Glycoimmunology in Xenotransplantation

Cheorl-Ho Kim

Glycoimmunology in Xenotransplantation

Springer

Cheorl-Ho Kim
Department of Biological Sciences
Sungkyunkwan University
Jangan-Gu, Suwon-Si, Korea (Republic of)

ISBN 978-981-99-7690-4 ISBN 978-981-99-7691-1 (eBook)
https://doi.org/10.1007/978-981-99-7691-1

This Springer imprint is published by the registered company Springer Nature Singapore Pte Ltd.
The registered company address is: 152 Beach Road, #21-01/04 Gateway East, Singapore 189721, Singapore

Keywords

EpitopesHistocompatibility antigens class II
Carbohydrate linkage
Endothelium
Vascular immunology
Glycosylation
Graft rejection immunology
Immune tolerance
Complement system
Xenotransplantation
Glycoimmunology
Glycan diversity
Xenoantigen
Pig
Red blood cell
Hyperacute xenograft rejection (HAR)
Acute humoral xenograft rejection (AHXR)
Cellular xenograft rejection (CXR)
Glycosylation in eukaryotes
Endoplasmic reticulum (ER)
Golgi apparatus
Glycoprotein
Glycolipid
Glycosaminoglycan

N-glycosylation
Glycosphingolipid
Ganglioside
Blood transfusion
ABO blood group system
Rhesus system (RH) and rhesus (D)
MNS antigen system
Lutheran system
Kell system
Duffy system
Kidd system
Dombrock blood group system
Gerbich blood group system
Knops blood group system
Allotransplantation
N-*glycolylneuraminic acid (neu5gc)*
Lewis histo-blood groupantigens
Accommodation
Immune incompatibility
Natural xenoantibody
Hyper acute rejection (HAR)
Galα1,3gal glycan
B-cell anticarbohydrate antibody production
A1,3gal-T KO pig
Non-α1,3gal epitope
Forssman antigen
Neu5gcNeuacα2,6galnacα1-R
A-lacnac
P1 antigen
Pk antigenI and i antigen system
Isoglobotrihexosylceramide (igb3; galα1,3galβ1,4glcβ1cer)
Hanganutziu-deicher (HD) antigen
Sd^a blood group antigen
Galnacβ1,4[neu5acα2,3]gal β1,4glcnacβ1,3gal glycan
Sd^a β1,4galnact-IISd^{a+} erythrocytic agglutination

Crispr9/cas-9
Triple α1,3gal-T
CMAH and β4galn-t2 KO pig
Swine leukocyte antigens (SLA)
*B-D-mannoside β-1,4-*N*-acetylglucosaminyl-transferase III (gnt-III)*
Complements system
Complement regulators (C regs)
Cd46 (MCPMembrane co-protein). cd55 (decay-accelerating factorDAF)
Cd59 (MAC-IP/MIRL)
Human CRP
Thrombomodulin
Delayed rejection of xenograft (DRX)
Blood coagulation
Endothelial protein C receptor (EPCR)
Cd36 (hcd36)
A1,3gal-T KO/hcd46/hepcr pig
Cd39
Cd73
Cd46
A1,3gal-T KO/hcd46/htm pig
C-reactive protein (CRP)
Resident memory T cells (trm)
Cellular xenogeneic rejection (CXR)
Cd200/cd200r
Transgenic hβ2m
HLA-E and HLA-G pig
Regulatory macrophages (mreg)
A20Tolerance
Mixed chimerism
Genome editing
Solid xenotransplantation
Islet xenotransplantation
Kidney xenotransplantation
Liver xenotransplantation
Heart and lung xenotransplantation
Corneal and tissue xenotransplantation

Infectious risk
Pig endogenous retrovirus (PERV)
Pig cytomegalovirus (PCMV)
Hepatitis E virus (HEV)
Pig lymphotropic herpesvirus (PLHV)
Chimeric organism
Human-non-human chimera (HNH-chimera)
Intra-bone bone marrow transplantation

Preface

The origin of life is depended on biodiversity, which is a pathway of evolution. Development, differentiation, growth, aging, apoptosis, xenotransplantation, oncogenesis, cancer, and diseases are represented by "cellular transformation." The answer to biodiversity-powering factor is originated from Glycan. One of the biological diversity types, named species diversity, is a phenomenon environmentally adapted from the evolutionary process for long period. Enforcing factor of different species, species difference, different phenotype, and different genotype between two species or each species is the acquisition of sugar utilization in different types of linkages. Sugars are photosynthesized from CO_2 and photonics by autotrophic plants and monomeric sugar residues are linked together to form their specific patterns in their organisms. Consequent diverse linkages specialize each organism, depending on residue-polymerized structures. If they are eukaryotes, the ER-Golgi apparatus is essentially synthesizing the glycan structures. The distinct structures of glycans discriminate each organism and are the essential molecular basis of the discrimination and difference between the organisms, giving an incompatibility between the different species.

Carbohydrate biosynthesis and expression on glycoproteins and glycolipids contribute to various biological processes such as protein stability, development, cell growth, differentiation, angiogenesis, neural maturation, remodeling, endocrinology, fertilization, and immunity. Diversity and variations in carbohydrate chain structures between family, species, kingdoms, and domains mark the global pattern and signs of immune self- and non-self-recognition. In evolution, the following next step of self- and non-self-recognition is operated by the specifically evolved self signs such as major histocompatibility complex (MHC) and human leukocyte antigen in human. For example, in human, diversity in ABH blood group antigens is observed in human family, and this type pattern distinguishes individuals from a pan-family to non-dividable unit of the family. Therefore, the utilization of carbohydrate structure to human basic diversity allowed human non-invasive boundary in order to protect boundary across from individuals. Resultantly, blood transfusion and organ transplantation are impossible even in the allogenic cross between humans. This explains how and how human beings are a lonely existence. From boned stage, human beings are lonely and thus seek the religions or lords such as God. If a human does not phenotypically express his (or her) specific blood type of ABH antigen, the individual produces antibody to react to the specific glycan. The antibody-producing mechanism is not well understood; however, immunoglobulin genes have probably memorized to produce each specific antibody before the era of evolutionary division from bacteria to human for a long time. Xenoantigens-expressing microflora or pathogens in human intestine may stimulate the production of antibodies. Such ABH-related antibodies induce hemolysis or hyperacute or allograft rejection due to incompatible graft property even between the same species. In mammals having the highly ordered and advanced immune system, the xeno-transplantation between the different species is not successfully settled down, from the very reason what mentioned above. The incompatibility is an immunologic rejection when the recipient host receives the tissues or organs from the different species of donors, as well-known in pig-to-human xenotransplantation.

A shortage and failure of human organs and tissues invite transplantation as the real solution. Xenotransplantation is regarded as the future therapeutic avenue for the organ failure and shortage in humans. However, the clinically available xenografts need to remove any issue of the anti-α-Gal antibody and complement. Development of available organs in human is a major part of future transplantation field of biology and medicine. Future xenotransplantation is directed to the use of animal tissues and organs. The miniature pigs are the major donor for human recipients and pig-to-human xenotransplantation needs to avoid the immune reactivity for the prevention of immune reaction with innate and adopted immune rejections. Understanding the glycan structures relevant to human-type linkages in glycan residues accelerates the application with genetic modification such as deletion and functional knock-out in pigs that overcome the issues of acute and hyperacute immune discrimination between the two species. Pig-to-human xenotransplantation is an attractive and replacement therapy to cover the lack of human organ donors. Xenotransplantation of shortage of organs needs an immunologically safe technology to save end-stage organ failure or hematopoietic cancer such as leukemia. There has recently been an increasing interest in xenotransplantation. The shortage of transplantable organs in human accelerates urgent supply of transplants and development of alternative transplantable organs. Revisiting issue in xenotransplantation is attributed to recent progress in information covering xenoantigens, immune surveillance, and immune protection to prevent xenograft rejection. In preclinical models, the circumstance made the feasible research direction to improve xenotransplantation efficacy. Such improved efficiency is visible from pancreatic islet xenograft survival of almost 3 years (950 days). Heart and kidney xenograft survivals have been succeeded for 945 days and 310 days, respectively. In addition, recent clustered regularly interspaced short palindromic repeats (CRISPR)/Cas9 as genome-editing technique has dramatically innovated the creation of donor pigs in short period. Also, information on porcine endogenous retrovirus (PERV) accelerated the application and challenges. Although tissue engineering and stem cell technology are under investigation, xenotransplantation technology is of future interests because animal tissues or organs are used to rescue failed or injured organs of the human recipient. Miniature pig (Sus scrofa) what has been genetically knocked out or modified are used as the source donor for xenotransplantation, simply due to its similarity to humans in sizes and immunity. The immunologic incompatibility between the donor pigs and the recipient human are based on the evolutionary distance between pigs and humans, as counted for about 4,000,000 years. This distance allows a xenograft rejection between the two mammals. Modification or deletion of the specific gene locus for immune rejection on genome of donor animals disrupts the immunological recognition ligands of the donor organs, consequently preventing the immune rejection of the human recipient and xenograft rejection. Xenotransplantation is a future promise to cover demand and supply in organ supplementation in human since beginning 2000s. The present conceptual xenotransplantation is the transplantation of pigs as animal organ source to humans. Basic preclinical studies have been initiated from the possibility of pig-to-nonhuman primate (NHP) transplantation although there was zoonosis issue, basically derived from PERV.

This book covers general xenoantigenic glycobiology in glycoprotein and glycolipid biosynthesis in ER and Golgi of eukaryotes. This book describes general glycobiology in emphasizing the structures, biosynthesis, glycosylation, and distribution of the glycans and xenogenic glycoantigens in eukaryotic cells of mammals including mouse, swine, chimpanzee, and human. In the middle, I have focused on topics in xenotransplantation glycobiology and expand descriptions of allogenic and xenoantigenic transplantation to open the dawn in insights into the origin of life. One of the biological diversity, named species diversity, is a phenomenon environmentally adapted from the evolutionary process for long period. The distinct structures of glycans discriminate each organism and are the essential molecular basis of the discrimination and difference between the organisms, giving an incompatibility between the different species. Diversity and variations in carbohydrate chain structures between family, species, kingdoms and domains mark the global pattern and signs of immune self- and non-self recognition. In human, diversity in ABH blood group antigens is observed in human family and this type pattern distinguishes individuals from a pan-family to non-dividable unit of the family. Blood transfusion and organ transplantation are impossible even in the allogenic cross between humans if carbohydrates are ignored. This explains how and how human beings are a lonely existence. ABH-related antibodies induce hemolysis or hyperacute or allograft rejection due to incompatible graft property even between the same species. The incompatibility is an immunologic rejection when the recipient host receives the tissues or organs from the different species of donors, as well-known in pig-to-human xenotransplantation. The immunologic incompatibility between the donor pigs and the recipient human are based on the evolutionary distance between pigs and humans. This distance allows a xenograft rejection between the two mammals. Modification or deletion of the specific gene locus for immune rejection on genome of donor animals disrupts the immunological recognition ligands of the donor organs, consequently preventing the immune rejection of the human recipient and xenograft rejection. This book helps undergraduate and graduate students, researchers, and professors who are involved in glycobiology and xenoantigenic biology with recent advances in the xenotransplantation basic and clinic.

The book also provides basic theories of autotransplantation, allotransplantation, and xenotransplantation, including immune system in mammals to protect self from invaders and incompatibility. Additionally, the book describes general blood coagulative biology in human and summarizes recent xenoantigen-removed KO swine (pig) construction and breeding with keywords of carbohydrate antigen, xenotransplantation, ER/Golgi apparatus, biological diversity, glycan-based evolution, and swine (pig) KO strain. Therefore, the book is also oriented to undergraduate and graduate students as well as glycobiology exam-based knowledge.

Suwon-Si, Republic of Korea
September 13, 2023

Cheorl-Ho Kim

Contents

Chapter 1
Origin of Life and Glycan

1.1 Origin of Life

The origin of life is dependent on biodiversity, which is the pathway of evolution. Organism evolution is intricately connected with glycan adaptation. In retrospect, the exhaustion time in biological evolution could not have been delayed by the human historic era, since BC1000. Development, differentiation, growth, aging, apoptosis, xenotransplantation, immunity, oncogenesis, cancer, and diseases are represented by "cellular transformation." The answer to biodiversity's powering factor originates from glycan. One of the biological diversities, named species diversity, is a phenomenon environmentally adapted from the evolutionary process for a long period. Enforcing factor of different species, species difference, different phenotype, and different genotype between two species or each species, is the acquisition of sugar utilization in different types of linkages. Sugars are photosynthesized from carbon dioxide CO_2, photonics by autotrophic plants, and monomeric sugar residues are linked together to form their specific patterns in their organisms. Consequent diverse linkages specialize each organism, depending on residue-polymerized structures. There is a big shift from ancient types of life, such as "bacteria" to "eukaryotes" by the formation of intracellular small organelles. If some life organisms form the "organelle," then it will become an eukaryote. If they are eukaryotes, the small organelle ER-Golgi apparatus essentially biosynthesize the glycan structures required for their life phenophena. The distinct structure of glycans discriminate each organism and are the essential molecular basis of the discrimination and difference between the organisms, giving an incompatibility between the different species.

Although glycans are biosynthesized and elaborated in ER-Golgi system by glycosyltransferases amd glycosidases, the means of glycobiology are glycans-based, which bear chemical structure, glycochemical reactivity, glycan diversity, and biosynthetic pathways in terms of glycan birth and life evolutional enigma. From a bioinformatics viewpoint, a glycosyltransferase superfamily, which was a

C.-H. Kim, *Glycoimmunology in Xenotransplantation*,
https://doi.org/10.1007/978-981-99-7691-1_1

sugar-nucleotide synthase once, has evolved to the present glycosyltransferases, which synthesize various glycans. Comparing glycan function between creatures may view some universality and diversity. Since genome analysis of *C. elegans* in 1998, Drosophila in 2002, and human in 2003, information from more than 30,000 genes is available in database banks, through systemic and genome wide tools. Genome DNA sequences of animals, including human, can be compared to comparatively solve the biological process and create new knowledge in physiological pathways. Using the limited number of human genes, phenotype diversity of cells, tissues, organs, organ systems, and individuals is constituted through glycan modification and transfer. From a small number of glycans, genes are concomitantly increased by intragenetic homologous recombination, specific mobile genetic elements (such as retrointegration and transposon), and infectious agents.

1.2 Cell Surfaced Xenoantigens

Cell surfaced carbohydrates are predominantly present as lipid- and protein-linked forms, which are known as glycolipids and glycoproteins, respectively. They are mostly front-end components present on the outer surface of cells regardless of being prokaryotes or eukaryotes. The surfaced glycosylmoieties reflect the signals and signs of the cells. Hence, such glycomoieties are acting as receivers of extracellular signals like cellular phone mechanism, and consequently transduce the received signals to intracellular accepting machines. Resultantly, the biological phenomena develop, particularly in multicellular organisms, as well-known modes in the cellular events including the cell-cell communication, recognition, binding, interaction, adhesion, fertilization, development, differentiation, growth and proliferation, aging, signal transduction, neurotransduction, innate immunity, adopted immunity, global immune systems, transformation, carcinogenesis, metastasis, and angiogenesis. Carbohydrate biosynthesis and functional expression, seen in glycoproteins and glycolipids, contribute to various biological phenomena and processes. The basic contribution of proteins is its stability.. Functionally, they contribute to cell proliferation and growth, development and differentiation, angiogenesis, neural maturation, tissue and bone remodeling, endocrinology, fertilization, and immunity. In addition to the above transducer functions, the glycans are the signs of each cell, tissue, organ, system, organism, and species. The signs mean the marker, symbol, or the face of each cell. This is the reason why surface glycosylated moieties should be predominantly located on the borders of cells.

All cells meet together in a face-to-face manner by means of recognition. In all places where cells existed, there are carbohydrates. Carbohydrates can function as a receptor and/or ligand upon the cell–cell interaction. For glycan communication with their counterparts even in a type of receptor or ligand, such counterparts play important roles through purely biophysical interactions. Thus, their synthesis, localization, and distribution in cells are strictly controlled by glycosylation, gene information, expression, and glycan-regulating machineries. These glycosylmoieties in glycoproteins and glycolipids determine the fate of the multicellular function, and eventually, cellular society formation. These multicellular function and sociality of cells depend on cell type, tissue type, organ type, organism, and species. Modifications of cellular glycosylation also contribute to phenotypic changes in cells. Indeed, most of the carbohydrate tumor antigens are differently glycosylated and terminally sialylated. Levels of several mucin-type sialylated Lewis antigens are representatively increased in the tumor cell glycoprotein *N*-glycan and *O*-glycan structures or glycolipids. For tumors, these glycosylated antigens can be used as tumor markers or tumor faces, especially for node-negative patients.

In recent viral Flu pandemics, human and avian influenza viruses are demonstrated to specifically bind sialyl motifs. Indeed, a carbohydrate antigen NeuAcα2-6Gal is a human type, whereas the NeuAcα2-3Gal is characterized as an avian type. In the case of the recently occurred human type H3N2 virus, the evolutionary transition in carbohydrate linkages and second dimensional structures was indicated to potentiate the viral transmission between the interspecies of humans and avians via the swine species. Therefore, each carbohydrate structure has been implicated as a face, sign, or signature of certain animals, organisms, organs, and cells, as well as biological states in cells and organs. The above glycan-specific Flu infection and carbohydrate-specific tumor transformation reflect the importance of carbohydrates in each species. If we expand this hypothesis to pigs or humans, pigs have their own specific face molecules as a form of carbohydrate, but not proteins or nucleic acids. Such molecules should be self-distinguishing, self-characterizing, and self-specialized, like the HLA molecules in humans. Those are the carbohydrate molecules and antigens. Thus, the carbohydrate molecules are not simply the only component in organisms. If it extends to the xenotransplantation, they are xenoantigenic carbohydrates. Paradoxically, the xenoantigens are carbohydrate xenoantigens.

1.3 Diversity and Variations in Glycans

Diversity and variations in carbohydrate chain structures between family, species, kingdom, and domains mark the global patterns and signs of immune system in self-immune recognition and nonself immune responses. In evolution, the following steps of self-recognition and non-self-recognition immune responses are operated by the specifically evolved self-signs as well recognized in major histocompatibility complex (MHC) and human leukocyte antigen (HLA). For example, human diversity in ABH blood group antigens is observed in human families, and this type of pattern distinguishes individuals from a pan-family to a nondividable unit of the family. Therefore, the human cell utilization of carbohydrate structure for basic human diversity allows human noninvasive boundary in order to protect boundary across from individuals. Resultantly, blood transfusion and organ transplantation are impossible even in the allogenic cross between humans. This explains why and how human beings are a lonely existence. From birth, human beings are lonely and thus seek religions or lords such as God. The human individuality further reflects the nondividable unit in their immunology. However, if God created humans, he may generously confer his creatures to share with the interspecies, namely human-to-human. That is indeed HLA allografts, allowing allograft change in allotransplantation among humans. Thus, the current allotransplantation is a clinically possible organ replacement. However, God made a big hurdle and barrier to keep the boundary between allografts of humans and ABO blood type that should be matched together before a large amount of perfusion. This is one example of tolerance generously provided by God. Then if we consider the relationship between different species, such as pig to human xenotransplantation, the border line to keep between the two species is established without the violation of the territorial body. The border line is performed by xenoantigenic carbohydrates. Therefore, to overcome this border, the removal of border lines is a prerequisite and the biologically protective behavior is the immunological rejection.

If a human does not phenotypically express his (or her) specific blood types of ABH antigen, the individual produces antibody to react to the specific glycan. The antibody-producing mechanism is not well understood; however, immunoglobulin genes have probably memorized to produce each specific antibody before the era of evolutionary division from bacteria to human. Or xenoantigens expressing microflora or pathogens in human intestine may stimulate the production of antibodies. Such ABH-related antibodies induce hemolysis, hyperacute, or allograft rejection due to incompatible graft property, even between the same species. In mammals having highly ordered and advanced immune systems, the xenotransplantation between the different species has not successfully settled down, for the very reason as mentioned above. The incompatibility is an immunologic rejection when the recipient host receives the tissues or organs from the different species of donors, well known for immunological rejection spectrum in pig-to-human xenotransplantation and summarized in the viewpoint of breeding rate, heart value, antigens, and xenoantigenic modification to genetically remove or overcome rejecting responses (Fig. 1.1).

1.4 Xenotransplantation of Immunologically Safe Technology

The 2011 USA Annual Organ Procurement and Transplantation Network (OPTN) report describes that one registered human donor is in every ten patients to await in order to organ transplantation in hospitals [1]. A shortage and failure of human organs and tissues invite transplantation as the real solution. Xenotransplantation is regarded as the future therapeutic avenue for organ failure and shortage in humans. However, the clinically available xenografts need to remove any issue of the anti-α 1,3Gal antibody and its complement. Development of available organs in humans is a major part of the future transplantation field of biology and medicine. Future

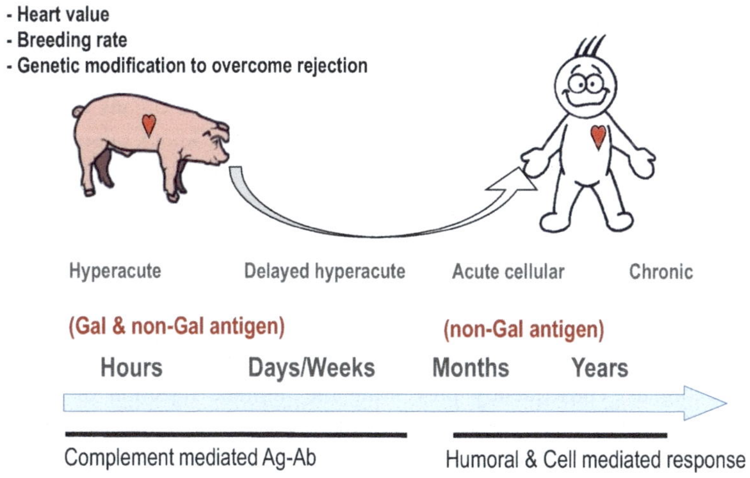

Fig. 1.1 Xenotransplantation and the rejection spectrum by α1,3Gal and non-α1, 3Gal antigenic epitopes

xenotransplantation is directed to the utilization of domestic animals for possible tissues and organs. Currently, developed miniature pigs are the major donors for human recipients, and pig-to-human xenotransplantation needs to avoid the immune reactivity for the prevention of immune reaction with innate and adopted immune rejections. Understanding the glycan structures relevant to human-type linkages in glycan residues accelerates its application with genetic modification such as deletion and functional knock-out in pigs that overcome the issues of acute and hyperacute immune discrimination between the two species. Pig-to-human xenotransplantation is an attractive and replacement therapy to cover the lack of human organ donors [2–4]. Xenotransplantation of shortage of organs needs an immunologically safe technology to save end-stage organ failure or hematopoietic cancer such as leukemia. There has recently been an increasing interest in xenotransplantation.

The shortage of transplantable organs in human accelerates urgent supply of transplants and development of alternative transplantable organs. Revisiting issues in xenotransplantation is attributed to recent progress in information covering xenoantigens, immune surveillance, and immune protection to prevent xenograft rejection. In preclinical models, the circumstance made the feasible research direction to improve xenotransplantation efficacy. Such improved efficiency is visible from pancreatic islet xenograft survival of almost 3 years (950 days). Heart and kidneys xenograft survivals have been succeessful for 945 days and 310 days, respectively [5]. In addition, recently developed CRISPR technology known as the clustered regularly interspaced short palindromic repeats-based genome-levelled technique, CRISPR/Cas9, has dramatically been innovated toward the generation and creation of donor pigs for a short period. This CRISPR gene editing technology is now a proper run in this field.

Also, information on porcine endogenous retrovirus (PERV) accelerated its application and challenges. Although tissue engineering and stem cell technology are under investigation, xenotransplantation technology is of future interest because animal tissues or organs are used to rescue failed or injured organs of the human recipient. Miniature pig (Sus scrofa) that has been a genetic knock out or modified is used as the source donor for xenotransplantation, simply due to its similarity to humans in size and immunity. On the basis of the relationship between pig donors and the human recipients, there is the immunologic incompatibility, basically caused by the evolutionary distance between pigs and humans, counted as about 4,000,000 years. This distance allows a xenograft rejection between the 2 mammals. Modification or deletion of the specific gene locus responsible for immune rejection on the genome of donor animals disrupts the immunological recognition ligands of the donor organs, consequently preventing the immune rejection of the human recipient, i.e., xenograft rejection [6]. Xenotransplantation is a future promise to cover demand and supply in organ supplementation in human since the beginning of the 2000s. The present concept of xenotransplantation is the transplantation of pigs as animal organ source to humans. Basic preclinical studies have been initiated from the possibility of pig-to-nonhuman primate (NHP) transplantation, although there was zoonosis issue, basically derived from porcine endogenous retrovirus (PERV).

1.5 Xenotransplantation of Future Prospect

Therefore, the present book focuses on carbohydrate xenoantigens and their recognition immunity. How to understand the xenoantigens and the recognition immunity? Then how to avoid the initial immunity upon encountering each counterpart. If the immunity is based on the preformed antibodies against xenoantigens, how to avoid the response from the antibody-mediated concerted reaction? If many different xenoantigens are associated with the total reactions, how to overcome such global immunity? In addition to the innate and adopted immunity, several blood disorders mediated by xenoantigen-operated blood diseases are also developed. Together with those, related carbohydrate xenoantigens are described. In addition, the book opens the detailed status of xenotransplantation in view of the future prospect. The xenotransplantation covers the immunity, sugar glycans, blood homeostasis, microbiology, and endogeneous harmful agents.

References

1. Zeyland J, Lipiński D, Słomski R. The current state of xenotransplantation. J Appl Genet. 2015;56:211–8. https://doi.org/10.1007/s13353-014-0261-6.
2. Yang YG, Sykes M. Xenotransplantation: current status and a perspective on the future. Nat Rev Immunol. 2007;7:519–31.

3. Ekser B, Ezzelarab M, Hara H, et al. Clinical xenotransplantation: the next medical revolution? Lancet. 2012;379:672–83.
4. Puga Yung G, Rieben R, Buhler LH, et al. Xenotransplantation: where are we standing in 2016? Swiss Med Wkly. 2016;147:w14403. (in press).
5. Cowan PJ, Tector AJ. The resurgence of xenotransplantation. Am J Transplant. 2017;17(10): 2531–6. https://doi.org/10.1111/ajt.14311.
6. Wolf E, Kemter E, Klymiuk N, Reichart B. Genetically modified pigs as donors of cells, tissues, and organs for xenotransplantation. Anim Front. 2019;9(3):13–20. https://doi.org/10.1093/af/vfz014.

Chapter 2
Current State of Xenotransplantation

2.1 Glycan Antigens of Pig Tissues and Red Blood Cells (RBC)

With advances in medical technology, even if serious organ damage occurs, it can be overcome through organ transplantation. But organ transplants basically require organ donors, and the donor organs are incredibly limited in their number when compared to the demanders. In fact, there are about 40,000 people waiting for organ transplants in Korea; however, only about 200 organ donors are registered on the official site [data available from URL: https://www.konos.go.kr/konosis (accessed September 20, 2020)]. When the organ supply and the demand were considered, there is a big gap, and it has led to the xenotransplantation idea. Using heterogeneous animals can easily solve the shortage of organs. However, organ transplants are cautious, even among the same species due to genetic differences in MHC-I and various immune responses. The barriers between xenotransplantation are much more complex than homogeneous transplantation. Therefore, in order to study the stabilization of xenotransplantation, model animals are used to specify and overcome the barriers of xenotransplantation to succeed in transplantation. Systematically, the most similar with human being is nonhuman primates (NHPs). They were first selected as a model animal of a study of xenotransplantation but was excluded for ethical issues.

Since pigs were the most suitable animals among other animals, the main target of the heterograft supplier was selected as pigs and conducted research. In homogeneous transplants, one of the main factors of the antigen is the MHC-I. Different individuals have different MHC-I, causing immune responses and failure in transplantation. However, not only is the structure of the protein MHC different in the interspecies, but the presence of other molecules causes a more serious immune response. That main molecule is glycan. All animals have their own unique glycans on the surface of the cell, which can act as an antigen. Therefore, glycan found in pigs, but not in humans, can act as an antigen in humans. This causes immunity

C.-H. Kim, *Glycoimmunology in Xenotransplantation*,
https://doi.org/10.1007/978-981-99-7691-1_2

rejection and acts as a barrier to xenotransplantation. But now, genetic modification can lead to pigs that can reduce or eliminate the expression of glycans by knock out the genes of the glycan synthesis enzyme that are expressed in pigs. It used to use Zinc finger or other inefficient techniques, but now it is possible to produce KO pigs quickly and effectively by using CRISPR-Cas9. By transplanting glycan-KO pig organs to NHP, data on the immune response and life expectancy of pigs are obtained. Then, it can be found out what kinds of glycans serve as antigens and what if they were removed and transplanted. It can also be considered whether there were any other barriers to fail the xenotransplantation.

2.2 Human-Animal Organ Exchange as Chimerism

Transplantation is currently facing shortage caused by limited supply problem of human donor allografts as functional materials and restriction by ethical issue. Human–animal exchange of organ is another type of anti-naturalism and anti-evolution. This is an opposition of the "God'-created order forwarded direction of evolution. Indeed the expanding gap between list of allograft-waiting person and the number of allotransplant receiving patients is a driving catalyst of research to the development of alternative source of grafts from house-cultivating pigs. This informs the birth of xenotransplantation in biological history. The xenotransplantation reflects a formation of chimera organisms, which are strictly prohibited in the immunology. In Greek mythology, a terrible monster chimera was killed by Bellerophon who ride a flying horse Pegasus after he used a magical golden bridle to catch Pegasus when he was a promised warrior the Mount Olympus. The mythology suggests that all the chimera organisms cannot be survived in the cosmic society. However, nevertheless, the xenotransplantation is to disobey the general philosophy in the biology and immunology.

Recently, the pig to human xenotransplantation has been regarded as an alternative strategy and relevant solution for shortage of the human organ with prolonged life span. In allotransplantation, human organs, tissues, and cells are under investigation to resolve the organ failure. Xenotransplantation is quite difficult rather than the allotransplantation, as it transplants living targets of one species, such as cells, tissues, or organs, to different species. Animal organs are used in the xenotransplantation, providing an endless or limitless sources to replace irreversibly dysfunctional, damaged, or impaired organs due to finally terminated step and failure in humans [1]. Xenotransplantation from pigs has now increasingly been supported with strong medical advocate for the substantial resolution to the continuous organ and tissue shortages from human donation, moreover, depending on the life span elongation. Although short-term xenograft transplantation is permissive for survival, however, long-term xenograft survival requires immunosuppression regimes, indicating the difficulty in successful long-term xenograft survival. In anatomic and immunologic view, the first accounting line of xeno-cells or xeno-tissues is vascular endothelial cells and endothelium in blood vessels. Therefore, vascular endothelial cells have

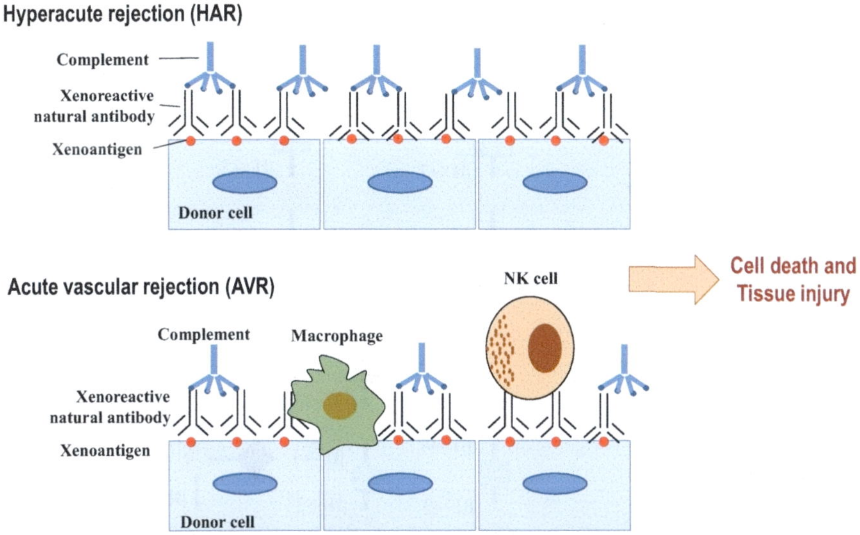

Fig. 2.1 Rejection profile in xenotransplantation. Acute vascular rejection (AVR) and hyperacute rejection (HAR) are the major types of rejection immune responses. The targeted cells or tissues are subjected to cell death and tissue injury. Various xenoreactive antibodies, complements, and innate immune cells are involved in the cell death and tissue injury via xenoantigenic recognition

mainly been subjected to combat xenografts to recipient host immune reactions, as known over the past three decades.

From the immune reaction through the vascular endothelial cells, several terms of acute vascular rejection (AVR), hyperacute xenograft rejection (HAR or HR), delayed antibody rejection (DAR or DR), delayed xenograft rejection (DXR), acute humoral xenograft rejection (AHR), acute cellular rejection (ACR), etc. have been classified to the current stage (Fig. 2.1). Currently, inhibition of HR and AVR is a critical issue for long-term pig xenograft survival in primates including humans. Such HR, DXR, and AVR are operated by the humoral immune system in the primates and are crucial barriers to solid organ transplantation. The HR, DXR, and AVR are based on the molecular recognition against non-self-antigens by host antibodies. Non-self-antigens such as proteins and carbohydrates present on the pig vasculatures act as xeno-antigens [2, 3]. In brief, the recognition responses of HR can be avoided by genetic or enzymatic removal of α1, 3-Gal-carrying carbohydrates in the pigs. Then, AVR or DXR becomes next response type of the known immunologic barriers to xenotransplantation. So far, two non-α1, 3Gal glycan xenoantigens of Sda and *N*-glycolylneuraminic acid (Neu5Gc) are known and characterized in levels of carbohydrate structure, glycosyltransferase-based biosynthesis and immunity from pigs (Fig. 2.2). These carbohydrate structures are partly similar to those structures linked to human blood group ABO glycans present on red blood cells (RBC) of human (Fig. 2.3).

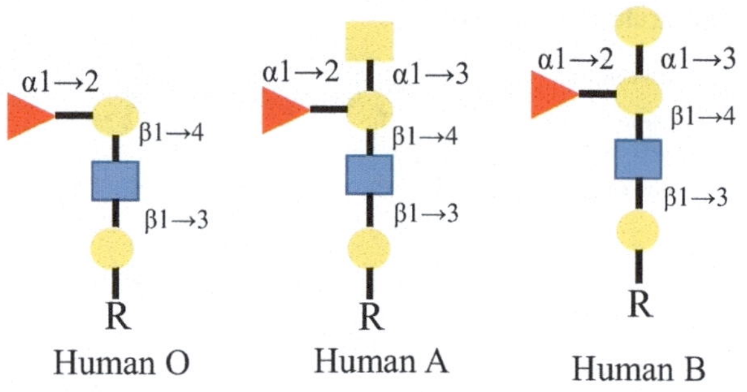

Fig. 2.2 Carbohydrate structure on human RBC ABO blood group glycans

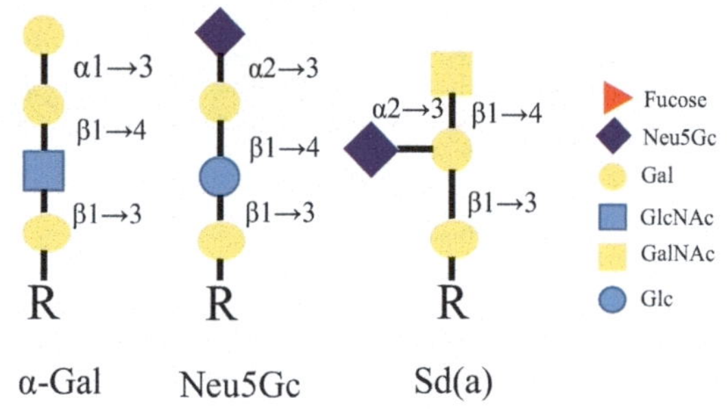

Fig. 2.3 Glycan antigens of porcine tissues or RBCs. Carbohydrate structure on glycan antigens of porcine RBCs. For example, RBCs of pig generate α1, 3-Gal antigenic epitopes linked to glycans, which are structurally similar to the B-type epitope of human blood group ABO

2.3 Emerging Xenoantigene Era

Here, the concept of cross-species and interspecies transplantation should be rearranged, since the terms are not new in the present xenoantigene era. In fact, despite the modern xenotransplantation, traditional transplantation has been clinically tried during the past 300 years [4, 5]. Moreover, with the recent progress in multiple genetic modification technology in donor pigs and immunosuppressive administration, cytotoxicity and survival rate of chronic graft rejections have been much improved. Also, the immunological tolerance has been clarified and used to induce it. However, such challenges of modern transplantation approaches are still insufficient to reach in better outcome and improve the successful xenotransplantation of xeno-tissues and xeno-organs to overcome short transplantation issue. Although relationship between the humans and non-human primates (NHPs) is phylogenetically and evolutionarily quite distant, the NHPs are used as a xenotransplantation model of humans, because the accumulated information of human immunity is still limited in view of xenotransplantation. Hence, the current utilization of NHPs as a model is also provoking another conflict to argue in xenotransplantation. In addition, the risk of xenozoonoses is equally dispensable in the NHPs and humans.

It is clear that xenotransplantation using different species is largely advantageous with a unique motif and benefit, inexhaustibly supplication of feasibly alternative xeno-organs. In addition, using the modern evolved technology, genetically modified pigs have been easily created for the alternative organ sources. There are also competent outcomes regarding lifespan-increasing transplantation using genetically modified pigs in NHPs, with prolonged survival rate. The recent trials combined with available approaches are under development. The combined strategies are the genetic modification of pigs, blockade-combined immunosuppression, and anti-inflammation approach. The most outstanding feature of the strategies is the α1, 3Gal-T-KO pigs combined with a T cell tolerance strategy, which exhibited prolonged survival of lifespan-increasing renal grafts for 3 months more [6]. The T cell tolerance strategy is involved with co-transplantation of a α1, 3Gal-T-KO thymus that was vascularized [7].

Among the known transplantation hurdles in clinical application, the following two main factors are the organ shortage and the lifelong immunosuppression using immunosuppressive drugs. These obstacles yield an applicable solution as a strategic approach to solve the two main obstacles. The use of xenogeneic organs supplied by genetically modified pigs is the solution strategy by triggering of immune tolerance across the xenoantigenic barrier between different species. Despite the above progress in xenotransplantation, currently, four weaknesses in the strategies have been suggested to prevent and avoid the rejection response to xenografts during the pig-to-human xenotransplantation. Among them, the most weak point is in the use of immunosuppressive drugs to induce the immune tolerance. These two strategies lead to noninfectious, opportunistic, and latent pathogens to become malicious to cause genuine pathogenicity. For example, α1, 3Gal-T-KO pig cells liberate and release pig-borne and pig-derived viruses, although the Galα1, 3-Gal xenoantigenic epitopes are absent in the viruses. Also, the transgenically enforced human complement regulatory protein (hCRP) expression on the pig cell surfaces leads to a complement defense defecting and weakening from pathogenic microorganisms including viruses. At the same time, transgenic hCRP expression rather facilitates such invasive pathogenic invaders to adopt hCRPs as the receptors for their binding and infection [8].

Recently, the human recipients waiting for transplantation of organs are gradually increasing in their number with lifespan elongation. In USA, over 130,000 patients are listed as the recipient patients waiting for life-saving organ transplants [9]. Strategic genetic modification is to produce "humanized" organ models from the pig for clinical application with safe relevance. The fundamental approaching milestone has been forwarded to the reduction of antigenic factors by the human gene transgenic expression in the donor pigs. Although the current xenotransplantation has practically several difficulties in combining multiple genetic KO modifications and random integration of human transgenes to a single pig, several technologies for pig genetic modification are recently developed. Among them, hopefully, CRISPR/Cas9-directed gene inactivation in experimental pigs has greatly improved the genetic modification and contributed to overcome xenograft rejection in pig-to-human transplantation.

Recently, the gene editing technology of CRISPR/Cas9 has an impact on the progressive modification of the pig genome to potentially match the human type. Now the CRISPR/Cas9 technology in a single step is a representative to produce multiple gene KO pigs [10, 11]. Genome editing technology or CRISPR/Cas9 technology can potentially be used for pig applications to DQ deletion from the pig or sequence alteration to remove immunogenic epitopes present in DQ genes of pigs. Because two DR and DQ in SLA are likely to function as xenoantigenic locus and most HLA antibodies contain the anti-DQ antibodies in recipient patients, as found in the renal allograft failure, the SLA DQ should be calarified for the importance in xenotransplantation [12]. The present CRISPR/Cas genome editing technology in the donor pigs is available to remove and eliminate the class type II antigenic epitope. Therefore, suitable donor pigs can be created without cross-match for all the patients with organ functional failure regardless of HLA sensitization level in the near future.

References

1. Ekser B, Ezzelarab M, Hara H, van der Windt DJ, Wijkstrom M, Bottino R, Trucco M, Cooper DK. Clinical xenotransplantation: the next great medical revolution? Lancet. 2012;379:672–83.
2. Ierino FL, Sandrin MS. Spectrum of the early xenograft response: from hyperacute rejection to delayed xenograft injury. Crit Rev Immunol. 2007;27:153–66.
3. Lin CC, et al. Recipient tissue factor expression is associated with consumptive coagulopathy in pig-to-primate kidney xenotransplantation. Am J Transplant. 2010;10:1556–68. https://doi.org/10.1111/j.1600-6143.2010.03147.x.
4. Zeyland J, Lipiński D, Slomski R. The current state of xenotransplantation. J Appl Genet. 2015;56:211–8. https://doi.org/10.1007/s13353-014-0261-6.
5. Cooper DKC, Gollackner B, Sachs DH. Will the pig solve the transplantation backlog? Annu Rev Med. 2002;53:133–47.
6. Lai L, Kolber-Simonds D, Park KW, et al. Production of alpha-1, 3-galactosyltransferase knockout pigs by nuclear transfer cloning. Science. 2002;295:1089–92.
7. Yamada K, Yazawa K, Shimizu A, et al. Marked prolongation of porcine renal xenograft survival in baboons through the use of alpha1, 3-galactosyltransferase gene-knockout donors and the cotransplantation of vascularized thymic tissue. Nat Med. 2005;11:32–4.
8. Ladowski JM, Reyes LM, Martens GR, Butler JR, Wang ZY, Eckhoff DE, Tector M, Tector AJ. Swine leukocyte antigen (SLA) class II is a xenoantigen. Transplantation. 2017;102(2):249–54. https://doi.org/10.1097/TP.0000000000001924.
9. Weiss RA. Transgenic pigs and virus adaptation. Nature. 1998;391(6665):327–8.
10. Reyes LM, et al. Creating class I MHC-null pigs using guide RNA and the Cas9 endonuclease. J Immunol. 2014;193:5751–7.
11. Li P, et al. Efficient generation of genetically distinct pigs in a single pregnancy using multiplexed single-guide RNA and carbohydrate selection. Xenotransplantation. 2015;22:20–31.
12. Tambur AR. HLA-DQ antibodies: are they real? Are they relevant? Why so many? Curr Opin Organ Transplant. 2016;21(4):441–6.

Chapter 3
Pig as Best Source for Clinical Xenotransplantation

Numerous patients in each country need their life-saving transplants. The current sources required for allogenic or xeno-antigenic transplantation are mainly the organs rather than tissues and cells. As mentioned earlier, genetically edited and modified pig organs with multiple genes are attracted to solve the issues of donor organ supply and shortage. However, hurdles still include natural antibodies reactive to xenoantigens in human, and therefore, elicited antibodies are the subjects to overcome the failure in current clinical xenotransplantation. Rhesus and cynomologus monkeys are used as models in the preventive rejection of allografts, but the baboon recipient is used for pig organ xenotransplantation [1]. However, for cell xenotransplantation, cynomolgus monkey has been used [2, 3].

Why pigs are selected as the most suitable source in the xenotransplantation? The answer is simple. Pigs have been considered to use as a model animal and applied for replacement of large organs toward human diseases [4, 5] and are the preferred species and donor animals as the current best source for clinical xenotransplantation. This is based on the facts that reasoned from their human-related anatomic size, nutrition, immunological, and physiologic aspects in similarity with humans, breeding characteristics, availability, multiplicity, and multiple genetic modifications to prevent immunological rejection, and overcome evolutionary immunologic obstacles [6–10]. Pigs are easily breedable with a short period of gestation and large size of a litter. In addition, pigs have the ease with genetic modification and cloning. Merits of pigs as nonhuman organ-tissue donors are in an unlimited and unrestricted availability, better breeding potential, reproductive maturity length, short pregnancy time, offspring number, anatomical and physiological aspect, relatively low infection risk, and easy dealing with nonpathogenic and pathogen-free pigs [7]. Moreover, when pigs are compared to NHPs or animals, ethical and economical advantageous animal is the pig, and they are the subject for biomedical investigation and development.

In the allograft transplantation, the human antigens of ABO blood group are a basic hurdle; however, in the xenotransplantation, currently several xenoantigens including α1,3-Gal, Neu5Gc and B4GalNT2-synthesized glycans are the major

C.-H. Kim, *Glycoimmunology in Xenotransplantation*,
https://doi.org/10.1007/978-981-99-7691-1_3

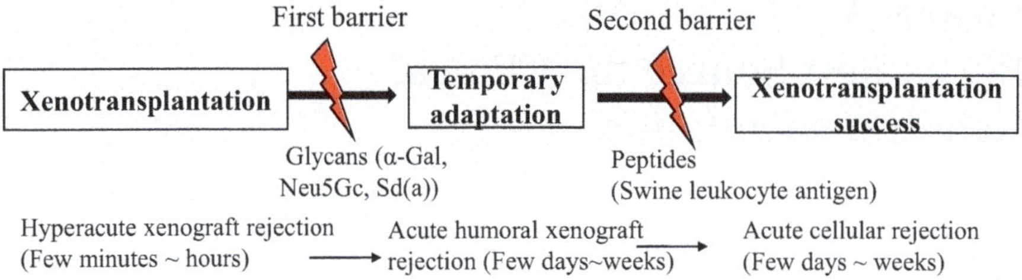

Fig. 3.1 Roadmap of xenotransplantation

hurdles, although complexed responses are followed [11]. Therefore, triple-negative deletion of xenoantigen genes such as α1,3Gal-T(GGTA1)/CMP-*N*-acetylneuraminic acid (NeuAc) hydroxylase (CMAH)/β1,4-*N*-GalNAc-T (β4GalNAc-T2, B4GALNT2) responsible for those productions have been attracted to target, and a targeted triple KO pig showed the eliminated barriers against the xenoreactive antibodies for patients [12]. Current roadmap of xenotransplantation is described in Fig. 3.1. Glycans (α1,3-Gal, Neu5Gc, Sda) and peptides (swine leukocyte antigen; SLA) are known as first and second barriers, respectively. HAR, AHR, and ACR are immunologically progressed upon encountering the xenoantigens. For their responding periods, HAR occurs within few minutes to hours, AHR and ACR occur within few days to weeks.

Nevertheless, currently, some hurdles have still to be overcome in the view of immune rejection responses, physiological incompatibility between human and pig, and transmission risk of pig-carrying parasites including microorganisms. Fortunately, through the recent breeding progress in multiple genetic modification of pigs, immunosuppressive drug development and safety solution, xenotransplantation is more closely approaching to our expectation to successful clinics [13].

References

1. Cooper DK, Satyananda V, Ekser B, van der Windt DJ, Hara H, Ezzelarab MB, et al. Progress in pig-to-non-human primate transplantation models (1998–2013): a comprehensive review of the literature. Xenotransplantation. 2014;21:397–419.
2. Ekser B, Ezzelarab M, Hara H, van der Windt DJ, Wijkstrom M, Bottino R, et al. Clinical xenotransplantation: the next medical revolution? Lancet. 2012;379:672–83.
3. van der Windt DJ, Bottino R, Kumar G, Wijkstrom M, Hara H, Ezzelarab M, et al. Clinical islet xenotransplantation: how close are we? Diabetes. 2012;61:3046–55.
4. Carter DB, Lai L, Park KW, Samuel M, Lattimer JC, Jordan KR, Estes DM, Besch-Williford C, Prather RS. Prather phenotyping of transgenic cloned piglets. Cloning Stem Cells. 2002;4:131–45.
5. Rogers CS, Stoltz DA, Meyerholz DK, Ostedgaard LS, Rokhlina T, Taft PJ, Rogan MP, Pezzulo AA, Karp PH, Itani OA. Disruption of the CFTR gene produces a model of cystic fibrosis in newborn pigs. Science. 2008;321:1837–41.
6. Cooper DK, Gollackner B, Sachs DH. Will the pig solve the transplantation backlog? Annu Rev Med. 2002;53:133–47.

7. Ekser B, Ezzelarab M, Hara H, et al. Clinical xenotransplantation: the next medical revolution? Lancet. 2012;379:672–83.
8. Ibrahim Z, Busch J, Awwad M, Wagner R, Wells K, Cooper DK. Selected physiologic compatibilities and incompatibilities between human and porcine organ systems. Xenotransplantation. 2006;13:488–99.
9. Ekser B, Markmann JF, Tector AJ. Current status of pig liver xenotransplantation. Int J Surg. 2015;23(Pt B):240–6.
10. Aigner B, Renner S, Kessler B, Klymiuk N, Kurome M, Wünsch A, Wolf E. Transgenic pigs as models for translational biomedical research. J Mol Med. 2010;88:653–64.
11. Byrne GW, McGregor CG, Breimer ME. Recent investigations into pig antigen and anti-pig antibody expression. Int J Surg. 2015;23(Pt B):223–8.
12. Estrada JL, Martens G, Li P, et al. Evaluation of human and non-human primate antibody binding to pig cells lacking GGTA1/CMAH/beta4GalNT2 genes. Xenotransplantation. 2015;22(3):194–202.
13. Niemann H, Petersen B. The production of multi-transgenic pigs: update and perspectives for xenotransplantation. Transgenic Res. 2016;25(3):361–74. https://doi.org/10.1007/s11248-016-9934-8.

Chapter 4
Glycan Antigens of Pig Interfering with Xenotransplantation: Three Immune Responses from the Glycans

4.1 Glycan Antigens

Human beings commonly understand carbohydrates or glycans only as a source of bioenergy. Because the proteins are biosynthesized in eukaryotic endoplasmic reticulum (ER) and Golgi apparatus, the *N/O*-glycans not only stabilize the protein synthesis process in ER and Golgi apparatus but also relates to cell-to-cell communication in cell membranes. Furthermore, glycans are also important to immune responses due to their interaction with lectins. All animals synthesize their unique glycans from each species and present them to the cell membrane. For example, human beings synthesize oligosaccharide from a dolichol (Dol) structure in the rough ER and do *N*-linked glycosylation. Then, *O*-glycan glycosylation take places in the Golgi apparatus, adding and removing various saccharides including galactose (Gal), mannose (Man), and *N*-acetylglucosamine (GlcNAc). Created glycans confer species specificity of each individual and play major antigen roles in xenotransplantation. Other types of glycans in the graft cell cause a more rapid immune response than protein antigens, causing the xenotransplantation to fail. In this regard, consider the glycans of pigs, which is the most suitable animal for xenotransplantation. α1, 3-Gal, Neu5Gc, and Sd(a), which are expressed in pigs but not in humans, cause various immune responses in humans to fail in xenotransplantation. Also, the major MHC or HLA should be considered crucial for transplantation. Their interspecies differences also act as barriers of xenotransplantation. If they can be removed from these interactions, it can be possible to success the xenotransplantation.

With the genetic modification method, CRISPR-Cas9, the genes of enzymes that make glycans in pigs can be Knocked Out (KO) to eliminate the glycan antigens itself. In addition, the immune reaction can be controlled by over-expression or suppression of control proteins that control each other's interaction. To identify these effects, the organs of verity of KO pigs will be transplanted to non-human-primate (NHP) to study the life span of transplanted organs. Then, we can check the changes in the life span of the organ to see if it will ultimately be possible to success

C.-H. Kim, *Glycoimmunology in Xenotransplantation*,
https://doi.org/10.1007/978-981-99-7691-1_4

of xenotransplantation. In xenotransplantation, glycan antigens raise three immune responses of acute humoral xenograft rejection (AHXR), HAR and ACR as the main forms.

4.2 Species-Specific Glycans

It is necessary to find out the order of the immune responses that occur in the human body during xenotransplantation. Basically, immunity occurs both innate immunes and acquired immunes. The immune system priority occurs by innate complements, antibodies, NK cells, and microphages. Macrophages are phagocytic cells as a heterogeneously functional population of mononuclear cells and fundamentally involve in inflammatory responses. As a similar heterogenous population, dendritic cells (DCs) in blood stimulate cytotoxic T cells and NK cells. They contribute to cellular leveled inflammatory process. But, DCs maturation mediated by their surrounded microenvironments decide either immunological progress or immune tolerance. NK cells are directly involved in cytotoxic cell death of donor cells in pig to human xenotransplantation. Pig ECs are directly subjected to the NK cell-mediated cell death even in the Ab absence, presumably due to human NK cell receptors that bind to α1, 3-Gal antigens. Naturally preformed or induced Abs bind to ABO blood group and xenograft Gal α1, 3Gal antigens. In transfusion or transplantation, the ABO antigen mismatching causes hyperacute rejection (HAR) by ABO antigens-reactive Abs. Similarly, the xenogeneic Galα1, 3Gal antigen induces HAR by Galα1, 3Gal-reactive Abs. For positive side of this phenomenon, this reaction can be applied to the cancer treatment through tumor-associated glycan-reactive Abs. Thus, the transplantation/transfusion and cancer immunotherapy are double-edged sword to occur via glycan antigens and Abs interaction. Therefore, α1, 3Gal epitope-downregulating strategy is also considered for the enzymatic competition between α1, 3Gal-T and other glycosyl-Ts in the trans-Golgi network by sialylation, GnT-III-ylation, and fucosylation (α1, 2FucT Tg pigs).

After a short period of time, chronic rejection occurs. This results in the coagulation dysregulation inflammatory response. Three stages of reaction occur when xenotransplantation is performed. These are HAR, AHXR, and ACR. These reactions are mediated by the types of glycans that people do not have. In people, the red blood cell (RBC) surfaced glycans representatively express each ABO blood type system. However, the pig's surface cells have glycans that act as epitopes because they are almost similar to human but slightly different structurally [1]. Among these groups, α1, 3-Gal has the fastest and greatest impact on the immune system [2]. It is made by α1, 3-GalT (or GGTA1) in pig cells, and this enzyme is held by most mammals. In the human and old monkey cases, the gene is deactivated by 92 bps deletion [3].

4.3 Hyperacute Xenograft Rejection (HAR)

The human's innate α1, 3-Gal-specific antibodies cause a sharp reaction to transplanted pig cells. This is called HAR (Fig. 4.1). This happens within minutes or hours of the transplantation by the naturally existing antibody (IgM) of the recipient, which catches the α1, 3Gal [4]. Complement proteins are a part of the innate immunity, and the activated complement proteins form complexes and attack cell membranes [5]. Serum antibodies that bind to α1, 3Gal epitope activate the complement system. The graft of NHP that contact with the pig's α1, 3Gal immediately activate the complement proteins, and they form a C3 complex and activate variety of complement protein complexes, destroying the cells by creating a membrane attack complex (MAC) [6]. Therefore, if HAR occurs, only the graft is lysis and destroyed. This complement reaction is controlled not only by antibodies but also by human complement regulatory proteins (CRPs). The known CRPs include CD46, CD55, and CD59 proteins. The human CRPs inhibit the complement system so do not become hemolysis even in the activation of the complement activation signal from the antibodies. HAR can be simply stopped by KO, the genes of these hCRPs. This reaction stops when transgenic pig expresses a human's CD46, CD55 or if the activation of antibodies [7] or complement proteins that have received signals for α1, 3-Gal is inhibited [8].

4.4 Acute Humoral Xenograft Rejection (AHXR)

However, the antibodies continue to reproduce in the body. The following reaction is caused by the antibodies that are regenerated within days and weeks later [9]. That response is the AHXR (Fig. 4.2). It is also called AVR, or DXR. This reaction is an immune response to newly created or recovered antibodies if the graft tissue is alive for more than a day. From this point on, not only the destruction of cells caused by antibodies and complement proteins occurs but also the combination of humoral and cellular immune response occurs. AHXR is accompanied by inflammation with a more powerful response than HAR. It is accompanied by mass reduction of

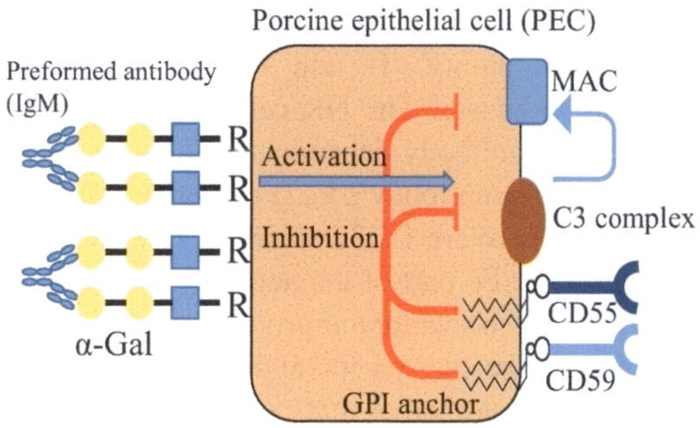

Fig. 4.1 Hyperacute xenograft rejection (HR)

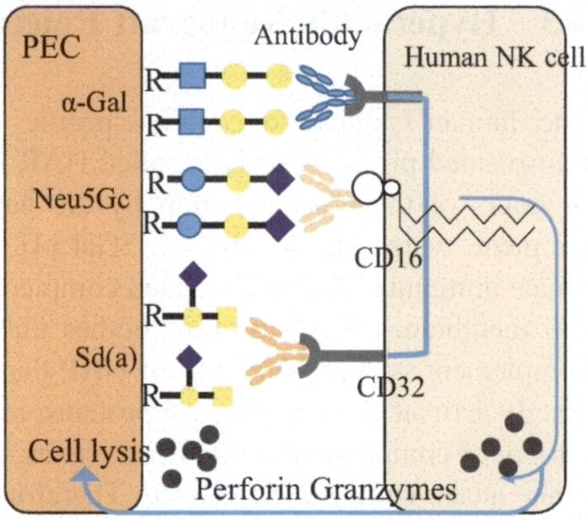

Fig. 4.2 Acute humoral xenograft rejection (AXR)

antibodies, complement proteins, and platelets. Also, cells go through necrosis, thrombosis, and infarct. It is found that AHXR appeared as an antigen other than α1, 3-Gal [10]. It turned out that GGTA1 KO (α1, 3GalT-KO) pig has antibody-meditated rejection within a few days. These antigens are Neu5Gc and Sd(a) blood group. These two glycans serve as antigens as they are not found in humans. Neu5Gc is made from Neu5Ac being hydroxylased by CMAH. In humans, they do not have Neu5Gc because the genes of the CMAH are deleted by the DNA mutation. The Sd (a) blood group is generated by β4GalNAc-T2 (B4GALNT2), which can also be a xenogenic antigen because it is not present in humans. Thus, these three glycan antigens act as the first generated barrier of xenotransplantation between pigs and humans.

4.5 Cellular Xenograft Rejection (CXR)

After the AHXR phase, various immune cells of microphages, mononucleose, NK cell, T-cells, and B-cells are all activated, so full-scale immune system begins to activate. This step is called acute CXR (ACXR), or simply CXR (Fig. 4.3). Previous HAR and AHXR were damaged only in graft and lysis at the tissue level, but from now on the whole organ is affected by the association of the vascular system. NK cells and microphages involve in the responses of the innate immunity. Also in adaptive immunity, T- and B-cells collaboratively induce the comprehensive immune responses. The NK cells can directly respond to cytotoxicity itself or go through the antibody-dependent cellular cytotoxicity pathway. The NK cell has a receptor of the antibody's Fc (FcR) on the surface and recognizes it and deliver lytic granules. There are CD32 which is FcγRII and CD16 which is FcγRIII, and they recognize the Fc part of the antibody that caught the pig antigens and activate NK cell. CD32 is a membrane protein and CD16 is a GPI-anchored protein, which is attached to the plasma membrane, transmitting intercellular signals [11]. If they recognize the antibodies that caught the glycan antigens, FcR and antibodies activate

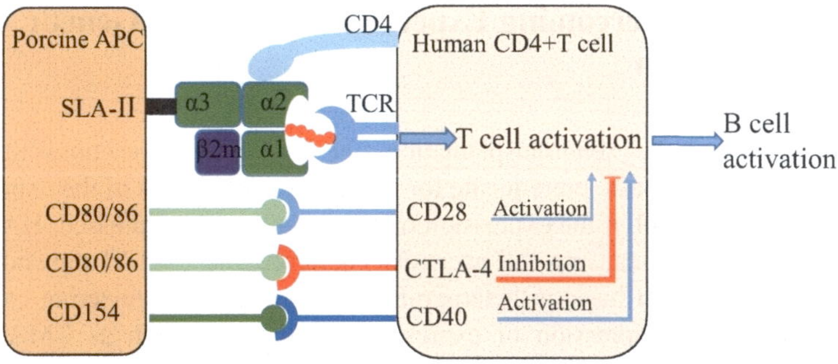

Fig. 4.3 Acute cellular rejection (ACR)

the NK cells to the release of the granzyme and perforin, which leads cell apoptosis [12].

The NK cell also recognizes MHC in addition to the previous three glycans. Naturally, the MHC molecular structure of pigs is different from that of humans. As a counterpart of human HLA or MHC, pig's MHC is named swine leucocyte antigen (SLA). Although they are homologous, they are different in the peptide chain structure. The human body distinguishes the differences of those and recognizes them as epitopes. Also, NK cell recognizes SLAs as epitopes itself. Thus, NK cells recognize the SLA class-I antibodies and cause antibody-dependent cellular cytotoxicity (ADCC). Furthermore, the key role of the NK cells is to recognize pig cells that are relatively deficient in HLA-I as non-self-cells and to apoptosis them. Also, T cell activation is accelerated as the microphage begins to produce proinflammatory tumor necrosis factor-α (TNF-α), interleukin-1(IL-1) and IL-6 cytokines. After these signals, adaptive immune begins to activate [12]. The pig antigen presenting cells (APC) recognize directly the T-cell receptor (TCR). At pig's APC, there are both SLA-I and SLA-II presenting the own peptide. The human TCR can interact with these SLA-peptide complexes especially more responsive to SLA-II. Not only this signal but also the interaction with the regulatory proteins should be considered. The pig's membrane proteins, CD80/86 and CD154 are expressed in the surface of pig's APC. CD80/86 is interacting with the human's CD28 and CD154 is interacting with the human's CD40. These signals activate the T-cell as a secondary regulator. The known surface antigen, CTLA-4, abbreviated from cytotoxic T-lymphocyte-associated antigen, interacts with CD80/86 and consequently, the interaction between CTLA-4 and CD80/86 inhibits the T-cell activation. Through these various regulation processes, T and B-cells are activated and humoral immunity begins and causes the entire organ to be damaged.

4.6 Rejection-Overcoming Experiments Through Genetic Manipulation

For the success of the xenotransplantation, the above three reactions must be suppressed. The most basic prerequisite for this is the suppression of the expression of the glycan. Above all, if the expression of α1, 3-Gal is not suppressed, HAR will appear within minutes, causing cell necrosis. Therefore, the GGTA1 pig should be assumed to be the most basic in heterograft. Further genetic engineering can then generate additional suppression of expression of enzymes such as CMAH and β4GalNAc-T2. Each of them is called GGTA1/CMAH [13] and GGTA1/B4GalNT2 double KO (DKO), and all three unexpressed objects are called GGTA1/CMAH/B4GalNT2 triple KO (TKO) [14]. The glycans least-expressed TKO has the highest success rate in xenotransplantation. In addition, as the initial immune response involves a complement system, the CRPs should also be considered. Since the CRPs of a pig does not prevent a human's compliment-mediated injury, the method of expression a human's CRPs (e.g., CD46, CD55, and CD59) is considered. In other words, it is more effective to make GTKO/hCRP once rather than to suppress GTKO or hCRPs one by one.

The examples of a real experiments are as follows. As a control group, the graft whose glycans are not removed at all is destroyed within 4–6 h after transplant. After transplant, the kidney from GTKO/hCD55 pigs survived for 125 days [15]. GTKO/CD46/CD55/thrombomodulin (TBM)/endothelial protein C receptor (EPCR)/CD39 pig's kidney survived for more than 136 days when it is transplanted to the baboon. The baboon died from infection rather than AHXR or ACXR in this experiment. And in an experiment in 2018, GGTA1/B4GalNT2 DKO pig's kidney was transplanted. At this time, the recipient survived for the longest time, as kidney functioned for 435 days [16]. In this experiment, death was caused by coagulation dysregulation. That is, in the above two experiments, the complement reaction produced by the glycan was perfectly suppressed. Additionally, it was found that the EPCR expression prevents coagulation dysregulation. EPCR importantly involves in anti-inflammatory responses and cytoprotection signaling in anticoagulation system [17]. It can also be seen that DKO pigs survive for a significantly longer period than GGTA1 pigs. It is the same with experiments on lung. The transplanted GTKO pig's lungs were out of HAR, but the function of xenograft was stopped in just 3.5 h by the coagulation dysregulation [18]. But, recipients, which received lungs from the GTKO/CD47/CD55 transgenic pig, saw a significant increase in survival dates upto 14 days [19]. This means that hCD47 has reduced the acute vascular adjustment.

4.7 Other Possibilities for Success Toward Rejection-Overcoming

The above experiments show that glycan expression suppression and hCRP expression can significantly inhibit HAR, AHXR, and even ACXR. In addition to the above methods, methods of inhibiting receptors in NK cells may also be valid. By using metalloprotease called ADAM17, activated NK cells lose their functional role via CD16 (FCγRIII) and CD62L downregulation, which are known as NK cell surfaced receptors. [20]. After passing through the xenotransplantation barrier of the above three reactions, the coagulation dysregulation becomes the main problem. In human, coagulation regulation proteins are known for tissue factor pathway inhibitor (TFPI), EPCR, TBM, and CD39. Overexpressed EPCR, TBM, EPCR, TFPI, and CD39 can protect the implanted grafts by prevention of the coagulation reactions [21].

References

1. Zhang R, Wang Y, Chen L, Wang R, Li C, Li X, Fang B, Ren X, Ruan M, Liu J, Xiong Q, Zhang L, Jin Y, Zhang M, Liu X, Li L, Chen Q, Pan D, Li R, Cooper DKC, Yang H, Dai Y. Reducing immunoreactivity of porcine bioprosthetic heart valves by genetically-deleting three major glycan antigens, GGTA1/β4GalNT2/CMAH. Acta Biomater. 2018;72:196–205. https://doi.org/10.1016/j.actbio.2018.03.055.
2. Huai G, Qi P, Yang H, Wang Y. Characteristics of α-gal epitope, anti-gal antibody, α1, 3 galactosyltransferase and its clinical exploitation (review). Int J Mol Med. 2016;37:11–20.
3. Perota A, Galli C. N-Glycolylneuraminic acid (Neu5Gc) null large animals by targeting the CMP-Neu5Gc hydroxylase (CMAH). Front Immunol. 2019;10:2396.
4. Platt JL, Fischel RJ, Matas AJ, et al. Immunopathology of hyperacute xenograft rejection in a swine-to-primate model. Transplantation. 1991;52(2):214–20.
5. Janeway CA Jr, Travers P, Walport M, et al. Immunobiology: the immune system in health and disease. In: The complement system and innate immunity. 5th ed. New York: Garland Science; 2001.
6. Yang YG, Sykes M. Xenotransplantation: current status and a perspective on the future. Nat Rev Immunol. 2007;7(7):519–31.
7. White DJG, et al. Expression of human decay accelerating factor or membrane cofactor protein genes on mouse cells inhibits lysis by human complement. In: Kootstra G, Opelz G, Buurman WA, van Hooff JP, MacMaster P, Wallwork J, editors. Transplant international official journal of the European Society for Organ Transplantation. Berlin/Heidelberg: Springer; 1992.
8. Brodsky RA. Paroxysmal nocturnal hemoglobinuria without GPI-anchor deficiency. J Clin Invest. 2019;129(12):5074–6.
9. Schuurman HJ, Cheng J, Lam T. Pathology of xenograft rejection: a commentary. Xenotransplantation. 2003;10:293–9.
10. Lin SS, Hanaway MJ, Gonzalez-Stawinski GV, Lau CL, Parker W, Davis RD, Byrne GW, Diamond LE, Logan JS, Platt JL. The role of anti-Galalpha1-3Gal antibodies in acute vascular rejection and accommodation of xenografts. Transplantation. 2000;70(12):1667–74.
11. Garman SC, Kinet JP, Jardetzky TS. Crystal structure of the human high-affinity IgE receptor. Cell. 1998;95(7):951–61.

12. Martens GR, Reyes LM, Butler JR, Ladowski JM, Estrada JL, Sidner RA, Eckhoff DE, Tector M, Tector AJ. Humoral reactivity of renal transplant-waitlisted patients to cells from GGTA1/CMAH/B4GalNT2, and SLA class I knockout pigs. Transplantation. 2017;101(4): e86–92.
13. Lutz AJ, Li P, Estrada JL, Sidner RA, Chihara RK, Downey SM, Burlak C, Wang Z-Y, Reyes LM, Ivary B, Yin F, Blankenship RL, Paris LL, Tector AJ. Double knockout pigs deficient in N-glycolylneuraminic acid and Galactose α-1, 3-Galactose reduce the humoral barrier to xenotransplantation. Xenotransplantation. 2013;20:27–35.
14. Li P, Estrada JL, Burlak C, Montgomery J, Butler JR, Santos RM, Wang Z-Y, Paris LL, Blankenship RL, Downey SM, Tector M, Tector AJ. Efficient generation of genetically distinct pigs in a single pregnancy using multiplexed single-guide RNA and carbohydrate selection. Xenotransplantation. 2015;22:20–31.
15. Higginbotham L, Mathews D, Breeden CA, Song M, Farris AB, Larsen CP, Ford ML, Lutz AJ, Tector M, Newell KA, Tector AJ, Adams AB. Pre-transplant antibody screening and anti-CD154 costimulation blockade promote long-term xenograft survival in a pig-to-primate kidney transplant model. Xenotransplantation. 2015;22:221–30.
16. Bertsimas D, Dunn J, Velmahos GC, Kaafarani HMA. Surgical risk is not linear: derivation and validation of a novel, user-friendly, and machine-learning-based predictive OpTimal trees in emergency surgery risk (POTTER) calculator. Ann Surg. 2018;268(4):574–83.
17. Bae JS, Yang L, Rezaie AR. Receptors of the protein C activation and activated protein C signaling pathways are colocalized in lipid rafts of endothelial cells. Proc Natl Acad Sci U S A. 2007;104(8):2867–72.
18. Nguyen B-NH, Azimzadeh AM, Zhang T, Wu G, Shuurman H-J, Sachs DH, Ayares D, Allan JS, Pierson RN. Life-supporting function of genetically modified swine lungs in baboons. J Thorac Cardiovasc Surg. 2007;133(5):1354–63.
19. Watanabe H, Sahara H, Nomura S, et al. GalT-KO pig lungs are highly susceptible to acute vascular rejection in baboons, which may be mitigated by transgenic expression of hCD47 on porcine blood vessels. Xenotransplantation. 2018;25:e12391.
20. Romee R, Foley B, Lenvik T, Wang Y, Zhang B, Ankarlo D, Luo X, Cooley S, Verneris M, Walcheck B, Miller J. NK cell CD16 surface expression and function is regulated by a disintegrin and metalloprotease-17 (ADAM17). Blood. 2013;121(18):3599–608.
21. Schmelzle M, Schulte Esch J 2nd, Robson SC. Coagulation, platelet activation and thrombosis in xenotransplantation. Curr Opin Organ Transplant. 2010;15(2):212–8.

Chapter 5
Glycosylation in Eukaryotes

5.1 Glycosylation via Eukaryotic Endoplasmic Reticulum and Golgi Apparatus

In eukaryotes from lower molds to terminally evolved humans, distinct biological event is so called glycosylation. Therefore, cell-specific and tissue-specific expression patterns of the glycans are crucial for their fate determination of roles to the organisms. This specific phenomenon is a complicated event specific only for the eukaryotic cells and populations who have the small organelle of endoplasmic reticulum (ER) and Golgi apparatus. This event is called complex pathway known as post-translational modification (PTM) of glycoproteins as well as glycolipid synthesis in eukaryotes. Glycans and carbohydrates are the third informative life chains in life science and biological system. The first and second life chains are the nucleic acids and proteins, respectively. They are synthesized in nuclear and ribosomes by strictly regulated signaling and ordered consequences in cells, tissues, organs, organ systems, and organisms. However, carbohydrates or glycans if they are coded life chains are synthesized or polymerized in ER.

In eukaryotes, cells regulate glycan synthesis depending on environmental stimulation from external and internal stimuli. The synthetic information is saved genes and totally encoded in their glycome. Thus, glycomes come from genomics, indicating the genomic regulation of glycomes. In the view of the first and second coding life chains, glycosylation is indeed considered to be belonged to just PTM event. Thus, some of the textbooks indicate that glycosylation is one of the PTMs in cells. However, it is precisely mentioned that general PTMs take place in the cytosol because the molecules or enzymes responsible for the protein modification are cytosolic enzymes, as the phosphorylases, kinases, methylases, and acetylases are localized in cytosols.

The fundamental question about glycosylation is raised for how to glycosylate in cells? The answer is very simple, and glycosylation occurs within the luminal ER compartments and Golgi apparatus because the enzymes responsible for the

C.-H. Kim, *Glycoimmunology in Xenotransplantation*,
https://doi.org/10.1007/978-981-99-7691-1_5

glycosylation are localized in the membrane-anchored or lumen forms in such organelle. The actors of the glycosylation are glycan synthetic and modifying enzymes such as glycosidases (for example, mannosidases or glucosidases) and glycosyltransferases (for example, Man-transferases or Gal-transferases) localized in the ER and Golgi apparatus. At the organelle, the transferases synthesize each specific glycan, depending on the linkage specificities of saccharides. Several nucleotide sugar substrates are utilized as donor sugars. As acceptor substrates, protein/lipids completely form glycoproteins and glycolipids. The enzymes are classified to type II transmembrane (TM) proteins by integral topologies. Their *N*-terminus are orientated within the cytoplasmic region. The enzymes are comprised of some structural domain regions with a short tail domain in cytoplasmic region, a stalk domain region, a transmembrane domain region and a COOH-terminal catalysis domain region. Information of glycan synthesis is transferred across the nuclear membranes to cytosol and again to the ER and Golgi machines. Glycan synthesis is sequentially progressed by the enzymatic actions of glycosyltransferases and glycosidases in the defined pathways of protein folding and quality control or biosynthesis of glycoproteins and glycosphingolipids [1].

A lot of glycosyltransferases catalyze the transferring reaction of a monosaccharide-linked donor to biphosphonucleotides activated, named nucleotide sugar donor substrate, to acceptor substrates such as glycoproteins and glycolipids. For example, UDP-Glc is subjected to enzymatic transfer of Glc to Glc2Man9GlcNAc2-dolichol acceptor substrate in ER, and consequently, Glc3Man9GlcNAc2-dolichol is formed.

5.2 Three Distinct Pathways for Glycoproteins, Glycolipids and Glycosaminoglycans

In eukaryotic glycosylation, three distinct pathways are well established for glycoproteins, glycolipids, and glycosaminoglycans. In the transplantation, glycan carbohydrates in glycoproteins and glycolipids are known as the xenoreactive antigens, while the glycosaminoglycans are known as extracellular matrix system for signal transduction and supporting bridges in cell architectures. In protein glycosylation, two different pathways of protein glycosylation are *O*-glycosylation and *N*-glycosylation pathways. GalNAc-T enzymes initiate the *O*-glycosylation on Ser/Thr residues in proteins. In humans, GalNAc-T enzymes are comprised of a family of about 20 members that transfer a monosugar from donor substrate GalNAc-UDP to the acceptor substrate Ser/Thr amino acids. GalNAc-T enzymes are ubiquitously present in all the animal cells, although they are not found in lower eukaryotes such as yeasts or non-animal, plants. To date, in Drosophila, 12 GalNAc-T enzymes are found, and in *Caenorhabditis elegans*, 9 enzymes are reported [2], giving an insight into evolution-based expansion of the enzymes in metazoans [3]. The forerunner GalNAc-Thr/Ser (Tn) serves as the next step of acceptor that catalyzes by three distinct additional glycosyltransferases that catalyze a Gal to yield the core 1. Seemingly, GlcNAc yields the core 3 and sialic acid (Sia) yields the sialyl-GalNAc, Core 1 sugar structure is abundantly expressed in most cells and tissues, while core 3 and its derivative structures are rarely expressed in the specific mucous epithelium or the salivary glands [4]. Structure of the core 1 sugar is synthesized by Core-1 GalNAc-T-1, known as T-synthase with molecular chaperone Cosmc [5]. Core 1 is further glycosylated to Core-2 structures by the β1,6-*N*-acetylglucosaminyltransferases (Core-2 GNTs) [1]. Core 2 glycan structures are also further glycosylated to poly-*N*-acetyllactosamine. O-glycan structures are terminally sialylated by sialyltransferases to transfer a terminator sugar, sialic acid residue.

5.3 Glycosylation of N-Linked Glycoproteins via ER and Golgi

Glycosylation of *N*-linked glycoproteins is found in all eukaryotes of three life domains. For the *N*-glycan role and function, it has been known that they regulate and modulate the surfaced protein functions of cells. But there is a big difference between prokaryotes and eukaryotes, as mentioned above. The *N*-glycosylation pathway of prokaryotes such as bacteria occurs at the periplasmic membrane of instead of the ER membrane, while that of eukaryotes take places at the ER membrane. Indeed, selectively targeted Asn residues of proteins are integrated to the bacterial periplasm or the eukaryotic ER lumen region. However, in the meaning of *N*-glycan itself, the *N*-glycosylation event is similarly operated in the two domains with a difference of ER or non-ER periplasmic membrane [6]. Therefore, the *N*-glycosylation pathways are well conserved across all domains of life, and this phenomenon in life and biological system reflects general concepts and fundamental aspects of *N*-glycosylation of glycoproteins. The reason why the eukaryotes utilize the ER system is described later somewhere.

The monosaccharide-based sugar blocks are assembled on isoprenoid and the dolichol-linked oligosaccharides are translocated to inner space such as ER lumen or periplasmic space is conserved in all systems. The oligosaccharide carrier is lipid moiety where bactoprenol is the carrier in the bacteria and dolichol (dolichol-pyrophosphate) in archaea and eukaryotes. Therefore, isoprenoid lipid, termed dolichol lipid, is the precursor oligosaccharide carrier of the ER glycosylation reaction, where the dolichol serves as the carrier of oligosaccharide moiety in the ER membrane. Dolichol synthesis is mediated by a specific enzyme, cis-prenyltransferase from initial substrate, farnesyl-pyrophosphate, via the step by step reaction that sequentially adds C5 isoprenoid units [7, 8]. Chain length of dolichols is different from each species, giving a species specificity, as the length fate of dolichol is depended on each distinct cis-prenyltransferase enzyme [9]. In the initial sugar synthesis of the building blocks, the three carbohydrate building blocks are GlcNAc, Man and Glc, supplied from as activated sugar nucleotides of donor substrates GDP-Man and UDP-GlcNAc in the ER membrane at the cytoplasmic side. However, dolichol-phospho-mannose (Dol-P-Man) and dolichol-phospho-glucose (Dol-P-Glc) are used for lumen mannosyltransferase (Man-T) and glucosyltransferase (Glc-T). Dol-P-Man and Dol-P-Glc are produced from Dol-phosphate (Dol-P), UDP-Glc and GDP-mannose (GDP-Man) in the cytosolic region and they are translocated across the ER membrane. Dol-P-Man synthase synthesizes donor substrate Dol-P-Man [10].

Oligosaccharyltransferase (OST), known as an *en block* enzyme, is a catalytic enzyme that transfers the reaction of the oligosaccharyl GlcNAc2Man9Glc3 moiety from the dolichyl-pyrophosphate (Dol-P-P) lipid carrier to the target of conserved protein Asn-X-Ser/Thr amino acid sequence at the amide group of Asn residue. The initially added *N*-glycan has 14 sugar residues of Glc3Man9GlcNAc2 biosynthesized in the ER region. They have a structure branched on a lipid dolichol-pyrophosphate anchor. The oligosaccharide is sequentially elongated by glycosyltransferases resident in lumen. However, those glycosyltransferases utilize Dol-P-bound monosaccharide substrates. The transferred core oligosaccharide structure from the Dol-P anchor to the consensus amino acid sequence of Asn-X-Ser/Thr is subjected to main stream of *N*-glycosylation via PTM pathway known as a protein quality control. At this step, the terminal Glc residue acts as an OST-recognizing signal in the luminal ER. OST is a membrane-bound complex enzyme and well conserved for its homology from protozoan to mammals. Translational products as the membrane-integrated proteins are subjected to co-translational insertion into the ER, the starting area for *N*-glycosylation. Folded proteins are in a sugar-delivering mode to export them to the next step in the Golgi apparatus. In the Golgi, protein *N*-glycan synthesis is terminated and also *O*-glycan synthesis is initiated. In eukaryotic *N*-glycosylation, OST initiates the *N*-glycosylation pathway to synthesize an *N*-glycan sequence. The OST initially adds the precursor sugars to the Asn amide group in the Asn-X-Ser/Thr consensus sequence [11]. In the highly conserved *N*-glycosylation site of the conserved Asn-X-Ser/Thr sequence in proteins, X is any a.a. residue but not Pro residue. It is co-translationally to Asn residue via "en bloc" transfer by OST and linked via GlcNAc to Asn residue specifically selected as the consensus Asn-X-Ser/Thr sequence of *N*-glycosylation a.a residue of the target proteins.

The glycosylation reactions are step-wisely and sequentially processed by "en bloc" transfer, leading to biosynthesis, Although the *N*-glycan sugar chain processing is well investigated, several questions are still raised how *N*-glycosylation influences the biological function and fate in the meaning of *N*-glycoproteins. In particular, the fate of sugar xenoantigenic chains is not well clarified to the extent to clinical application. Initially, *N*-glycans have been crucial for precise protein folding formation and quality control by the ER chaperone. The relevant protein folding and quality control (QC) process are regulated by the ER-resident chaperone and lectin cycle named "calnexin/calreticulin cycle" known as the term of ER-associated degradation (ERAD). The ERAD event is progressed by unfolded protein response (UPR). To do such mission in protein quality control, accessory proteins including ERp57, GRP170, DnaJ-like cofactor (ERdj1–5), HSP70 family BiP (GRP78),

HSP90 member GRP94, calnexin, calreticulin, and protein disulfide isomerase (PDI) with ER-resident proteins are harmonist. If the folding is succeeded, then they are transported to R-to-Golgi intermediate compartment (ERGIC) toward Golgi. However, if they are misfolded or incompletely folded, they transport to specific pericentriolar subcompartment toward ER-derived QC compartment (ERQC) and finally to ERAD and ubiquitination process or apoptosis.

5.4 Evolutionally Conserved N-Glycosylation

In eukaryote cells, the co-translationally transferred *N*-linked oligosaccharides are commonly conserved, although protozoa or lower eukaryotes show "high mannose" typed oligosaccharide structures in the transferred proteins. Therefore, the "high mannose" typed oligosaccharides are considered to be intermediates during evolution of the protein *N*-glycosylation [6]. The Glc3Man9GlcNAc2 block is a common assembly of all pathways in eukaryotes [12]. Regardless of the biological diversity and evolutional origin, the oligosaccharides transferred to con-translationally synthesized proteins are commonly and evolutionarily conservative in the glycoprotein *N*-glycosylation pathway occurred in ER of all eukaryotic organisms.

Glycoprotein *N*-glycosylation in the evolution of eukaryotic species is one of the wide processes and modifications of proteins. For example, as calculated from the mouse glycoproteomic analysis, over 2300 *N*-glycosylated proteins having more than 5000 glycosylation amino acid sites were theoretically known [13]. The huge *N*-glycosylation sites indicate that the transferring enzyme OST is very unique in its substrate specificity affordable for such a wide distribution and diversity. Considering the short and linear sequence of the theoretical *N*-glycosylation site of NxS/T acceptor amino acids, it is potentially assumed to insert at many amino acid sites linked in the proteins. For example, results of the *N*-glycoproteomic analysis from differently independent species indicate a highly diverse number of *N*-glycosylation sites on glycoproteins [14], giving a conclusion of the highly flexible adaptation of this type of *N*-glycosylation modification. Moreover, the *N*-glycosylation event of glycoproteins is most frequently occurring in multicellular eukaryote cells rather than in unicellular eukaryotic species [14]. Therefore, this strongly suggests the hypothesis that diversification in *N*-glycoprotein glycosylation is a driving and dynamic force of biological evolution, functioning in the multicellular evolutionary adaptation [15]. Furthermore, the results also strengthen that the course of eukaryotic evolution needs a selective pressure to enhance the protein *N*-glycosylation degree in secreted proteins of eukaryotic cells. N-glycosylated glycans in proteins can lead to and precisely direct the protein folding and maturing machinery to defined locations of the polypeptides. Each proteins specifically and differentially processed N-glycosylated glycan in protein is marked as a folding and maturing marker to be used for a covalently attached signaling molecule, which controls and processes the nonmatured polypeptide in the ER, and consequently, depending on the modified structures of glycans, the same protein *N*-glycan goes through a polypeptide-, cell type, or species-dependent manner in the Golgi apparatus, leading to *N*-glycosylation structure diversity towards each different functional activity [16].

5.5 Glycan Chain-Functional Specificity

The glycan-specific functions of *N*-glycosylate oligosaccharides in proteins confer functionally diverse eukaryotic *N*-glycomes. The target-interacting characters of glycans and the versatile specificity of OST in substrate recognition determine the fate of *N*-glycosylation in glycoproteins in eukaryotic cells. Totally, *N*-glycosylation takes place through the two membrane-bound organelles of ER–Golgi, which are systemically compartmentalized in ER and Golgi apparatus. In the ER processing followed by the Golgi apparatus, the *N*-glycosylated oligosaccharide structures are gradually modified by "trimming" pathway with the removal of unneeded saccharide and then re-addition of needed saccharides by the "processing" pathway with saccharides such as Gal, Fuc, or Sia. Currently, the Golgi is an organelle to edit the structures of *N*-glycosylated saccharides in *N*-glycoprotein trafficking. Functional deletion or mutation of Golgi apparatus-resident enzymes and transporting proteins cause defections in *N*-glycan modification and processing. The Golgi glycosyl enzymes are enriched in cisternaes like complex compartmentation uits [17, 18]. When glycoproteins across the ER membrane and ER lumen traverse the Golgi apparatus, Golgi-trafficked *N*-glycoproteins encounter Golgi-localized glycosyltransferases in a mode that early glycosylation-related enzymes are localized in the cisternae nearest the ER compartment or termed cis-Golgi and late-resident glycosylation enzymes are resident in the trans side of Golgi stack termed trans-Golgi compartment. As mentioned before, another *O*-glycan glycosyltransferase resident in the Golgi apparatus generates *O*-glycan structures if glycoproteins [19]. Golgi-resident glycosyltransferases also operate the retrograde transport reversely acting within the trans- to cis-cisternae direction in Golgi [20].

5.6 Glycosylation in Glycosphingolipids and Gangliosides

Finally, in glycolipid biosynthesis, glycosyltransferases also transfer monosaccharide residues to ceramide-based sphingolipids composed of sphingosine and fatty acids. Glycosphingolipids are synthesized from ceramide by ER- or Golgi-localized glycosyltransferases [21, 22]. The substrate specificities of such glycosyltransferases are distinguished from those of *N*-glycoprotein-specific and O-glycoprotein-specific glycosyltransferases. Basically, ceramides are attached with glucose (Glc) and galactose (Gal) to form lactosylceramide (LacCer). Upon Sia linking to the LacCer substrate by GM3 synthase, the simplest ganglioside GM3 is made, naming a ganglioside in the condition of sialic acid residues in LacCer. Gangliosides are synthesized by three different pathways, named ganglioside a-, b-, and c-series

[23]. Indeed, when *N*-acetylgalactosamine (GalNAc) is attached to GM3 by GM2/GD2 synthase, which is the a-series ganglioside pathway. In contrast, when additional Sia is attached by GD3 synthase, the b- or c-series ganglioside is in the biosynthetic pathway. Therefore, the disialyl GD3 synthase and GM2/GD2 synthases determine the ganglioside pathways. Therefore, when the GM2/GD2 synthase initially act to their acceptor substrates, ganglioside a-series are synthesized, while a stepwise catalysis of the initial GD3 synthase and the next GM2/GD2 synthase leads to synthesis of the ganglioside b-series pathway. Likewise, when GT3 synthase catalyzes to disialyl acceptor GD3 before the catalysis of GM2/GD2 synthase, ganglioside c-series pathway is operated [24, 25].

5.7 Glycan Antigen-Therapeutic Applications

To regulate and control carbohydrates function in therapeutic applications, glycosyltransferases can be manipulated by gene KO, transgene, and alteration of substrate specificities. Indeed, the cell surface carbohydrates can be modified to a specific phenotype of cells to display their altered interaction between the glycan epitope and lectins and/or antibodies. If the sugar structures are some markers of xenoantigenic determinants, xenotransplantation needs the removal of the antigens. For example, α1,3Gal-T gene has been targeted in pigs to delete the Galα1,3Gal xenoantigenic epitope, and α1,2-fucosyltransferase (α1,2Fuc-T) is rather transduced to decrease the Galα1,3Gal level [26].

References

1. Bard F, Chia J. Cracking the glycome encoder: signaling, trafficking, and glycosylation. Trends Cell Biol. 2016;26(5):379–88.
2. Tian E, et al. Recent insights into the biological roles of mucin-type O-glycosylation. Glycoconj J. 2009;26:325–34.
3. Bennett EP, Mandel U, et al. Control of mucin-type O-glycosylation: a classification of the polypeptide GalNAc-transferase gene family. Glycobiology. 2012;22(6):736–56.
4. Iwai T, Inaba N, et al. Molecular cloning and characterization of a novel UDP-GlcNAc: GalNAc-peptide beta1,3-Nacetylglucosaminyltransferase (beta 3Gn-T6), an enzyme synthesizing the core 3 structure of O-glycans. J Biol Chem. 2002;277(15):12802–9.
5. Ju T, Cummings RD. A unique molecular chaperone cosmc required for activity of the mammalian core 1 beta 3- galactosyltransferase. Proc Natl Acad Sci U S A. 2002;99:16613–8.
6. Schwarz F, Aebi M. Mechanisms and principles of N-linked protein glycosylation. Curr Opin Struct Biol. 2011;21:576–82.
7. Swiezewska E, Danikiewicz W. Polyisoprenoids: structure, biosynthesis andfunction. Prog Lipid Res. 2005;44:235–58.
8. Welti M. Regulation of dolichol-linked glycosylation. Glycoconj J. 2012;30(1):51–6.
9. Schenk B, Fernandez F, Waechter CJ. The ins(ide) and out(side) of dolichyl phosphate biosynthesis and recycling in the endoplasmic reticulum. Glycobiology. 2001;11:61R–70R.

10. Orlean P, Albright C, Robbins PW. Cloning and sequencing of the yeast gene for dolichol phosphate mannose synthase, an essential protein. J Biol Chem. 1988;263:17499–507.
11. Aebi M. N-linked protein glycosylation in the ER. Biochim Biophys Acta. 1833;2013:2430–7.
12. Samuelson J, Banerjee S, Magnelli P, Cui J, Kelleher DJ, Gilmore R, Robbins PW. The diversity of dolichol-linked precursors to Asn-linked glycans likely results from secondary loss of sets of glycosyltransferases. Proc Natl Acad Sci U S A. 2005;102:1548–53.
13. Zielinska DF, Gnad F, Wisniewski JR, Mann M. Precision mapping of an in vivo N-glycoproteome reveals rigid topological and sequence constraints. Cell. 2010;141:897–907.
14. Zielinska DF, Gnad F, Schropp K, Wisniewski JR, Mann M. Mapping N-glycosylation sites across seven evolutionarily distant species reveals a divergent substrate proteome despite a common core machinery. Mol Cell. 2012;46:542–8.
15. Varki A. Biological roles of oligosaccharides: all of the theories are correct. Glycobiology. 1993;3:97–130.
16. Helenius A, Aebi M. Roles of N-linked glycans in the endoplasmic reticulum. Annu Rev Biochem. 2004;73:1019–49.
17. Klumperman J. Architecture of the mammalian Golgi. Cold Spring Harb Perspect Biol. 2011;3: a005181.
18. Papanikou E, Glick BS. Golgi compartmentation and identity. Curr Opin Cell Biol. 2014;29: 74–81.
19. Röttger S, et al. Localization of three human polypeptide GalNAc-transferases in HeLa cells suggests initiation of O-linked glycosylationthroughout the Golgi apparatus. J Cell Sci. 1998;111:45–60.
20. Lowe M. Structural organization of the Golgi apparatus. Curr Opin Cell Biol. 2011;23:85–93.
21. Gault CR, Obeid LM, Hannun YA. An overview of sphingolipid metabolism: from synthesis to breakdown. Adv Exp Med Biol. 2010;688:1–23.
22. Maccioni HJ, Quiroga R, Spessott W. Organization of the synthesis of glycolipid oligosaccharides in the Golgi complex. FEBS Lett. 2011;585(11):1691–8.
23. Yu RK, Tsai YT, Ariga T. Functional roles of gangliosides in neurodevelopment: an overview of recent advances. Neurochem Res. 2012;37(6):1230–44.
24. Bieberich E. Synthesis, processing, and function of N-glycans in N-glycoproteins. Adv Neurobiol. 2014;9:47–70. https://doi.org/10.1007/978-1-4939-1154-7_3.
25. Kim CH. Ganglioside Biochemistry. Singapore: Springer; 2022. https://doi.org/10.1007/978-981-15-5815-3.
26. Milland J, Christiansen D, Sandrin MS. Alpha1, 3-galactosyltransferase knockout pigs are available for xenotransplantation: are glycosyltransferases still relevant? Immunol Cell Biol. 2005;83(6):687–93.

Chapter 6
Human Red Blood Cell (RBC) Blood Groups System

Understanding of the blood group system of human gives benefits in transplantation and transfusion medicine. Karl Landsteiner received Noble Prize in 1930, and Jan Jansky classified human blood groups to four different blood group types. The human blood group genes are present in allelic or located on the identical chromosomal region. Blood type indicates a specific reaction pattern to antiserum in a given system. Blood group is antigenic determinant antigens expressed on cellular surfaces of RBC. Human blood group system thus contains several distinct antigenic determinants and expanded to transfusion complications [1]. Currently, 33 blood group systems classified show over 300 antigens [2, 3], and each coding gene is autosomal. Only XG and XK genes are located on sexual chromosome X. Both X and Y chromosomes carry MIC2 gene. Amino acid sequence variation causes the antigen polymorphisms in Rh, Kell, and glycolipids as well as glycoproteins in ABO type (Table 6.1). Currently, four main types of A, B, AB, and O blood groups commonly present in human are expressed due to their inherited determination. Each group is either Rhesus (Rh) D positive (+) or RhD negative (−), possibly giving eight blood groups. The Rh antigens are only present on RBCs. Each blood group antigen has been identified by human natural antibodies and blood antigen molecules on the RBC surfaces. Currently, ISBT termed from the International Society of Blood Transfusion has a definition of 33 independent systems of human blood group (Table 6.1). The known specific antigens, but not RhD, are expressed on normal blood cells and tissues. Apart from ABO and Rhesus systems, the RBC membrane antigens represent the blood groups. The ABO antigens widely present in the body refer to histo-blood group antigens. Some systems have only one determinant, and other systems have on more multiple determinants such as MNS and Rh. The MNS blood group consists of three genes, while the Chido/Rodgers blood group and Rh group are comprised of two genes. Some systems consist of a single gene. In addition, more 30 RBC antigens present in very high or very low prevalent levels are not characterized yet. Blood groups are inherited, and some groups are environmentally, developmentally, or by disease changed.

C.-H. Kim, *Glycoimmunology in Xenotransplantation*,
https://doi.org/10.1007/978-981-99-7691-1_6

Table 6.1 ISBT blood group system classification of human

Blood group (symbol)	ISBT no.	Antigen no.	Gene(chromosome)	Cluster of differentiation
(a) Major human blood group systems				
ABO (ABO)	1	4	ABO(9q34.2)	
MNS (MNS)	2	46	GYP-A,-B,-E (4q31.22)	CD235(glycophorin)
Pk (P1PK)	3	2	A4GALT(22q13.2)	CD77 (Gb3)
Rh (RH)	4	52	RhD, RhCE (1p36.11)	CD240
Lutheran (LU)	5	20	LU(BCAM) (19q13.32)	CD239
Kell (KEL)	6	32	KEL(7q34)	CD238
Lewis (LE)	7	6	FUT3(19p13.3)	
Duffy (FY)	8	6	FY(DARC) (1q23.2)	CD234
Kidd (JK)	9	3	SLC14A1(18q12.3)	
(b) Minor human blood group systems				
Diego (DI)	10	22	SLC4A1(17q21.31)	CD233(anion exchanger)
Yt (YT)	11	2	ACHE(7q22.1)	(acetylcholinesterase)
Xg (XG)	12	2	XG(Xp22.33)	CD99(MIC2)
Scianna (SC)	13	7	ERMAP(1p34.2)	(ERMAP)
Dombrock (DO)	14	7	ART4(12p12.3)	CD297 (GPI-Ap)
Colton (CO)	15	4	AQP1(7p15.1)	(aquaporin-1)
Landsteiner wiener (LW)	16	3	ICAM4(19p13.2)	CD242(ICAM-4)
Chido/Rodgers (CH/RG)	17	9	C4A, C4B (6p21.32)	(C4 protein)
Hh (H)	18	1	FUT1(19q13.33)	CD173 (Le^y)
Kx (XK)	19	1	XK(Xp21.1)	(membrane transporter)
Gerbich (GE)	20	11	GYPC(2q14.3)	CD236(glycophorin C)
Cromer (CROM)	21	16	DAF(1q32.2)	CD55(complement DAF)
Knops (KN)	22	9	CR1(1q32.2)	CD35(C3-binding protein)
Indian (IN)	23	4	CD44(11p13)	CD44(surface glycoprotein)
Ok (OK)	24	3	BSG(19p13.3)	CD147(basigin)
Raph (RAPH)	25	1	CD151(11p15.5)	CD151(tetraspanin)
John Milton Hagen (JMH)	26	6	SEMA7A(15q24.1)	CD108 (GPI-semaphorin)
I (I)	27	1	GCNT2(6p24.2)	(β1,6GlcNAc-T)
Globoside (GLOB)	28	1	B3GALNT1 (3q26.1)	(β1,3GalNAc-T1)
Gill (GIL)	29	1	AQP3(9p13.3)	(aquaporin 3)
RhAG (RHAG)	30	3	RHAG(6p12.3)	CD241(ammonia transporter)

6.1 Major and Minor Blood Group Systems of Human

In humans, major blood group systems include ABO (ABO), MNS (MNS), D235 (glycophorin), Pk (P1PK), Rh (RH), Duffy (FY), Kidd (JK), Kell (KEL), Lewis (LE), and Lutheran (LU). *Minor human blood group systems include* Dombrock (DO), Gerbich (GE), Knops (KN), Globoside (GLOB), Diego (DI), Yt (YT), Xg (XG), CD99(MIC2), Chido/Rodgers (CH/RG), Cromer (CROM), Scianna (SC), Hh (H), Kx (XK), Colton (CO), Landsteiner Wiener (LW), Indian (IN), Ok (OK), Raph (RAPH), Gill (GIL), John Milton Hagen (JMH), I (I), and RhAG (RHAG).

6.2 Known Function of Blood Groups

Among the 33 antigens of human blood group, five antigens contain the glycan antigens (ABO, H, P1Pk, I, and GLOB). The two antigens are present in blood plasma (LE and CH/RG). The other 23 antigens contain the membrane proteins of RBCs [4, 5] with 5 major proteins of DI, Rh, RhAG, MNS, GE, and CO known as membrane transporter proteins, and the remaining 17 antigens are not functionally known. The other antigens are known to function as receptor and ligand, enzyme, and glycocalyx [6]. The antigen-null phenotypes do not show immune responses [7]. But Knops involves in complement receptor (CR)-1 [8], and Cromer is related to decay acceleration factor (DAF) [9].

Human ABO groups play mainly a role in blood hemostasis [10], exerting quantitative blood plasma levels of von Willebrand factor (vWF) and coagulation factor VIII (FVIII). They are associated with ischemic stroke, myocardial infarction, and venous thromboembolism by A and AB blood groups [11] via ABO glycosyl transferases modulation of thrombosis. Cerebral venous thrombosis is associated with non-O blood types [12]. In fact, human blood group ABO is involved in the severe preeclampsia, where AB group shows a risk with 2.1-folds rather than non-AB group [13]. ABO system is also associated with malignancies. A group shows a potent chronic hepatitis-B infection and pancreatic cancer phenotypes [14]. B group shows a potent ovarian cancer phenotype [15]. O blood group can protect against falciparum malaria and falciparum malaria infection by blocking rosette formation [16], but enhances the *Vibrio cholerae* infection in the O1 El Tor and O139 strains.

6.3 Blood Grouping, Cross-Matching Requisition, and Antibody Screening

Pregnant women routinely receive a blood group test to prevent any complications when the mother blood type is RhD negative, but the child shows father-inherited RhD-positive type. The pregnant RhD-negative women must receive only RhD-negative blood. Agglutination and transfusion are our understanding of RBC antigens associated with diseases. Screening, typing, and cross-matching of blood groups are emphasized. Before blood transfusion, blood group test and cross-matching requisition are required. Blood requisition form should include blood group testing, typing, and cross-checking. Because the ABO incompatibility causes complement-derived intravascular hemolysis, ABO group typing is tested using blood RBCs. The A and B antigens as well as blood plasma serum A and B antibodies are used for the typing. Then, the Rh type is tested using by Rh-negative. Cross-matching is mixed with donor RBCs and the recipient serum [17] via three steps of ABO incompatibility and additional MN, P, and Lewis antigenic detection. The second step checks Rh antibodies using the first step reactants. The third step detects Rh, Kidd, Kell, and Duffy antibodies by adding antiglobulin sera to the second step reactants. The first two steps detect the fatal HTR. For antibody screening, commercialized RBCs that carry all the antigens capable of production of hemolytic antibodies are used to detect those serum antibodies present in the recipient and donor.

6.4 Solution of Blood Transfusion Replacement

During alloimmunization, immune recognition of incompatible RBCs and hemolysis are problematically fats to the recipients, needing antigen modulation. The ABO blood system can be enzymatically converted to distinct antigens of blood group. Goldstein and Lenny developed the "enzyme conversion of the ABO group to O-RBC group, calling ECO-RBC. For example, galactosidase enzyme converts the B antigen to the O antigen [18]. The developed treatment does not affect the RhD, RhC, and RhE types as well as Duffy, Kell, Kidd, Lewis, MNS, and Lutheran groups due to Gal residue. However, conversion of A antigen by enzyme is not easy because of two A2 and A1 type sugar structures [19]. ECO-RBCs were also generated from donors of A1, A2, B, or AB types [20]. The enzymic antigen conversion can be also applied to allotransplantation [21]. As another strategy, RBC antigen masking by polyethylene glycol treatment is known. The last strategy is in vitro RBC generation via genetically manipulated stem cells [22].

6.5 Blood ABO Group System of Human

Human ABO blood group system is currently known as the most dominant blood group and is important for transplantation and transfusion. Normally, from individuals above the age of 6 months, their A antigen-specific or B antigen-specific antibodies are generated. Blood A antigen and B-specific antibody are co-presented vice-versa in sera, while group O antigen is present without serum A and antigens but with both anti-A- and anti-B-specific antibodies. The ABO's precursor form is *H-antigen* expressed in RBC surfaces in the ABO blood antigens. Rare Bombay phenotype individual carries the homozygous H gene but does not synthesize H-antigen. The Bombay phenotype individuals have isoantibodies to H-antigen and also to A/B antigens.

6.5.1 Historical Progress in ABO Blood Type

The most general blood groups are the Rhesus blood and ABO blood in transfusion and plantation in humans. The barrage of the A blood type (and/or B) antigenic RBC cells is present as an enzyme action consequence of each specific glycosyltransferase that attach distinct monosaccharides of conformation-depended sugar structures with the protein Rh (D) from D antigen. A lot of different systems of blood group are decided by different RBC antigens. However, clinically important blood systems are reported to only blood group ABO system and Rh system. Four major blood groups are listed, namely O, A, B, and AB blood group in the human ABO blood system [23]. Among them, the O, A, and B types were discovered by Austrian pathologist Karl Landsteiner, a Viennese medical doctor (M.D), as the ABO blood group of human in 1900, which is the very year rediscovered for the Mendel's finding. The Karl Landsteiner's ABO blood group of human has been accepted as one of the successful outcomes in the human blood transfusion and medicine field and also in the medical history. The chronological history of human blood group ABO system has greatly been traced to:

1. History of blood group ABO, No. 1: The first milestone as in the first description on 1666 year on animal transfusions when Mr. Boyle published the animal transfusions in the journal, Philosophical TRANSACTIONS (1665-1666, No 1, p385-388). His description dealt with the animal transfusions with a thematic title of Tryals suggested by Mr. Boyle to Dr. Lower for the blood transfusion improvement from live animals into other animals [24].
2. History of blood group ABO, No. 2: The second milestone of the progress in ABO discovery is the question and performance of the question "Can a fierce dog become more tame if transfused with blood of a cowardly dog? (this sentence is the copied sentence of the cited reference)."
3. History of blood group ABO, No. 3: The third milestone of the progress of ABO discovery arises to the actual performance of the historic year 1666, which

treatment by the lamb blood transfusion was banned for insanity issue by national congress in year 1676.

4. History of blood group ABO, No. 4: The fourth milestone of the blood group ABO progress is the events that in the years 1900–1901, Austria Viennese pathologist Karl Landsteiner found the first human ABO blood group. In June 14, 1868, Karl Landsteiner born in Vienna Baden, Austria, investigated human medicine during 1885 and 1891 in Vienna, Austria. In 1896, he obtained a medical assistantship as appointed at the Austria Hygiene Institute. Then, he moved to the Pathological Anatomical Institute of the University of Vienna as a medical assistant in 1897, and in 1911, he was nominated as a pathological anatomy professor. During 1908–1919, Karl Landsteiner was again appointed as a Vienna municipal institute director. Karl Landsteiner also stayed as a medical prosector in den Haag (The Haugu), Holland, until 1921. During 1923–1939, Karl Landsteiner moved to the Rockefeller Institute for Medical Research, New York, USA, and also an emeritus researcher since 1939 to his death. His work represents that human individual's blood serum agglutinates other individual's RBCs through finding the A-specific and B-specific agglutinins currently known as agglutinable antibodies and agglutinogens as agglutinable antigens. In 1900, thus, Karl Landsteiner described the agglutination of healthy individual's serum agglutinate ..... blood cells from independent individual' [25]. He designated human bloods to A, B, and C groups (but not O group) [26]. The groups represent indeed the individually inherited characteristics of each RBC cells in human. From the findings, he received Nobel medicine and physiology prize in 1930. Karl Landsteiner was deceased in 1943 in New York, USA. Since his accomplishment, it marked the beginning of safe transfusion [27].
5. History of blood group ABO, No. 5: In 1990, Von Dungerne and Hirschfeld discovered that human blood type A and type B are genetically inherited by a simple way of the Mendelian heredity as Mendelian dominant traits, showing each dominant inheritance, but blood group O type is recessive [25].
6. History of blood group ABO, No. 5: The fifth milestone of the blood group ABO progress is the events that in 1990, the blood group ABO characteristics were further categorized in the 1990s by Yamamoto et al. [28]. Yamamoto and Hakomori [29] found the gene sequences of glycosyltransferases α1,3GalNAc-transferase (GTA) and α1,3Gal-transferase (GTB). Interestingly, analysis of the gene sequences exhibited only simple changes in 4 a.a residues between glycosyltransferases GTA and GTB enzyme, indicating that these two GTA and GTB glycosyltransferases share high homologous characters with the two enzymes with a difference at only 4 a.a residues from the whole polypeptide with 354 amino acid residues. Amino acid alterations of these four residues include Arg176Gly, Gly235Ser, Leu266Met, and Gly268Ala and converts the enzyme's substrate specificity from GTA to GTB. Therefore, the human blood type A is directed by α1,3GalNAc-specific GTA and blood type B is by α1,3Gal-specific GTB in the enzyme biochemistry and genetics of the blood groups. Also, the blood type AB is expressed by both two enzyme genes GTA and GTB as inherited by a classical Mendelian way. In addition, as known for the single

cis-AB protein, the O blood group type reflects the DNA sequence substitution or deletion, which is related for activity absence in functional enzymes of GTA and GTB [30–35]. Approximately natural mutations are detected in 200 alleles, giving an insight of "nature's gift to structure-function studies."

6.5.2 The ABO Blood Group Discovery Exploits the Human Genetic Polymorphism Studies and Additional Discoveries

Multiple polymorphic system has been understood through better human genetic definition and variation of population. One of them is the ABO group, as it is an elucidated case of the first human polymorphism. Historically, Landsteiner had not gained any human inheritance and not considered about the genetic inheritance trait of the human blood type ABO markers. However, his A and B blood types are well fit to inherit for the simple inherited Mendelian traits of a dominant character. Except for blood groups, Karl Landsteiner was also a pioneer of the hematology, and he studied poliomyelitis, paroxysmal hemoglobinuria, and syphilis [36].

In 1902, 2 years later of his ABO discovery (current O was formerly C), the fourth AB type was discovered by his coworkers Alfred Von Decastello and Adrian Sturli, and consequently, the human blood ABO blood system was completed [37]. Eight years later from the AB type discovery, von Dungern and Hirschfeld established the capital letters such as the A, B, O, and AB, which is currently generalized for the official blood group nomenclature in 1910 [38, 39]. The reason why A type is designated is based on the most abundantly occurred blood group in Europe. The rare group observed in Europe was designated as B. Finally, the unagglutinable RBC group was automatically designated as O. The simultaneously co-existence of A and B group was designated as AB. Soon, the A1 and A2 known as the A subgroups were found by two scientists of Von Dungern and Hirschfeld [38–40] and many other rare markers resembled from the ABO system have been discovered since the mid-1930s [41–43]. For example, later Landsteiner studied blood transfusions, genetic variation and inheritance in humans [44]. The inheritance was suggested by Epstein and Ottenberg in 1908 [20] with Mendel's laws and 2 years later in 1910, Emil von Dungern and Ludwig Hirschfeld suggested its genetic theory [38].

For the inheritance study on the ABO groups, Emil von Dungern and Ludwig Hirschfeld established the scientific center in 1910 [46]. Thereafter, from the discovery of ABO groups, RBC membrane has been a major subject to further discover until the middle of the twentieth century. The Emil von Dungerne and Hirschfeld (1884–1954, also called Hirszfeld) studied the familial inheritance and found that blood groups of A and B types are dominantly inherited, but O blood type is recessive. The blood groups of A type and B type linked to familial genetic variation has largely influenced Hirschfeld and colleagues during the World War I [27].

When during 1907–1911, Hirschfeld studied in the von Dungern department of the Institute for Cancer Research, Heidelberg, Hermany. He also elucidated the distribution of the human ABO antigens and their caused immunological conflicts [46]. In 1928, he published a monograph of "Constitution Serology and Blood Group Research in a German language of 'Konstitutionsserologie und Blutgruppenforschung" in Berlin. Since the 1950s, the biochemistry of the human ABO system was outlined, confirming that the membrane terminal *N*-acetyl-D-GalNAc directs A type, D-Gal directs type B, and L-Fuc directs type O of human blood group [43, 47]. ABO blood group has been geographically related when compared to human races, the theory of the two different human races was raised, proposing that the population with A types is come from central or northern Europe, while the B type is come from India. Those two human races are together mixed in the middle region between Indai and Europe. Today, this suggestion is recognized to be too ridiculous, not easily acceptable [48].

As mentioned earlier, Karl Landsteiner was an opener of dawn of a breakthrough in the blood type diversity of hematology. By his pioneered great discovery of approximately 700 human RBC antigens, the human blood group is now subgrouped into 30 systems and documented by ISBT, International Society of Blood Transfusion [49]. Each individual has unique blood groups, except for absolutely the same blood groups such as the identically same twins or triplets.

6.5.3 Hardy-Weinberg Law Has Been Applied to Three ABO Alleles

Historically, Mendelian dominant has been attractively expanded to the study of population distribution due to its application interests. In population, generation to generation change was subject of genetic discussion since the rediscovered outcomes of Mendel's finding. Simply applied theory named "binomial" calculation law and later, this is also so-called "the Hardy-Weinberg (HW) law" in university text biology book was proposed by mathematician Hardy [50] and Weinberg (1908) [51]. They proposed random mating equation in the condition of selection absence of nonselection environments. Calculated frequency of the 3 genotypes that are derived from two alleles located on a single locus is that *A* (its frequency p) and a (its frequency $q = 1\text{-}p$), giving the sum of the calculation, p^2 *AA*, $2pq$ *Aa*, and q^2 *aa*, after one random mating generation, regardless of starting frequencies, without allele frequencies change in random mating generations. Thus, population distribution speculates to multiple genetic alleles located on a single locus. Hardy wrote his "Pure mathematics I," a student textbook at Cambridge University in 1950s. Bernstein [52] applied the Hardy-Weinberg law to three alleles of the ABO [52, 53].

References

1. Reid ME. Blood group systems. In: Reference module in life sciences, Brenner's Encyclopedia of genetics. 2nd ed; 2013. p. 351–2. Blood Group Systems.
2. Lögdberg L, Reid ME, Zelinski T. Human blood group genes 2010: chromosomal locations and cloning strategies revisited. Transfus Med Rev. 2011;25:36–46.
3. Westhoff CM. The rh blood group system in review: a new face for the next decade. Transfusion. 2004;44:1663–73.
4. Anstee DJ. The functional importance of blood group-active molecules in human red blood cells. Vox Sang. 2011;100:140–9.
5. Daniels G, Reid ME. Blood groups: the past 50 years. Transfusion. 2010;50:281–9.
6. Denomme GA. The structure and function of the molecules that carry human red blood cell and platelet antigens. Transfus Med Rev. 2004;18:203–31.
7. Luo H, Chaudhuri A, Zbrzezna V, He Y, Pogo AO. Deletion of the murine Duffy gene (Dfy) reveals that the Duffy receptor is functionally redundant. Mol Cell Biol. 2000;20:3097–101.
8. Rao N, Ferguson DJ, Lee SF, Telen MJ. Identification of human erythrocyte blood group antigens on the C3b/C4b receptor. J Immunol. 1991;146:3502–7.
9. Telen MJ, Hall SE, Green AM, Moulds JJ, Rosse WF. Identification of human erythrocyte blood group antigens on decay-accelerating factor (DAF) and an erythrocyte phenotype negative for DAF. J Exp Med. 1988;167:1993–8.
10. Zhang H, Mooney CJ, Reilly MP. ABO blood groups and cardiovascular diseases. Int J Vasc Med. 2012;2012:641917.
11. Wiggins KL, Smith NL, Glazer NL, Rosendaal FR, Heckbert SR, Psaty BM, et al. ABO genotype and risk of thrombotic events and hemorrhagic stroke. J Thromb Haemost. 2009;7: 263–9.
12. Tufano A, Coppola A, Nardo A, Bonfanti C, Crestani S, Cerbone AM, et al. Non-O blood group as a risk factor for cerebral vein thrombosis. Thromb Haemost. 2013;110:197–9.
13. Hiltunen LM, Laivuori H, Rautanen A, Kaaja R, Kere J, Krusius T, et al. Blood group AB and factor V Leiden as risk factors for pre-eclampsia: a population-based nested case-control study. Thromb Res. 2009;124:167–73.
14. Wang DS, Chen DL, Ren C, Wang ZQ, Qiu MZ, Luo HY, et al. ABO blood group, hepatitis B viral infection and risk of pancreatic cancer. Int J Cancer. 2012;131:461–8.
15. Gates MA, Wolpin BM, Cramer DW, Hankinson SE, Tworoger SS. ABO blood group and incidence of epithelial ovarian cancer. Int J Cancer. 2011;128:482–6.
16. Anstee DJ. The relationship between blood groups and disease. Blood. 2010;115:4635–43.
17. Miller RD. Transfusion therapy. In: Miller RD, Ericksson LI, Fleischer LA, Weiner-Kronish JP, Young LA, editors. Miller's anesthesia. 7th ed. Philadelphia: Churchill Livingstone Elsevier; 2010. p. 1739–66.
18. Goldstein J, Siviglia G, Hurst R, Lenny L, Reich L. Group B erythrocytes enzymatically converted to group O survive normally in A, B, and O individuals. Science. 1982;215:168–70.
19. Goldstein J. Conversion of ABO blood groups. Transfus Med Rev. 1989;3:206–12.
20. Liu QP, Sulzenbacher G, Yuan H, Bennett EP, Pietz G, Saunders K, et al. Bacterial glycosidases for the production of universal red blood cells. Nat Biotechnol. 2007;25:454–64.
21. Kobayashi T, Liu D, Ogawa H, Miwa Y, Nagasaka T, Maruyama S, et al. Alternative strategy for overcoming ABO incompatibility. Transplantation. 2007;83:1284–6.
22. Hashemi-Najafabadi S, Vasheghani-Farahani E, Shojaosadati SA, Rasaee MJ, Armstrong JK, Moin M, et al. A method to optimize PEG-coating of red blood cells. Bioconjug Chem. 2006;17:1288–93.
23. Mahapatra S, Mishra D, Sahoo D, Sahoo BB. Study of prevalence of A2, A2B along with major ABO blood groups to minimize the transfusion reactions. Int J Sci Res. 2016;5(3):189–905.
24. Boyle M. Tryals proposed by Mr Boyle to Dr. Lower, to be made by him, for the improvement of transfusing blood out of one live animal into another; promised to Number 20. P 357. Phil Trans R Soc. 1665–1666;1:385–8.

25. Landsteiner K. Zur Kenntnis der antifermentativen, lytischen und agglutinierenden Wirkungen des Blutserums und der Lymphe. Centralbl Bakteriol. 1900;27:357–62.
26. Landsteiner K. Ueber Agglutinationserscheinungen normalen menschlichen Blutes. Wien Klin Wschr. 1901;14:1132–4.
27. Hirschfeld L, Hirschfeld H. Serological differences between the blood of different races. The result of researches on the Macedonian front. Lancet. 1919;2:675–9.
28. Yamamoto F, Clausen H, White T, Marken J, Hakomori S. Molecular genetic basis of the histo-blood group ABO system. Nature. 1990;345(6272):229–33.
29. Yamamoto F, Hakomori S. Sugar-nucleotide donor specificity of histo-blood group A and B transferases is based on amino acid substitutions. J Biol Chem. 1990;265(31):19257–62.
30. Yamamoto F, McNeill PD, Hakomori S. Human histo-blood group A2 transferase coded by A2 allele, one of the A subtypes, is characterized by a single base deletion in the coding sequence, which results in an additional domain at the carboxyl terminal. Biochem Biophys Res Commun. 1992;187(1):366–74.
31. Joziasse DH, Shaper JH, Jabs EW, Shaper NL. Characterization of an alpha 1-3-galactosyltransferase homologue on human chromosome 12 that is organized as a processed pseudogene. J Biol Chem. 1991;266(11):6991–8.
32. Tsuji T, Hakomori S, Osawa T. Identification of human galactoprotein b3, an oncogenic transformation-induced membrane glycoprotein, as VLA-3 alpha subunit: the primary structure of human integrin alpha 3. J Biochem. 1991;109(4):659–65.
33. Yamamoto F, McNeill PD, Hakomori S. Identification in human genomic DNA of the sequence homologous but not identical to either the histo-blood group ABH genes or alpha 1-3 galactosyltransferase pseudogene. Biochem Biophys Res Commun. 1991;175(3):986–94.
34. Tsuji T, Yamamoto F, Miura Y, Takio K, Titani K, Pawar S, Osawa T, Hakomori S. Characterization through cDNA cloning of galactoprotein b3 (Gap b3), a cell surface membrane glycoprotein showing enhanced expression on oncogenic transformation. Identification of Gap b3 as a member of the integrin superfamily. J Biol Chem. 1990;265(12):7016–21.
35. Yamamoto F, Marken J, Tsuji T, White T, Clausen H, Hakomori S. Cloning and characterization of DNA complementary to human UDP-GalNAc: Fuc alpha 1-2Gal alpha 1-3GalNAc transferase (histo-blood group a transferase) mRNA. J Biol Chem. 1990;265(2):1146–51.
36. Speiser P, Smekal FG. Karl landsteiner. 2nd ed. Wien: Brüder Hollinek; 1975.
37. von Decastello A, Sturli A. Ueber die Isoagglutinine im Serum gesunder und kranker Menschen. Münch Med Wschr. 1902;49:1090–5.
38. Von Dungern E, Hirschfeld L. Ueber Vererbung gruppenspezifischer Strukturen des Blutes. Z Immun Forsch. 1910;6:284–92.
39. Von Dungern E, Hirschfeld L. Ueber gruppenspezifische Strukturen des Blutes. Z Immun Forsch. 1911;8:526–62.
40. Thomsen O, Friedenreich V, Worsaae E. Über die Möglichkeit der Existenz zweier neuer Blutgruppen; auch ein Beitrag zur Beleuchtung sogenannter Untergruppen. Acta Pathol Microbiol Scand. 1930;7:157–90.
41. Prokop O, Uhlenbruck G. Lehrbuch der menschlichen Blut- und Serumgruppen. 2nd ed. Leipzig: Georg Thieme; 1966.
42. Race RR, Sanger R. Blood groups in man. 6th ed. Oxford London Edinburgh Melbourne: Blackwell Scientific Publications; 1975.
43. Prokop O, Göhler W, Mayr W, Geserick G, Radam G. Human blood groups. Montreal: D. J. Paradis Editions Inc.; 1986.
44. Apecu RO, Mulogo EM, Bagenda F, Byamungu A. ABO and rhesus (D) blood group distribution among blood donors in rural south western Uganda: a retrospective study. BMC Res Notes. 2016;9(1):513. https://doi.org/10.1186/s13104-016-2299-5.
45. Epstein AA, Ottenberg R. A simple method of performing serum reactions. Proc N Y Pathol Soc. 1908;8:117–23.
46. Geserick G, Wirth I. Genetic kinship investigation from blood groups to DNA markers. Transfus Med Hemother. 2012;39:163–75. https://doi.org/10.1159/000338850.

47. Dahr W. Alloantigens of proteins and glycoproteins in membranes of human red blood cells. In: Mayr WR, editor. Advances in forensic Haemogenetics 2. Berlin Heidelberg: Springer; 1988. p. 8–15.
48. Bodmer W. Genetic characterization of human populations: from ABO to a genetic map of the British people. Genetics. 2015;199:267–79.
49. Society ARC (n.d.). Blood types-Australian red cross blood service. All about blood. www.donatedblood/com.all/aii-aboutblood/blood-types.
50. Hardy GH. Mendelian proportions in a mixed population. Science. 1908;28:49–50.
51. Weinberg, W., 1908 Uber den Nachweis der Vererbung beim Menschen. Jahresh. Ver. Vaterl. Naturkd. Wurttemb. 64: 369–382. English translations in Edwards 2008 Boyer, S. H. (Editor) Papers on human genetics, Prentice Hall, Englewood Cliffs.
52. Bernstein F. Zusammenfassende Betrachtungen uber die erblichen Blutstrukturen des Menschen. Z Indukt Abstamm Vererbungsl. 1925;37:237–370.
53. Crow JF. Felix Bernstein and the first human marker locus. Genetics. 1993;133:4–7.

Chapter 7
Non-ABO Blood Group Systems

7.1 MN and P Blood Group Secretors and Rhesus System (RH) and Rhesus (D), Lutheran, Kell, Duffy, and Kidd Systems

After the ABO blood group discovery, Karl Landsteiner and his colleagues further discovered additional markers of blood groups of *MN and P*, and a third, Secretor, was found in 1930.

7.1.1 MNS Antigen System

MNS antigens are expressed by glycophorin (GP) A and GPB on RBSs. Landsteiner and Levine described the MNS antigens for the first time in 1927. Two glycoprotein genes of GPA and GPB are involved and regulated by co-dominant LM and LN alleles on an autosomal chromosome 4. Two anti-M and anti-N IgM can cause transfusion reactions (TR). As mentioned above, the MN group consisting of the S and s factors was reported in 1947, and the basic biochemistry of the MNS antigen system and MNSs, based upon two genes of GPA and GPB, were outlined in the 1970s. The MNSs antigens expressed on two sialylglycoproteins of GPA and GPB in RBC. Although the GPA gene, called GYPA, and GPB gene, called GYPB, encode their proteins, but the genes GYPA and GYPB are very similar, enabling gene crossing-over or conversion. This similarity frequently forms hybrid genes to generate variant glycophorins [1]. The generated glycophorin variants are serologically distinguished by Miltenberger diagnosis subsystem. Now, the Miltenberger diagnosis nomenclature system has been further progressed to utilize terms of GP system by the first propositus [2]. Therefore, the Miltenberger subtype III termed the name of Mi.III has been replaced by the name GP.Mur to denote the glycophorin generated by a hybrid GYP gene (B-A-B). Through the world, GP.Mur population

C.-H. Kim, *Glycoimmunology in Xenotransplantation*,
https://doi.org/10.1007/978-981-99-7691-1_7

frequency is very low under about 5.0%. For example, Caucasian Miltenberger antigens and GP.Mur antibodies are rarely present, but in Far east Asian, such as Taiwanese frequency, the ratio is relatively high [3]. The GP.Mur sera contain diversely polyclonal antibodies known for Hil-specific, Mia-specific, Mur-specific, MINY-specific, and MUT-specific antibodies. Thus, they are totally called anti-"Mia'. St^a antigen [4] is another variant type caused by a hybrid GYP gene (B-A) having a single crossing over at each No. 3 intron of GYPA gene and GYPB gene with expression of a hybrid protein of the *N*-terminal GYPB and C-terminal GYPA. To date, 4 A, B, C, and D distinct alleles for St^a are found depending on the crossing-over location [5].

The polymorphism in the coding amino acids of the specific GPA and GPB glycoproteins produces MN system [6, 7]. The strongest Rh antigen factor is the factor D. Rh system is very complex because the RH system consists of other additional antigen factors of C, Cw, c, E, and e and several genetic variants, indicating the difficulty in understanding. For example, a blood scientist, Wiener, found multialleles present at 1 locus on chromosome. Similarly, two biologists, Fisher and Race, found 3 gene loci closely adjacent each other in the D-C-E order [6, 7]. The known RH antigens are free of carbohydrate antigens, but palmitolyated polypeptides in the RH antigens, as studied biochemically. As described earlier, the Kell, Duffy, and Kidd systems consist of glycoproteins as their antigens. Likely to the human ABO groups, the human RH system is very important for blood transfusion and perinatal care medicine. Additionally, in forensic science, the RH and MNSs blood systems are highly recognized to be very valuable for their criminal discrimination.

In 1969, after the end of the World War II, general guidelines of human blood transfusion were initially documented, as described for the German language written to "Bundesgesundheitsblatt" (in English, corresponding to Federal Health Gazette), and recent version was born in 2002 [8]. The Federal Health Office defined the following systems from 1990 [9]. They are (1) ABO, Duffy (FY), Kell (K), MNSs, Rhesus (RH); (2) GC, GM, HP, INV; (3) ACP, ADA, AK, ESD, GLO, PGD, PGM1; and (4) HLA-A, B, C, and DR locus biomarkers, which are serologically defined. Among the 4 blood marker systems, antigens present in erythrocyte membrane are ABO, Duffy (FY), Kidd (JK), MNSs, and RH.

7.1.2 Rhesus System

Again continuously, 40 years later, the Rhesus system (RH) and Rhesus (D) antigen were discovered by Landsteiner and Weiner, separately in 1927 [10, 11]. The newly discovered groups were the Rhesus (Rh) factors [12] by an unusually displayed disease caused in the hemolytic newborn. A Rh factor-lacked mother, which the strong Rh factor is D antigen, reacting to the D expressed on the RBCs of the mother's child, causing the hemolytic disease of the newborn (HDN). This is because the mother produced the antigen D-specific antibodies that destroyed her fetal RBCs. Like this, several Rh determinants were newly detected through modified assaying techniques of RBC agglutinations. Importantly, R. A. Fisher [13] well described the Rh determinants for closely linked three loci. Interestingly, each locus has two independent alleles of Cc, Dd ND Ee, which determine each corresponded antigens C and c, etc. However, only d-specific antibodies were not confirmed yet, although the reason is now known. It was because of that the reason was known much later, simply due to its lacking or the absence of D.

At that time, a Swiss geneticist Ernst's explication largely influenced the Fisher's interpretation, where Ernst explicated that the colorful flower, Primula incompatibility system shows complex-typed with a close linkage. In such Rhesus or D system, human blood groups are either Rh-negative of Rh-positive, which is decided by the absence or presence of D antigens expressed on RBC surfaces. Genes of the Rhesus (Rh) or D antigens are determined by Chromosome 1, which has three pairs of close-linked alleles. All human population shares with the same Rhesus and ABO systems. But in human populations, the two systems differ only in the specific type of distributions and frequencies in independent ethnic groups and different races [14]. In the meaning of clinical importance, the distribution of ABO system is crucial with the Rh system, as the knowledge is directly linked to the blood transfusion. In fact, the Rh identification in the human bloods is essential for the prevention of erythroblastosis fetalis, the abnormally present erythroblasts in the fetal bloods. This erythroblastosis fetalis occurs in cases of Rh-positive fetus-bearing Rh-negative mother.

The most prevalent blood group is ABO system, and Rhesus (Rh) system is the second one. Red blood cells have frequently a distinct protein RhD antigen. Among the 50 defined Rh blood group antigens, 5 antigens are clinically regarded. Rh factor as an immunogenic D-antigen is found on individual RBC surface, referring to as Rh + (or D-antigen condition). If not, Rh- (D-antigen absence) is referred. Each person carries 1 group among the 8 groups of A RhD+ (A+), A RhD-e (A-), B RhD+ (B+), B RhD- (B-), O RhD+ (O+), O RhD- (O-), AB RhD+ (AB+), and AB RhD- (AB-). Of interests, the O RhD- blood (O-) can safely transfuse to others, effective for medical emergencies. Most recipients are thus safe because the RhD O- blood is compatible with all ABO and RhD groups. Anti-Rh antibodies are absent in the RhD-negative RBCs of individuals. Only RhD-positive RBCs-exposed individuals can produce anti-Rh IgG antibodies and cross the pregnant placenta. Rh-positive child-born pregnant Rh-negative mothers can display prophylaxic shock against Rh immunization.

7.1.3 Lutheran System

Lutheran system contains four allelic antigen pairs with single a.a substituted glycoprotein antigen in the Lutheran group coded on chromosome 19. Lutheran-specific antibodies are not frequent and clinically important.

7.1.4 Kell System

The Kell system is originated from Mrs. Kellacher serum that causes hemolysis with her newborn infant's erythrocytes. Kell erythrocyte antigens are reactive with anti-K antibodies and the third immunogenic antigen with the first ABO and second Rh antigens. Twenty-five Kell antigens are known, and anti-K antibodies severely induce the hemolytic TR and hemolysis disease of fetus and newborn (HDFN).

7.1.5 Duffy System

Haemophilia patient Duffy contains Duffy-antigen known as Fy glycoprotein on the RBC surfaces. Duffy-antigen known as a nonspecific chemokine receptor is a human malarial receptor. Antigens of Fy^a and Fy^b are expressed on the Duffy glycoprotein, and they form combinatory four different Fy(a + b−), Fy(a + b+), Fy(a − b+), and Fy(a − b−) phenotypes. They cause IgG types to induce hemolytic TR.

7.1.6 Kidd System

Antigen Kidd is called Jk antigen, known as a membrane urea transporter glycoprotein of renal endothelial cells and RBCs. Rare Kidd-specific antibodies cause severe TR. The Kidd antigens were discovered in the Mrs. Kidd serum in the case she delivered a HDFN-bearing baby and later defined by anti-Jka antibody. After the first discovery of Jk^a, the second and third Jk^b and Jk3 were discovered.

7.1.7 Dombrock Antigens in Dombrock System

The Dombrock antigens (Do) consist of two Do^a and Do^b antithetical antigens. In addition, Dombrock antigens have five highly prevalent antigens of Gy^a, Hy, Jo^a, DOYA, and DOMR. Thus, Dombrock antigens by ISBT terminology include Do^a (DO1), Do^b (DO2), Gy^a (DO3), Hy (DO4), Jo^a (DO5), DOYA (DO6), and DOMR (DO7). Gy(a–) phenotype is the defected Dombrock system. Do^a-specific antibody was found in 1965 and Do^b-specific antibody binds the antithetical antigen [15, 16]. The glycoprotein Do has 5 N-glycans. The two antibodies specify three phenotypes (Table 7.1) [17]. Human *Dombrock system* is based on the RBC glycoprotein Do antigens. Do^a-specific antibodies were found in a blood-transfused patient. Do antigen names are based on blood donors. Second Dombrock antigen, Do^b, is Do^a's antithetical antigen. Three antigens of Gy^a (calling for Gregory antigen), Hy (calling for Holley antigen), and Jo^a (calling for Joseph antigen) are

Table 7.1 The Dombrock antigen system. RBCs are Do(b–) if *JO/JO* encodes the Do antigen but are Do(b+^{W}) if *HY/JO* encodes it

Antigen type	Reaction by Do^{a}-specific (sp) Ab	Do^{b}-sp Ab	Gy^{a}-sp Ab	Hy-sp Ab	Jo^{a}-sp Ab
Do(a + b–)	Yes	0	Yes	Yes	Yes
Do(a + b+)	Yes	Yes	Yes	Yes	Yes
Do(a–b+)	No	Yes	Yes	Yes	Yes
Gy(a–)	No	No	No	No	No
Hy–	No	Low	Low	No	Low
Jo(a–)	Low	Low	Yes	Low	No

present on the Do antigen on glycosylphosphatidylinositol (GPI)-anchored ADP-ribosyltransferase-4 (ART-4) but not on a GPI-paroxysmal nocturnal hemoglobinuria type III RBCs [18]. The gene (DO, ART4) is located on the chromosome 12. *ART4* gene variations generate the different Do antigens. The Do^{b} is different from the Do^{a}. Do^{b} antigen has an RGD motif. Do^{a} and Do^{b} are prevalent and genetic markers. Anti-Do^{a} and anti-Do^{b} cause hemolytic TR. The Gy^{a}, Hy, and Jo^{a} antigens are 99% more of individuals. Do^{a} is mainly expressed on RBCs as well as minorly on bone marrow, lymphocytes, lymph nodes, spleen, intestines, fetal heart, ovaries, and testes. The expression level of Dombrock antigens is high in the fetal liver.

DO gene located on the chromosome 12 and *ART4* gene generates several Do phenotypes and consists of 3 exons, 314 amino acids, and GPI-anchored protein. The DO*B allele encodes an RGD motif. *DO*A* allele is similar to its chimpanzee *DO* homolog, which encodes Asn instead of Asp (D), the *DO*B* allele RGD motif seems is normal phenotype. Donor typing for RBC DO*A, DO*B, and Do(b–) is crucial for blood transfusion to anti-Do^{b} patients. More reliable test is the DNA analysis than hemagglutination due to not easy availability of anti-Do^{a} and anti-Do^{b} antibodies. No disease is related with the Do^{a} nor the [Gy(a–)] null. Higher prevalent Gregory (Gy^{a}) and Holley (Hy) antigen phenotypes are co-related each other. Gy(a–) phenotype is Hy– RBC, while the Hy– is Gy(a+^{W}). Gy^{a} and Hy located on the same glycoprotein are upgraded to the ISBT High Incidence Antigen Gregory Collection [19]. The higher prevalent Jo^{a} is associated with Gy^{a}. Additional high prevalent antigen is Jc^{a} as the Jo^{a} is related to Gy^{a} and Hy. RBC Gy(a–) are Do(a–b–), indicating that the Gy(a–) is the Dombrock-null phenotype.

*Among phenotypes of the antigen Do^{a} (DO1) and Do^{b} (DO2), DO*A* and *DO*B* alleles show 3 nucleotide difference in exon 2 and determine the Do type. Do(a–) or Do(b–) bloods are determined by hemagglutination detection. In the Gy^{a} (DO3) antigen, Dombrock Gy^{a} absences is the Do$_{null}$ [Gy(a–)] phenotype, caused by *DO* gene silencing. Regarding the Hy (DO4) antigen, the Hy+/Hy– phenotypes is generated by the nucleotide change in exon 2 to create Gly108Val as the Hy null on an allele. The 793G (265Asp) indicates the invariable Do(a–b+) Hy– RBC phenotype. The 898G allele is found in the original Hy– proband, terming the HY1. For the Jo^{a} (DO5), nucleotide 350 T generates the Jo^{a} null phenotype. The Jo (a–) RBC is the *DO*JO/JO* RBC or *DO*HY/JO* RBC. The 793A (265Asn) indicates the Jo(a–) RBC Do(a + b–). *DO*HY/JO* RBC genotype is the Do(a+b+^{W}). For the Jc^{a} RBC, Jc(a–) is a combined form of *DO*HY* and *DO*JO* allele. In the DOYA (DO6) and the DOYA− proband, the silenced Do^{a} leads to Do(a–b–) with weak Gy^{a}, Hy, and Jo^{a} RBC. The DOMR (DO7) is caused by the nucleotide differences with DOMR+/DOMR− and show the DOMR antigen null. This explains the DOMR− with weak Do(b+) RBC. DOMR− RBCs show a weak Gy^{a}, Hy, and Jo^{a} RBC. For additional DO allele forms including DO*A-HA/SH/WL, DO*B-SH/BWL/B-SH-Q149K, and DO*B-I175N alleles are found. Thus, the DO is useful for anthropology studies, as the DO allele of chimpanzee is similar to the wild-type human.

Do^{a}, Do^{b}, Hy, and Jo^{a} are weakly immunogenic, as Dombrock antibodies are well studied. Although hemagglutination is classically based on the antigen and antibody reaction, Dombrock antigen-reactive antibody is not easy to determine. Anti-Hy and anti-Jo^{a} are not easily differentiated for each other. Hy and Jo^{a} are not easy to explain. Do does not recognize complement. Antibody production is activated by transfusion and pregnancy and normally found in serum. The anti-Gy^{a} antibody is specific for only one epitope. HDFN is not caused but hemolytic transfusion reaction (TR) is raised. Anti-Do^{a} or anti-Do^{b} causes TR. Anti-Hy causes biphasic destruction of Hy+ RBCs [20]. Hy-specific, Gy^{a}-specific, and Jo^{a}-specific antibodies also cause TR. In DO system, antibodies do not induce HDN. If DO gene is artificially expressed, they act as immunogen [21] to strongly agglutinate human RBCs, but not Gy(a–). Two Mabs agglutinate other great apes, not other organisms including lesser apes, old and new world monkeys, dogs, mice, prosimians, sheep, or rabbits [22]. Because mice bear a DO homolog, mice RBCs are not reacted, indicating that the anti-Do is restricted to only apes [23].

7.1.8 Gerbich System of GPC and GPD Antigens

Gerbich system consists of glycophorin A-D. Among them, GPC (CD236C) and GPD (CD236D) antigens carry the Gerbich antigens as minor sialoglycoproteins present on RBCs [24]. GPC expression is found in earlier stage in normal erythroid differentiation and leukemic differentiation as well as in T- and B-cells, monocytes, and myeloid precursor cells [25, 26]. Multifunctional sialoglycophorin C (GPC) and GPD function as the antigens of Gerbich. Gerbich antigens attached on GPC and GPD of human RBCs form RBC shape. *The Gerbich antigens consist of high- and low-frequency GPC and GPD antigens. The human RBC glycophorins anchor the cytoskeleton to give RBC stability. Type I transmembrane (TM) GPC and GPD proteins* consist of 3 extracellular *N*-terminal domains, a domain in TM region, and a C-terminal domain in cytoplasmic region. GYPC gene encodes GPC and GPD. But a leaky translation event is started from two initiation codons and consequently produces GPD form as a short-truncated GPC part [27]. The GPC is a glycoprotein with 128 a.a. residues having 1 *N*-glycosylation site and 12 *O*-glycosylation site [28]. GPD started from GPC Met22 has a protein with 107 a.a. residues without the *N*-glycosylation site. CPC and GPD complex with proteins 4.1 and p55 to form a spectrin-actin structure in membrane [29].

Location of the GYPC gene is on chromosome 2q14-q21, and it has a 13.5 kb length in DNA sequence and 4 exons [27]. Exons 2 and 3 are highly homologous, and consequently, contribute to unequal crossing over with exon 2/3 loss, giving Gerbich-negative phenotypes. These RBCs loss the high-prevalent Ge2 (exon 2), Ge3 (exon 3), and Ge4 (exone1) antigens. The exon 2 loss causes the Yussef (Yus) phenotype, occurred in the Mediterranean and Middle East region. Exon 3 loss causes the Gerbich phenotype occurred in Melanesians [28]. Ge1 is caused by lack in a high-prevalent Gerbich antigen, which is observed in Melanesians and now obsolete. Ge2 has first been known as the Gerbich antigen. The Gerbich variants are emerged from *P. falciparum* infected survival and selective pressure. Deletion event of alleles occur in endemic regions of *P. falciparum* malaria infection, like the Middle East region and Papua New Guinea. The known RBCs such as rare Gerbich and Yus have short exon deleted variants in GPC and GPD proteins with different glycosylation [30]. Four alleles in the Yus show exon 2 deletion with deletion of its adjacent introns. The Gerbich type-specific 3 alleles similarly show exon 3 deletion and its adjacent introns [31]. Deletions of exon 3 and exon 4 are known as Leach type I [28], and Exon 3 nucleotide changed type is known as Leach type II, which causes a frame-shift mutation and generates a prematured stop codon. Both Leach type I and II generate the null RBC called Leach or Gerbich null [32]. The rare Leach phenotype is found in Western European ancestry. Complete GPC and GPD loss display elliptocytic RBCs and mild anemia. Six high-prevalent form of Gerbich antigens are such Ge2, Ge3, Ge4, GEPL (or GE10), GEAT (or GE11), and GETI (or GE12). Five low-prevalent forms of Gerbich antigens are Ls^a (or GE6), An^a (or GE7), Dh^a (GE8), Wb (or GE5), and GEIS (or GE9). Other antigenic forms of Ge4, Wb, Dh^a, and GEAT antigens are present on GPC. Gerbich Ge2 and An^a antigens are present on GPD. Gerbich Ge3, Ls^a, GEIS, GEPL, and GETI antigens are present on GPC and GPD. Gerbich Ge2 is present at the GPD *N*-terminus, and Gerbich Ge4 is present on the GPC *N*-terminus. Gerbich Ge3 is present in GPC and GPD.

For host function of Gerbich glycophorin C and D antigens in RBC structure and hemolysis, *6 high-prevalent Gerbich Ge2, Ge3, Ge4, GEPL (or GE10), GEAT (or GE11), and GETI (or GE12) antigens are known. Five low-prevalent Gerbich Wb (or GE5), Ls^a (or GE6), An^a (or GE7), Dh^a (or GE8), and GEIS (or GE9) antigens are also known. Four Gerbich Ge4, Wb, Dh^a, and GEAT antigens are present on GPC.* Gerbich *antigens of Ge2 and An^a are present on GPD.* Gerbich *antigens of Ge3, Ls^a, GEIS, GEPL, and GETI are present on GPC and GPD. Gerbich GPC and GPD-reacting antibodies are natural-preformed antibodies but do not induce hemolytic TR. Ge2- or Ge3-specific autoantibodies induce autoimmune hemolytic anemia (AIHA).* Gerbich antigen-reactive antibodies occurs in pregnancy or transfusion [33]. In alloimmunization to Gerbich antigens, rare Leach phenotype individual produces Ge2-specific antibody, Ge3-specific antibody, or Ge4-specific antibody. Rare Yus individual produces Ge2-specific antibody due to the high Ge2 immunogenicity rather than Ge3 and Ge4. Gerbich antigens-specific

Table 7.2 Ligands of *P. falciparum* merozoites recognize human erythrocyte receptors

Malaria merozoite ligand	Receptors of human RBC
P. falciparum EBA-175	GPA
P. falciparum EBL-1	GPB
P. falciparum EBA-140	GPC
P. falciparum Rh4	Complement receptor-1
P. falciparum Rh5	Basigin

antibodies lead to hemolytic TR or HDFN. The Gerbich-associated HDFN are destruction of macrophage-dependent cells and proliferation blocking of early erythroid progenitor cells. Ge2-specific antibody or Ge3-specific antibody cause AIHA [34].

Gerbich GPC and GPD antigens function as malaria *Plasmodium* merozoites infectious receptor. For protozoa Plasmodium causing malaria disease, 5 human-susceptible malaria protozoa are known for *P. knowlesi, P. falciparum, P. malariae, P. ovale,* and *P. vivax.* Among them, protozoa parasite *P. vivax* is widespread, while *P. falciparum* causes the severe outcome such as death [35]. Host RBC receptors are GPC and GPD required for a ligand of *P. merozoites* EBA-140, where EBA indicates the erythrocyte-binding antigen. The *P. merozoites* EBA-140 ligand has 1 *N*-glycan and several *O*-glycan structures. GPC mediates rosetting in malaria *P. falciparum* and *P. vivax* infection through binding to the STEVOR protein. For *P. falciparum* EBA-140 merozoite infection, its ligand recognizes the *N*-terminal No. 36–63 a.a sequence of GPC. EBA-140 ligand homologs are also present in chimpanzee parasite *P. reichenowi* [24]. Chimpanzee erythrocyte GPD is also the ligand-binding receptor required for infection of *P. reichenowi* EBA-140, whereas GPC of human RBC acts as the receptor required for human parasite *P. falciparum* EBA-140. GPC is also target for a rodent parasite *P. berghei* infection. *GYPC gene mutations are related to host antimalarial pathway resistant to P. falciparum interaction and cause the lost high-prevalence Gerbich antigenicity and consequently decrease the susceptible RBC level to infection of P. falciparum merozoite. The Gerbich antigens are therefore Plasmodium infectious actors to human RBCs and allow possible preventive tool for partial malaria infection in malaria-endemic regions.* Blood-stage malaria parasites invade the host RBCs using by interaction of merozoite ligands with host RBC receptors. Duffy binding-like (DBL) protein and reticulocyte binding-like (RBL) protein involve in *Plasmodium* merozoite invasion [36–40]. The known DBL family of *P. falciparum* strain is a form of erythrocyte binding-ligand (EBL) capable of binding to distinct RBC receptors (Table 7.2), allowing invasion of malaria merozoites. Four known *P. falciparum* EBLs are EBA-175, -140, -181, and -1 [41]. EBA-175 and EBA-140 [42, 43] recognize sialyl *O*-glycans attached to GPA and GPC, respectively [44]. EBA-140 binds to the GPC sialyl *N*-glycan only [45].

Sialyl N-glycan of GPC bind to EBA-140. GPC *O-/N*-glycans have the EBA-140-recognizing site. Sialylglycans on GPC and GPC bind to EBA-140 [45, 46]. The known homologs of EBA-140 as EBL merozoite ligands are also present in *P. reichenowi* and chimpanzee-infectious species *P. falciparum*

[47]. Because GPC is human protein, GPD is suggested to be an RBC receptor of chimpanzee for infection of *P. reichenowi* EBA-140 strain [48]. In the experimental rodent model of *P. berghei* malaria, RBC receptors for *P. berghei* merozoites that was discovered from mouse embryonic stem cells [49]. In the mouse stem cell-derived RBCs, deletion of GYPC without GPC caused a decreased invasive capacity by *P. berghei* strain. Malaria *P. berghei* strains depend on mouse RBC GPC more than that *P. falciparum* strain depends on RBCs of human [49]. Parasites prefer to invade RBCs, and they release variant surface antigens (var) antigens on their surfaces. The known variant surface antigens (var) are STEVOR, RIFIN, and EMP1 (known as an erythrocyte membrane protein 1), which easily bind to the RBC membranes and emerging on the RBC surfaces. The known var-encoded variable proteins recognize uninfected RBCs and also endothelium for infected RBC sequestration to save the malaria invaders from splenic destruction [50]. Infected RBCs bound clumps with uninfected RBCs form rosettes, and consequently, shielding the infected RBCs to evade the parasites from immunity [51]. *P. falciparum* STEVOR protein-caused rosetting is intermediated by GPC of RBCs [52]. Therefore, the rosetting event is prevented by soluble GPC. In addition, soluble GPC inhibits STEVOR interaction with RBCs.

7.1.9 Knops Blood Group System

The Knops antigen has been discovered by specific antibodies characteristic of high titer or low avidity with weak reactions by antihuman globulin (AHG) sera [53]. Later, one of them defines a blood group Knops by the ISBT as the 22nd blood system [54]. Knops polymorphisms are important for antigen typing, although Yka is not identified yet. CR1 expression as Knops antigens in RBC by inheritance or acquisition can be identified by genotyping, like to the Dombrock system. The Knops polymorphisms have been known, and they are found in short consensus repeats (SCR) 25 of the CR1. Additionally, other SNPs are also existed in the CR1 gene, and new Knops antigens can be discovered. Caucasian sera obtained from three independent patients of Copeland, Stirling, and Wainright, identified in 1965, the showed antibodies with similar specificity. The antibody names of Co and St (Cost) were given for the first two patients and termed Cs^a. Again, a new antigen York (Yk^a) was found and related to Cs^a, but YK^a was termed as the Knops system [55]. Similarly, anti-Kn^a discovered was saline-reactive anti-K and antiglobulin antibodies from Caucasian [56]. The anti-Kn^a-opposed Kn^b has been later found in the anti-Kp^b-containing Hall sera [57]. A new antigen McC^a related to Kn^a was found and half percent of McC(a–) sera are Kn(a–). Another allele Sl^a and Vil pair have been also discovered. Initially, the McC^c terminology was used for Sl^a antigen. Similarly, McC^d was used for Vil antigen [58, 59]. The Sl term indicates the antibody-bearing patient names (Swain and Langley). Because this protein is the Knops-carrying protein, Sl^a re-termed S11. Vil also renamed S12 [60]. As the last type, the high-prevalent Knop is KAM discovered in the Helgeson phenotype

Table 7.3 Knops antigens and genes. A of the start codon AUG is the starting number

ISBT (antigen name)	Nucleotides	Amino acid encoded
KN1(Kn^a)	4708G,4681G	Val1561
KN2(Kn^b)	4708A,4681A	Met1561
KN3(McC^a)	4795A,4768A	Lys1590
KN4(S11	4828A,4801A	Arg1601
KN5(Yk^a)	Unknown	Unknown
KN6(McC^b)	4795G,4768G	Glu1590
KN7(S12)	4828G,4801G	Gly1601
KN8(S13)	4828A,4855 T*,4801A,4828 T*	Arg1601, Ser1610
KN9(KCAM)	4870A,4843A	Ile1615

Caucasian (serologic Kn deficient) [61], currently KCAM recognized by the ISBT. Like Sl^a, KCAM antigen is highly expressed in Caucasians.

The Knops antigen-carrying protein and its gene was identified in 1991, and CR1 protein carries the Kn^a, McC^a, Yk^a, and Sl^a antigens [62]. The Helgeson phenotype is characteristic of low copy numbers of CR1 on the erythrocytes. Using mAb-specific for erythrocyte antigens (MAIEA), Cs^a was found not to be located on CR1. Therefore, Cs^a and Cs^b are kept as ISBT collection. Frequency of the Knops antigens (Kn^a, Kn^a, McC^a, McC^b, S11, S12, Yk^a, KCAM and Refs) (Table 7.3) is in Caucasians, Africans, African Americans, and Brazilians. CR1 carries antigens of Kn^a, McC^a, and Sl^a. CR1 also carries Yk^a. The term Helgeson phenotype is also attributed to low copy number of the erythrocyte CR1. Using MAIEA, Cs^a was demonstrated not to be carried on CR1. Cs^a and Cs^a are remained for ISBT collection. Knops antigens follow mendelian-dominant trait inheritance with properties of high-frequent RBC antigens (but not for Kn^b and McC^b), weakened development of cord blood cells, and weakened of stored RBCs. They are not cleaved by papain or ficin, but cleaved by trypsin. And they are absent on platelets and in low plasma level/absence in urine or saliva. For the antigenic distribution, the Knops are diverse in their protein strength but weakened when the cells are stored or bear low CR1 numbers. S-S bond disrupting agents break Knops, McCoy, York, and Swain-Langley antigens. Knops-specific antibodies are nonreactive to trypsin-digested RBCs because a trypsin-susceptible site is located on CR1 SCR 28 position. Excessive immune diseases such as systemic lupus erythematosus or HIV infection can impair CR1 and weakening in Knops antigen strength [63]. From the stored RBC membranes, CR1-bearing vesicles are budded. Therefore, in the stored RBCs, the Knops antigens are weak in the strength. Thus, this is the reason why In (Lu) RBCs have weakened Knops antigens [64]. Two mutations in SCR 25 were associated with the S11 (Sla) and McCoy antigens. The McC^a and McC^b polymorphisms are present at DNA sequence of 4795 bp, in which nucleotide A encodes Pro (McC^a), while nucleotide G encodes Asp (McC^b). Mutations of S11 and S12 are only 11 a.a away from its relation. An A at 4828 bp encodes Arg and G encodes Gly. The final Knops antigen is the KCAM, which was initially known as KAM. The

KCAM-generating SNP is found but not involved in a specificity of blood group system until a McCoy-like antibody was identified in an ethnic Caucasian.

With disease association, in 1997, CR1 has been recognized as a rosetting ligand of *P. falciparum*-infected RBCs with uninfected RBCs [65]. *P. falciparum*-infected erythrocytes generate rosettes in severe malaria diseases. CR1 on uninfected erythrocytes forms rosettes and CR1-defected erythrocytes (called Helgeson phenotype) form weak rosettes and solubilized CR1 inhibits resetting formation. Sl:-1 phenotype RBCs weakly bind to PfEMP1 known as the parasite rosetting ligand [66]. Therefore, this is hypothesized that the polymorphism is easily selected in malaria-endemic regions due to prevention of severe malaria. The hypothesis theory has been examined in African peoples in Kenya and Mali, Africa. The Knops haplotype combination of Kn(a+)/McC(a–)/Sl:-1 showed the protection effect against the malaria infection. Currently, the known CR1 and other CRs are pathogenic infectious receptor. CR1-using pathogens are *Leishmania major* (as host of monocyte-macrophage) [67], *Legionella pneumophila* (as host of monocyte-macrophage) [68], *L. panamensis* [69], and *Mycobacterium tuberculosis* (as host of monocyte-macrophage) [70]. McCa and McCb-heterozygous individuals are resistant to infection of *M. tuberculosis* [71]. McCb has been speculated to evolve among African populations by *M. tuberculosis*. McCb also gave Africans survival advantages due to its heterozygous property.

References

1. Rearden A, Phan H, Dubnicoff T, Kudo S, Fukuda M. Identification of the crossing-over point of a hybrid gene encoding human glycophorin variant Sta. J Biol Chem. 1990;265(16):9259–63.
2. Tippett P, Reid ME, Poole J, Green CA, Daniels GL, et al. The Miltenberger subsystem: is it obsolescent? Transfus Med Rev. 1992;7(3):170–82.
3. Chen TD, Chen DP, Wang WT, Sun CF. MNSs blood group glycophorin variants in Taiwan: a genotype-serotype correlation study of 'Mi(a)' and St(a) with report of two new alleles for St(a). PLoS One. 2014. 2014;9(5):e98166.
4. Cleghorn TE. Two human blood group antigens, St-a (Stones) and Ri-a (Ridley), closely related to the MNSs system. Nature. 1962;21(195):297–8.
5. Suchanowska A, Smolarek D, Czerwiński M. A new isoform of Sta gene found in a family with NOR polyagglutination. Transfusion. 2010;50(2):514–5.
6. Race RR, Sanger R. Blood groups in man. 6th ed. Oxford London Edinburgh Melbourne: Blackwell Scientific Publications; 1975.
7. Prokop O, Göhler W, Mayr W, Geserick G, Radam G. Human blood groups. Montreal: D. J. Paradis Editions Inc.; 1986.
8. Bundesärztekammer. Richtlinien für die Erstattung von Abstammungsgutachten. Dtsch Ärztebl. 2002;99:C509–11.
9. Hoppe J-D. Neufassung der Richtlinien für die Erstattung von Blutgruppengutachten. Bundesgesundheitsbl. 1990;33:264–9.
10. Garratty G, Dzik W, Issitt PD, Lublin DM, Reid ME, Zelinski T. Terminology for blood group antigens and genes-historical origins and guidelines in the new millennium. Transfusion. 2000;40:477–89.

11. Landsteiner K, Wiener AS. An agglutinable factor in human blood, recognized by immune sera for rhesus blood. Proc Soc Exp Biol Med. 1940;43:223–4.
12. Levine P, Stetson RE. An unusual case of intragroup agglutination. JAMA. 1939;113:126–7.
13. Fisher RA. The rhesus factor: a study in scientific method. Am Sci. 1947;35(95–102):113.
14. Apecu RO, Mulogo EM, Bagenda F, Byamungu A. ABO and rhesus (D) blood group distribution among blood donors in rural South Western Uganda: a retrospective study. BMC Res Notes. 2016;9(1):513. https://doi.org/10.1186/s13104-016-2299-5.
15. Lomas-Francis C, Reid ME. The dombrock blood group system: a review. Immunohematology. 2010;26(2):71–8.
16. Molthan L, Crawford MN, Tippett P. Enlargement of the dombrock blood group system: the finding of antiDob. Vox Sang. 1973;24:382–4.
17. Lewis M, Allen FH Jr, Anstee DJ, et al. ISBT working party on terminology for red cell surface antigens: Munich report. Vox Sang. 1985;49:171–5.
18. Telen MJ, Rosse WF, Parker CJ, Moulds MK, Moulds JJ. Evidence that several high-frequency human blood group antigens reside on phosphatidylinositol-linked erythrocyte membrane proteins. Blood. 1990;75:1404–7.
19. Lewis M, Anstee DJ, Bird GWG, et al. Blood group terminology 1990. ISBT working party on terminology for red cell surface antigens. Vox Sang. 1990;58:152–69.
20. Beattie KM, Castillo S. A case report of a hemolytic transfusion reaction caused by anti-Holley. Transfusion. 1975;15:476–80.
21. Chapel-Fernandes S, Callebaut I, Halverson GR, Reid ME, Bailly P, Chiaroni J. Dombrock genotyping in a native Congolese cohort reveals two novel alleles. Transfusion. 2009;49:1661–71.
22. Halverson GR, Schawalder A, Miller JL, Reid ME. Production of a monoclonal antibody shows the conservation of epitopes on the dombrock glycoprotein in the great apes (abstract). Transfusion. 2001;41(Suppl):24S.
23. Glowacki G, Braren R, Cetkovic-Cvrlje M, Leiter EH, Haag F, Koch-Nolte F. Structure, chromosomal localization, and expression of the gene for mouse ecto-mono(ADP-ribosyl) transferase ART5. Gene. 2001;275:267–77.
24. Jaskiewicz E, Peyrard T, Kaczmarek R, Zerka A, Jodlowska M, Czerwinski M. The Gerbich blood group system: old knowledge, new importance. Transfus Med Rev. 2018;32(2):111–6. https://doi.org/10.1016/j.tmrv.2018.02.004.
25. Colin Y, Joulin V, Le Van Kim C, Roméo PH, Cartron JP. Characterization of a new erythroid/megakaryocyte-specific nuclear factor that binds the promoter of the housekeeping human glycophorin C gene. J Biol Chem. 1990;265:16729–32.
26. Le Van Kim C, Colin Y, Mitjavila MT, Clerget M, Dubart A, Nakazawa M, et al. Structure of the promoter region and tissue specificity of the human glycophorin C gene. J Biol Chem. 1989;264:20407–14.
27. Le Van Kim C, Mitjavila MT, Clerget M, Cartron JP, Colin Y. An ubiquitous isoform of glycophorin C is produced by alternative splicing. Nucleic Acids Res. 1990;18:3076.
28. Colin Y, Le Van Kim C, Tsapis A, Clerget M, d'Auriol L, London J, et al. Human erythrocyte glycophorin C. Gene structure and rearrangement in genetic variants. J Biol Chem. 1989;264:3773–80.
29. Salomao M, Zhang X, Yang Y, Lee S, Hartwig JH, Chasis JA, et al. Protein 4.1R-dependent multiprotein complex: new insights into the structural organization of the red blood cell membrane. Proc Natl Acad Sci U S A. 2008;105:8026–31. https://doi.org/10.1073/pnas.0803225105.
30. Chang S, Reid ME, Conboy J, Kan YW, Mohandas N. Molecular characterization of erythrocyte glycophorin C variants. Blood. 1991;77:644–8.
31. Gourri E, Denomme GA, Merki Y, Scharberg EA, Vrignaud C, Frey BM, et al. Genetic background of the rare Yus and Gerbich blood group phenotypes: homologous regions of the GYPC gene contribute to deletion alleles. Br J Haematol. 2017;177:630–40. https://doi.org/10.1111/bjh.14578.

32. McShane K, Chung A. A novel human alloantibody in the Gerbich system. Vox Sang. 1989;57: 205–9.
33. Daniels G. Human blood groups. Wiley-Blackwell; 2013. https://doi.org/10.1002/9781118493595.ch1.
34. Göttsche B, Salama A, Mueller-Eckhardt C. Autoimmune hemolytic anemia associated with an IgA autoanti-Gerbich. Vox Sang. 1990;58:211–4.
35. Crompton PD, Moebius J, Portugal S, Waisberg M, Hart G, Garver LS, et al. Malaria immunity in man and mosquito: insights into unsolved mysteries of a deadly infectious disease. Annu Rev Immunol. 2014;32:157–87. https://doi.org/10.1146/annurev-immunol-032713-120220.
36. Malpede BM, Tolia NH. Malaria adhesins: structure and function. Cell Microbiol. 2014;16: 621–31. https://doi.org/10.1111/cmi.12276.
37. Weiss GE, Gilson PR, Taechalertpaisarn T, Tham W-H, de Jong NW, Harvey KL, et al. Revealing the sequence and resulting cellular morphology of receptor-ligand interactions during Plasmodium falciparum invasion of erythrocytes. PLoS Pathog. 2015;11:e1004670. https://doi.org/10.1371/journal.ppat.1004670.
38. Beeson JG, Drew DR, Boyle MJ, Feng G, Fowkes FJI, Richards JS. Merozoite surface proteins in red blood cell invasion, immunity and vaccines against malaria. FEMS Microbiol Rev. 2016;40:343–72. https://doi.org/10.1093/femsre/fuw001.
39. Satchwell TJ. Erythrocyte invasion receptors for Plasmodium falciparum: new and old. Transfus Med. 2016;26:77–88. https://doi.org/10.1111/tme.12280.
40. Tham WH, Healer J, Cowman AF. Erythrocyte and reticulocyte binding-like proteins of Plasmodium falciparum. Trends Parasitol. 2012;28:23–30. https://doi.org/10.1016/j.pt.2011.10.002.
41. Adams JH, Blair PL, Kaneko O, Peterson DS. An expanding ebl family of Plasmodium falciparum. Trends Parasitol. 2001;17:297–9. https://doi.org/10.1016/S1471-4922(01)01948-1.
42. Wanaguru M, Crosnier C, Johnson S, Rayner JC, Wright GJ. Biochemical analysis of the plasmodium falciparum erythrocyte-binding Antigen-175 (EBA175)-Glycophorin-a interaction: implications for vaccine design. J Biol Chem. 2013;288:32106–17. https://doi.org/10.1074/jbc.M113.484840.
43. Salinas ND, Paing MM, Tolia NH. Critical glycosylated residues in exon three of erythrocyte glycophorin a engage Plasmodium falciparum EBA-175 and define receptor specificity. MBio. 2014;5:e01606–14. https://doi.org/10.1128/mBio.01606-14.
44. Thompson JK, Triglia T, Reed MB, Cowman AF. A novel ligand from Plasmodium falciparum that binds to a sialic acid-containing receptor on the surface of human erythrocytes. Mol Microbiol. 2001;41:47–58.
45. Rydzak J, Kaczmarek R, Czerwinski M, Lukasiewicz J, Tyborowska J, Szewczyk B, et al. The Baculovirus-expressed binding region of Plasmodium falciparum EBA-140 ligand and its glycophorin C binding specificity. PLoS One. 2015;10:e0115437. https://doi.org/10.1371/journal.pone.0115437.
46. Malpede BM, Lin DH, Tolia NH. Molecular basis for sialic acid-dependent receptor recognition by the Plasmodium falciparum invasion protein erythrocyte-binding antigen-140/BAEBL. J Biol Chem. 2013;288:12406–15. https://doi.org/10.1074/jbc.M113.450643.
47. Martin MJ, Rayner JC, Gagneux P, Barnwell JW, Varki A. Evolution of human-chimpanzee differences in malaria susceptibility: relationship to human genetic loss of N-glycolylneuraminic acid. Proc Natl Acad Sci U S A. 2005;102:12819–24. https://doi.org/10.1073/pnas.0503819102.
48. Zerka A, Kaczmarek R, Czerwinski M, Jaskiewicz E. Plasmodium reichenowi EBA-140 merozoite ligand binds to gycophorin D on chimpanzee red blood cells, shedding the light on origins of Plasmodium falciparum. Parasit Vectors. 2017;10:554. https://doi.org/10.1186/s13071-017-2507-8.
49. Yiangou L, Montandon R, Modrzynska K, Rosen B, Bushell W, Hale C, et al. A stem cell strategy identifies glycophorin C as a major erythrocyte receptor for the rodent malaria parasite

Plasmodium berghei. PLoS One. 2016;11:e0158238. https://doi.org/10.1371/journal.pone.0158238.
50. Wahlgren M, Goel S, Akhouri RR. Variant surface antigens of Plasmodium falciparum and their roles in severe malaria. Nat Rev Microbiol. 2017;15:479–91. https://doi.org/10.1038/nrmicro.2017.47.
51. Sherman IW, Eda S, Winograd E. Cytoadherence and sequestration in Plasmodium falciparum: defining the ties that bind. Microbes Infect. 2003;5:897–909.
52. Niang M, Bei AK, Madnani KG, Pelly S, Dankwa S, Kanjee U, et al. STEVOR is a Plasmodium falciparum erythrocyte binding protein that mediates merozoite invasion and resetting. Cell Host Microbe. 2014;16:81–93. https://doi.org/10.1016/j.chom.2014.06.004.
53. Moulds JM. The knops blood-group system: a review. Immunohematology. 2010;26(1):2–7.
54. Daniels GL, Anstee DJ, Cartron JP, et al. Blood group terminology 1995. Vox Sang. 1995;69: 265–79.
55. Molthan L, Giles CM. A new antigen, Yka, and its relationship to Csa (cost). Vox Sang. 1975;29:145–53.
56. Helgeson M, Swanson J, Polesky HF. Knops-Helgeson (Kna), a high frequency erythrocyte antigen. Transfusion. 1970;10:737–8.
57. Mallan MT, Grimm W, Hindley L, Knighton G, Moulds MK, Moulds JJ. The Hall serum: detecting Knb, the antithetical allele to Kna (abstract). Transfusion. 1980;20:630.
58. Lacey P, Laird-Fryer B, Block U, Lair J, Guilbeau L, Moulds JJ. A new high incidence blood group factor, Sla; and its hypothetical allele. (abstract). Transfusion. 1980;20:632.
59. Molthan L. Expansion of the York, cost, McCoy, knops blood group system: the new McCoy antigens McCc and McCd. Med Lab Sci. 1983;40:113–21.
60. Daniels GL, Cartron JP, Fletcher A, et al. International Society of Blood Transfusion Committee on terminology for red cell surface antigens: Vancouver report. Vox Sang. 2003;84:244–7.
61. Moulds JM, Pierce S, Peck KB, Tulley ML, Doumbo O, Moulds JJ. KAM: a new allele in the knops blood group system (abstract). Transfusion. 2005;45(Suppl):27A.
62. Rao N, Ferguson DJ, Lee SF, Telen MJ. Identification of human erythrocyte blood group antigens on the C3b/C4b receptor. J Immunol. 1991;146:3501–7.
63. Holme E, Fyfe A, Zoma A, Veitch J, Hunter J, Whaley K. Decreased C3b receptors (CR1) on rythrocytes from patients with systemic lupus erythematosus. Clin Exp Immunol. 1986;63:41–8.
64. Daniels GL, Shaw MA, Lomas CG, Leak MR, Tippett P. The effect of in(Lu) on some high-frequency antigens. Transfusion. 1986;26:171–2.
65. Rowe JA, Moulds JM, Newbold CI, Miller LH. P. Falciparum rosetting mediated by a parasite-variant erythrocyte protein and complement-receptor 1. Nature. 1997;388:292–5.
66. Moulds JM, Kassambara L, Middleton JJ. Identification of complement receptor one (CR1) polymorphisms in West Africa. Genes Immun. 2000;1:325–9.
67. Da Silva RP, Hall BF, Joiner KA, Sacks DL. CR1, the C3b receptor, mediates binding of infective leishmania major metacyclic promastigotes to human macrophages. J Immunol. 1989;143:617–22.
68. Payne NR, Horwitz MA. Phagocytosis of legionella pneumophila is mediated by human monocyte complement receptors. J Exp Med. 1987;166:1377–89.
69. Robledo S, Wozencraft A, Valencia AZ, Saravia N. Human monocyte infection by Leishmania (viannia) panamensis. J Immunol. 1994;152:1265–75.
70. Schlesinger LS, Bellinger-Kawahara CG, Payne NR, Horwitz MA. Phagocytosis of Mycobacterium tuberculosis is mediated by human monocyte complement receptors and complement component C3. J Immunol. 1990;144:2771–80.
71. Noumsi GT, Tounkara A, Diallo D, Moulds JM. Knops blood group polymorphism and protection from Mycobacterium tuberculosis (abstract). Transfusion. 2006;46(Suppl):19A.

Chapter 8
Conceptual Onset of Xenotransplantation from ABO Blood Type-Incompatible Organ Allotransplantation

8.1 Sialic-Acid (Sia)-Attached Blood Group Determinants

Although regular AB(H) blood group system depicts non-SA-containing erythrocyte differentiation, another example of the Neu5Ac-attached to specific antigens is so-called human blood group MN, raised by genetic variations in the N-terminal regions of glycophorin in RBCs, which yield different and small O-linked sialylation from each individual [1, 2]. Such variations in the antigens also produce the different sialylglycans and consequently generate the specific binding antibodies against human RBCs to affect failure during blood transfusion. Similarly, some antibodies against other blood groups are also considered [3]. For example, the M and N blood group antigens of GPA consist of Siaα2,3Gal- and Siaα2,6GalNAc- residues, linked to *O*-glycan saccharides. These two M and N blood group antigens are bound by M/N-specific antibodies. The GPA-M and GPA-N antigens are easily confirmed by M-specific and N-specific Mab bindings. The GPAs with SA-Gal called GPA2,3 can be produced by asialoGPAs resialylation reaction using α2,3Sia-transferase. The GPAs with GalNAc-linked Sia (GPA2,6) are also produced, indicating the distinct presence of only one of two SA residues linked to Gal or GalNAc in the epitope [4].

8.2 *N*-Glycolylneuraminic Acid (Neu5Gc)-Based Blood Groups in Cats

In the early 1900s, it was identified that domestic cats have only one common blood group system [5, 6], whereas humans, dogs, horses, and other mammal species have multiple systems [7]. Except for primates, gene mutations are not found to generate different types of blood group in animals. If one major antibody system is present, blood transfusion within the species is restricted, as expected in cats. The cats have one blood group with two major blood types of A and B serotypes, where blood A

C.-H. Kim, *Glycoimmunology in Xenotransplantation*,
https://doi.org/10.1007/978-981-99-7691-1_8

type is dominant to blood B type. Recognition of the two blood group A and B types in cats is caused by cats B antibodies, which recognize a sialyloglycolipid on the RBCs of A cats. For the type A (NeuGc) and B (NeuAc) difference, the only presence or absence of Neu5Gc residue has been known, as the NeuGc residue is attached to the GD3 ganglioside of cat RBC [8]. The NeuGc-synthesizing enzyme gene, CMAH gene, is the factor of the Cat A and B type, which the CMAH gene promoter region differentially synthesize the NeuGc in cat RBC, indicating the change in the CMAH gene promoter region [9]. Interestingly, the similar polymorphisms in Neu5Gc synthesis of dog erythrocytes are also found [10]; however, the dog Neu5Gc-recognizing antibodies are unknown yet. Thus, Sias act as alloantigens within the same species. Cats blood group antigens have blood type A, B, and AB on cat erythrocytes, indicating their incompatibility in blood group system also cause cat blood transfusion failures and cat isoerythrolysis. AB group as a rare type is also allelic but recessive to A blood type, but dominant to B blood type in cats. Therefore, in cats, blood group NeuGc antigens are well identified. NeuGc antigen determines the blood A type, while NeuAc antigen determines the blood B type.

Type A group has two distinct isotypes of NeuGc,NeuGc,Gal,Glc,Ceramide, termed (NeuGc)2-GD3 and NeuAc,NeuGc,GD3. In addition, (NeuGc)2-disialoparagloboside and NeuAc,NeuGc-disialoparagloboside are also found in cat erythrocytes. However, type B blood group has only (NeuAc)2-GD3. Type AB group has (NeuGc)2-GD3, NeuAc, NeuGc-GD3, and (NeuAc)2-GD3. Those antigens are resided in the plasma membrane protein, 50-kDa of cats. The B group is only the form of (NeuAc)2-GD3 ganglioside and also with the NeuAc residue in a plasma membrane 50 kDa protein. As the blood type A antigen determinant, NeuGc residue is also specifically resided as glycolipid (NeuGc)2-GD3 [8] and the A blood type antigen is also found in a plasma membrane 50-kDa protein.

8.3 General Aspects of ABO Blood Type Carbohydrates in Transplantation

During allograft transplantation, preformed antibodies as an important barrier raise unusual immune responses to graft rejection, as those antibodies have been known in kidney transplantation in 1964 [11]. In the allograft transplantation, the blood group antigens of ABO type were recognized as a rejection barrier. In contrast, in the xenograft transplantation, sugar antigens such as α-Gal, Neu5Gc, and the B4GalNT2-generated saccharide(s) are the rejection barriers. As blood type system of ABO blood has been discovered by Karl Landsteiner, the blood group antigenic structures were elucidated in 1950s by three blood scientists, Elvin Kabat, Walter J. Morgan, and Winifred M. Watkins [12–14]. The ABH antigens are greatly different from each other in antigenic properties [15]. Among the three sugar epitopes of antigen A, B, and H (O), H (O) antigen is composed of two saccharides of L-Fuc and D-Gal, while A and B antigens carry trisaccharides, composed of the common antigen H plus D-GalNAc for A antigen or D-Gal for B antigen. Therefore, the ABO antigenic glycans are based on fucosylated oligosaccharides linked to glycolipids and glycoproteins on the RBC surfaces. The ABO saccharide determinants are critical factors harmful in clinical transfusion of human blood and transplantation of human organs [16, 17]. In human, the surfaces of RBCs are major and intrinsic distribution with substantially in body fluids and tissues as rare distribution. The blood group antigens in close proximity to GM1 gangliosides are expressed on the epithelial cell surfaces in intestines, functioning as the infection sites of intestinal infectious bacteria like enterotoxigenic *E. coli* (ETEC) and *V. cholerae* species [18]. Additionally, they also function as infectious receptors of pathogenic bacteria of *Campylobacter jejuni* and *Heliobacter pylori* as well as Noroviruses that cause vomiting and diarrhea [19]. In the view of structural mimetics, along with the ABO blood group oligosaccharides, the αGal1,3βGal1,4GlcNAc, and fucosyl α1,3Gal-Le^X of Galα1,3Galβ1,4(Fucα1,3)GlcNAc structure are also considered as xenoantigenic structures in pig-to-human organ xenotransplantation. In humans, Lewis y (type II chain) is recognized as a tumor-associated antigen (TAA) [20]. Lewis y contains two Fuc residues with structure of Fucα1,2Galβ1,4 (Fucα1,3)GlcNAcβ1-R, belonging to the A, B, and H(O). Lewis antigen family has distinct terminal fucosyl residue catalyzed by the α1,2-Fuc-transferase. Lewis y (type 2 chain) antigen as a difucosyl tetrasaccharide is expressed on the blood group Type 2 oligosaccharides of glycoproteins and glycolipids. Lewis y (type II chain) as TAA is expressed in some cancers such as colorectal carcinomas and large bowel tumors and therefore useful for the classification of human bladder and human renal tumor cells. The Lewis y antigen is also a diagnosis and prognosis biomarker of breast cancer, cholangioma and hepatoma cells. In addition, the A antigenic oligosaccharides of blood group attached on the core tetrasaccharide of Lewis y are suggested as a ligand for B-subunit of *E. coli* heat-labile toxin (LTB) due to its analogous property to a LTB-derived hybrid protein and B-subunit of CTB cholera toxin.

8.4 Lewis Histo-Blood Group Antigens and Their Comparison with ABO(H) Group

Apart from the blood group ABO types, Lewis histo-blood group oligosaccharides are general barriers in transplantation medicine and blood transfusion. The ABO locus alleles encode the antigen-synthesis genes of glycosyltransferases. Human ABO blood type system has three major alleles of O, A, and B antigens. In contrast, pig carries only O and A allele for each antigen. In the pig AO blood type, the pig A antigen-coding gene shows its homology to the human ABO blood antigenic genes as well as the ABO genes present in other species. The porcine A gene encodes an α1,3-GalNAc-transferase enzyme that generates blood group A antigenic sugar. But the O gene in pig is deleted; therefore, a silent O antigenic gene is not activated in pig

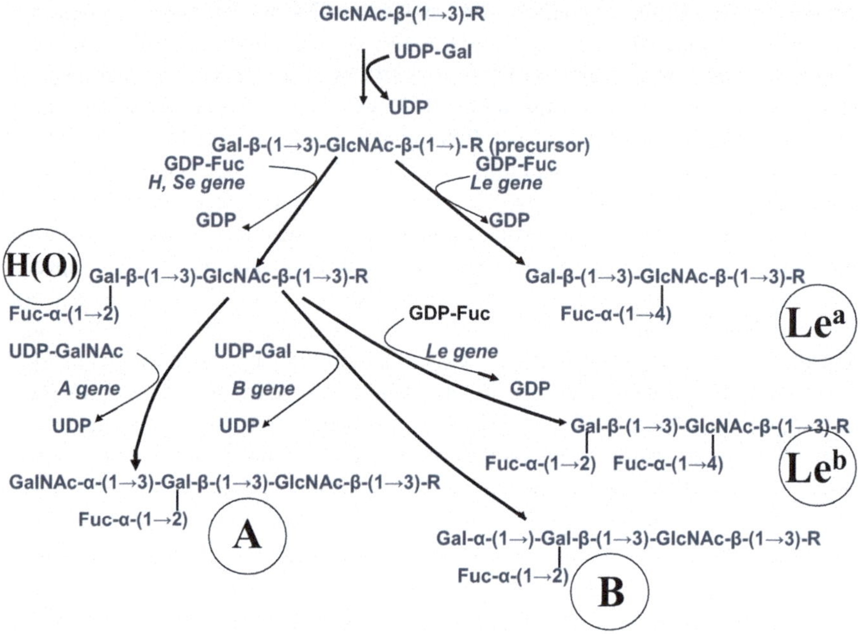

Fig. 8.1 Synthetic pathway of blood types A, B, H and lewis (Le^a, Le^b) type

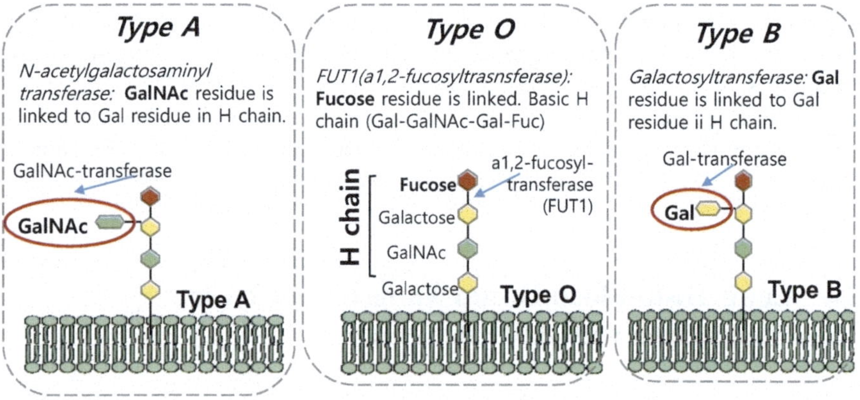

Fig. 8.2 Carbohydrate antigens known for human ABO blood type antigens

organs during xenotransplantion to human recipients [21]. Figure 8.1 shows the synthetic pathway of A, B, H antigens and Lewis (Le^a and Le^b) type. Carbohydrate antigens known for human ABO blood type antigens has been described (Fig. 8.2).

As described earlier, the immunologically dominant carbohydrates of the A and B antigenic saccharides have each different structure of GalNAcα1,3(Fucα1,2)Gal-R and Galα1,3(Fucα1,2)Gal-R saccharide, respectively. The A antigen gene encodes the A glycan-transferase, and the A saccharide transferase catalyzes the enzymatic reaction that GalNAc residue as a donor-transferring substrate to the H acceptor substrate Gal residue (Fucα1,2Gal-R). Similarly, the B saccharide transferase gene encodes B transferase, where Gal residue as a donor substrate is transferred to the same the acceptor H substrates. Group O blood type is just the status of H substrates, because O genes are absent. For the past 2 decades, the Lewis histo-blood group ABO polymorphism in human has been known as the best-known example of terminal saccharide variation [22]. Experimentally, human B transferase cDNA and nonfunctional O allelic cDNA were isolated from cDNAs information deduced by the a.a sequence of the human A antigenic saccharide-transferase [23]. The fine ABO blood group gene locus has been known from the molecular analysis. The known O alleles in humans carry an N-terminal single base deletion in the gene, and some O alleles are substituted in an amino acid [24]. For example, between A antigen transferase and B antigen transferase, only four amino acid substitutions are found. Apart from the three major alleles, subgroup alleles and cisAB and B (A) gene alleles required for the A and B antigenic synthesis are mutated [22]. The human blood group ABO alleles are officially listed at the Website, http://www.bioc.aecom.yu.edu/bgmut/index.htm, known for the Blood Group Antigen Gene Mutation Database Web site. Defective expression of FUT1 gene, which is inherit in humans, rarely represents so-called "O Bombay phenotype".

8.5 Human O Bombay Phenotype Group

The human O Bombay phenotype is not capable of the H antigen synthesis and consequently produces serum anti-H antibodies. Rejection of an organ graft xenotransplanted from a related primate species fast occurs rather than those of allo-transplantated grafts. However, the rejection response can be delayed when general immunosuppression agents or immune therapeutic application are made. Although pigs are considered to be preferable organ and cell sources compared to nonhuman primates, a pig organ to baboon transplantation was reported to be fast rejected within minutes. From the antigenic rejection properties of ABO-incompatibility in organ transplantation, the organ recipients recognize saccharide epitopes present in the pig cell surfaces.

8.6 Glycoantigenic Differences in Primates Such as Baboons or Old World Monkeys

Initial studies of carbohydrate antigenic rejection have likely been started using baboons because baboons are readily available from Africa area such as South Africa. In addition, baboons have only the blood group AB type glycan structures among A, B, and AB antigens. Baboons lack the O antigen, and this is highly similar to the case of humans. Thus, baboons can be readily used for heart transplantation model in humans for the ABO-compatible or -incompatible transplants. About 30% of AB-incompatible transplants showed hyperacute rejection in baboons (within 24 h), while about 60% hyperacute rejection is reported in case of humans [24]. These indicate that transplantation of an ABO-incompatible organ is dangerous, even if ABO-incompatibility weakly rejects the grafts between closely related species [25]. Synthetic A or B oligosaccharides, which are intravenously (i.v.) infused, are bound by the serum antibodies of A- or B-specific antibodies, and the antibody–antigen complex is cleared in the blood. This reduces the serum levels of antibody and no-antibody means no-fear of hyperacute rejection. Thus, transplantation of ABO-incompatible organ graft is possible.

There were a few other carbohydrate targets, but α1,3Gal was the most crucial. It is surprising to note that the α1,3Gal structure and the ABO group antigenic carbohydrates are together similar [26]. The carbohydrate structures of α1,3Gal and blood group B are mostly similar [27], indicating that the immune responses may be highly resembled together in human. For example, α1,3GalT KO may accumulate precursors, increasing several carbohydrate chains including lactosamine (Galβ1,4GlcNAc-) precursors, H antigenic blood group, glycolipid P1 and x2. However, because such increased carbohydrate chains are ubiquitously expressed in humans, the human immune system does not recognize the sugars. Actually, in the α1,3GalT-KO endothelial cells, such terminal α1,3Gal epitopes were disappeared, and terminal GalNAc and Gal β1,3GalNAc were also reduced [28]. Instead, the lectin-binding reactions against the α2,3-Sia and α2,6-Sias were increased.

Dr. Uri Galili discovered that all mammals, which are evolutionarily occupied below Old World monkeys, have the expressed structures of α1,3Gal carbohydrate. But humans and Old World NHPs lack the α1,3Gal carbohydrate due to genetic deletion of α1,3Galactosyltransferase (GT), and they produce anti- α1,3Gal antibodies to protect from the α1,3Gal-carriers [29–31]. According to the Galili and his colleagues [32], it has been hypothesized that Old World monkeys, apes, and humans generate α1,3Gal-specific antibodies as "preformed or natural" forms in new-born infancy. The role of the α1,3Gal-specific antibodies are speculated as "defensive" protector against the α1,3Gal-expressing microflora or viruses inhabited in gastrointestinal tracts. In new born and infant baboons and humans, antiswine antibodies known as anti-α1,3Gal antibodies, specifically blood A type-specific antibody or B-specific antibody, are not expressed at least for the first 3 months [33, 34]. Then they start to produce the antibodies presumably after inhabitation in the gastrointestinal tracts. Almost since 30 years, as the reality and existence of α1,3Gal as the main antigen for anti-swine antibodies, it presence was firstly reported in 1991 at the first International Congress on Xenotransplantation (Minneapolis, Minnesota, USA) [35], lighting that the progress in α1,3Gal study is extraordinarily made. Recent advances of CRISPR/Cas-based genome editing technology has successfully potentiated the triple KO pig production (GGTA1/CMAH/B4GALNT2) targeting the three xenoantigens. The CRISPR/Cas technology gives a possible cross-matching compatibility for xenotransplantation of allo-unsensitized recipients.

8.7 Allotransplantation's Major Barrier Is the Blood Group ABO System

Because transplantation or transfusion with blood in ABO difference causes rejection or hemodynamic alterations, the complete concordance of ABO antigens has been assessed in allotransplantation. Human allotransplantation uses allografts medically supplied from human population with free ranged subjects when the organs and tissues are clean or free from known pathogens.

Humoral immune response is increasingly understood on the incompatible antigens such as blood group carbohydrates. In addition, histopathological progression of antibody-dependent immune rejection of organ allografts in ABO-incompatible responses is also elucidated in terms of the cell type-specific ABH antigenic expression, as are understood with the carbohydrate chain type-specific immune response. Indeed, blood group A kidney-transplanted group O patient has an antibody in the A-specific antibody repertoire in the A antigen expressed on a chain type 1 [36]. Apart from xenotransplantation, the major allotransplantation barrier is the ABO group, as A and B donor grafts express A and B antigenic carbohydrates to cause HR. Since A2 donor with low graft A antigen expression [37], xenograft settle of A1/B kidney donors was successful in the ABO barriers [38]. ABO-incompatible renal transplantation is clinically under trial [39]. The anti-class II antibodies in allotransplantation are important, as anti-class II antibodies are major crucial factors to induce transplanted glomerulopathy and loss of graft [40]. AHR seems to be like to the vascular rejection after allotransplantation in incompatible recipients [41]. The antibody subtype of hyperacute rejection (HR) is the preexisting IgM as the primary Ig class, whereas the AHR antibody subtype is the elicited IgM and IgG. In the allotransplantation over ABO or HLA barriers, several plasmopheresis, preformed natural antibody absorption, and complement inhibition are strategies to avoid HR of a solid organ xenograft [42].

Although human complement regulating CD55, CD46, and CD59-transgenic pigs overcome HR, the pigs cannot avoid the AHR development because of the classical and alternative complement pathways. Donor antigens-specific antibodies cause two distinct rejection types of HR and AHR in the solid organ allotransplantation. HR in the recipients occurs rapidly within minutes to hours upon transplantation because of preformed antibodies against the donor antigens [43]. Complements and antibodies are accumulated as deposits in the vascular wall and they provoke tissue damage and injury to endothelium with fibrin-platelet thrombosis together with neutrophil homing. Terminally, endothelium on arterial and capillary walls is progressed to a rapid and irreversible graft failure toward phenotypic rejection [43]. AHR occurs slowly within several days mainly due to the elicited antibody reactions to the xenograft. In the solid organ xenograft-derived AHR, many molecules including IgG, IgM, C5b-9, and C4d are detected with the signs of capillary integrity reduction, death of endothelial cells, and accumulation of fibrins [42]. AHR is the second immunological injury, and the damage is displayed after HR because xenograft is still live.

Allo-HLA antigens are also the immune elicitor to the allotransplantation hosts, and there are three cases of immune-reactive mechanisms in hosts. They are the transplantation described above, pregnancy or blood transfusion. The T cell repertoire as preexisted is well known to function as alloreactive holder to allo-antigens. The memory T cells respond to allo-HLA antigens, and its high memory T cell frequency is also attributed to cross-reaction potentials of virus-specific memory T cells [44]. Pathogenic infection and antigen vaccination are also known to relate with HLA-specific antibody production. For example, infected pathogenic agents lead to cause directly allo-antigenic reaction by molecular mimics, and this is called the response by co-stimulatory factors of alloreactive leukocytes or "heterologous immunity." Such direct alloreactivity takes place when the mismatched HLA proteins present in donor cells are directly bound by host T cells, resulting in the acute rejection mediated by T cells. Direct responses of memory T cells to alloantigenic MHC induce acute rejection and chronic inflammation such as the well-known allograft nephropathy as transplantation tolerance.

From the immunological patterns and responses to allo- and xenografts, it is clear that immunological responses between xenotransplantation and allotransplantation are quite similar, although xenotransplantation is largely complex. Indeed, surface proteins of pig islets may display the vigorous immunologic rejection rather than allografts [45]. Antibody binding to ABH blood group antigens or α1,3Gal antigens contribute to endothelium activation, as known in the pig endothelial cells and α1,3Gal antigenic rejection. When pig endothelial cells are incubated with the tetrameric isolectin B4, known as an α1,3Gal-recognizing lectin from plant *Griffonia simplicifolia* or the decameric α1,3-Gal IgM, activate endothelial cells [46, 47]. HR process requires complement activation. However, anti-α1,3Gal IgG and IgM mixture activates pig endothelium without complement but increasing adhesion of human neutrophils [48]. Anti-α1,3Gal IgG accelerates migrative potentials of NK cells of human across pig endothelial vessel and enhances motility of the NK cells [49]. Antibodies-recognizable carbohydrates hence induce immune reaction. Endothelial integrin family and von Willebrand factor are reported as porcine endothelium [α1,3]-Gal-attached protein [50]. Phenotype pig blood antigens such as ABO groups in pigs are variable and thus frequently assessed by reverse phenotyping, hemagglutination, or tissue ABO antigen immunohistochemical test. Currently, quick and simple phenotyping microtyping can assess ABO groups in pigs. Another important factor in the immune responses involved in antigen-specific antibodies against blood group ABH and α1,3Gal antigens are their molecular densities expressed on the donor cells. In fact, Galili group has calculated the estimated numbers of α1,3Gal antigenic epitopes, which calculated to $1–35 \times 10^6$ epitopes in each cell. However, the A epitope number present in the RBC A1 cells was estimated to $0.81–1.17 \times 10^6$ in each cell [51].

8.8 Similarity in Antibody-Driven Rejection Between Allotransplanted Incompatible ABO Grafts and Xenotransplanted Vascularized Grafts

In the carbohydrate antigens, α1,3Gal determinant and blood group ABH antigens are distinct in their carbohydrate chain structures. The xenoreactive α1,3Gal or Galα1,3Galβ1-R epitope antigen is loaded on the type 2 chain known as Galβ1,4GlcNAc disaccharide. However, blood group A type of the GalNAcα1,3 (Fucα1,2)Galβ1-R structure, blood group B type structure of Galα1,3(Fucα1,2) Galβ1-R, and blood group H type of Fucα1,2Galβ1-R structure are loaded on the type 1 chain known as the Galβ1,3GlcNAc, type 2 and 3 chains known as Galβ1,3GalNAc[α], and type 4 chain known as Galβ1,3GalNAcβ (Fig. 8.1). Tissue distribution of ABH is not well known due to its trace amounts of expression, although some ABH distribution has been known. For example, in individuals of blood group A1 type, type A glycolipids attached on the type 4 chain are rich in kidney organs, while RBCs express mainly type 2 chain A glycolipids attached on the type 4 chain [36]. Thus, it has been suggested that the immune response is selective to the differently expressed blood group A and B types. For an instance, monoclonal Abs produced mouse are specific for type 1, 2, 3, and 4A blood group antigens are known.

In terms of action mechanism(s) of the Ab-mediated rejections between ABO-incompatible allografts of human organs and xenografts swine organs, those rejection types show the similarity because donor organ target carbohydrate antigens and donor carbohydrate-specific naturally preformed Abs are similar. The immunological rejection of ABO-incompatible allografts shares the similar mechanism with the similar mode with the vascular-transplanted xenografts [52]. Consequently, the prevention of Ab-mediated rejection in allograft transplantation of the ABO-incompatible organs is similarly applicable for the beneficial prevention of the Ab-mediated xenograft rejection in xenotransplantation. Therefore, successful strategy for transplantation across the ABO-incompatible rejection barrier can be expanded to that of xenograft rejection. The molecular mechanistic characters between the allografts of ABO-incompatible organs and the xenografts of xenotransplantation are well known in the immune reaction. Success in clinical blood ABO group-incompatible allotransplantation reflects progress in humoral immunity [53].

Several approaches including the removal of ABO-specific antibodies during transplantation by plasmapheresis or immunoadsorption, impairing of the B cell population by rituximab or splenectomy, use of immunomodulators by intravenous Ig (IVIG), and a synthetic drug Losartan known as an endothelial cell protective agent and a hypertension preventive agent are used for the success of transplantations. From the similar mechanism of Ab-mediated rejecting modes between the allograft transplantation of ABO-incompatible organs and vascular xenotransplantation of xenograft organs [54], Ab-mediated rejection of xenografts can be curbed. ABO-incompatible organ allotransplantation is used as the factors for accommodation [55] that is in similar state of long-term survival of xenograft organs in vascular xenotransplantation. In addition, blood group A/B-specific antibodies and α1,3Gal-specific antibodies are similar origins and function, in terms of age, gender, and total IgM [56]. α1,3Gal antigen-specific antibodies consist predominantly of IgM and IgG2 in non-sensitized and naive recipients or patients, while recipients or patients exposed to pig tissues express various subtypes of antibodies such as anti-α1,3Gal IgG2, IgG1, and IgM [57]. If such preformed antibodies against A/B antigens are removed off before allotransplantation of ABO-incompatible organs, patient recipients become immunologically tolerant to the blood group-incompatible antigens and terminate the A- or B-specific Ab production, or such accommodation status is activated [58]. This fact has been also understood by the results that the α1,3Gal-specific Ab titer in α1,3Gal-T KO mice is highly produced when xenoantigens are immunized compared to α1,3Gal-carrying allogenic peptides. The reason has been supported by an enhanced activation of helper T cells by xenoantigenic peptides when compared to allogenic peptides [59].

However, molecular antigenic incompatibilities of individual species and the antigenically loaded sizes contribute to the immunological rejection strength of xenografts in the xenotransplantation of xenografts. In the HR, HR response in allografts of ABO-incompatible organs occurs when Abs preformed for A or B antigenic determinants bind to the endothelium and consequently activate complementation reaction [52]. Seemingly, HR of xenografts occurs when Abs preformed specifically for α1,3Gal antigen bind to the endothelium and activate complementation. Antibody binding and complement fragmentation lyse the endothelium and raise prothrombosis. Hemorrhage and platelet aggregation observed in the microvasculatures induce the termination within short period from minutes to hours. Thus, HR of xenografts or allografts of ABO-incompatible organ is similar.

In the AVR, AVR slowly occurs from several hours to several days after transplantation and displays diffused microvascular thrombosis, endothelium swelling, and focal ischemia with fibrin deposition in vascular vessel. AVR of xenografts are characterized by cellular infiltrate of innate immune cells such as neutrophils, monocytes/macrophages, and NK cells [11]. A crucial and driving factor is the activation cellular event in the AVR, which is known as a type II transcriptional upregulation in endothelial cells in order to transcribe genes encoded for cellular adhesion molecules (CAMs), chemokines, cytokines, and prothrombosis molecules. Naturally preformed (anti-α1,3Gal) and induced (anti-α1,3Gal and anti-non-α 1,3Gal) xenoantigen-specific Abs are crucial for the AVR of xenografts [60]. Direct binding of α1,3Gal by the innate immunity provokes to AVR. Currently, human leukocyte antigen (HLA), blood group A- and B-binding antibodies, endothelial cell-specific antigens, and non-HLA molecules trigger the allograft-specific AVR responses in ABO-incompatible organs. In the ACR, the cellular rejections of organ xenografts are not well understood in NHPs because the organ xenografts are preferably damaged in HR or AVR. However, T-cell activation is driven mostly through the indirect activation mode, where the T-cell activation responses are superior to drive in a xenoantigenic recognition reaction due to the antigenic differences [54]. In addition, the known A/B or α1,3Gal antigens for incompatible carbohydrates also lead to vigorous immune response of T-cell activation through antigen uptake and preformed or elicited Abs, which are specific for the antigenic carbohydrates.

8.9 Accommodation

The conceptual definition of original accommodation is simply depicted to resistance to immune response-mediated damage and injury [61]. Dr. Jeffrey Platt initially presented tolerance and the term "accommodation", as the terminology of "accommodation" described by Dr. Platt indicates resistance of endothelial cells against antibody-driven immune rejection during kidney transplantation of ABO-incompatible organ or experimental xenotransplantation of xenografts [62] Accommodation responses have been initially observed during allotransplantation of ABO-incompatible kidney organs when recipients of kidney transplants have antibodies in their blood as well as and heterotopic cardiac xenografts [62].

Now, accommodation events occur in considerable rates occupied with approximately 10–30% of common transplantation of ABO-compatible organs [63]. He originally named the accommodation event as a state that transplanted grafts show immune-resistant status with immune response-driven injuries and damages. Such "accommodation" term has been ascended to the success in the allotransplantation of HLA-incompatible organ grafts as well as allotransplantation of ABO-compatible HLA-incompatible organ grafts. Presently, clinical accommodation indicates the successful ABO-incompatible renal allograft transplantation [40]. The term "accommodation" is therefore referred to indicate patient recipients who bear donor organ-recognizing ABO allogenic Abs or HLA-specific allogenic Abs without any damage to the allograft with the unknown mechanism or a phenotype responsible for accommodation [64]. Although the accommodation state of nonexistence of acute rejection is resembled to tolerance, tolerance is the status of nonresponsiveness of donor-dependent immune response. Since transplanted recipients bear normally B cell-derived immunity to their donors, the accommodation events are commonly found in transplantation of organs.

Such similar phenomenon is also observed in discordant xenotransplants, in which grafts are survived without injury even in the presence of α1,3Gal epitope-recognizing antibodies where carbohydrate structure of α1,3Gal epitopes is similar to carbohydrate structures of the ABO antigens [65]. Bach et al. [66] suggested that the xenoantigenic accommodation model is a hamster-to-rat cardiac xenotransplantation, where HR was prevented by treatment with recombinant cobra venom factor (CVF) known as C3b (Cobra) and the nontoxic and complement-activating cobra venom component with the immune-suppressor cyclosporine [66]. The CVF depletes complement. Then, although IgM type of anti-hamster Abs was elicited, the heart-derived transplanted grafts survived without any signal of acute antibody-mediated damage or graft rejection. Presumably, one possible explanation has been made that endothelial cells become resistant to tissue injuries caused by antibody-mediated responses and complement-mediated responses as well as damages caused by the production of protective proteins such as a potent NF-κB signaling-inhibitory anti-inflammatory A20 protein, antiapoptotic Bcl-2, and antiapoptotic Bcl-xL, and heme oxygenase-1 (HO-1) known as anti-inflammatory, antiapoptotic, and antiproliferative enzyme. As an interesting point, there is a similarity between experimental survival of discordantly incompatible xenografts and clinical survival of allografted ABO-incompatible kidneys [67]. Commonly in both xenografts and allografted transplantation, the antibody-recognizing endothelial antigens are mainly carbohydrates. Carbohydrates-specific immune responses involve preexisting B cells and mainly IgM, whereas protein antigens-specific responses need B cells maturation with T-cell cooperation and IgG antibodies. Therefore, the accommodation of ABO-incompatible allografts is not the same as the accommodation of HLA-incompatible grafts. The reason why ABO-incompatible renal transplantation is more successful than HLA-incompatible may be the difference between protein antigen-specific antibody responses and carbohydrate antigen-specific antibody responses [64].

C4d known as a split fragment of C4 complement component contributes to the antibody-mediated rejection (AMR). Although cellular rejection-based organ loss is not frequently observed with immunosuppressive regimens, AMR of humoral rejection type clinically causes cellular injury and damage in transplantation, as known for transplanted renal and cardiac grafts. C4d is an AMR marker and activated by the classical complement and lectin complement pathways of complement activation, as complement activation activates effector cells (B/T-cells, macrophages/monocytes, and neutrophils) and endothelial cells as target cells during AMR of allografts. In fact, complement C4d observed in HLAi-grafted patients is regarded as a signal of AMR [68], because C4d is detected in accommodated ABO-incompatible grafted recipient patients [69], indicating the accommodation take places in ABO-incompatible allotransplantation, but not in HLA-incompatible allotransplantation. Without graft injury and damage, the donor HLA-specific Abs or C4d deposition reflects the accommodation state. Because antigenic carbohydrates respond differently to recognition and binding of Abs than HLA, accommodation event is easily existed in ABOi but not in HLAi transplantation. This indicates that transplanted grafts can be accommodated to Abs against antigenic carbohydrates but not antigenic proteins. Therefore, it is suggested that the term "accommodation" is not automatically used in patient recipients with donor-specific anti-HLA Abs or intragrafted complement deposits [64].

The term accommodation does not mean a non-injured or non-damaged grafts in the existing condition of Abs. Instead, it rather means a protected phenotype, indicating the accommodated status. Therefore, those phenotype has been suggested to include the following proteins of A20 protein, Bcl-2/Bcl-xL, or HO-1, which were mentioned above [66, 70], and also complement regulatory proteins, complement receptor proteins of CD46, CD55, and CD59 or decay-accelerating factor (DAF) [71]. If accommodation is the resistant status of a graft to acute damage and rejection by donor-specific Abs or other blood factors [72], accommodation would be explained by the following conditions of nonrejection enhancement, grafts adaptation of transplants, and immune tolerance. Accommodation reflects changes in endothelial cells of transplanted grafts or other related cells resistant to cell toxicity [73]. Therefore, before xenograft operation, preformed anti-ABO antibodies should be removed from recipient to keep the non-antibody titer, called "accommodation". If pig α1,3Gal xenograft survival time persists for 3–4 weeks in humans by the antibody removal using such absorption, hCRP-TG organs or immunosuppression, this state of accommodation is similar to the status of renal allografts with blood group ABO-incompatibility. At present, this is suggested to be impossible due to all other xenogenic barriers [74].

References

1. Varki A, Gagneux P. Multifarious roles of sialic acids in immunity. Ann N Y Acad Sci. 2012;1253(1):16–36.
2. Sadler JE, Paulson JC, Hill RL. The role of sialic acid in the expression of human MN blood group antigens. J Biol Chem. 1979;254:2112–9.
3. Uemura K, Roelcke D, Nagai Y, Feizi T. The reactivities of human erythrocyte autoantibodies anti-Pr2, anti-Gd, Fl and Sa with gangliosides in a chromatogram binding assay. Biochem J. 1984;219:865–74.
4. Duk M, Sticher U, Brossmer R, Lisowska E. The differences in significance of alpha 2,3Gal-linked and alpha 2,6GalNAc-linked sialic acid residues in blood group M- and N-related epitopes recognized by various monoclonal antibodies. Glycobiology. 1994;4(2):175–81.
5. Ingebrigsten R. The influence of isoagglutinins on the final results of homoplastic transplantation arteries. J Exp Med. 1912;16:169–77. https://doi.org/10.1084/jem.16.2.169.
6. Ottenburg R, Thalhimer W. Studies in experimental transfusion. J Med Res. 1915;28:213–29.
7. Bell K. The blood groups of domestic mammals. In: Agar NS, Board PG, editors. Red blood cells of domestic mammals. Elsevier Science Publishers; 1983. p. 133–64.
8. Andrews GA, Chavey PS, Smith JE, Rich L. N-Glycolylneuraminic acid and N-acetylneuraminic acid define feline blood group A and B antigens. Blood. 1992;79:2485–91.
9. Bighignoli B, Niini T, Grahn RA, Pedersen NC, Millon LV, Polli M, Longeri M, Lyons LA. Cytidine monophospho-N-acetylneuraminic acid hydroxylase (CMAH) mutations associated with the domestic cat AB blood group. BMC Genet. 2007;8:27.
10. Yamakawa T, Suzuki A, Hashimoto Y. Genetic control of glycolipid expression. Chem Phys Lipids. 1986;42:75–90.
11. Bach FH, Robson SC, Winkler H, et al. Barriers to xenotransplantation. Nat Med. 1995;1:869.
12. Watkins WM, Morgan WTJ. Specific inhibition studies relating to the Lewis blood-group system. Nature. 1957;180:1038–40.
13. Morgan WTJ, Watkins WM. Genetic and biochemical aspects of human blood-group A-, B-, H-, Le-a- and Le-b-specificity. Br Med Bull. 1969;25:30–4.
14. Kabat EA. Immunochemical studies on the carbohydrate moiety of water soluble blood group A, B, H, Lea and Leb substances and their precursor I antigens. In: Isbell H, editor. Carbohydrates in solution (Adv chemistry series 117). Washington: American Chemical Society; 1973. p. 334–61.
15. Karki G, Mishra VN, Mandal PK. An expeditious synthesis of blood-group antigens, ABO histo-blood group type II antigens and xenoantigen oligosaccharides with amino type spacer-arms. Glycoconj J. 2016;33(1):63–78.
16. West LJ, Pollock-Barziv SM, Dipchand AI, Lee KJ, Cardella CJ, Benson LN, Rebeyka IM, Coles JG. ABO-incompatible heart transplantation in infants. N Engl J Med. 2001;344:793–800.

17. Breimer ME. Tissue specificity of glycosphingolipids as expressed in pancreas and small intestine of blood group A and B human individuals. Arch Biochem Biophys. 1984;228:71–85.
18. Finne J, Breimer ME, Hansson GC, Karlsson KA, Leffler H, Vliegenthart JF, Halbeek HV. Novel polyfucosylated N-linked glycopeptides with blood group A, H, X, and Y determinants from human small intestinal epithelial cells. J Biol Chem. 1989;264:5720–35.
19. Holmner A, Lebens M, Teneberg S, Angstrom J, Ökvist M, Krengel U. Novel binding site identified in a hybrid between cholera toxin and heat-labile enterotoxin: 1.9 Å crystal structure reveals the details. Structure. 2004;12:1655–67.
20. Yamamoto F, Yamamoto M. Molecular genetic basis of porcine histo-blood group AO system. Blood. 2001;97(10):3308–10.
21. Yamamoto F, Clausen H, White T, Marken J, Hakomori S. Molecular genetic basis of the histo-blood group ABO system. Nature. 1990;345:229–33.
22. Yamamoto F, McNeill PD, Yamamoto M, Hakomori S, Bromilow IM, Duguid JK. Molecular genetic analysis of the ABO blood group system, 4: another type of O allele. Vox Sang. 1993;64:175–8.
23. Kelly RJ, Ernst LK, Larsen RD, Bryant JG, Robinson JS, Lowe JB. Molecular basis for H blood group deficiency in Bombay (Oh) and Para-Bombay individuals. Proc Natl Acad Sci U S A. 1994;91:5843.
24. Cooper DK. Clinical survey of heart transplantation between ABO blood group-incompatible recipients and donors. J Heart Transplant. 1990;9:376–81.
25. Cooper DK, Human PA, Rose AG, Rees J, Keraan M, Reichart B, Du Toit E, Oriol R. The role of ABO blood group compatibility in heart transplantation between closely related animal species. An experimental study using the vervet monkey to baboon cardiac xenograft model. J Thorac Cardiovasc Surg. 1989;97:447–55.
26. Stussi G, West L, Cooper DK, Seebach JD. ABO-incompatible allotransplantation as a basis for clinical xenotransplantation. Xenotransplantation. 2006;13:390–9.
27. Galili U. Xenotransplantation and ABO incompatible transplantation: the similarities they share. Transfus Apher Sci. 2006;35:45–58.
28. Miyagawa S, Takeishi S, Yamamoto A, et al. Survey of glycoantigens in cells from alpha1-galactosyltransferase knockout pig using a lectin microarray. Xenotransplantation. 2010;17:61–70.
29. Galili U, Shohet SB, Kobrin E, Stults CL, Macher BA. Man, apes, and Old World monkeys differ from other mammals in the expression of alpha-galactosyl epitopes on nucleated cells. J Biol Chem. 1988;263:17755–62.
30. Galili U. Evolution of alpha 1,3galactosyltransferase and of the alpha-Gal epitope. Subcell Biochem. 1999;32:1–23.
31. Galili U, Rachmilewitz EA, Peleg A, Flechner I. A unique natural human IgG antibody with anti-alpha-galactosyl specificity. J Exp Med. 1984;160:1519–31.
32. Galili U, Mandrell RE, Hamadeh RM, Shohet SB, Griffiss JM. Interaction between human natural anti-alpha-galactosyl immunoglobulin G and bacteria of the human flora. Infect Immun. 1988;56:1730–7.
33. Rood PP, Tai HC, Hara H, Long C, Ezzelarab M, Lin YJ, van der Windt DJ, Busch J, Ayares D, Ijzermans JN, et al. Late onset of development of natural anti-nonGal antibodies in infant humans and baboons: implications for xenotransplantation in infants. Transpl Int. 2007;20:1050–8.
34. Dons EM, Montoya C, Long CE, Hara H, Echeverri GJ, Ekser B, Ezzelarab C, Medellin DR, van der Windt DJ, Murase N, et al. T-cell-based immunosuppressive therapy inhibits the development of natural antibodies in infant baboons. Transplantation. 2012;93:769–76.
35. Cooper DK. Depletion of natural antibodies in non-human primates – a step towards successful discordant xenografting in humans. Clin Transpl. 1992;6:178–83.
36. Holgersson J, Breimer ME, Samuelsson BE. Basic biochemistry of cell surface carbohydrates and aspects on the tissue distribution of histo-blood group ABH and related glycosphingolipids. APMIS. 1992;100(suppl 27):18.

37. Breimer ME, Molne J, Norden G, et al. Blood group A and B antigen expression in human kidneys correlated to A1/A2/B, Lewis, and secretor status. Transplantation. 2006;82:479–85.
38. Squifflet JP, De Meyer M, Malaise J, et al. Lessons learned from ABO-incompatible living donor kidney transplantation: 20 years later. Exp Clin Transplant. 2004;2:208–13.
39. Norde'n G, Breimer ME. ABO-incompatible kidney transplantation: overview and new strategies. Trends Transplant. 2007;2:35–43.
40. Montgomery RA, Locke JE, King KE, et al. ABO incompatible renal transplantation: a paradigm ready for broad implementation. Transplantation. 2009;87:1246–55.
41. Shimizu A, Colvin RB, Yamanaka N. Rejection of peritubular capillaries in renal Allo- and xeno-graft. Clin Transpl. 2000;14(Suppl 3):6–14.
42. Vadori M, Cozzi E. The immunological barriers to xenotransplantation. Tissue Antigens. 2015;86(4):239–53. https://doi.org/10.1111/tan.12669.
43. Ierino FL, Sandrin MS. Spectrum of the early xenograft response: from hyperacute rejection to delayed xenograft injury. Crit Rev Immunol. 2007;27:153–66.
44. D'Orsogna L. Infectious pathogens may trigger specific allo-HLA reactivity via multiple mechanisms. Immunogenetics. 2017;69:631–41.
45. Bottino R, Trucco M. Use of genetically-engineered pig donors in islet transplantation. World J Transplant. 2015;5:243.
46. Vanhove B, de Martin R, Lipp J, Bach FH. Human xenoreactive natural antibodies of the IgM isotype activate pig endothelial cells. Xenotransplantation. 1994;1:17.
47. Palmetshofer A, Galili U, Dalmasso AP, et al. [alpha]-Galactosyl epitope-mediated activation of porcine aortic endothelial cells. Transplantation. 1998;65:971.
48. Ehrnfelt C, Serrander L, Holgersson J. Porcine endothelium activated by anti-[alpha]-Gal antibody binding mediates increased human neutrophil adhesion under flow. Transplantation. 2003;76:1112.
49. Hauzenberger E, Klominek J, Holgersson J. Anti-Gal IgG potentiates natural killer cell migration across porcine endothelium via endothelial cell activation and increased natural killer cell motility triggered by CD16 cross-linking. Eur J Immunol. 2004;34:1154.
50. Holzknecht ZE, Platt JL. Identification of porcine endothelial cell membrane antigens recognized by human xenoreactive natural antibodies. J Immunol. 1995;154:4565.
51. Economidou J, Hughes-Jones NC, Gardner B. Quantitative measurements concerning A and B antigen sites. Vox Sang. 1967;12:321.
52. Holgersson J. Can ABO-incompatible organ transplantation pave the way for clinical xenotransplantation? Transplantation. 2007;84(12 Suppl):S48–50.
53. Tydén G, Kumlien G, Genberg H, et al. ABO incompatible kidney transplantations without splenectomy, using antigen-specific immunoadsorption and rituximab. Am J Transplant. 2005;5:145.
54. Auchincloss H Jr, Sachs DH. Xenogeneic transplantation. Annu Rev Immunol. 1998;16:433.
55. Koch CA, Khalpey ZI, Platt JL. Accommodation: preventing injury in transplantation and disease. J Immunol. 2004;172:5143.
56. Parker W, Lundberg-Swanson K, Holzknecht ZE, et al. Isohemagglutinins and xenoreactive antibodies. Hum Immunol. 1996;45:94.
57. Yu PB, Parker W, Everett ML, et al. Immunochemical properties of anti-Gal[alpha]1–3Gal antibodies after sensitization with xenogeneic tissues. J Clin Immunol. 1999;19:116.
58. Ishida H, Tanabe K, Ishizuka T, et al. Differences in humoral immunity between a nonrejection group and a rejection group after ABO-incompatible renal transplantation. Transplantation. 2006;81:665.
59. Tanemura M, Yin D, Chong AS, et al. Differential immune responses to [alpha]-Gal epitopes on xenografts and allografts: implications for accommodation in xenotransplantation. J Clin Invest. 2000;105:301.
60. Gollackner B, Goh SK, Qawi I, et al. Acute vascular rejection of xenografts: roles of natural and elicited xenoreactive antibodies in activation of vascular endothelial cells and induction of procoagulant activity. Transplantation. 2004;77:1735.

61. Platt JL, Kaufman CL, de Mattos G, Barbosa M, Cascalho M. Accommodation and related conditions in vascularized composite allografts. Curr Opin Organ Transplant. 2017;22(5): 470–6.
62. Platt JL, Vercellotti GM, Dalmasso AP, et al. Transplantation of discordant xenografts: a review of progress. Immunol Today. 1990;11:450.
63. Dijke EI, Platt JL, Blair P, et al. B cells in transplantation. J Heart Lung Transplant. 2016;35: 704–10.
64. Rose ML, West LJ. Accommodation: does it apply to human leukocyte antigens? Transplantation. 2012;93(3):244–6.
65. Bach FH, Turman MA, Vercellotti GM, et al. Accommodation: aworking paradigm for progressing toward clinical discordant xenografting. Transplant Proc. 1991;23:205.
66. Bach FH, Ferran C, Hechenleitner P, et al. Accommodation of vascularized xenografts: expression of "protective genes" by donor endothelial cells in a host Th2 cytokine environment. Nat Med. 1997;3:196.
67. Stussi G, West L, Cooper DK, et al. ABO-incompatible allotransplantation as a basis for clinical xenotransplantation. Xenotransplantation. 2006;13:390.
68. Stewart S, Winters GL, Fishbein MC, et al. Revision of the 1990 working formulation for the standardization of nomenclature in the diagnosis of heart rejection. J Heart Lung Transplant. 2005;24:1710.
69. Haas M, Rahman MH, Racusen LC, et al. C4d and C3d staining in biopsies of ABO- and HLA-incompatible renal allografts: correlation with histologic findings. Am J Transplant. 2006;6:1829.
70. Williams JM, Holzknecht ZE, Plummer TB, Lin SS, Brunn GJ, Platt JL. Acute vascular rejection and accommodation: divergent outcomes of the humoral response to organ transplantation. Transplantation. 2004;78(10):1471–8.
71. Ding JW, Zhou T, Ma L, et al. Expression of complement regulatory proteins in accommodated xenografts induced by anti-alpha-Gal IgG1 in a rat-to-mouse model. Am J Transplant. 2008;8: 32.
72. Cascalho MI, Chen BJ, Kain M, Platt JL. The paradoxical functions of B cells and antibodies in transplantation. J Immunol. 2013;190:875–9.
73. Jindra PT, Hsueh A, Hong L, et al. Anti-MHC class I antibody activation of proliferation and survival signaling in murine cardiac allografts. J Immunol. 2008;180:2214–24.
74. Breimer ME. Gal/non-gal antigens in pig tissues and human non-gal antibodies in the GalT-KO era. Xenotransplantation. 2011;18:215–28.

Chapter 9
Classification of Rejection in Host Recipients in Xenotransplantation

9.1 Xenoantigenic Carbohydrate Antigens and Immune Incompatibility

Glycome difference between human and non-human species is the most basic concept as a main barrier of xenotransplantation. Determination of xenogeneic carbohydrate antigens different from interspecies and different species is the first and the crucial step because cell surfaced carbohydrates are biosynthesized by diverse and various glycosyltransferases. They are bound by preformed natural Abs or de novo biosynthesized Abs. They are subjects of the reaction to incompatible blood products, tissues, and organs. In the vascular endothelium, the transplanted endothelial cells of the donor organ are the first lines and contacting sites with the host recipient's blood in organ transplantation. In vascular endothelium, endothelial cells regularly express their self-protecting substances in terms of anti-inflammatory molecule and anticoagulative agents on cell surfaces, which are naturally crucial for the blood anticoagulation and for homeostatic self-protection. Genetically distinct differences are well known between the donors and host recipients, which are the basic causing factors for the rejection during cell, tissue, and organ xenotransplantation in the recipient immune system, consequently resulting in complete failure of the transplantation. Several biological incompatibilities between pigs and humans are the fundamental basis to prevent clinical xenotransplantation. Indeed, pig xenogeneic organs are immunologically rejected by the immune system of human by the most basic AMR. Xenoantigenic grafts are the major subject of immune rejections in mode of species- and tissue-specific carbohydrate antigens.

C.-H. Kim, *Glycoimmunology in Xenotransplantation*,
https://doi.org/10.1007/978-981-99-7691-1_9

9.2 Diverse Xenograft Rejection in Host Recipient

In xenotransplantation, the most key hurdle is the antibody-driven cell death due to cell destruction of the xenografts. Therefore, antibody-removing trials are technically performed in clinical prevention of antigen-Ab interaction. For example, in human host patients, two carbohydrate antigen barriers of the pig α1,3Gal antigen and NeuGc-linked glycoantigen can be easily avoided by such plasmapheresis for simple antibody removal. Interestingly, after perfusion, antibody responses are differed with the subtype of anti-α1,3Gal IgG. However, blood group type A-specific anti-A Abs are IgM [1]. In the transplantation, carbohydrate residues and protein epitopes are targeted by the humoral antixenograft immunity of antibody-mediated rejection. The genetic incompatibility between human cellular adhesion molecules (CAMs) and their interacting pig ligands is a rejection-driving force. Moreover, additional incompatibility between complement proteins and coagulation factors of human and their regulatory proteins is also a contribution factor to xenograft rejection in xenotransplantation.

In blood harmonist, as thromboregulation is lost in the organ xenografts of AVR, α1,3GalT KO pigs hearts and kidneys transplanted into baboons are readily rejected after thrombotic microangiopathy caused by induced deposition of non-α1,3Gal-specific IgG Abs in vascular vessels [2, 3]. Pig cells can also directly activate human prothrombin [4]. In the coagulation cascade, tissue factor pathway inhibitor (TFPI) of pigs is known to partly be incompatible with coagulation factor Xa (FXa) of humans [5], where FXa is a serine protease, known as a target candidate for the antithrombotics. Structure of ligand/FXa complexes is found. Functional incompatibility in pig TFPI hinders recognition of human factor Xa and also impedes coagulation reaction drived by human tissue factor. The pig anticoagulant activity is corresponded to that of the human protein without apparent incompatibilities between human tissue factor pathway and pig TFPI. Additional coagulation factors can explain for the dysregulated thromboregulation and endothelium function as well as dysfunctional tissue factor in xenograft rejection. These two properties of pig cells also contribute to dysregulation of coagulation, accelerating the rejection. Therefore, in organ xenotransplantation, species incompatibilities are carefully treated and should be analyzed more importantly rather than just antigenic differences.

Genetically modified pig organs having the Galα1,3Gal carbohydrate xenoantigen deficiency do not exhibit HR in NHPs; however, acute humoral rejection is observed during 3–6 months regardless of the use of immunosuppression [6, 7]. The xenoantigen-specific antibody response is known to be encoded by germline progenitors located in V3-21 region of the VH3 family [8]. The IgH chain variable regions are assembled from variable (VH), diversity (D), and joining (J) genes. Among the human VH complex consisting of 100 gene segments in 7 families, the largest VH3 family has 22 functional genes. For example, the human α1,3Gal-specific antibody is encoded by specific heavy chains of the VH3 Ig gene family.

9.3 Natural Xenoantibodies

Natural xenoantibodies reject wild-type pig organs by binding to the Galα1,3Gal carbohydrate xenoantigen and trigger to initiate complement-mediated cytotoxicity (CDC). As complement regulatory proteins, human DAF-transgenic pigs are resistant against HR, but susceptible of acute humoral xenograft rejection [9–11]. Xenoreactive Abs produced from human DAF transgenic pigs and pig xenografts of wild type are encoded by restricted progenitors of germline [12–16]. The xenoantibodies encoding IgVH genes are linked with IGHV alleles [9–12], which are regarded as the germline progenitors for xenoantibodies on human bioartificial liver [14–16]. Although later described, the unexpected anti-non-α1,3Gal xenoantibodies are largely produced upon α1,3Gal-T KO xenografts transplantation. The anti-non-Gal antibodies obtained from humans and NHPs with immunization of α1,3Gal-T KO pig cells have been found to bind the α1,3Gal-T KO pig cells with the enhanced binding capacity. This suggests that carbohydrate xenoantigens also induce the xenoantibody response to α1,3Gal-T KO pig cells.

Carbohydrate-specific antibody production in evolution is well conserved in humans and NHPs as innate immune responses [17]. Canonical structure of anti-carbohydrate antibodies is known to determine the genetic selection of VH3 genes of the immune repertoire during xenotransplantation [14–16]. Continued antibody selection for xenoantigens restricts behaviors of germline progenitors. Thus, pig cells are considered to be altered after genetic modification to induce selection of specific xenoantibodies. This antibody-mediated rejection is coupled with complement binding to antibodies against xeno antigens-cell activation-cell lysis axis. The event occurs in the absence condition of pig CD55-controlled human complement [18]. Major issue is destroyed damage and injury of pig tissues by naturally preformed Abs, the cellular rejection, complement system, and coagulative dysregulation. Those species incompatibilities are caused by multiple factors. Among them, human IgGs likely involve in their function in each.

In summary, as described above, based on the severity and prolonged rejection period, the four classes of rejection such as (1) hyperacute rejection (HR), (2) delayed rejection (DR), (3) acute rejection (AR), and (4) chronic rejection (CR) that are in collaboration.

References

1. Rydberg L, Molne J, Strokan V, et al. Histo-blood group A antigen expression in pig kidneys–implication for ABO incompatible pig-to-human xenotransplantation. Scand J Urol Nephrol. 2001;35:54–62.
2. Kuwaki K, Tseng Y-L, Dor FJMF, et al. Heart transplantation in baboons using [alpha]1,3-galactosyltransferase gene-knockout pigs as donors: initial experience. Nat Med. 2005;11:29.

3. Yamada K, Yazawa K, Shimizu A, et al. Marked prolongation of porcine renal xenograft survival in baboons through the use of [alpha]1,3-galactosyltransferase gene-knockout donors and the cotransplantation of vascularized thymic tissue. Nat Med. 2005;11:32.
4. Siegel JB, Grey ST, Lesnikoski B-A, et al. Xenogeneic endothelial cells activate human prothrombin. Transplantation. 1997;64:888.
5. Kopp CW, Siegel JB, Hancock WW, et al. Effect of porcine endothelial tissue factor pathway inhibitor on human coagulation factors. Transplantation. 1997;63:749.
6. Yamada K, Yazawa K, Shimizu A, Iwanaga T, Hisashi Y, Nuhn M, O'Malley P, Nobori S, Vagefi PA, Patience C, Fishman J, Cooper DK, Hawley RJ, Greenstein J, Schuurman HJ, Awwad M, Sykes M, Sachs DH. Marked prolongation of porcine renal xenograft survival in baboons through the use of alpha1,3-galactosyltransferase gene-knockout donors and the cotransplantation of vascularized thymic tissue. Nat Med. 2005;11(1):32–4.
7. Kuwaki K, Tseng YL, Dor FL, Shimizu A, Houser SL, Sanderson TM, Lancos CJ, Prabharasuth DD, Cheng J, Moran K, Hisashi Y, Mueller N, Yamada K, Greenstein JL, Hawley RJ, Patience C, Awwad M, Fishman JA, Robson SC, Schuurman HJ, Sachs DH, Cooper DK. Heart transplantation in baboons using alpha1,3-galactosyltransferase gene-knockout pigs as donors: initial experience. Nat Med. 2005;11(1):29–31.
8. Kiernan K, Harnden I, Gunthart M, Gregory C, Meisner J, Kearns-Jonker M. The anti-non-gal xenoantibody response to xenoantigens on gal knockout pig cells is encoded by a restricted number of germline progenitors. Am J Transplant. 2008;8(9):1829–39.
9. Lam TT, Paniagua R, Shivaram G, Schuurman HJ, Borie DC, Morris RE. Anti-non-Gal porcine endothelial cell antibodies in acute humoral xenograft rejection of hDAF-transgenic porcine hearts in cynomolgus monkeys. Xenotransplantation. 2004;11(6):531–5.
10. Ramirez P, Montoya MJ, Rios A, Garcia Palenciano C, Majado M, Chavez R, et al. Prevention of hyperacute rejection in a model of orthotopic liver xenotransplantation from pig to baboon using polytransgenic pig livers (CD55, CD59, and H-transferase). Transplant Proc. 2005;37(9): 4103–6.
11. Chen G, Sun H, Yang H, Kubelik D, Garcia B, Luo Y, et al. The role of anti-non-Gal antibodies in the development of acute humoral xenograft rejection of hDAF transgenic porcine kidneys in baboons receiving anti-Gal antibody neutralization therapy. Transplantation. 2006;81(2): 273–83.
12. Nozawa S, Xing P-X, Wu GD, Gochi E, Kearns-Jonker M, Swensson J, et al. Characteristics of immunoglobulin gene usage of the xenoantibody binding to Gal-α(1,3) Gal target antigens in the Gal knockout mouse. Transplantation. 2001;72:147–55.
13. Kearns-Jonker M, Fraiman M, Chu W, Gochi E, Michel J, Wu G-D, Cramer DV. Xenoantibodies to pig endothelium are expressed in germline configuration and share a conserved immunoglobulin VH gene structure with antibodies to common infectious agents. Transplantation. 1998;65:1515–9.
14. Kearns-Jonker M, Swensson J, Ghiuzeli C, Chu W, Osame Y, Baquerizo A, Demetriou A, Cramer DV. The human antibody response to porcine xenoantigens is encoded by IGHV3-11 and IGHV3-74 IgVH germline progenitors. J Immunol. 1999;163:4399–412.
15. Kleihauer A, Gregory CR, Borie D, Kyles AE, Shulkin IS, Patanwala I, Zahorsky-Reeves J, Starnes VA, Mullen Y, Todorov I, Kearns-Jonker M. Identification of the VH genes encoding xenoantibodies in non-immunosuppressed rhesus monkeys. Immunology. 2005;116:89–102.
16. Zahorsky-Reeves J, Gregory C, Cramer DV, Kyles AE, Borie DC, Christe KL, Starnes VA, Kearns-Jonker M. Similarities in the immunoglobulin response and VH gene usage in rhesus monkeys and humans exposed to porcine hepatocytes. BMC Immunol. 2006;7(1):3.
17. Nguyen H, Sato N, MacKenzie C, Brade L, Kosma P, Brade H, Evans S, Oliva B, Bates P, Querol E, Aviles F, Sternberg M. Automated classification of antibody complementarity determining region 3 of the heavy chain (H3) loops into canonical forms and its application to protein structure prediction. J Mol Biol. 1998;279(5):1193–210.
18. Cozzi E, White DJ. The generation of transgenic pigs as potential organ donors for humans. Nat Med. 1995;1(9):964–6.

Chapter 10
Hyper Acute Rejection (HAR)

10.1 Introduction

Hyper acute rejection (HAR) occurs 48 h immediately after transplantation. HAR is caused by cytotoxic antibody against donor's HLA or ABO antigen. Its symptoms include high fever, edema, complement activation, and blood coagulation. Blood coagulation is also caused by bleeding from endothelial cells, and its symptoms are heart attack, stroke, pulmonary embolism, and thrombosis. In blood vessel, blood clot (thrombus) formation induces endothelial cell injury and abnormal blood stream. During HAR, antibody-mediated rejection (AMR) occurs by donor antigen-induced Abs, which were directed against ABO antigenic carbohydrates, donor-specific molecule HLA proteins, antigens of endothelial cells, or surfaced antigens on pig cells. Despite technology development, there are still obstacles and hurdles of HAR and cellular xenogeneic rejection (CXR). First, as a representative xenoantigen, GGTA1 (α1,3-galactosyltransferase) generates a major α1,3-Gal epitope as xenoantigen. Its synthesized α1,3-Gal residue-linked epitopes are present on surfaces of most mammal cells in synthetic mechanism of the donor substrates α1,3-Gal+Galβ1,4GlcNAc (*N*-acetyllactosamine; LacNAc) to yield Galα1,3Galβ1,4GlcNAc-R carbohydrate structures, known as α1,3-Gal antigenic epitope. For specific aspects of α1,3-Gal epitope, first, the α1,3-Gal antigenic epitope is not existed in humans and several species including Old World monkeys and apes, indicating that α1,3-Gal epitope is a major xenoantigen. Second, cytidine monophosphate (CMP)-*N*-acetylneuraminic acid (NeuAc) hydroxylase (CMAH) hydroxylates sialic acid Z(Sia), Neu5Ac. The CMAH gene is widely expressed in endothelial cells of mammals, except highly evolved species, humans. Neu5Gc is one of non-Gal xenoantigens. Third, β4GalNT2 (β1,4*N*-acetylgalactosaminyl transferase) synthesizes the Sia-α2,3-(GalNAc-β1,4)Gal-β1,4-GlcNAc structure known as Sd(a) antigen, transferring a donor substrate of β1,4-GalNAc residue to the Gal residue of an α2,3-sialylglycan structure. Sd(a) is one of non-α1,3Gal xenoantigens (Fig. 10.1). Fourth, human HLA-compatible swine leukocyte antigen (SLA) is a

C.-H. Kim, *Glycoimmunology in Xenotransplantation*,
https://doi.org/10.1007/978-981-99-7691-1_10

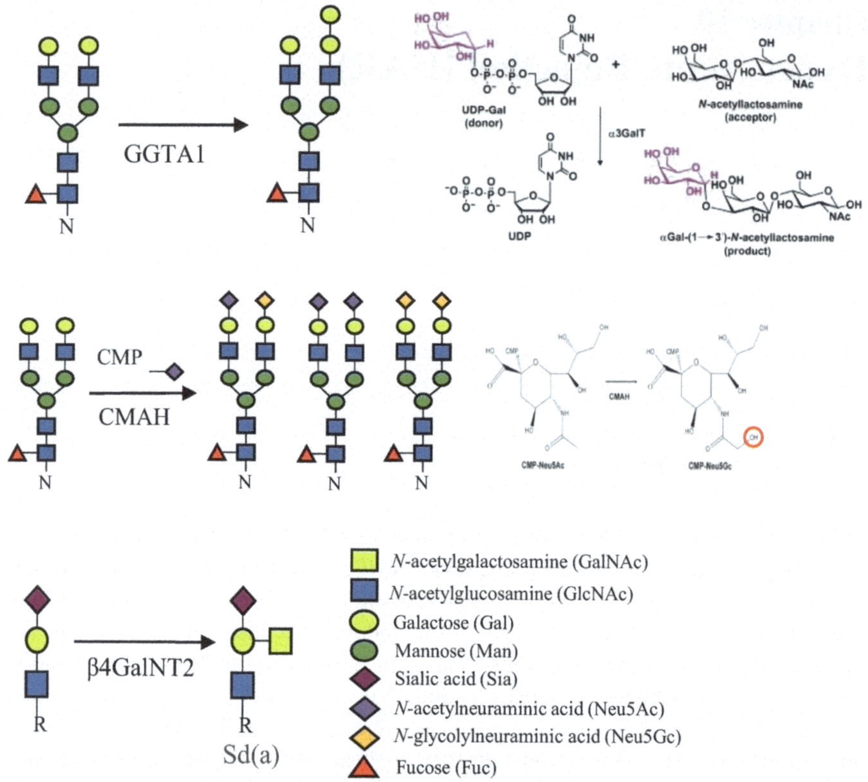

Fig. 10.1 Xenoantigen synthesis, which causes antibody-mediated rejection (AMR). GGTA1 (α1,3-galactosyltransferase) generation of α1,3-Gal antigenic epitope, α1,3Gal + *N*-LacNAc (Galβ1,4GlcNAc) transferring to Galα1,3Galβ1,4GlcNAc-R structure (α1,3-Gal epitope). Second, CMP-*N*-acetylneuraminic acid (Neu5Ac) hydroxylase (CMAH) hydroxylates Neu5Ac. *N*-glycolylneuraminic acid (NeuGc) is a non-α1,3-Gal xenoantigen. Third, β4GalNT2 (β1,4N-acetylgalactosaminyl transferase) synthesizes the Sd(a) antigen of Sia-α2,3-[GalNAcβ1,4] Galβ1,4GlcNAc-R, transferring an donor β1,4GalNAc residue to the acceptor Gal residue of an α2,3-sialyl chain. Sd(a) is also non-α1,3Gal xenoantigens

swine type MHC protein, which is highly polymorphic. SLA of pigs, like HLA of humans, is directly associated with the immune responses of pigs to microbial and viral infections as well as vaccinations.

It has also a polymorphic diversity in the classical class I (C-I) and class II (C-II) gene clusters and also in the conservative clusters of class III (C-III) gene. Expression of class I gene cluster of SLA is found in most pig cells. SLA and HLA are 75–80% identical in amino acid levels and 3D structures. Technically, efforts have been tried to overcome the AMR in swine (Fig. 10.2). For example, NeuGc/α1,3-Gal-KO pigs and GPI-anchored CD55-transgenic pigs are known to reduce xenograft rejection in pig to human xenotransplantation [1].

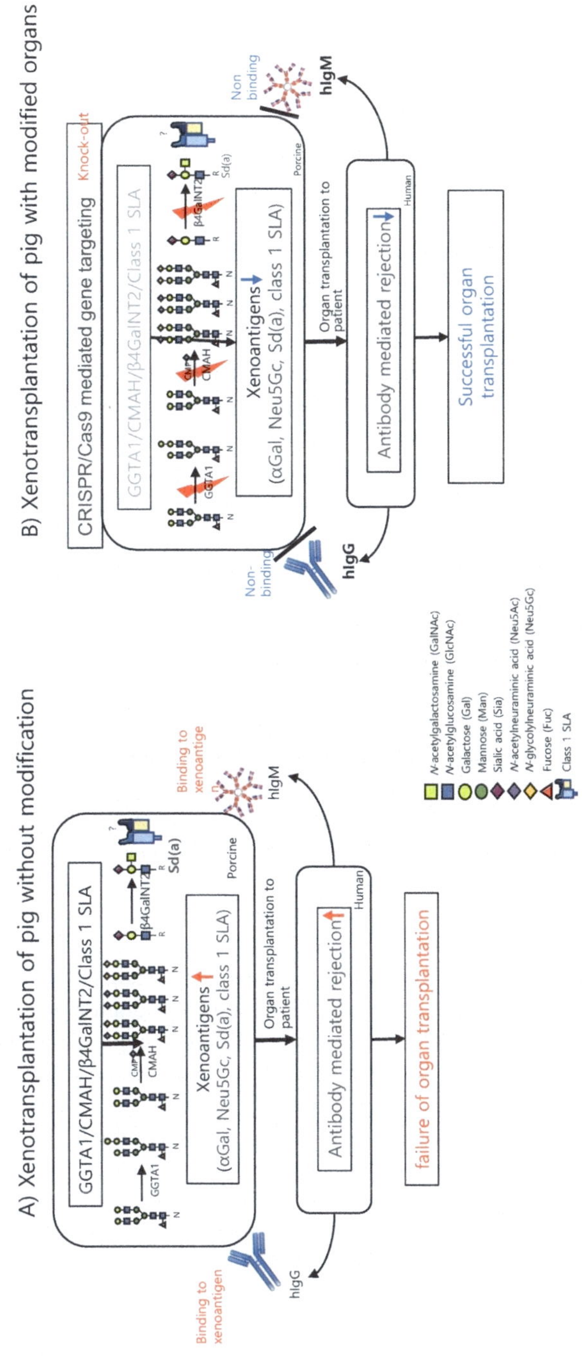

Fig. 10.2 Comparison of xenoantigenic modification

10.2 History, Property, and General Aspects in HR

Recognition, binding, and interaction of anti-α1,3-Gal Abs of primates such as humans with porcine α1,3-Gal epitopes (Galα1,3Gal antigen) elicit HR, as HR is explained by definition of early graft failure or grafts failure within short times from 24 h after transplantation. Thus, naturally pre-formed or preexisting antibodies against endothelial α1,3-Gal epitopes are a factor of HR, contributing to complement activation and rapid destruction of transplanted grafts. However, the graft failure or hurdles are currently overcome by gene disruption of the GGTA1 (α1,3-Gal-trasnferase or α1,3-Gal-T) gene, which inactivate the gene expression, or enhanced overexpression of complement regulatory factor genes of human including DAFs of CD46, CD55, and CD59. During xenotransplantation, α1,3-Gal antibodies raise HR. Human natural antibodies or xenoantigenic antibodies can directly recognize the α1,3-Gal epitopes. Three major components involved in HR are known: (1) specific antigens, (2) antigen-recognizing Abs, and (3) complements.

HR is specific for the fast onset and development during 2–3 min after organ transplantation, by direct endothelial injured damage in blood capillaries. In parallel to the antibody response, the continuing immune response develops, and the event is similar to the inflammatory process, exhibiting vascular blood coagulation, neutrophil infiltration to the inflamed endothelium and tissue. The process involves the cellular apoptosis and necrosis of the transplanted organs collaboratively by innate immune cells and NK cells. Why does it so fast in the HR of the xenograft? This is initially caused due to the host's donor antigen-specific antibodies, which are naturally produced. The α1,3-Gal epitope of Galα1,3Gal is an immunologically major factor in HR induced by natural Abs in pig-to-human xenotransplantation or pig-to-primate xenotransplantation using xenografts.

In the beginning of 1990s, the α1,3-Gal epitope, Galα1,3Gal carbohydrate structure, has been recognized as the major xenoepitope causing acute rejection (AR). Thereafter, during 2000s, α1,3-galactosyltransferase (GGTA1 transferase is α1,3Gal-T) KO pigs have been created for the purpose of overcoming the HR issue. Dr. Sandrin group in Australia is leading to understanding the basic functions of the carbohydrate antigens with endless enthusiasm. After HR is overcome or prevented, the other rejection response in xenotransplantation is consequently followed by the types of AVR, AXR, and delayed xenograft rejection (DXR) [2]. AVR is involved in anti-Galα1,3Gal Abs [3] to activate endothelial cells [4].

In hyperacute antibody-mediated xenograft rejection, the immunological obstacle of HR is overcome using by the α1,3Gal-T KO pigs and multiple KO and transgenic pigs. They avoid immunological and coagulation barriers. Galα1,3Gal-specific IgG Abs induce and mediate a cellular migration that human NK cells translocate across pig aortic endothelial cells, as this event is separately occurred from antibody-dependent cellular cytotoxicity (ADCC) [5]. NK cells recognize the Galα(1,3)Gal [6] although the receptor is not understood. Disseminated intravascular coagulation is also caused by xenoreactive Abs when they activate endothelial cells. The interaction of the α1,3Gal-specific ligands recruits the NK cells or classical complement components, triggering to antibody-ADCC and complement pathway to induce the donor cell death and cell lysis [7]. The recognition of α1,3Gal-specific antibodies to α1,3-Gal antigenic epitopes present in the xeno-grafts stimulates the complement activation cascade, inducing the disruption of the xenograft transplanted by HR as the initial and major hurdle. Therefore, the genetic removal of the organ cell-surfaced Galα1,3Gal antigens in the donor animals is prerequisite for the next step of successful transplantation and immunotolerance induction. Because pigs as NHP are closely related to humans, they are potential usable sources of organ grafts in xenotransplantation. Apart from pig xenografts, alternative chimpanzees have also been considered for the potential source due to the similar organ size. In addition, chimpanzees share compatibility of blood group with humans and therefore they are regarded as an endangered species.

Old World primates and human do not express the Galα-1,3Gal carbohydrate epitopes in terminal glycan structures of glycolipids and *N-/O*-glycans. However, all other mammalian species except for the Old World primates and humans express α1,3Gal glycan structures in the terminal residues [8]. The glycosyltransferase known as α1,3Gal-T enzymatically biosynthesizes the α1,3-Gal epitope with Galα1,3Galβ1,4GlcNAc-R structure [9]. The first target α1,3Gal epitopes, which are recognized by human preformed and natural anti-pig Abs, are abundantly produced on many porcine cells, tissues, vascular endothelium, and intestinal barriers. The human natural Abs of IgG and IgM types bind to the α1,3Gal antigenic epitopes to elicit the HR responses to destruct the porcine xenografts within short times from minutes or hours. As a consequence, humans express a large amount of anti-α1,3Gal antibody in circular system as is an immune barrier to xenotransplantation. In human serum antibodies, several naturally produced antibodies can bind the Galα1,3Gal antigens expressed on the pig-derived cell surfaces, especially they are largely present in endothelial cells in vascular vessels. The α1,3-Gal carbohydrate epitopes are unique in their carbohydrate structures. A1,3-Gal carbohydrate epitopes, anti-α1,3-Gal Abs and α1,3Gal-T activities are distributed in diverse patterns of mammal species. The genetic deletion and enzymatic inactivation found in the α1,3Gal-T are characteristic of ancestral primates, accounting for the α1,3Gal-T deletion and dysfunctionality in primate is originated from physical geographical boundaries, figuring adaptation and natural selection.

The three parameters of α1,3-Gal epitopes, α1,3Gal-T enzymatic activities, and anti-α1,3Gal Abs are also diverse in ancestral species primates. The natural production of preformed anti-α1,3-Gal Abs is an apparent hurdle in the pigs to human xenotransplantation because xenoantigenic immune responses is continuously operated upon transplantation of xenografts. Therefore, genetically constructed α1,3Gal-T KO pigs are now a basic means of rejection solution. Reversely, the anti-α1,3-Gal Abs are also usable for the immunogenicity induction of vaccines and tissue regeneration. The demanding direction to fulfill the organ shortage and failure are depended on the lifespan elongation, and the importance of xenotransplantation is increasingly recognized through the world, especially in the developed countries. Such situation and circumstances are well expressed and described by a surgical medical doctor Cooper. Indeed, in a recent essay, Dr. DK Cooper, a transplant surgeon, who was interested in heart transplantation during his 3 years PhD course, depicted his interest and passions in glycobiology since his joint action on ABO-incompatible allotransplantation to expand and overcome the shortage and failure of organs. He pursued the xenotransplantation and cross-species xenotransplantation of porcine tissues and organs to human patient recipients [10].

There are also additionally known carbohydrate antigens, which are known as antigenic targets expressed on pig cells, and they are specifically recognized by human Abs as anti-pig Abs: (1) *N*-glycolylneuraminic acid (NeuGc) residue generated by the CMAH enzyme and (2) the Sd(a) pentasaccharide structure of GalNAcβ1,4(Neu5Acα2,3)Galβ1,4GlcNAcβ1,3Gal antigenic residues synthesized by β1,4*N*-acetylgalactosaminyltransferase (β4GalNT2) enzyme. These two enzyme genes have experimentally been deleted in pigs to overcome primate complement and coagulation activation. Using those genetically manipulated pigs to delete the genes, the pig organs that were xenografted to non-human primate (NHP) recipients exhibited the prolonged survival for several months longer than the wild type. Toward clinical transplantation of such pig cells and organs, xenotransplantation of pig organs to humans are largely expected within the coming future. Recently, longer xenograft survival has been obtained from genetically engineered pigs in NHPs having 10 months more survival in life-span kidney xenotransplantation [11], and more than years for nonlife span heart [12] xenotransplantation as heterotopic xenograft organs.

10.3 Complement Cascade in HR

In vascularized organ xenotransplantation of wild-type pig to native NHPs, naturally preformed xenoantigen-specific Abs raise HR. The xenoantigen-specific Abs also activate the complement reaction cascade. To overcome these responses, pigs have been genetically modified and generated to induce complement regulatory protein (CRP) expression of human. These genetically mutated organs of the pigs efficiently protect the HR [13]. In HR, pre-formed natural Abs against α1,3Gal epitopes present in the endothelial cells (EC) surfaces of pig activate the complement cascade reaction by binding to α1,3Gal epitopes [14]. Type I (nontranscriptional or intrinsically acting) complement cascade activation is fast and induces the downstream reactions such as EC retraction, P-selectin (CD62P) expression, platelet-activating factor (PAF) secretion and the decreased antithrombotic homeostasis [15]. When IgGs recognize the target α1,3-Gal antigens, IgGs serially activate complement cascade of C3b production and consequently complement mediated vascular collapse or destruction. In the rabbit-to-swine multivisceral xenotransplantation, it was reported that the liver moderately develops HR, but the other organs raise severe HR. Then, IgGs accumulated in the endothelium of vascular vessels cause various pathogenic events such as edema, fibrinoid degeneration, hemorrhage, myositis, necrosis and thrombosis [16].

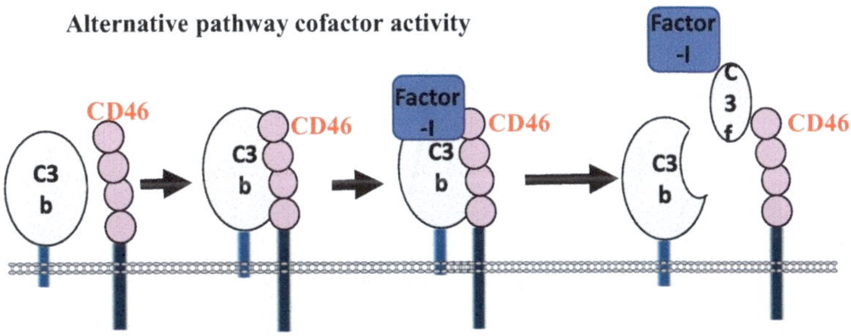

Fig. 10.3 CD46 function in alternative pathway of complement activation

Nitric oxide species (NOS) and reactive oxygen species (ROS) are known as free radicals additionally cause tissue damages, injuries, xenograft rejection and dysfunction. Complement system enhances the HR progression occurred in a xenoantigenic perfused liver by collaboration with leukocytes and free radical formation [17]. If Gal-mimetic, named GAS914, as a soluble polymeric form of a Galα1,3Gal trisaccharide is intravenously administered, levels of xenoreactive Abs are significantly reduced and therefore such administration can prevent HR and also AVR at the initial stage [18]. Transplanted grafts, which are vascularized, are damaged and lost due to activation of type II (transcriptional or secondly induced) EC that is AXR-characteristic with fibrin-deposited thrombosis and immune cell infiltration. Neutrophils, monocyte/macrophages and NK cells are specific for AXR. Among them, the only NK cells have well been elucidated for their immunological role [19, 20].

In the α1,3Gal-T KO cardiac pig to human xenotransplantation, HR has well been described, but not in α1,3Gal-T KO-hCD55-tarnsgenic (TG) donor pigs. This indicates that non-α1,3-Gal-specific Ab-dependent complement activation is presumably related to the HR event. However, local CD55 expression negatively regulates the response at early times [21]. Also, multiple genetic modification-combined α1,3GalT-KO pigs have been generated with hCD55 [21] or hCD46 [22] (Fig. 10.3). Hearts from α1,3Gal-T KO-hCD55 or hCD46 TG pig organs were grafted heterotopically into baboons and the recipient animals showed their survival length between 15 days and 52 days [21]. Their survivals show more moths to 1 year when the host B cells were depleted [22]. In the xenotransplantation of α1,3Gal-T KO kidney of pigs to baboons, multiple TG manipulations have been tried, as transgened for human CD55 and CD59 to reduce complement cascade and transgened for human CD39 to downregulate thrombosis and purinergic receptor signaling as well as α1,2-Fuc-T (or HT) as a H-transferase. In addition, the baboon recipients were also subjected to combined plasma exchanges and received bortezomib and human C1 inhibitor (hC1-INH), which inhibit complement activation with antibody-generating B cell downregulation (or plasma cells). The generated donor pig is referred to α1,3Gal-T KO.hCD55.hCD59.hCD39.hHT TG. In the α1,3Gal-KO.hCD55.hCD59.hCD39.hHT TG pig to primate renal

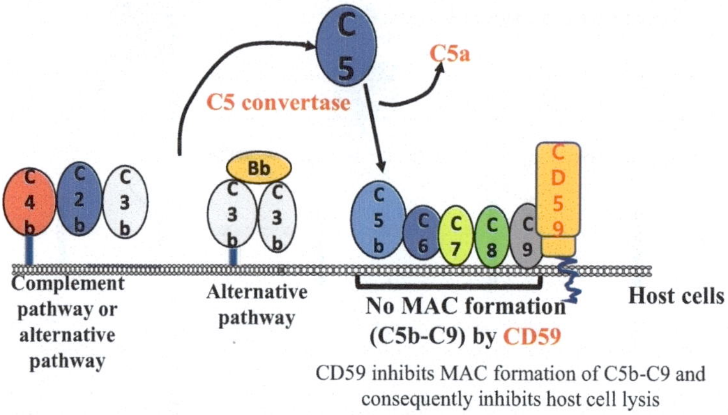

Fig. 10.4 CD59 function in inhibition of complement or alternative pathway. CD59 inhibits MAC formation of C5b-C9 and consequently inhibits host cell lysis

xenotransplantation, the donor pigs did not exhibit a sign of HR [22, 23]. CD59 function in (Fig. 10.4).

In kidney xenotransplantation into baboons utilizing the human CD55/CD59/CD39/HT transgenic α1,3GalT-KO pigs, when baboons received immunosuppression drugs such as corticosteroids, mycophenolate mofetil, tacrolimus and human recombinant C1 inhibitor in the combined administration with bortezomib or cyclophosphamide. The results showed that several parameters such as C3, IgM and C5b-C9 were accumulated in the rejection xenografts, but not C4d accumulation. This suggested that only alternate pathway of complementation was activated, but classical complement pathway was blocked. Moreover, the rejected organs of baboon showed remarkable large infiltration of monocyte/macrophage cells with minimal infiltration of lymphocytes [24]. Therefore, it has been considered that several factors of non-α1,3Gal Abs, innate immune responses with monocyte activation, alternative pathway of complement and PCMV replication are possibly concerted to lead to graft rejective response. The survived length of grafted xenoantigenic kidney was short compared to that of heterotopic xenoantigenic hearts due to loss of grafts displayed by thrombotic microangiopathic pathology and AXR driven by immunoglobulin Abs and deposition of complements [23].

Despite the immune reaction, toward T cell tolerance, thymo-kidney transplants survived for 3 months [21]. The elicited anti-non-α1,3Gal Abs showed cytotoxicity of type of AXR. Therefore, bortezomib known as the B cell-specific blocker and plasma cell-specific inhibitor efficiently reduce the levels of active plasma cells in bone marrow (BM). Structural characterization of non-α1,3Gal antigen is a next step of identification in pig-to-baboon combination because the non-α1,3Gal antigen-responses are specially interested in the complete success in xenotransplantation. For example, in human, some arguments between anti-Neu5Gc antibodies [25] and double α1,3Gal-T and Neu5Gc-KO pig [26] are raised. Pig donors are advantageous when hCRP (CD55, CD59) is combined, as the combination was efficient in avoiding the rejection, however, the complement activation protection is observed in the early transplantation.

In addition, the usage of hC1-INH as a recombinant form (rhC1-INH) is further recommended, as it is effective for allo-specific Abs-driven rejective response. The rhC1-INH inhibits the classical complement pathway activation. Unfortunately, it still allows xenotransplantation rejection response, because of occurring in activation of alternative complement pathway, as reported in rodent models of xenotransplantation [27], although not evidenced in case of NHPs. Thus, more wide ranged blockades of complement activation, including anti-C5 monoclonal antibody (mAb), have been suggested for use in macrophages. Deposition of complement factor iC3b, which binds to CD11b/18 (CR3), is also found. This macrophage's xenoantigen-specific cell response invites the regulator development in the inhibition pathway of transgenic CD47-SIRPα [28]. As known, rejection induces coagulation leading to consumptive coagulopathy.

Historically, the physiological and ethical conditions offered pigs as the appropriate donor sources for xenotransplantation. However, in the experiment that the pig organs to Old World monkey transplantation, the α1,3Gal antigenic epitope was known to activate the Old World monkey complement system, activating complement C5a and C3b, where the anti-α1,3Gal Abs recognize the α1,3Gal epitope, consequently stimulates the pathways of complement cascade. This further activates a cascade of activation. Opsonic C3b complement fragment recognizes the pathogens to internalize them by phagocytes. C5a fragment as a chemotactic protein recruits macrophages and neutrophils as inflammatory phenotypes [29]. Once activated complement pathway affects the downstream activation of chemotactic complement fragmentation to yield C5a and C3b from C5 and C3. Then general inflammatory responses initiated by the produced chemotactic peptides are activated to recruit innate immune cells of leukocytes including macrophages. Therefore, when the endothelial α1,3-Gal epitopes of the xenograft vascular vessels recognize the anti-α1,3Gal Abs to activate platelet aggregation pathway. Consequently, the xenograft vascular transplants are urged to the disruption and collapse as a pattern of the rapid HR [30].

For the humoral antibody-triggered damage to epitopes on endothelial cells, it is strengthened that the complement activation is the mostly crucial mediator and other events including ADCC, inflammation, and endothelial coagulative and thrombotic activation are also known. Naturally preformed or de novo synthesized Abs also raise the immune response during implantation of xenoantigenic ligaments of pig extracellular matrices, bioprosthetic heart valves, intestinal submucosa, and cartilages, eliciting the preformed or de novo Abs interaction, deposition, complement activation, chemotactic complement, and macrophagic cell activation. Abs of IgM and IgG activate macrophages, while and IgM and iC3b activate neutrophils in the bioprosthetic heart valves. In clinical islet transplantation, antibody binding, classical complement activation, IgM, IgG, C4, C3, C5, and C9 deposition, membrane attack complex (MAC) formation damage the early allo and xenogeneic islet injury [29] (Fig. 10.5).

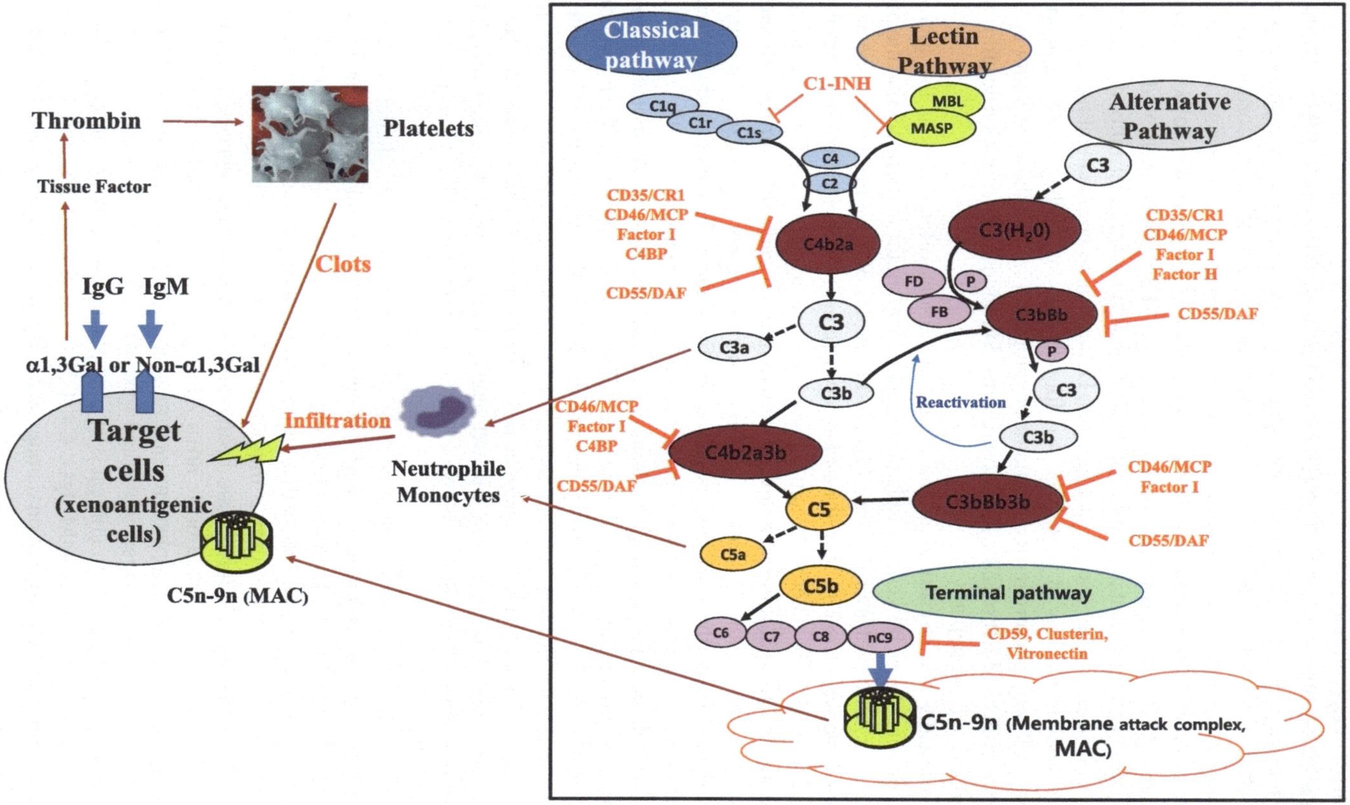

Fig. 10.5 Complement activation cascade to xenoantigenic cells. α1,3Gal or non-α1,3Gal-epitopes are recognized for the three different complement cascade pathways of classical, lectin, and alternative pathways. Human complement regulatory proteins (CRP) such as CD55 or CD59 are enforced for expression in pigs to protect them from human complements

10.4 Galα1,3Gal Glycan Xenoantigens and Its Biosynthesis

The α1,3Gal or Galα1,3Galβ1,4GlcNAc-R glycan epitope is the first target of the mammalian humoral immunity, which is a type of natural and elicited immunity against pig-primate solid organ transplantation. The recognition and binding of the naturally formed α1,3Gal-specific Abs to the α1,3Gal epitope indicates the anti-α 1,3Gal reactivity of human as the HR reaction form during α1,3-Gal epitope-bearing organ transplantation. The Galα1,3Galβ4GlcNAc-R epitope is significantly present on the EC surfaces of pig endothelium and epithelial cells, calculated to approximately 1–3 × 10^7 epitope numbers per cell [31]. Higher 1–8% IgM levels than 1–2.4% IgG levels have been calculated to be specifically bind to the Galα1,3Galβ4GlcNAc-R glycans [32]. In addition, the Galα1,3Galβ4GlcNAc-R glycan epitope develops the AHR by xenotransplanted grafts-elicited anti-α1,3Gal Abs including IgG and IgM types.

Considering those facts such as the α1,3-Gal glycan epitope, anti-α1,3Gal Abs, and α1,3Gal-T enzyme activities, α1,3Gal-T enzyme is considered to be lost in ancestral primates of Qld-World primates. A major and commonly observed xenoantigen Galα1,3Gal causes HR in pig-to-primate xenotransplantation including pig-to-human xenograft transplantation. α1,3Gal-T enzyme acts on β-galactosyl-1,4-*N*-acetylglucosaminyl (β-Gal1,4GlcNAc-) terminus, *N*-acetyllactosamine (LacNAc), nonreducing terminal LacNAc residues of glycoproteins, and asialo-α1-acid glycoprotein, but not to 2′-fucosyl-LacNAc (FucLacNAc) [33]. Because asialofetuin has the terminal Gal residue (Galβ1,4GlcNAc) in N-glycoprotein glycans, it can be used.

In humans, α-1,3Gal epitope is synthesized by the catalytic reaction that Gal residue is attached to Galβ1,4GlcNAc glycans linked to glycosphingolipids (GSLs) or glycoproteins [34] (Fig. 10.1). The Galα1,3Gal residues are the specific glycan linkage structure of Galα1,3LacNAc. In the Golgi apparatus, a glycosyltransferase named α1,3galactosyltransferase (α1,3Gal-T) catalyzes the biosynthetic pathway, which catalyzes the transferring reactions of a terminal Gal residue in α1,3Gal-linkage from uridine diphosphate (UDP)-Gal to NAcLac glycans attached to the *O*-linked or *N*-linked glycoproteins or GSLs, producing oligosaccharide Galα1,3Galβ1,4GlcNAc-R structures, termed as α-1,3Gal epitopes.

The anti-α1,3Gal-specific recognizing lectin GS-IB4 is also known, which isolated from plant *Griffonia simplicifolia* isolectin B4 [35]. In addition, purified monoclonal Abs and purified polyclonal Abs are also available for the specific recognition of the Galα1,3Galβ4GlcNAc-R glycan epitope. The GS-IB4 lectin recognizes Galα1-terminal glycans, but the lectin is impossible to distinguish the structure difference between the terminal Gal residue attached to different carbon atoms of the Galα1,3 residues and Galα1,4 residues. Alternative α1,3Gal-recognizing lectin MOA is from the *Marasmius oreades* agglutinin, as abbreviated to MOA because it binds to Galα1,3-terminal glycan. MOA is also impossible to distinguish linkage structures of the α1,2-fucosyl B antigen of B blood type 2, which is the Galα1,3(Fucα1,2)Galβ1,4-R glycan structure and the non-fucosylated α-1,3Gal epitope (Galα1,3Galβ1,4GlcNAcβ1,3-R) known as Galili antigen [36]. Thus, the two MOA and GS-IB4 lectins are specific for terminal Galα1,3 epitopes in GSL glycolipids and in the fucosyl blood group B antigenic determinant. However, although these two lectins recognize specifically the Gal α1,3LacNAc4 on immune-affinity TLC plates, they recognize weakly the iGb3 and Gb3. GS-IB4 binds to α1,3Gal epitopes in the order of Galα1,3Gal epitope > Galα1,3(Fucα1,2) Gal epitope > Galα1,4-Gal epitope [37, 38]. In contrast, GS-IB4 recognizes the terminally extended α1,4Gal-R structures such as NOR1 glycan of Galα1,4GalNAcβ1,3Gal α1,4Galβ1, 4Glcβ1Cer and NOR2 glycan of Gal α1,4GalNAcβ1,3Gal α1,4GalNAcβ1,3Gal α1,4Galβ1,4Glcβ1Cer, but not the Gb3 on TLC plate [37, 38]. The anti-α1,3Gal monoclonal antibody has also the binding specificity of α1,3Gal of the xenoantigen and glycans, while the polyclonal anti-α 1,3Gal antibody is based on the xenoantigen prepared and obtained from the exclusion of other specific Ans through immunoadsorption column [39].

From the substrate specificity of α1,3-Gal-T enzyme, it is interested in the enzymatic comparison between α1,3-Gal-T and other galactosyltransferases. The similarities in amino acid sequence of pig α1,3Gal-T were high between pig α1,3Gal-T and other family of α1,3Gal/GalNAc transferases such as the human blood A and B type glycosyltransferases and Forssman synthase. The α1,3Gal glycan structure is also resembled with those of human blood group B determinants (B antigen), with only unique difference of the existence of a Fuc-residue attached to the penultimate Gal residue. Thus, it is possible that certain anti-α1,3Gal Abs can react to the terminal Galα1,3Gal epitopes present on A and B antigens [40]. Therefore, human humoral immunity occurred in xenotransplantation contributes to current understanding and mechanistic explanation of the blood group ABO rejection in clinical allotransplantation.

The α1,3Gal-T was once regarded as a single α1,3GalT, which synthesizes the xenoepitope α1,3-Gal. Interestingly, the iGb3 synthase and α1,3GalT are responsible for synthesis of α-1,3Gal [41] because rat iGb3 synthase also synthesizes α1,3Gal epitope as the GSL iGb3 structure and rats have two distinct enzymes of α1,3Gal-T [42] and iGb3 synthase. Interestingly, rats carry two distinct glycosylation pathways for α1,3Gal epitopes. From the study on the rat iGb3 (isoglobotriaosylceramide) synthase (iGb3S), it was demonstrated that the rat iGb3 synthase (iGb3S) [43] also synthesizes the Galα1,3Gal. Thus, the Galα1,3Gal structure is simply biosynthesized by the α1,3Gal-T and iGb3 synthase enzymes. However, the amino acid similarity of the rat iGb3S sequence is low to be 40% when compared to other α1,3Gal-T enzymes of pig, ox, and mouse. In addition, iGb3S is solely specific for the glycolipid synthesis of the iGb3 from UDP-Gal, where lactosyl-ceramide is used as an acceptor substrate [41]. The iGb3 is an isoglobo-series glycolipid class, which is used as the synthetic precursor of isoForssman and isogloboside iGb4 [43]. The most interesting aspect of the iGb3S is that the enzyme is only found from rat. From the substrate usage specificity and glycan structure synthesized by iGb3S, the term of α1,3Gal-T generally referred to glycoprotein-specific enzyme, but not the iGb3 enzyme. Therefore, the description of iGb3 synthase is moved to the following subtitle in this text.

The α1,3Gal epitope with Galα1,3Galβ1,4GlcNAc-R structure or Galα1,3Gal (α-1,3Gal) is a disaccharide linked to LacNAc residues ubiquitously found in marsupials, New World Monkeys, pigs, and non-primate mammal (NPM). In contrast, the structures are not present in apes, Old World monkeys, and humans [44]. Its expression is present in thyroglobulins of bovine, canine, and pig, while it is found in glycoproteins of cell surfaces in mouse lymphoma cells and Ehrlich ascites. In cows and rabbits, α-1,3Gal epitopes are abundantly expressed on surface membrane glycolipids on RBCs [45]. The α-1,3Gal epitopes of Galα1,3Galβ1,4GlcNAc-R are unique carbohydrate attached to glycoproteins and glycolipids [46]. The α-1,3Gal epitope expression is depended on each organ, being high level of the α-1,3Gal epitope in the liver hepatocytes. The expression of α-1,3Gal epitope is also high in the proximal tubules of kidney, but lowly in the distal tubules of kidney and glomeruli. For the glycosylated molecules, the α-1,3Gal epitope is found in α2- and β3 integrins, fibrinogen, and von Willebrand's factor, which are related with blood coagulation of platelets. Twenty or more glycoproteins including the several integrins such as α1, αv, α3/α5, β1, and β3 integrin as well as DM-GRASP and von Willebrand's factor (vWF) present on endothelial cells have α-1,3Gal epitope [47].

10.5 Strategies to Overcome the HR

HR process occurs by three major components in immune system: (a) antigens, (b) Abs, and (c) complements [47]. The first way to overcome the HR is the genetic inactivation of the active α1,3Gal-T enzyme gene. There are two common methods are known. First, one strategy is inactivation of the α1,3Gal-T enzyme expression. Technologically, the transgenic animal production to compete or add another glycosyltransferases to diverge the glycan-biosynthetic pathway from synthesis of α-1,3Gal epitopes. This strategy contributes to the synthesis of different carbohydrates from the α-1,3Gal epitopes not recognized by natural or preformed ABs. The α1,2fucosyl-transferase (Fuc-T) and α2,3 sialyl-transferase (ST) or α2,6ST are the subjects known as differently replaceable glycosyltransferases. These different glycosyltransferases are competitive for the same substrate usage. Through blocking, competing, or activating the specific carbohydrate synthesizing enzyme, the Galα(1,3)Gal expression level is decreased or absent.

Although the α-1,3Gal epitope synthesis was reduced rather than controls, the effect is not complete. The second is genetic deletion strategy, which refers to as the deletion of the pig α1,3Gal-T gene, as reported for the exon 9-targeted pig α1,3Gal-T. The exon 9 region of the pig α1,3Gal-T gene has been constructed in a pPL657 vector, which is the region responsible for the α1,3Gal-T catalytic domain [48]. The two alleles of the α1,3Gal-T genes were subjected to the homozygous inactivation, which disrupts the enzyme function. The KO strategy of the α1,3Gal-T gene completely disrupts the HR. From the anti-α1,3Gal positive B-cell in xenograft α-Gal epitope of α1,3Gal-T KO mice to xenograft α1,3-Gal epitope, the mice showed the lack of α1,3-Gal epitope antigens due to the α1,3Gal-T gene disruption [49]. The α1,3Gal-T KO animals are applicable to transplantation of the α1,3Gal-T KO pigs to humans without complement-mediated HR, and HR was eliminated. The α1,3Gal-T KO pig to human organ transplantation becomes a clinically acceptable treatment. Another minor strategy is the galactosidase treatment to the grafts, which can also remove the α1,3Gal residue. However, it yields the non-α1,3Gal carbohydrate antigens, and they consequently recruit macrophages/neutrophils, NK cells, and T cells.

10.6 Genetic Background of α1,3Gal-T Gene

The mammalian α1,3-Gal-T genes are evolved to the species-specific mode of α1,3Gal-T and α1,3Gal epitope that are equally conserved in lemurs and New World monkeys. Lemurs are a unique primate group native in east African coast Madagascar island and related to apes and monkeys with two superfamilies of the distinct aye-ayes (*Daubentonia madagascariensis*) and ring-tailed lemurs (*Lemur catta*). The α1,3Gal-T has been suggested to be came out with beginning of mammalian evolution [50]. Ancestral Old World primates are estimated to be selectively evolved to inactivate their own α1,3Gal-T gene since two continents are separated from the South American and African continents. Five genetic mutations were found in the α1,3-Gal-T-coding region of the Rhesus macaques and humans.

The Rhesus macaques are known as Asian and Old World monkeys. In the orangutan species, three genetic mutations were also found toward the α1,3Gal-T gene inactivation, giving α1,3Gal-T gene with a specific evolution trend through the mammals. From the species of Old World primates, the α-1,3Gal saccharide antigenic gene was started to interrupt toward evolution with selective pressure and adaptation to survive. At preset in the apically evolved humans, the human α1,3Gal-T gene exists as a form of pseudogenes. Location of the human α1,3Gal-T pseudogene is found in Chromosome 9 with two different point mutations, and these mutations lead to a frame shift, which consequently results in genesis of a prematured stop codon [51]. The same exon region of the chimpanzee has been known to have further the same two point mutations, while the gorilla species have the frame-shift mutation to yield a stop codon as a premature form.

The Old World monkey α1,3Gal-T has also the common mutations. Starting from these evolutionary inactivation process and events in ancestral Old World primates, thereafter continuously diverse evolutionary direction of apes and monkeys has been progressed. Hence, the evolution has been estimated to be started almost within ~20–25 million years [52]. Consequently, the genetic α1,3Gal-T gene inactivation and disruption evolutionarily lead to the immunological failure of tolerance to the α-1,3Gal epitope and the antigen-specific synthetic onset of the anti-α1,3Gal Abs. The mouse α1,3Gal-T gene has been known to have the 9 exon DNA sequences, but consists of only six exon encoding regions. Mouse (*Mus musculus*) α1,3Gal-T exon regions 4–9 are translated to the 1500 bp-conding α1,3Gal-T mRNA (GenBank: M26925.1) [41]. In the α1,3Gal-T gene of pigs (*Sus scrofa*), the α1,3Gal-T gene is located on swine chromosome 1 having an 1269 bp for the α1,3-Gal-T mRNA (GenBank: L36152.1) [53]. The bovine α1,3Gal-T mRNA has an 1828 bp for the α1,3-Gal-T mRNA (GenBank: J04989.1) [54]. The mRNA conding loci of the α1,3Gal-T genes are similarly found to the known primates such as disambiguation *Gorilla gorilla* (GenBank: M73304.1), rhesus macaque *Macaca mulatta* (GenBank: M73306.1), and Bornean orangutan *Pongo pygmaeus* (GenBank: M73305.1) [55].

Considering the expression and distribution of the α-1,3Gal antigenic epitope and serum anti-α-1,3Gal Abs and also the α1,3Gal-T, the α1,3Gal-T is not effective in ancestral primates. In addition, from the human α1,3-Gal-T genomic organization and mammalian nucleotide sequence, the evolutionary linkage has been deduced, indicating that in ancestral primates, the mutations figure out genetic inactivation and deletion of the coding region of α1,3Gal-T gene. When the DNA sequence of the α1,3-Gal-T pseudogene of humans has been compared to those of other mammal species with the evolutionary tree figured by phylogenetic systematics from various animal species, the evolutionary α1,3Gal-T gene inactivation is attributed to the genetic mutations operated by strong selective pressures as natural selection and adaptation in ancestral primates. Partial sequences of the corresponding part of α1,3Gal-T coding region are conserved in humans. For example, such overlapping human clones named clone HGT-2 and HGT-10 were isolated using a bovine homolog.

Among them, HGT-2 bears a predicted coding region, but the region was multiply mutated with frame-shifts and nonsense codons. HGT-2 clone located on 1.5 Kb uninterrupted DNA sequence is similar to bovine α1,3-Gal-T known as a processed pseudogene and partially flanked by Alu family repeats. The HGT-2 sequence has a short 5′- and 3′-untranslated region, and a coding DNA sequence is homologous to bovine α1,3-Gal-T sequence. The sequence has multiple frame-shifted DNA mutations and nonsense codons. Predicted peptide is 68% similar to the bovine enzyme protein. In addition, another clone HGT-10 has the *N*-terminal region-resembling sequence of bovine α1,3-Gal-T. Location of HGT-10 sequence is on chromosome-9 and -12 with pseudogene and nonfunctional gene [52]. The α1,3Gal-T gene is located on house mouse chromosome 2 (*Mus musculus*) [56], swine chromosome 1 (*Sus scrofa*), bovine cattle chromosome 11 (*Bos taurus*) [57], dog (*Canis lupus familiaris*) chromosome [58], and human (*Homo sapiens)* chromosome 9 pseudogene. Like the α1,3Gal-T pseudogene of human, it has been demonstrated to be inactivated for functional α1,3Gal-T in ancestral primates, by DNA sequence deletion, which raises the enzyme truncations and stop codon mutations [59].

When human α1,3Gal-T pseudogene is compared with other species in the nucleotide sequence level, natural selective deletion of evolution-based genes occurred with protection by gene product-specific antibody production. Thus, several evolved animals including apes, humans, and Old World monkeys are negative for the α1,3Gal epitope products. Instead, the naturally preformed α1,3Gal epitope-reactive Abs are generated in a large quantity, and the species are not immunotolerant to them, in order to defend the xeno-species from possible pathogenic infections, if they produce the α1,3Gal epitopes. To conclude the biological meaning of the genetic mutation, α1,3Gal-T gen is inactivated to protect themselves from parasitic pathogenic infections, immunologic totality, and related diseases. The possible systems are predicted to be those endemically connected continents such as Africa and South America of the Old World [52].

10.7 The Production of α-1,3Gal Carbohydrate Epitope-Specific Antibodies

The humoral immunity as a basic defenser acts as hurdle and barrier in short-term survival of transplanted xenografts and also long-term survival of xenografts. Additionally, humoral immune responses directly reject tissue and cellular xenografts. Xenoantigen-specific xenoantibodies directly reject xenografted solid organs. Because human and Old World primates genetically lack the α1,3Gal antigenic epitopes, they are consequently immunologically intolerant to α1,3Gal antigenic epitopes. Old World primates and humans therefore will produce anti-α1,3Gal Abs [60, 61].

In humans, the α1,3Gal-T gene and enzyme are not active, and humans contain anti-α1,3Gal Abs when they are exposed to α1,3Gal antigenic epitopes. The α1,3Gal epitope can be measured by the anti-α1,3Gal epitope-recognizing Abs or the α1,3Gal glycan-recognizing lectin named *Bandeiraea simplicifolia* (*Griffonia simplicifolia*) IB4. α1,3Gal antigen expression is broad in erythrocytes and nucleated in non-primate mammal, New World monkey, and prosimian cells including marmoset, monkey, spider monkey, and squirrel monkey [52]. Among them, prosimians as primates include strepsirrhines such as adapiforms, lemurs, and lorisoids as well as the haplorhine tarsiers, the omomyiforms, i.e. prosimians include all primates but exclude the simians. Prosimians are characteristics of more primitive, ancestral, or plesiomorphic than the present simian monkeys as well as apes and humans.

In contrast, ancestral primates including Old World monkeys, apes, and humans have no α-1,3Gal glycans as surfaced epitopes. However, the α1,3Gal-specific Abs, which are known to hold approximately 1% of immunoglobulins, is naturally producing fetal-borne type of antibody like anti-ABO blood type in human. Human intestinal microbial flora covers various commensal bacteria and opportunistic bacteria, even in the form of pathogenic or non-pathogenic phenotypes, and they produce α-Gal epitope. Human and old world exhibit the lacked α1,3Gal antigenic epitopes through the whole bodies. These current facts explain why Old World primates and humans are immunologically intolerant to the α1,3Gal antigenic epitopes. Humans and Old World primates consequently generate α1,3Gal antigen-specific Abs [62].

Primates what lost the α1,3Gal epitope are urged to survive by means of the α1,3Gal-specific antibody production to defend against α1,3Gal antigen-expressing pathogens. Anti-α1,3Gal Abs are preformed in a natural generation mode in humans. Their anti-α1,3Gal Abs are belonged to the Ig isotypes of IgA, IgG, and IgG. Human anti-α1,3Gal Abs have the selective binding properties to the α1,3Gal antigenic epitopes, and this characteristic reactivity of anti-α1,3Gal Abs is attributed to HR of α1,3Gal organ transplants. Molecular property of anti-carbohydrate Abs is well conservative in primates and humans during natural evolution as an immune functional outcome in innate immunity [49, 63]. In human case, the carbohydrate Gal α1,3Gal-specific natural antibodies are IgM with a less amount of IgG [64, 65]. If α1,3Gal-T1 gene is deleted in mice and pigs, anti-α1,3Gal antibodies are induced in both KO animals, and the KO animals do not develop autoimmune attacks. From the facts, it has been concluded that the generation of anti-α1,3Gal Abs is tolerant to iGb3 synthase-based function [66, 67]. The most well-explaining finding to support the above conclusion is reported by papers below. The α1,3Gal-T1 KO mice produced cytotoxic α1,3Gal-specific antibodies of IgG and IgM upon oral inoculation of *Escherichia coli* O86:B7, as enteric bacterial flora having α1,3Gal-like carbohydrates stimulate anti-α1,3Gal antibodies in humans [66]. Similarly, α1,3Gal-T1 KO pigs produced the anti-α1,3Gal antibodies [67].

10.8 B-Cells and Anti-carbohydrate Antibody Production

In 1930s, Gorer [68–70] found that allotransplantation produces donor-recognizing Abs, and that to do the production, a genetic locus present in the MHC gene region including the human HLA proteins is related. This genetic locus decides acceptance or rejection of transplants [71]. In general, immune responses of B cells underlying the organ rejection in the transplantation are the most potent immunological barriers, as elucidated from studies of the molecular responses of B cells to antigen, potentially applicable for therapeutics [68]. However, there are still covered questions how functions of B cells determine the fate of transplanted organs. In addition, there is still a fundamental question of when, how, or whether potentially developed and potent therapeutic agents can optimally be administered for use [68]. B cells regulate cellular immunity, mediate immune tolerance, and accommodation. B cells express IgM type as the first isotype during rearrangements of light and heavy light chains, which are essentially required for the development of B cells [72, 73].

It has been known that IgM gene-lacking or disrupted animals were demonstrated to be B cells-deficient as well as Abs-deficient [74, 75]. In the case of miniature pigs, different from other large farm animals like cattle, goats, and sheep, pigs are known to bear only one heavy chain locus for Ig gene that is required to inactivate for complete replacement of Ig [76, 77]. B-cell populations to secrete xenoantigen-specific antibodies are important to understand the mechanisms how they can recognize the xenoantigenic carbohydrates because the information can deduce the therapeutic avenue in xenotransplantation [78]. Thus, many attempts to understand the B-cell population have been made. Upon pig antigenic immunization, the anti-α 1,3Gal Ig-producing cells are proliferated in the lymph nodes, BM and spleen for

6 months and antibody-producing plasma cells, which are terminally differentiated, are predominantly located in the bone marrow region.

The naturally preformed and induced α1,3Gal-specific Abs are T-cell-dependently secreted. This simply indicates that the interaction of CD154/CD40 between B-cells and T-cells elicits the humoral immunity, and the interaction inhibition leads to the diminish of antibody production applicable to xenotransplantation [78]. In the lymph nodes and spleens, both CD20+CD138+Ig+B-cell population and immatured CD138+Ig+DR+ plasma cells produce natural anti-α1,3Gal IgM. Human immature CD138+Ig+DR+ plasma cells secrete anti-α1,3Gal IgG, while matured plasma CD138+Ig−DR cells in NHP secrete anti-α1,3Gal IgG. With respect to production of the non-α1,3Gal-specific xenoantibody production, splenic marginal zone B-lymphocyte IgDloCD23lo IgMhiCD21hi are known to T-cell-independently secrete IgM subtype, instead, in a NK cell-dependent manner [79].

Apart from B cell repertoire, CD4+ T cells population among various T cells are characteristically cytotoxic rather than CD8+ T cells [80]. From the tissue and organ rejection caused by Galα1,3-Gal-mediated reaction, there is a basic question of how the immune response of the Galα1,3-Gal-recognizing Abs is generated. Therefore, an immunological mechanism depicted from experimental analysis using the α1,3-Gal KO mouse model has been proposed to answer the Galα1,3-Gal-specific response of IgM Abs, which response is independently induced of cooperative help from T-cells, calling T-cells independent. However, the production of Galα1,3-Gal-reactive IgG Abs is totally dependent on T-cells, as reported by Tanemura et al. [81]. Galα1,3-Gal epitope-specific B cells bind to Galα1,3-Gal xenoepitope antigen on glycoproteins and glycolipids, and they present xenoepitope-linked peptides to MHC C-II and T-cells, consequently, eliciting downstream B cell activation and switching of isotypes [82].

Among circulating B cells in human immune system, approximately 1% of B cell population seems to generate the α1,3Gal epitope-specific Abs, as similar mode of the human ABO blood type antibodies. The B cell population is the B1b-like CD5-CD11b + population, as it produces anti-α1,3Gal antibodies in the analysis of the resident B-cells population capable of the xenoantigen Galα1,3Galβ4GlcNAc-R glycan antigen (or α1,3Gal)-specific Abs production in human and NHPs [83]. α1,3Gal epitope-specific Abs are not observed at fetal and birth, although it starts to produce it after few months of birth. The isotypes of α1,3Gal epitope-recognizing Abs are diverse for the IgA, IgM, and IgG [84, 85], and the human α1,3Gal-specific Abs are encoded by different heavy chain genes basically located in the Ig gene VH3 family [86]. As B-cell-targeting agents to reduce the xenoantibodies, anti-CD20 agents deplete B-cells [87]. In addition, anti-CD19 mAbs are known to completely deplete memory B-cells or short term-survived plasma B cells [25]. As an immunosuppressive agent, anti-CD20 mAbs prolonged the survived period of renal and cardiac xenografts with length of more than 1-year survival period in a model xenotransplantation of pig-to-primate heart xenografts [88]. To target the antibody-secreting cells, plasma cell-specific agent bortezomib is also reported in xenotransplantation of NHPs and humans [89].

10.9 The Generation of α1,3Gal-T KO Pigs

Gal-T-KO mice examined by Sandrin et al. excluded the α1,3Gal-antigen synthesis. Before producing α1,3Gal-T KO pigs, α1,3Gal-T KO mouse was produced [1, 2]. Historically, the named female "Dolly" of domestic sheep was created as the first cloned mammal of the world by the development of the duplication technology. Dolly was the first mammalian clone generated by using an adult somatic cell and the nuclear transfer technology in a method first used in human in vitro fertilization in the 1970s. Duplicated sheep Dolly mated to David, a male sheep and four lambs were born. Thus, we learned that the animal can be cloned from a mammary cell. Accordingly, the pig α1,3Gal-T KO clone became possible by the duplication technology, and the first cloned pig was born. The first α1,3Gal-T KO pigs did not become available until 2003 [90].

The heart and kidney transplantations of α1,3Gal-T KO pig organs remarkably prolonged survived period of the transplanted grafts. Among them, the heart xenografts gave a particularly prolonged and functional survival for 6 months, indicating a potentially advanced development of xenotransplantation technology. This also indicates that the absence of α1,3Gal epitope expression on pig tissues reduces the cellular responses in primates as well as the constitutive humoral immune responses to the xenografts. Thus, the stepwise removal of the α1,3Gal epitopes is beneficial and needed to overcome the immune rejective hurdles and rejection barriers for upcoming xenotransplantation. Subsequently, for the next step, α1,3Gal-T KO pigs were genetically engineered for further genetic manipulation, for example, to express hCRP. These additionally combined technologies to generate relevant pigs as graft supplier have further increased the survival of xenografts [91].

The α1,3Gal-T KO pigs experimentally was negative for HR [92]. In tumor study, the α1,3Gal antigenic epitopes are also possibly applied for the elevated immunogenicity to use in cancer immunotherapy. α1,3Gal-T gene has long been targeted to pursue this issue to remove the Galα1,3Gal antigenic epitopes. In 2001 and 2002, miniature pigs which α1,3-GalT gene has heterozygously been deleted and thus inactivated in enzyme activity were produced. Using the Galα(1,3)Gal or Galα1-3Galβ1-4GlcNAc (α1,3Gal) epitope-targeted KO pigs, the immune-suppressed baboons heart has been xenotransplanted in 2005, giving a graft survival period of 92–179 days in the recipient baboons [10]. α1,3-Gal-T null animals generated from genetic manipulation have shown the eliminated epitopes, and consequently HR elimination. However, the α1,3-GalT gene-targeted pigs are not sufficient for the complete removal of xenoantigens, requiring further discovery, identification, characterization, and elimination of other surfaced xenoreactive antigens from pig cells. From the analysis using anti-α1,3Gal antibody and lectin, any α1,3Gal antibody-reacting glycolipids were detected in α1,3Gal-T KO pig tissues. Hence, the biosynthesis of Fucosyl-iGb3 in pig organs and tissues is issued. The issue whether iGb3 expressed in pig still acts as antigen is questioned as raised in this chapter.

10.10 Clinical Use of α1,3Gal Epitope toward Human Diseases

Apart from application to the xenotransplantation of α1,3Gal epitope, the α1,3Gal antigens can clinically be applied by chemical, biochemical, and biotechnological application to tumor and viral vaccine developments as well as other treatment such as wound healing. Given the basis of the xenoantigen structure and the binding specificity of the α1,3Gal antigenic epitopes and anti-α1,3Gal Abs, these information is applicable to clinical exploitation. Indeed, in human deficiency of the α1,3Gal antigenic epitopes, the clinical application of α1,3Gal antigenic epitope is considered in a manner of targeting tumor vaccines. Tumor cells derived from humans commonly synthesize α1,3Gal antigenic epitopes, and this potentiates the immunogenicity to induce immune response toward the tumor regression. Because human cancer cells display the α1,3Galantigenic epitopes, the α1,3Gal antigens expressed on autologous cancer antigens as vaccines are possibly presented to antigen-presenting cells (APCs) as innate immune cells. The anti-α1,3Gal antigen-based autologous cancer vaccines lead to increase in their immunogenicity to effectively elicit their anticancer immune responses, contributing to eradication of the residual tumor cells.

Considering the anti-α1,3Gal Abs, the α1,3Gal-T KO pigs have a medium to study on the immunogenicity induction of tumor or viral vaccines. Recent pandemic Flu or HIV vaccine is a good candidate, these α1,3Gal epitope-carrying vaccines are expected to have higher anti-α1,3Gal antibody-binding capacities. In cardiovascular section, the whole heart organs are believed to be required for whole heart xenotransplantation, as human cardiac heart valves are substituted by surgical operation in cardiovascular disease and the bioprosthetic heart valves constructed by glutaraldehyde-treated bovine pericardium or pig aortic valves are commercially available [93]. They are, however, the following issues including calcification, antibody infiltration to the xenografts, and DXR or AVR are raised in glutaraldehyde-linked xenografts. For example, pig to monkey xenotransplantation using the α1,3Gal-T KO heart and kidney showed that pig hearts and kidneys can survive for ≤6 months and 8–32 days, respectively. To apply in cancer immunotherapy of the α1,3-Gal epitope, tumor cells are transfected with the α1,3Gal-T gene. The α1,3Gal-T gene-transfected tumor cells are reacted with human IgG and IgM, although human tumor cells are weakly reacted to complement-mediated lysis in human. Tumor-associated antigens (TAAs) are recognized by innate immune cells such as APCs. For the immune responses of cancer vaccines, tumor vaccines should be effectively uptaken by APCs to transport them to the adjacent lymph nodes where TAAs taken to APCs are intracellularly processed and the processed antigenic fragments are presented to MHC-II. In α1,3Gal epitope-carrying TAA-binding α1,3Gal-specific antibodies recognize α1,3Gal epitope antigens. The anti-α1,3Gal antibody Fc region recognizes the Fcα receptors present in the APCs to elicit the uptake of the α1,3Gal epitope-carrying TAA vaccine by the APCs.

Like general tumor vaccines, the enhanced immunogenicity, and immunoreactivity of constructed viral vaccines, as tried to the viral vaccines of influenza viral Flu and human immunodeficiency virus (HIV) are easily measured, calling vaccination technology. Similar to the tumor vaccine technology, the targeted anti-α1,3Gal antigens complexed to the APCs can be expanded to the above viral vaccines of HIV and influenza Flu [94]. The Flu vaccine is currently based on surfaced coat glycoprotein hemagglutinin (HA) in the viral envelop, and this is an applicable instance for the anti-α1,3Gal antigenic vaccines. The Flu-coated HA glycoprotein contains six to eight complex typed *N*-glycosylated glycan chains. Enforced α1,3Gal epitope expression on the HA and sialic acid removal by viral neuraminidase generated the terminally *N*-acetyllactosamine-based HA [95]. The *N*-acetyllactosamine residues have been α1,3Galactosylated by transferring α1,3Gal from the UDP-Gal to the LacNAc. In fact, α1,3Gal antigenic epitope-carrying HA has easily been created. The anti-Flu virus Abs and responses of host T cells, when immunized with the α-1,3Gal antigenic epitope-carrying vaccine, were significantly increased. Furthermore, when α1,3Gal-T KO mice vaccinated with α1,3Gal antigenic epitopes were experimentally infected with Flu virus, the survivals were greatly increased [96]. This indicates the anti-Gal antibody compatibility with antigen presentation enhances immunogenicity of flu vaccine. In case of the HIV vaccination, gp20 having 24 N-linked glycosylation chains has terminally Sia-Galα1-4GlcNAc-R. Upon neuraminidase digestion and glycosylation with UDP-gal and α1,3Gal-T, gp120 is α-galactosylated to form α1,3Gal epitope. The levels of T cell responses and anti-gp120 antibody were also enhanced during the immunizing booster of α1,3Gal epitope-carrying gp120. Thus, the immunogenicity of the α1,3Gal epitope-expressing HIV and Flu vaccines can be enhanced, allowing expansion to application.

Recently, several utilization technologies and application of α1,3Gal epitopes have been developed. For example, nanoparticles combined with α1,3Gal epitopes have been examined for tissue regeneration of internal injuries and damages. Using α1,3Gal-based nanoparticles, α1,3Gal antigenic epitopes are also applicable for brain motor neurons, ischemic myocardium, skin burns, skin remodeling, nerve regeneration, and wound healing because the anti-α1,3Gal antibody interacts with α1,3Gal antigenic epitopes to activate the host complement system. Thus, the interaction rapidly recruits macrophages as innate immune cells. In fact, host-protecting responses such as wound healing need the cooperation of the locally recruited macrophages and the activated macrophages. Then, such activated macrophage cells produce growth factors and IL-1 and TNF-α cytokines, colony-stimulating factors (CSFs) such as macrophage CSF, granulocyte macrophage (GM) CSF (sargramostim), granulocyte CSFs (G-CSFs, filgrastim), and platelet-derived growth factor (PDGF) and vascular endothelial growth factor (VEGF). The cytokines and growth factors stimulate the healing responses of tissue damages and injuries. If the α1,3Gal antigens are artificially integrated on cell surface's glycolipids or glycoproteins, the α1,3Gal antigen–antibody interaction induces proinflammatory and cell lysis responses. However, the proinflammatory responses can be avoided by the α1,3-Gal epitope-encapsulated nanoparticles, which can be complexed in the form of cholesterol, phospholipids, and α1,3Gal-carrying glycolipids.

In α1,3Gal-T KO pigs, α1,3Gal nanoparticle-treated wounds recruited macrophages in the α1,3Gal-T KO pigs. Angiogenesis responses are also enhanced in the wounds treated with α1,3Gal epitope-based nanoparticles, and scar or keloid formation is reduced [97]. Healing time of α1,3Gal liposome-treated skin wounds in α1,3Gal-T KO mice was rapidly progressed. Similarly, the α1,3Gal-specific antigenic Abs are easily applied to healing of the skin burns, as done in α1,3Gal-T KO mice. α1,3Gal-based liposome-treated burns greatly recruited neutrophils and macrophages in α1,3Gal liposome-treated burns. Therefore, the α1,3Gal-based nanoparticle attraction of macrophages is based on interaction between anti-α1,3Gal antibody and α1,3Gal antigen, which accelerates skin burns and tissue regeneration and wound healing. α1,3Gal nanoparticles can regenerate injured nerves and ischemic myocardium in a similar mode, attracting and recruiting macrophages and stem cells. The cells express and secrete the produced cytokines, chemokines, and growth factors. The stem cells stimulate the extracellular matrixes and microenvironments to induce their differentiation to cardiomyocyte cells, allowing tissue regeneration.

References

1. Martens GR, Reyes LM, Li P, Butler JR, Ladowski JM, Estrada JL, Sidner RA, Eckhoff DE, Tector M, Tector AJ. Humoral reactivity of renal transplant-waitlisted patients to cells from GGTA1/CMAH/B4GalNT2, and SLA class I knockout pigs. Transplantation. 2017;101(4): e86–92.
2. Sandrin MS, McKenzie IF. Recent advances in xenotransplantation. Curr Opin Immunol. 1999;11:527–35.
3. Lin SS, Hanaway MJ, Gonzalez-Stawinski GV, Lau CL, Parker W, Davis RD, Byrne GW, Diamond LE, Logan JS, Platt JL. The role of anti-Gal1–3Gal antibodies in acute vascular rejection and accommodation of xenografts. Transplantation. 2000;70:1667–74.
4. Xu H, Yin D, Naziruddin B, Chen L, Stark A, Wei Y, Lei Y, Shen J, Logan JS, Byrne GW, Chong AS. The in vitro and in vivo effects of anti-galactose antibodies on endothelial cell activation and xenograft rejection. J Immunol. 2003;170:1531–9.
5. Hauzenberger E, Klominek J, Holgersson J. Anti-Gal IgG potentiates natural killer cell migration across porcine endothelium via endothelial cell activation and increased natural killer cell motility triggered by CD16 crosslinking. Eur J Immunol. 2004;34:1154–63.
6. Inverardi L, Clissi B, Stolzer AL, Bender JR, Sandrin MS, Pardi R. Human natural killer lymphocytes directly recognize evolutionarily conserved oligosaccharide ligands expressed by xenogeneic tissues. Transplantation. 1997;63(1318–1330):1.
7. Platt JL, Fischel RJ, Matas AJ, Reif SA, Bolman RM, Bach FH. Immunopathology of hyperacute xenograft rejection in a swine-to-primate model. Transplantation. 1991;52:214–20.
8. Galili U, Shohet SB, Kobrin E, et al. Man, apes, and old world monkeys differ from other mammals in the expression of alpha-Galactosyl epitopes on nucleated cells. J Biol Chem. 1988;263:17755–62.
9. Galili U, Clark MR, Shohet SB, et al. Evolutionary relationship between the natural anti-Gal antibody and the Gal alpha 1,3Gal epitope in primates. Proc Natl Acad Sci U S A. 1987;84:1369–73.

10. Cooper DK. Modifying the sugar icing on the transplantation cake. Glycobiology. 2016;26 (6):571–81.
11. Iwase H, Hara H, Ezzelarab M, Li T, Zhang Z, Gao B, et al. Immunological and physiological observations in baboons with life-supporting genetically engineered pig kidney grafts. Xenotransplantation. 2017;24:e12293.
12. Mohiuddin MM, Singh AK, Corcoran PC, Thomas ML 3rd, Clark T, Lewis BG, et al. Chimeric 2C10R4 anti-CD40 antibody therapy is critical for long-term survival of GTKO.hCD46.hTBM pig-to-primate cardiac enograft. Nat Commun. 2016;7:11138.
13. Cozzi E, White DJ. The generation of transgenic pigs as potential organ donors for humans. Nat Med. 1995;1(9):964–6.
14. Lexer G, Cooper DKC, Rose AG, et al. Hyperacute rejection in a discordant (pig to baboon) cardiac xenograft model. J Heart Transplant. 1986;5:411–8.
15. Cho JH, Kim JS, Lim MU, et al. Human leukocytes regulate ganglioside expression in culturedmicro-pig aortic endothelial cells. Lab Anim Res. 2012;28(4):255–63.
16. Galvao FH, Soler W, Pompeu E, et al. Immunoglobulin G profile in hyperacute rejection after multivisceral xenotransplantation. Xenotransplantation. 2012;19(5):298–304.
17. Ngo BT, Beiras-Fernandez A, Hammer C, Thein E. Hyperacute rejection in the xenogenic transplanted rat liver is triggered by the complement system only in the presence of leukocytes and free radical species. Xenotransplantation. 2013;20(3):177–87. https://doi.org/10.1111/xen.12035.
18. Zhong R, Luo Y, Yang H, et al. Improvement in human decay accelerating factor transgenic porcine kidney xenograft rejection with intravenous administration of gas914, a polymeric form of alphaGAL. Transplantation. 2003;75:10–9.
19. Rieben R, Seebach JD. Xenograft rejection: IgG1, complement and NK cells team up to activate and destroy the endothelium. Trends Immunol. 2005;26:2–5.
20. Fox A, Mountford J, Braakhuis A, et al. Innate and adaptive immune responses to nonvascular xenografts: evidence that macrophages are direct effectors of xenograft rejection. J Immunol. 2001;166:2133–40.
21. McGregor CG, Ricci D, Miyagi N, Stalboerger PG, Du Z, Oehler EA, et al. Human CD55 expression blocks hyperacute rejection and restricts complement activation in Gal knockout cardiac xenografts. Transplantation. 2012;93(7):686–92.
22. Mohiuddin MM, Corcoran PC, Singh AK, Azimzadeh A, Hoyt RF Jr, Thomas ML, et al. B-cell depletion extends the survival of GTKO.hCD46Tg pig heart xenografts in baboons for up to 8 months. Am J Transplant. 2012;12(3):763–71.
23. Kuwaki K, Tseng YL, Dor FJ, Shimizu A, Houser SL, Sanderson TM, et al. Heart transplantation in baboons using alpha1,3-galactosyltransferase gene-knockout pigs as donors: initial experience. Nat Med. 2005;11(1):29–31.
24. Le Bas-Bernardet S. Bortezomib, C1-inhibitor and plasma exchange do not prolong the survival of multi-transgenic GalT-KO pig kidney xenografts in baboons. Am J Transplant. 2015;15 (2):358–70. https://doi.org/10.1111/ajt.12988.
25. Scobie L, Padler-Karavani V, Le Bas-Bernardet S, Crossan C, Blaha J, Matouskova M, et al. Long-term IgG response to porcine Neu5Gc antigens without transmission of PERV in burn patients treated with porcine skin xenografts. J Immunol. 2013;191(6):2907–15.
26. Lutz AJ, Li P, Estrada JL, Sidner RA, Chihara RK, Downey SM, et al. Double knockout pigs deficient in N-glycolylneuraminic acid and Galactose alpha-1,3-Galactose reduce the humoral barrier to xenotransplantation. Xenotransplantation. 2013;20(1):27–35.
27. Romanella M, Aminian A, Adam WR, Pearse MJ, d'Apice AJ. Involvement of both the classical and alternate pathways of complement in an ex vivo model of xenograft rejection. Transplantation. 1997;63(7):1021–5.
28. Ide K, Wang H, Tahara H, Liu J, Wang X, Asahara T, et al. Role for CD47-SIRPalpha signaling in xenograft rejection by macrophages. Proc Natl Acad Sci U S A. 2007;104(12):5062–6.

29. van der Windt DJ, Marigliano M, He J, et al. Early islet damage after direct exposure of pig islets to blood: has humoral immunity been underestimated? Cell Transplant. 2012;21:1791–802.
30. Byrne GW, Stalboerger PG, Du Z, Davis TR, McGregor CG. Identification of new carbohydrate and membrane protein antigens in cardiac xenotransplantation. Transplantation. 2011;91:287–92.
31. Galili U, Shohet SB, Kobrin E, Stults CL, Macher BA. Man, apes, and Old World monkeys differ from other mammals in the expression of alpha-galactosyl epitopes on nucleated cells. J Biol Chem. 1988;263:17755–62.
32. McMorrow IM, Comrack CA, Sachs DH, DerSimonian H. Heterogeneity of human anti-pig natural antibodies cross-reactive with the Gal(alpha1,3)Galactose epitope. Transplantation. 1997;64:501–10.
33. Scheer M, Grote A, Chang A, et al. BRENDA, the enzyme information system in 2011. Nucleic Acids Res. 2011;39(Database issue):D670–6.
34. Cooper DK. Xenotransplantation–state of the art. Front Biosci. 1996;1:d248–65.
35. Peters BP, Goldstein IJ. The use of fluorescein-conjugated Bandeiraea simplicifolia B4-isolectin as a histochemical reagent for the detection of alpha-D-galactopyranosyl groups. Their occurrence in basement membranes. Exp Cell Res. 1979;120:321–34.
36. Winter HC, Mostafapour K, Goldstein IJ. The mushroom Marasmius oreades lectin is a blood group type B agglutinin that recognizes the Galalpha 1,3Gal and Galalpha 1,3Galbeta 1,4GlcNAc porcine xenotransplantation epitopes with high affinity. J Biol Chem. 2002;277:14996–5001.
37. Wu AM, Lisowska E, Duk M, Yang Z. Lectins as tools in glycoconjugate research. Glycoconj J. 2009;26:899–913.
38. Diswall M, Gustafsson A, Holgersson J, Sandrin MS, Breimer ME. Antigen-binding specificity of anti-αGal reagents determined by solid-phase glycolipid-binding assays. A complete lack of αGal glycolipid reactivity in α1,3GalT-KO pig small intestine. Xenotransplantation. 2011;18 (1):28–39.
39. Liu J, Weintraub A, Holgersson J. Multivalent Galalpha1,3Gal-substitution makes recombinant mucin-immunoglobulins efficient absorbers of anti-pig antibodies. Xenotransplantation. 2003;10:149–63.
40. Galili U. Xenotransplantation and ABO incompatible transplantation: the similarities they share. Transfus Apher Sci. 2006;35:45–58.
41. Taylor SG, McKenzie IF, Sandrin MS. Characterization of the rat alpha(1,3) galactosyltransferase: evidence for two independent genes encoding glycosyltransferases that synthesize Galalpha(1,3)Gal by two separate glycosylation pathways. Glycobiology. 2003;13:327–37. https://doi.org/10.1093/glycob/cwg030.
42. Osman N, McKenzie IF, Mouhtouris E, Sandrin MS. Switching amino-terminal cytoplasmic domains of alpha(1,2) fucosyltransferase and alpha(1,3)galactosyltransferase alters the expression of H substance and Galalpha(1,3)Gal. J Biol Chem. 1996;271:33105–9.
43. Keusch JJ, Manzella SM, Nyame KA, Cummings RD, Baenziger JU. Expression cloning of a new member of the ABO blood group glycosyltransferases, iGb3 synthase, that directs the synthesis of isoglobo-glycosphingolipids. J Biol Chem. 2000;275:25308–14.
44. Rydberg L, Holgersson J, Samuelsson BE, Breimer ME. Alpha-Gal epitopes in animal tissue glycoproteins and glycolipids. Subcell Biochem. 1999;32:107–25.
45. Galili U. The alpha-gal epitope and the anti-Gal antibody in xenotransplantation and in cancer immunotherapy. Immunol Cell Biol. 2005;83:674–86.
46. Boix E, Zhang Y, Swaminathan GJ, Brew K, Acharya KR. Structural basis of ordered binding of donor and acceptor substrates to the retaining glycosyltransferase, alpha-1,3-galactosyltransferase. J Biol Chem. 2002;277:28310–8.
47. Sandrin MS, Osman N, McKenzie IF. Transgenic approaches for the reduction of Galalpha(1,3) Gal for xenotransplantation. Front Biosci. 1997;2:e1–11.

48. Galili U. Anti-Gal: an abundant human natural antibody of multiple pathogeneses and clinical benefits. Immunology. 2013;140:1–11.
49. Galili U. Interaction of the natural anti-Gal antibody with alpha-galactosyl epitopes: a major obstacle for xenotransplan-tation in humans. Immunol Today. 1993;14:480–2.
50. Galili U. Significance of the evolutionary α1,3-galactosyltransferase (GGTA1) gene inactivation in preventing extinction of apes and old world monkeys. J Mol Evol. 2015;80(1):1–9.
51. Larsen RD, Rivera-Marrero CA, Ernst LK, Cummings RD, Lowe JB. Frameshift and nonsense mutations in a human genomic sequence homologous to a murine UDP-Gal:beta-D-Gal(1,4)-D-GlcNAc alpha(1,3)-galactosyltransferase cDNA. J Biol Chem. 1990;265(12):7055–61.
52. Huai G, Qi P, Yang H, Wang Y. Characteristics of α-Gal epitope, anti-Gal antibody, α1,3 galactosyltransferase and its clinical exploitation (review). Int J Mol Med. 2016;37(1):11–20.
53. Strahan KM, Gu F, Preece AF, Gustavsson I, Andersson L, Gustafsson K. cDNA sequence and chromosome localization of pig alpha 1,3 galactosyltransferase. Immunogenetics. 1995;41:101–5.
54. Joziasse DH, Shaper JH, Van den Eijnden DH, Van Tunen AJ, Shaper NL. Bovine alpha 1→3-galactosyltransferase: isolation and characterization of a cDNA clone. Identification of homologous sequences in human genomic DNA. J Biol Chem. 1989;264:14290–7.
55. Galili U, Swanson K. Gene sequences suggest inactivation of alpha-1,3-galactosyltransferase in catarrhines after the divergence of apes from monkeys. Proc Natl Acad Sci U S A. 1991;88:7401–4.
56. Church DM, Goodstadt L, Hillier LW, Zody MC, Goldstein S, She X, Bult CJ, Agarwala R, Cherry JL, DiCuccio M, et al. Mouse genome sequencing consortium: lineage-specific biology revealed by a finished genome assembly of the mouse. PLoS Biol. 2009;7:e1000112. https://doi.org/10.1371/journal.pbio.1000112.
57. Lindblad-Toh K, Wade CM, Mikkelsen TS, Karlsson EK, Jaffe DB, Kamal M, Clamp M, Chang JL, Kulbokas EJ III, Zody MC, et al. Genome sequence, comparative analysis and haplotype structure of the domestic dog. Nature. 2005;438:803–19.
58. Lander ES, Linton LM, Birren B, Nusbaum C, Zody MC, Baldwin J, Devon K, Dewar K, Doyle M, FitzHugh W, et al. International human genome sequencing consortium: initial sequencing and analysis of the human genome. Nature. 2001;409:860–921.
59. Galili U. Evolution and pathophysiology of the human natural anti-alpha-galactosyl IgG (anti-Gal) antibody. Springer Semin Immunopathol. 1993;15:155–71.
60. Larsen RDR-MC, Rivera-Marrero CA, Ernst LK, Cummings RD, Lowe JB. Frameshift and nonsense mutations in a human genomic sequence homologous to a murine UDP-Gal:beta-D-Gal(1,4)-D-GlcNAc alpha(1,3)-galactosyltransferase cDNA. J Biol Chem. 1990;265:7055–61.
61. Rodriguez IA, Welsh RM. Possible role of a cell surface carbohydrate in evolution of resistance to viral infections in old world primates. J Virol. 2013;87:8317–26.
62. Galili U, Andrews P. Suppression of a-galactosyl epitopes synthesis and production of the natural anti-Gal antibody: a major evolutionary event in ancestral Old World primates. J Hum Evol. 1995;29:433–42. https://doi.org/10.1006/jhev.1995.1067.
63. Tahiri F, Li Y, Hawke D, Ganiko L, Almeida I, Levery S, Zhou D. Lack of iGb3 and Isoglobo-series Glycosphingolipids in pig organs used for xenotransplantation: implications for natural killer T-cell biolog. J Carbohydr Chem. 2013;32(1):44–67.
64. Galili U, Rachmilewitz EA, Peleg A, Flechner I. A unique natural human IgG antibody with antialpha-galactosyl specificity. J Exp Med. 1984;160(5):1519–31.
65. Sandrin MS, Vaughan HA, Dabkowski PL, McKenzie IF. Anti-pig IgM antibodies in human serum react predominantly with Gal(alpha 1-3)Gal epitopes. Proc Natl Acad Sci U S A. 1993;90(23):11391–5.
66. Posekany KJ, Pittman HK, Bradfield JF, Haisch CE, Verbanac KM. Induction of cytolytic anti-Gal antibodies in alpha-1,3-galactosyltransferase gene knockout mice by oral inoculation with Escherichia coli O86:B7 bacteria. Infect Immun. 2002;70(11):6215–22.
67. Dor FJ, Tseng YL, Cheng J, Moran K, Sanderson TM, Lancos CJ, Shimizu A, Yamada K, Awwad M, Sachs DH, Hawley RJ, Schuurman HJ, Cooper DK. alpha1,3-Galactosyltransferase

geneknockout miniature swine produce natural cytotoxic anti-Gal antibodies. Transplantation. 2004;78(1):15–20.
68. Dijke EI, Platt JL, Blair P, Clatworthy MR, Patel JK, Kfoury AG, Cascalho M. B cells in transplantation. J Heart Lung Transplant. 2016;35(6):704–10.
69. Gorer PA. The genetic and antigenic basis of tumour transplantation. J Pathol Bacteriol. 1937;44:691–7.
70. Gorer PA. The antigenic basis of tumour transplantation. J Pathol Bacteriol. 1938;47:231–52.
71. Gorer PA. The significance of studies with transplanted tumours. Br J Cancer. 1948;2:103–7.
72. Tomizuka K, Shinohara T, Yoshida H, Uejima H, Ohguma A, Tanaka S, Sato K, Oshimura M, Ishida I. Double trans-chromosomic mice: maintenance of two individual human chromosome fragments containing Ig heavy and κ loci and expression of fully human antibodies. Proc Natl Acad Sci U S A. 2000;97:722–7.
73. Kitamura D, Roes J, Kühn R, Rajewsky K. AB cell-deficient mouse by targeted disruption of the membrane exon of the immunoglobulin mu chain gene. Nature. 1991;350:423–6.
74. Mendicino M, Ramsoondar J, Phelps C, Vaught T, Ball S, LeRoith T, Monahan J, Chen S, Dandro A, Boone J. Generation of antibody-and B cell-deficient pigs by targeted disruption of the J-region gene segment of the heavy chain locus. Transgenic Res. 2011;20:625–41.
75. Tesson L, Usal C, Ménoret S, Leung E, Niles BJ, Remy S, Santiago Y, Vincent AI, Meng X, Zhang L. Knockout rats generated by embryo microinjection of TALENs. Nat Biotechnol. 2011;29:695–6.
76. Kuroiwa Y, Kasinathan P, Sathiyaseelan T, Jiao JA, Matsushita H, Sathiyaseelan J, Wu H, Mellquist J, Hammitt M, Koster J. Antigen-specific human polyclonal antibodies from hyperimmunized cattle. Nat Biotechnol. 2009;27:173–81.
77. Ramsoondar J, Mendicino M, Phelps C, Vaught T, Ball S, Monahan J, Chen S, Dandro A, Boone J, Jobst P. Targeted disruption of the porcine immunoglobulin kappa light chain locus. Transgenic Res. 2011;20:643–53.
78. Vadori M, Cozzi E. The immunological barriers to xenotransplantation. Tissue Antigens. 2015;86(4):239–53. https://doi.org/10.1111/tan.12669.
79. Li S, Yan Y, Lin Y, et al. Rapidly induced, T-cell independent xenoantibody production is mediated by marginal zone B cells and requires help from NK cells. Blood. 2007;110:3926–35.
80. Davila E, Byrne GW, La Breche PT, McGregor HCJ, Schwab AK, Davies WR, et al. T-cell responses during pig-to-primate xenotransplantation. Xenotransplantation. 2006;13(1):31–40.
81. Tanemura M, Yin D, Chong AS, Galili U. Differential immune responses to αGal epitopes on xenografts and allografts: implications for accommodation in xenotransplantation. J Clin Invest. 2000;105:301–10.
82. Wang L, Radic MZ, Galili U. Human anti-Gal heavy chain genes. Preferential use of VH3 and the presence of somatic mutations. J Immunol. 1995;155:1276–85.
83. Fischer-Lougheed J, Gregory C, White Z, Shulkin I, Gunthart M, Kearns-Jonker M. Identification of an anti-idiotypic antibody that defines a B-cell subset(s) producing xenoantibodies in primates. Immunology. 2008;123:390–7.
84. Fang J, Walters A, Hara H, Long C, Yeh P, Ayares D, Cooper DK, Bianchi J. Anti-gal antibodies in α1,3-galactosyltransferase gene-knockout pigs. Xenotransplantation. 2012;19:305–10.
85. Galili U, Rachmilewitz EA, Peleg A, Flechner I. A unique natural human IgG antibody with anti-a-galactosyl specificity. J Exp Med. 1984;160:1519–81.
86. Hamanova M, Chmelikova M, Nentwich I, Thon V, Lokaj J. Anti-Gal IgM, IgA and IgG natural antibodies in childhood. Immunol Lett. 2015;164:40–3.
87. Reddy V, Jayne D, Close D, Isenberg D. B-cell depletion in SLE: clinical and trial experience with rituximab and ocrelizumab and implications for study design. Arthritis Res Ther. 2013;15 (Suppl 1):S2.
88. Iwase H, Liu H, Wijkstrom M, et al. Pig kidney graft survival in a baboon for 136 days: longest life-supporting organ graft survival to date. Xenotransplantation. 2015;22:302–9.

89. Le Bas-Bernardet S, Tillou X, Branchereau J, et al. Bortezomib, C1-inhibitor and plasma exchange do not prolong the survival of multi-transgenic GalT-KO pig kidney xenografts in baboons. Am J Transplant. 2015;15:358–70.
90. Kolber-Simonds D, Lai L, Watt SR, Denaro M, Arn S, Augenstein ML, Betthauser J, Carter DB, Greenstein JL, Hao Y, et al. Production of alpha-1,3-galactosyltransferase null pigs by means of nuclear transfer with fibroblasts bearing loss of heterozygosity mutations. Proc Natl Acad Sci U S A. 2004;101:7335–40.
91. Azimzadeh A, Kelishadi S, Ezzelarab MB, Singh A, Stoddard T, Zhang T, Burdorf L, Avon C, Laaris A, Cheng X, et al. Early graft failure of GalTKO pig organs in baboons is reduced by expression of a human complement-regulatory protein. Xenotransplantation. 2015;22:310–6.
92. Park S, Kim WH, Choi SY, Kim YJ. Removal of alpha-Gal epitopes from porcine aortic valve and pericardium using recombinant human alpha galactosidase A. J Korean Med Sci. 2009;24:1126–31. https://doi.org/10.3346/jkms.2009.24.6.1126.
93. Choi SY, Jeong HJ, Lim HG, Park SS, Kim SH, Kim YJ. Elimination of alpha-gal xenoreactive epitope: alpha-galactosidase treatment of porcine heart valves. J Heart Valve Dis. 2012;21:387–97.
94. Galili U, Repik PM, Anaraki F, Mozdzanowska K, Washko G, Gerhard W. Enhancement of antigen presentation of influenza virus hemagglutinin by the natural human anti-Gal antibody. Vaccine. 1996;14:321–8.
95. Henion TR, Gerhard W, Anaraki F, Galili U. Synthesis of alpha-gal epitopes on influenza virus vaccines, by recombinant alpha 1,3galactosyltransferase, enables the formation of immune complexes with the natural anti-Gal antibody. Vaccine. 1997;15:1174–82.
96. Abdel-Motal UM, Guay HM, Wigglesworth K, Welsh RM, Galili U. Immunogenicity of influenza virus vaccine is increased by anti-gal-mediated targeting to antigen-presenting cells. J Virol. 2007;81:9131–41.
97. Galili U. Discovery of the natural anti-Gal antibody and its past and future relevance to medicine. Xenotransplantation. 2013;20:138–47.

Chapter 11
Non-α1,3Gal Carbohydrate Antigenic Epitopes

11.1 Introduction

Non-Gal (Non-α1,3-Gal) antigenic epitope is also targeted by the cellular immune system. Preformed natural non-Gal Abs do not cause HR in pig-to-human xenografts. However, non-α1,3Gal Abs can induce graft-injured damages to cells in both vascularized organs and tissues. Transplanted xenografts induce production of non-α1,3Gal-specific Abs in the hosts. Likely to the human allograft rejection, the anti-non-α1,3Gal Abs control survivals of xenografts in the host. The pig MHC molecules-specific Abs of humans preferably bind to the glycan structures, but not proteins. Elucidation on the relationship between non-α1,3Gal antigenic epitope structures of pigs and the reactive non-α1,3Gal Abs of humans is interesting in understanding of the Abs-mediated graft damages in near future. Relevant understanding of downstream reaction of non-α1,3Gal antigens in pig xenografts to human Abs should be systematically explained and edited in future. When α1,3-Gal antigen-based HR is overcome, AHR is elicited by low levels of natural Abs to α1,3Gal epitopes, which occur within 3 days to 3 weeks. However, AHR is poorly understood. As a similar expression, AVR or DXR is known. Non-α1,3Gal antigenic epitopes and non-α1,3Gal Abs are suggested to involve in such AHR, AVR, or DXR. To date, a representative and predominant non-α1,3Gal antigen is Neu5Gc as the major non-α1,3Gal xenoantigen.

AHR, AVR, or DXR occurs within a few days, and Abs to non-α1,3Gal antigenic epitopes, but not α1,3Gal epitopes, are important [1, 2] and thus represent procoagulant changes in the pig endothelium. These events consequently activate complement system with additional activation of blood coagulation system, which result in apoptosis, edema, platelet aggregation, and thrombosis in the transplanted grafts [3]. Thus, in this status, α1,3Gal-T gene (*GGTA1*) KO or regulators of complement system are not effective for inhibition of AVR. Binding of Abs to non-α1,3Gal antigens in endothelial cell membranes still leads to both chronic rejection and DXR, and consequently resulting in loss of grafts. Therefore, identified

C.-H. Kim, *Glycoimmunology in Xenotransplantation*,
https://doi.org/10.1007/978-981-99-7691-1_11

discovery of such non-α1,3Gal and non-Neu5Gc-based glycan antigens of pigs will open new creation and future opportunity for generation of xenoantigen-modified pigs to prevent antigen–antibody-driven tissue and organ injuries of the grafts. Aberrantly-modified monosaccharide Neu5Gc is the currently known human target for most non-α1,3Gal xenoantigenic Abs [4].

Further identification of the new pig gene products may provide clues for genetically-modified organ sources of pigs to prevent and reduce the xenografted antigenicities as well as to increase immunological resistant capacity to DXR [5], inviting better prolonged and long-term survivals of xenografts. In the recent study to search immunologic target genes involved in DXR, potential antigens involved in DXR have been searched in a GGTA-1/CMAH double KO animals in pig-to-monkey xenotransplantation model [6]. To search the potential antigens involved in DXR in the α1,3Gal-antigen removed condition, two antigenic deficient double KO pigs of the α1,3Gal and Neu5Gc epitopes were examined and the double KO pigs exhibited the reduced DXR in the presence of non-α1,3Gal antigens in endothelial cell membranes, but these pigs exhibited thrombotic microangiopathy or chronic rejection, resulting in loss of grafts in an immunization model of pig endothelial cells to monkey as hosts [6].

When cynomolgus monkeys were transplanted with aortic endothelial cells (AEC) as primary cells and renal microvascular endothelial cells (RMEC) as primary cultured cells which isolated from a pig GGTA1/CMAH double-KO strain, the binding of monkey Abs and cytotoxic activity to RMEC were increased. RMEC showed higher immunogenic capacity rather than that of AEC. From transcriptome analysis of cells isolated from the GGTA1/CMAH double KO pigs, the number of transcribed genes (1500 gene expressions) was higher than that AEC. Among them, in the pig RMEC, 101 candidate genes were specifically expressed in the cells, not expressed in AEC of pig and also in RMEC of monkeys or humans. Then, several immunologic target genes have been demonstrated to involve in DXR. Discovery and identification of new targeted antigens are crucial for newly and genetically-modified pigs in xenotransplantation. Their modified pig organs are expected to be resistant to chronic antibody-driven vascular endothelial activation. Also, the pigs are important for establishment of antigen-specific tolerance [7]. Some successful examples as follows: A20 known as TNF-α-stimulated protein 3 and heme oxygenase 1 (HO-1) are those known as the anti-apoptotic and anti-inflammatory molecules. They are reported to suppress activation of AVR and endothelial cells [8, 9].

11.2 Non-α1,3Gal Xenoreactive Antibodies Recognize α-Lactosamine, Forssman Antigen, Neu5Gc, Tn-, T-, Sialosyl-Tn, NeuAcα2,6GalNAcα1-R, α-LacNAc, P1 Antigen, and Pk Antigen

The α-1,3Gal antigenic epitopes are not the sole carbohydrate xenoepitopes for the naturally preformed and elicited xenoreactive Abs in human. Human immunoglobulins bind to pig α1,3Gal-T-KO endothelial cells. In addition, non-α1,3Gal-reactive preformed and induced IgG and IgM Abs are found in primates which are xenografted, and develop the AHR, even in the condition of α1,3Gal epitope absorption [10] or in the grafted α1,3Gal-T-KO pigs to primates. Thirteen percent of IgM and 36% of IgG in the healthy serum of human recognize non-α1,3Gal antigenic epitopes expressed in pig endothelial cells causing cell injury and damage by complement-dependent cytotoxicity (CDC) or ADCC [11]. Such non-αGal xenoreactive antibodies are known to recognize the carbohydrate antigens, such as α-lactosamine, Forssman antigen of carbohydrate structure of GalNAcα1,3GalNAcβ1,3Galα1,4Galβ1,4Glcβ1-R, Neu5Gc known as H-D, Tn-, T-, sialyl-Tn, NeuAcα2,6GalNAcα1-R, α-LacNAc, P1 and Pk antigen from the results of immunized baboons, healthy human blood donors, or neonatal porcine islets-grafted patients [12].

The presence of α-lactosamine in human serum was suggested from a cross-reaction property of anti-α1,3Gal antibody. The blood sera of the clinical patients intraportally injected with islet-like clusters of pig showed the terminal α-linked GalNAc-specific Abs and the Abs specific for the above linear and branched oligomannose glycans, Forssman epitopes, Galα1,3 Lewis^{X}, Neu5Acα2,3Galβ1-3GlcNAc, and various α2,3-NeuGc and α2,6-NeuGc terminal glycans [13] were found. Among them, Forssman antigen does not cause xenograft rejection of pig xenografts despite the recognizing properties of baboon and human Abs to the Forssman antigens [13]. Indeed, pigs are Forssman-negative, but produce anti-Forssman antibodies. The α1,3Gal epitope-disrupted α1,3Gal-T KO pigs produce the β-LacNAc structures which are unmasked by sialic acids. Although two P1 and x2 as rare antigens of blood groups are unexpressed in normal wild pig types, the α1,3Gal-T KO pigs produce them [14].

In the AHR, human TNF receptor (TNFR) and HO-1 are known for their anti-apoptotic and anti-inflammatory properties and they suppress activation of endothelial cells and AHR [8, 9], as shown in the results of the human HA-tagged hTNFR-I IgG1-Fc (TNFRI-Fc) and HO-1 in pig organs [15, 16]. Recently, various xenotransplantation-compatible pigs have been created by multiple genetic modification using the embryo engineering by a TALEN-coupled with SCNT. The TALEN system combined with transgenic expression for multiple genes is now under application for xenotransplantable pigs. Therefore, using TALEN system, a pig line has been designed to establish the quadruple phenotype of the NeuGc-synthesizing CMAH gene KO in α1,3Gal-KO pigs expressing double transgenic soluble human TNFRI IgG1-Fc (shTNFRI-Fc) and hHA-tagged-HO-1 (hHAHO-1).

The quadruple CMAH-KO/GT-KO/HO-1/shTNFRI-Fc genetically mutant pigs were genetically modified [16]. The human serum antibody-driven immune cytotoxicity to the endothelial cells of the CMAHKO/GTKO/hHAHO-1/shTNFRI-Fc mutant pigs was created. Therefore, the quadruple-modified pig line is much progressed for testing in xenotransplantation [17].

11.3 Definition of Non-Gal (Non-α1,3Gal) Antigen and Antibody

The known anti-non-α1,3Gal Abs influence xenograft survival of pigs in xenotransplanted recipients, if a xenograft recipient's blood contains high amounts of natural anti-non-α1,3Gal Abs, which is preformed, prior to xenotransplantation operation [18]. The recognition level of anti-pig Abs to aortic endothelial cells (AECs) of α1,3Gal-T KO pig is weak compared to wild-type (WT) AECs of pig [19]. For example, when the monkey received pig AEC or pig renal microvascular endothelial cells (pRMEC) as a kidney transplant in the kidney of pig organ to monkey xenotransplantation, the antibody recognition with AECs of pig is increased with the cells injured and damaged in vivo, and the fibrin-platelet thrombosis and ischemic injuries were developed to the xenografted organ. Such similar outcomes are obtained for the microangiopathy thrombosis and consumptive micropathy coagulation during kidney transplantation of pig compared to heart transplantation of pig [20]. pRMEC is suggested to be highly immunogenic compared to the endothelial cells in endothelium of the coronary vascular system. The thrombotic microangiopathy is correlated with the levels of Igs and complement depositions occurred in the grafted organs. For example, the α1,3Gal-T KO pig kidneys-transplanted baboon recipients showed the rapid rejection of the transplanted xenogenic grafts due to the cytotoxic Abs induced by non-α1,3Gal antigenic epitopes by day 16 [21].

The harmful function for anti-non-α1,3Gal Abs in the α1,3Gal-T KO pig organs is not yet clear. Antibody concentration of anti-non-α1,3Gal Abs in serum is regarded as a determination factor for serum measurement of antibody levels. In the human sera, antibody types of IgG and IgM, which are specific for non-α1,3Gal and non-NeuGc antigens present on pig AECs are occupied for almost 100% of the serum Abs. Those Abs are considered to be the SDa-specific antibodies and other minority is estimated to be unknown pig antigens-specific. Actually, from the results obtained from pig AECs binding to anti-non-α1,3Gal antibody of IgG and IgM types and to anti-non-α1,3Gal/non-Neu5Gc antibody IgG and IgM types, the SDa has been suggested to be a major antigen for anti-pig Abs of humans [22]. Because the β1,4 *N*-acetylgalactosaminyltransferase synthesizes the SDa antigenic determinant, the gene KO pigs reduce the antigen production. At present, α1,3Gal and Neu5Gc (or SDa)-deficient pigs are most likely to be the subjects for clinical xenotransplantation.

Non-α1,3Gal antigens are composed of donor pig's proteins and also carbohydrates. However, the precise antigenic and molecular structures are not well defined in terms of chemistry. In definition of non-α1,3Gal Abs of humans, non-α1,3Gal-recognizing Abs indicate naturally occurring antibodies and they increase in WT or α1,3Gal-T-KO pig grafts-receiving humans/NHPs. In reactivity of human natural Abs with embryonic cells of pigs, both native human serum, which is non-depleted and Galα1,3Gal-epitopes-depleted serum of human blood group AB type showed the cytotoxicities to ventral mesencephalon cells of pigs. Although immunoadsorption of Galα1,3Gal epitope removes the Galα1,3Gal epitopes from the xenograft of pig lungs, the xenografts have still capacities of HR to even lower levels against human blood-perfused pig lungs. Thus, non-α1,3Gal antigen-reacting Abs are important in the immune xenograft rejection. Non-α1,3Gal-specific Abs can be stimulated in WT or α1,3Gal-T-KO pig grafts-received humans/NHP. Non-α1,3Gal-specific Abs are easily reactive with both proteins and carbohydrates of glycoproteins, which carbohydrates chains are attached.

The terminologies of the "non-α1,3Gal-specific antigens" and "non-α1,3Gal-specific Abs" are currently described in the immunological xenotransplantation. Despite their frequent uses of the "non-α1,3Gal-specific antigens" and "non-α 1,3Gal-specific Abs," the precise terminological definitions of them are not definitive. According to Breimer [23], "Non-α1,3Gal-specific antigens" are newly defined, which is based on the following conditions: (1) biochemically perspective to "xenoantigens expressed in α1,3Gal-T KO pigs" or a (2) immunobiologically perspective to "Pig-derived antigens specifically recognized by non-α1,3Gal-specific Abs." To date, totally 15 non-α1,3Gal antigenic glycans, reactive against xenoreactive Abs of humans or naturally occurring Abs of humans have been known (Table 11.1) [12].

11.3.1 Human Blood Group System P Antigen (P1PK)

Among non-α1,3Gal antigenic glycans, P antigen (now, P1PK) as a human blood group system is formed by the A4GALT (Fig. 11.1). The P1 antigen (previously P antigen) historically was first discovered by Landsteiner K and Levine P. in 1927 [24]. The P (P1PK) group mainly has three GSLs (P^k, P1 and NOR) [25, 26] as well as terminal Galα1,4Galβ on N-glycoproteins [27]. P antigens are biosynthesized by globotriaosylceramide Gb3 (CD77) synthase (it is termed an α1,4-Galactosyltransferase or P1/Pk synthase). The Globoside (GLOB) antigen (previously, P) is currently one of the members of the independently classified GLOB blood group system. Other publications are also found with difference [12]. The P antigen or globoside-4 (Gb4) is a GSL on human erythrocytes. P-related Abs are associated with hemolytic TRs, hemolytic disease of the newborn (HDN), and spontaneous abortion. Anti-P Abs are also involved in viral paroxysmal cold hemoglobinuria (PCH) in children. In PCH, anti-P is called the Donath–Landsteiner antibody. Globoside system also known as globe disease is associated with

Table 11.1 Non-α1,3Gal antigenic glycans reactive against xenoreactive Abs or naturally occurring Abs in human

1. Neu5Gc (Hanganutziu-Deicher; HD antigen)
2. Thomsen-Friedenreich antigen (T or TF): Galβ1,3GalNAcα1-Ser/The as core 1 structure in O-linked glycosylation
3. Tn (TF precursor) antigen: GalNAcα1-Ser/Thr
4. Sialosyl-Tn: NeuAcα2,6GalNAcα1-R (Ser/Thr)
5. P^k: Gb3, Galα1,4Galβ1,4Glcβ1-R[a]
6. P-antigens[a]
7. Forssman antigen: GalNAcα1,3GalNAcβ1,3Galα1,4Galβ1,4Glcβ1-R
8. I: Galβ1,4GlcNAcβ1,3Galβ1,4GlcNAcβ-R
9. I: Galβ1,4GlcNAcβ1,3[Galβ1,4Glcβ1,6]Galβ1,4GlcNAcβ-R
10. αRhamnose-containing oligosaccharides, L-Rhamnose-α-Rhamnose and L-Rhamnose-α1-3GlcNAcβ1-2Rhamnose-α-R
11. Sulphatide I: SO_4-3Gal-R
12. βGlcNAc-carrying oligosaccharides, GlcNAcβ-R and GlcNAcβ1,4GlcNAcβ-R
13. Galα1,3Le^x: Galα1,3Galα1,4(Fucα1,3)GlcNAcβ1,3Galβ1,4Glcβ1-R
14. A type blood group: GalNAcα1,3(Fucα1,2)Galβ1,4GlcNAcβ-R
15. B type blood group: Galα1,3(Fucα1,2)Galβ1,4GlcNAcβ-R

[a]P antigen (now, P1PK) is formed by the A4GALT. Blood group, Globoside System is associated with B3GALNT1, β1,3-*N*-Acetylgalactosaminyltransferase 1 (GalNAc-T1, Globoside blood group antigen)

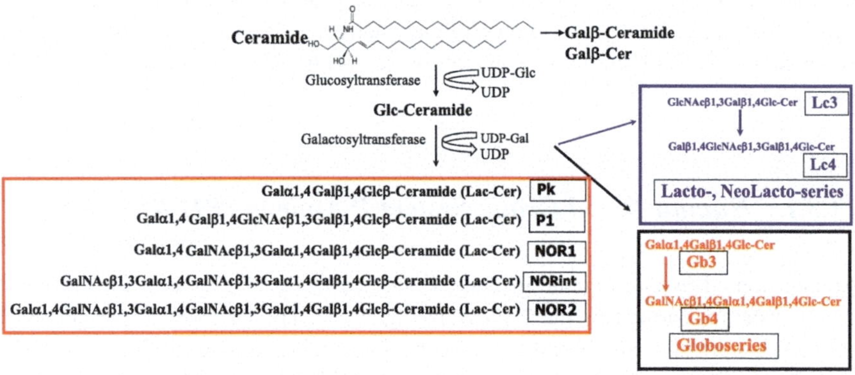

Fig. 11.1 Glycan structures of P1PK blood group antigens. Mammal glycosphingolipid biosynthetic pathway is started from Gal-Cer or Glc-Cer, Globo-series, and Lacto- and Neolacto-series

pulmonary disease, chronic obstructive and deficiency anemia. Blood group, Globoside System is associated with B3GALNT1, β1,3-*N*-Acetylgalactosaminyltransferase 1 or Globoside Blood Group is associated with ERK Signaling and TGF-Beta Pathway. P (P1PK) antigens are composed of carbohydrates and have P^k (Gb3), P1 and NOR1, NORint, and NOR2, which Gb3/CD77 synthase (α1,4-Gal-T or P1/Pk synthase) synthesize them. Among them, Pk antigen acts as a receptor for Shiga toxins and zoonotic meningitis-causing *Streptococcus* species.

P, P^k, PI, and LKE are a receptor for P-fimbriae UPEC *E. coli*. P1 antigen level is regulated by the A4GALT gene transcription via early growth response 1 (EGR1) and runt-related transcription factor 1 (RUNX1) to the SNP rs5751348 genomic location. A4GALT gene has two P^1 and P^2 genotypes [28]. P (P1PK) is determined by antibody reactions of P1-, P-, P^k-, and $PP1P^k$-specific antibodies and NOR-specific Abs. P1 includes (1) anti-P1(+), (2) anti-P (+) and anti-$PP1P^k$ (+) and (3) anti-P^k (−). P2 phenotype includes (1) anti-P1 (−), (2) anti-P (+), (3) anti-$PP1P^k$ (+), and (4) anti-P^k (−). Rare p phenotype has not the P antigens due to A4GALT gene defection and thus includes (1) anti-P1 (−), (2) anti-P (−), (3) anti-$PP1P^k$ (−), and (4) anti-P^k (−). They strongly react to anti-PP1Pk, causing for early spontaneous abortions, HDFN, and delayed hemolytic TRs. The A4GALT gene has 34 mutations in 37 alleles with dysfunctional enzyme activity and risk to the rare p. Also, the rare NOR is defined by NOR1 and NOR2 of terminal Galα1,4GalNAcβ1-attached GSL antigens, present only in *Rana ridibunda* by the A4GALT gene mutation. That is p.Q211E substituted mutation of the Gb3/CD77 synthase gene (rs397514502) [29]. For P (P1PK) antibody, anti-P1 Abs are detected in liver flukes and hydatid cyst disease as well as P1-like substance-bearing bird excrement exposure. Anti-$P1P^k$ Abs contain a mixture of GLOB-, P1-, and P^k-specific Abs in the p person sera. Allo-anti-GLOB is in the $P1^k$ and $P2^k$ person sera (as preformed IgM isotype with minor IgG). The Abs cause hemolytic transfusion reactions and HDFN (if IgG isotype crosses the placenta). anti-PP1Pk is related to early spontaneous abortion (the placental antigen P^k and GLOB are reacted by IgG isotype Abs). Preformed anti-NOR Abs [30] bind to NOR1 and NOR2 antigens, inducing NOR poly-agglutination.

11.3.2 Human Blood I and I Antigen System

The human blood I and i antigen gene are located on chromosome 6. The I antigen is found on the RBC membrane in adults, while the i antigen is found in newborns and fetuses. Unlike AB(O)H antigens, the antigens I and i are found in RBC membrane, human cells and body fluids of amniotic fluid, milk, ovarian cyst fluid, plasma, saliva and urine [31]. The antigens I and i are the repeated units of *N*-acetyllactosamine (LacNAc) on the ABH and Lewis antigens [31, 32]. LacNAc repeats are synthesized by the B3GNT1 and B4GALT1. The i antigen is linear form, whereas the I antigen is branched form [32]. The blood group I, branching enzyme *I* gene encodes a β1,6-*N*-acetylglucosaminyltransferase (GalNAc-T1; IGnT). But the I gene is not identified yet. The I and i antigens may be associated with hematopoiesis [33]. Cold agglutinin disease-related autoantibodies are reactive to I antigen as IgM (kappa subtype) isotype. But transient cold agglutinin disease-related autoantibodies are IgG isotype. In contrast to IgG, cold-reactive IgM Abs as cold agglutinins recognize RBC I antigen and cause RBC agglutination with complement activation, hemolysis and

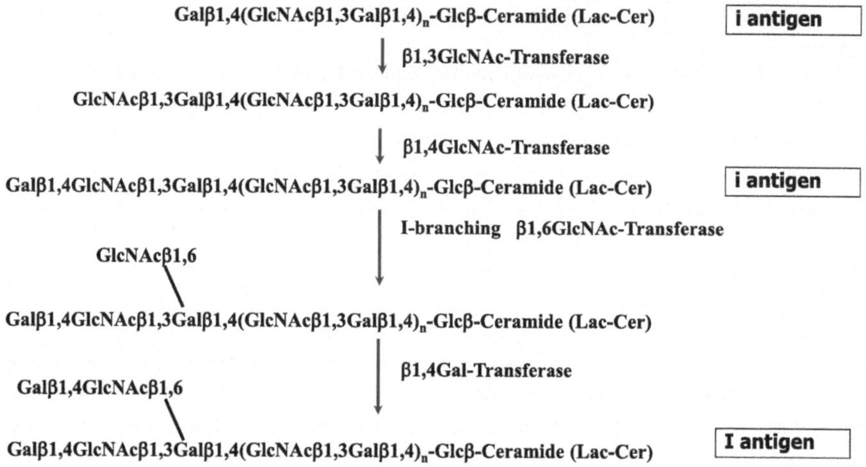

Fig. 11.2 Glycan structures and biosynthesis of the i and I

anemia. Limitedly and rarely, certain adults have the RBC i antigen, calling the adult i phenotype due to the I-branching enzyme, GCNT2 gene mutation. They have the I antigen-specific alloantibodies or cold agglutinin. The adult i antigen is a recessive inheritance involved in congenital cataracts, where cataracts occur when i antigen is present on the lens epithelium due to the I-branching enzyme mutation known as IGNTB gene. The I antigen and i antigen were first described in 1956 and 1960, respectively [31]. I and i antigens are changed during human development [33]. The letter I indicates the I antigen-lacking individuality.

Human Ii system-similar antigens are found in primates such as chimpanzees and monkeys, not in non-primates such as cats, dogs, or guinea pigs. The I antigen reached with a cold agglutinating anti-I autoantibody is present on RBCs and only limited population are not reactive to anti-I, calling I-negative. Cord blood cells have a weak I antigen. Cold agglutinating antibody anti-i detected the i antigen. Adult human RBCs generate antigen I, but weakly antigen I. The antigen I is predominant in neonatal and fetal human RBCs. Since human birth, the I antigen level is increased but the level of i antigen is decreased, until 18 months after birth life [34]. Similar to ABH antigens, antigens Ii resemble histo group blood antigens. I and i antigens are also expressed in oncogenesis [34] as oncofetal antigens [35]. The Ii antigen determinants are glycolipids and glycoproteins glycans and the interior structures of the ABH and Lewis antigen carbohydrate chains.

From type 2 chain of Galβ1-4GlcNAc structure, the antigens i and I structures are linear forms with repeated branches of LacNac, Galβ1-4GlcNAcβ1-3Galβ1-4GlcNAc-R, and Galβ1-4GlcNAcβ1-3(Galβ1-4GlcNAcβ1-6)Galβ1-4GlcNAc-R (Fig. 11.2). The LacNAc repeats are sequentially formed by the β1,3-GlcNAc-T and β1,4-Gal-T enzymes. The blood group *I* locus encodes a β6GlcNAc-T as antigen conversion of i antigen to I antigen is based on a third enzyme, named I-branching enzyme of β1,6-GlcNAc-T (β6GlcNAc-T) [36].

11.4 Non-α1,3Gal Antibodies Cause Cytotoxicity, Damages, and Injuries of Xenoorgans, But Not Hyperacute Xenorejection

In addition, he added the following definitions as non-α1,3Gal-specific Abs, which the Abs are reactive with α1,3Gal-T KO organs of pig and also as Abs responsible for antibody reaction with pig cells treated with α-galactosidase enzymes. Such Abs are not reactive with Galα1,3-terminating carbohydrate antigens. In Abs against non-α1,3Gal antigens, xenoreactive IgG1 subclass Abs are specific for non-α1,3Galα3Gal-antigenic epitopes whereas Abs of the IgG2 and IgM subclass are reactive for Galα1,3Gal epitopes. The non-α1,3Gal-specific Abs also induce ADCC.

In xenotransplantation, many interests have emergingly been increased during the beginning 1980 and scientific approaches are started in 1995, the very year when the transgenic pigs were for the first time, developed with the transgenic technology applied for complement regulatory protein (CRP) of human known as decay acceleration factor (DAF) of human. The HR of hearts from the pig was avoided when xenografted into NHP. Using the xenografts, approximately 100 days xenograft was survived in the immunosuppressed condition. Simultaneously, the α1,3Gal antigen of Galα1,3Galβ1,4GlcNac-R epitopes in glycoprotein and glycolipid carbohydrate core chains was identified as HR factor. The α1,3Gal antigen-lacking species have naturally preformed anti-α1,3Gal antibodies, as historically predated according to Landsteiner's Law.

The elimination of the α1,3Gal-T1 gene was reported in 2002. Using the xenograft, survival was slightly longer lasted rather than the above hCRP transgenic pigs. However, the xenografts still raised the acute humoral xenograft rejection or thrombopathy. The α1,3Gal-T KO and hCRP TG transgenic organs xenografting in NHP suggested the future systemic search and discovery of any potential candidates to provoke the remaining rejection responses in humans. Thus, the term of "non-α1,3Gal" antigens has been emerged in the human and NHP immune system. After α1,3Gal-based HR, non-α1,3Gal-mediated DXR and the thrombotic microangiopathy are major in organ xenografts, where the thrombotic microangiopathy reflects microvasculatured fibrin-platelet thrombosis, which causes ischemic damage and injury of the transplanted grafts. The microvascular circulation comprises vessels includes arterioles, capillaries, and venules to deliver oxygens and nutrients to cells in organ tissues and also to maintain body hydrostatic pressures. The microvascular circulation is crucial to avoid chronic rejection of allograft or graft failure, and to preventively protect, attenuate or suppress chronic rejection responses.

The endothelial cells in lining of the vasculatured grafts are mainly targeted during the immune responses of recipient hosts, which are progressed by antibody-dependent rejections like ADCC or thrombotic microangiopathy. Preformed natural Abs and elicited Abs against the endothelium of vascular vessels are the major mechanistic mediators of the primary immune response in the DXR.

This mechanism(s) explain the responses as depicted from the results that antibody or complement activation lead to the chronically activated damages or injuries to the endothelial cells of vascular endothelium. The events stimulate the vasculature thrombopathies, leading to microvascular thrombotic damages and ischemic injuries. In organ transplantation, chronic vasculopathy of allografts is thus a main causing factor of transplantation limitation for the patient survival for long-term period after transplantation.

It has been suggested that the serum levels of antibody types of IgM and IgG specific for α1,3Gal and anti-nonα1,3Gal antigens are affected by the several parameters of human ABO type blood group, age, diet, gender, individual history of vaccination and childhood era's geographic location from the cohort analysis. Human non-α1,3Gal-specific Abs activate in vitro complement system and cause damages and injuries of cells as well as cell damages mediated by ADCC [37] without any knowledge and information on which xenoantigen(s) are potential for in vivo cell damages of transplanted grafts. Human HLA-specific Abs cross-reacting with cells of pig are cytotoxic to pig cells. From results obtained from NHP models, non-α1,3Gal antibodies display to induce graft rejection by the immune system of hosts [10, 38]. Hence, the issue and information of non-α1,3Gal antibodies-regulated xenograft rejecting response should be a future subject to solve from the threat mediated by the host immunity in pig-to-human xenograft transplantation. This is also simultaneous issue even in the α1,3Gal-T KO pig organs. The current knowledge and information on the non-α1,3Gal antigens and Abs in pig-to-human xenotransplantation is depicted. In 1990s' era, major interest was in the α1,3Gal antigen. Then, at the present 2017, the pig non-α1,3Gal glycan and protein antigenic epitopes and anti-non-α1,3Gal Abs of human are strengthened.

11.5 Conceptional Difference Between Naturally Preformed Antibodies and Induced Antibodies

In the xenorejection reaction, the issue whether non-α1,3Gal antibodies are preformed or not is crucial. Immunization with bacterial carbohydrates during infection induces the most anti-glycan Abs in human, as naturally occurred. This immune response has also been confirmed by the cases of the anti-ABO blood group and anti-α1,3Gal epitope-specific Abs. Cytotoxicity of anti-α1,3Gal-specific Abs in human is known to occur in the first 3 months of fetal development, whereas cytotoxic non-α1,3Gal epitopes-specific antibodies are not occurred in the first year of fetal development [23]. For protein antigens, humans are known to have less naturally preformed foreign antigens (like SLA)-specific antibodies, although the immune system in human encountered with new antigenic proteins drives the powerful immune reaction to generate antibodies, as confirmed in the incompatible HLA allograft-receiving humans.

The issue of the non-α1,3Gal antigenic epitope has been raised as a remaining hurdle to overcome xenotransplantation. In 1993, a blood AO group of pig blood system has been reported by Oriol et al. [39]. The blood type A antigen of pig possibly cause an immune reaction to the blood O type or B type in human hosts. Concerning non-α1,3Gal antigen, there are several candidates of non-α1,3Gal xenoantigenic epitopes in addition to the α-GalNAc or β-GalNAc, β3-Gal, Galα1,3-Lew^x, NeuGc antigen known as Hanganutziu–Deicher (H-D) epitope, Forssman antigen, NeuAc-linked glycans (Neu5Acα2,3Galβ1,3GlcNAc), and Sid blood group (Sd^a) antigens. Despite the diverse antigens to date, these antigens are controversially not well defined for their xenoantigenicities even in part levels, requiring additional studies. Surfaced glycolipids in mammal cells are much richer than glycoproteins in molecular levels. In the pig kidney, as the α-1,3Gal type, the Galα1,3 Lew^x attached in glycolipids has been reported to be a human natural antibody-reactive epitope [40].

Gal α1,3Lew^x and Gal α1,3nLc4 (neolactotetraosylceramide) were also reactive to human natural antibody [41]. In addition, NeuGc-Gal-GlcNAc as the H-D antigen type and Galα1,3Lew^x as the α-1,3Gal type attached to *N*-glycans also in the miniature pig kidney function as xenoantigens due to their reactivity with human natural antibody [42]. Regardless to α1,3Gal and H-D antigens, antibodies from a patient who intraportally received pig islet-like clusters have been reacted with α-linked GalNAc and Galβ1,3GlcNAc. The patient's antibodies reacted the above carbohydrate structures even in the forms of terminally sulfated or sialylated as well as β-GlcNAc. However, the antibodies denied the β1,3-linked, oligomannosyl and Gal α1,3-Lew^x sugar structures [13].

With regard to the Forssman antigen in pigs, the terminal GalNAc residue of the Forssman antigen seems to be somewhat crucial. Although the terminal GalNAc, which is a residue of the Tn-antigen such as GalNAcα-O-Ser/Thr, is found in glycoproteins from pig, the expression level is decreased in the α1,3Gal-T-KO pig with unknown reason. Breimer group [23] reported that Thomsen–Friedenreich antigen (T-antigen; Galβ1,3GalNAcα-O-Ser/Thr), Tn antigen, and sialyl-Tn antigen (NeuAc α2,6GalNAcα-O-Ser/Thr) also function as xenoantigens.

In pig cells, terminal α1,3-Gal or β-Gal and α-GalNAc residues are not expressed, as confirmed by experimental results obtained using α-galactosidase or β-galactosidase-digested pig red cells [43]. The α- or β-Gal and α-GalNAc are the basic carbohydrate structures observed during de novo biosynthesis of *N*-glycan or *O*-glycans in mammals. α- or β-Gal and α-GalNAc residues are also demonstrated not to have the reactivity with non-α1,3Gal antibody [44]. Therefore, among them, the terminal GalNAc may function as a non-α1,3Gal antigen. Some Sia-carrying antigenic epitopes except for the H-D antigen in adult pig islets is suggested to be the non-α1,3Gal antibody-reactive epitope [45]. For example, Neu5Acα2,3Galβ1,3GlcNAc is the candidate [46]. As an interesting case, Sid blood group (Sd^a)/CAD antigen synthesized by pig β1,4*N*-acetylgalactosaminyl transferase 2 gene is also xenoantigenic [47], although, human and primate express the B4GALNT2. To the present point, α1,3Gal-KO pigs synthesize many non-α1,3Gal antigen epitopes on glycolipids that can be reactant with human serum even regardless the Forssman, H-D antigen and iGb3 antigens [48].

Before α1,3Gal-T KO pigs were developed using an initial and conventional technology of homologous recombination, enzyme competition to control sugar addition and processing by glycosyltransferases has been applied and basic insights into importance of α1,3-Gal carbohydrate were obtained in vitro. To prevent, eliminate or decrease the pig α1,3Gal xenografts, enzymatic competitions in usage of sugar substrates of α1,3Gal-T with ER-Golgi glycosyltransferases has been tried [49, 50]. In the modification of α1,3-Gal carbohydrate antigen, the enhanced expression of GnT-III exhibited the suppressed α1,3-Gal expression in the *N*-linked sugar antigen of glycoproteins [48]. Regarding to cellular receptors of innate immune cells, which are related to interacting glycans as ligands, the glycan receptors of NK cells or other neutrophils such as monocytes and macrophages are the target progress to understand the immune rejection in the part of non-α1,3Gal epitope rejection. Also, PERV issue is a part due to its ligand properties with *N*-linked sugar.

In case of the non-α1,3Gal antigens such as the NeuGc residue epitope as H-D sugar and other non-α1,3Gal antigens, enzymatic competition methods were in part effective to reduce the reactivity of the competitively-modified pig cells with human sera in in vitro level. For alternative way to regulate the functions in the cell receptors of innate immune cells such as Mannose lectins and Siglecs, the carbohydrate antigenic epitopes are targeted to eliminate the downstream signaling pathway to reject to immunity. However, they are also not satisfactorily progressed in further studies, although these basic immune studies are important for future in biology. For better understanding of acting mechanism(s) of non-α1,3Gal natural antibody-driven xenograft rejection, the Chinese tree shrews (CTS) as a rejection model of discordant xenografts has been developed [13]. A wild-type animal or CTS has been widely used in research of NHP for several decades since its finding in Southeast Asian part of the Tupaiidae family.

11.6 Isoglobotrihexosylceramide or Isoglobotriaosylceramide (iGb3; Galα1,3Galβ1,4Glcβ1Cer) Glycan Xenoantigens

11.6.1 Background of iGb and iGb3 Synthase

The α1,3-galactosyltransferase (α1,3Gal-T1, GGTA1) (EC 2.4.1.51) is not the sole enzyme, which can biosynthesize the Galα1,3Gal saccharide. Another α1,3-glycosyltransferase family Galα1,3 transferase member is the isoglobotriaosylceramide (iGb3) synthase (iGb3S) (EC 2.4.1.87), which is called GGTA2. The iGb3 synthase, iGb3S, biosynthesizes the known isoglobo-series GSLs (Fig. 11.1). The xenoantigenic Galα1,3Gal-R saccharide epitope has initially regarded as the glycan determinant catalyzed by a single α1,3Gal-T enzyme. Over the long evolution time, primates and humans have been subjected to gene inactivation of the GGTA1 or α1,3Gal-T enzyme gene and also generate specific Abs to the α1,3Gal epitope (Galα1,3Galb4GlcNAc-R). To obtain the "humanizing" pig organs for xenotransplantation, the biallelic (GGTA1) KO pig has been developed as the basic step. Indeed, during the pig production of homozygously disrupted GGTA1 gene for α1,3Gal-T toward the clinical xenotransplantation, the immunogenic Galα1,3Gal glycan epitopes were still detected in GGTA1$^{-/-}$ KO pigs (α1,3Gal-T KO). This disrupts the xenotransgenic central dogma that α1,3GalT-derived epitope is the sole the Galα1,3Gal glycan.

In pig-baboon xenotransplantation using cells derived from GGTA1 KO pig (α1,3Gal-T KO), a residual antibody generation against α1,3Gal antigens and the residual Abs-associated chronic cytotoxic activity were raised. After biotechnological development of α1,3Gal-T-KO pigs by knocking out the α1,3Gal-T enzyme, GGTA1, additional type of glycosyltransferase, named GGTA2, has become important in pig. Thus, HR response has been overcome by the mutant pig generation, which the mutated pigs lack the immunogenic α1,3Gal-T enzyme as GGTA1. However, issue of antibody-mediated immune rejection has still been raised. From GGTA1 gene deleted KO pigs, at present, it is still unclear how much residual amounts of α1,3Gal antigenic epitopes are present in GGTA1$^{-/-}$ KO pigs. For example, Diswall et al. [51] demonstrated synthesis of some glycolipids as a type of α1,3-Gal epitope from α1,3-Gal-T-KO pig tissues. That was a fucosylated iGb3, but not iGb3. Although the Galα1,3Gal antigenic determinants are not observed from the cells, tissues and organs of α1,3Gal-T KO pigs, the existence of fucosylated iGb3 was a new finding. For the iGb3-synthesizing enzyme, many researchers have been tried to isolate the iGb3 product and iGb3 synthesizing gene [52]. From the basis of α1,3-GalT (GGTA1) enzyme, the second α1,3-Gal-specific enzyme is named to the above GGTA2. The GGTA2 gene was not identified yet in α1,3-Gal-T KO pig cells.

The GGTA2 (it is also described as A3Gal-T2) synthesizes α1,3Gal antigens such as isoglobotriaosylceramide (iGb3; Galα1,3Galβ1,4GlcNAcβ1-ceramide) on lactosylceramide (Lac-Cer) cores [53]. Isoglobotrihexosylceramide synthase (iGb3S, A3GalT-2) catalytically biosynthesizes isoglobo-series GSLs, which have a terminal α-1,3Gal disaccharide. This is termed as iGb3 and iGb3S is thus the α1,3-Gal epitope-synthesizing enzyme (Fig. 11.3) because iGb3S synthesizes the Galα1,3Gal saccharide antigens on GSLs by the attachment of Gal residue to Lac-Cer (Fig. 11.3). It was shown that pig GGTA1$^{-/-}$ KO line-derived cells still generate even lower amounts of Galα1,3Gal epitopes than the normal cells, as found in the cells during staining with anti-Galα1,3Gal-specific monoclonal Abs.

Despite construction of the α1,3-GalT null animals, however, a residual α1,3Gal antigenic levels were reactive in biallelic α1,3Gal-T KO pig cells and induced chronic rejection of α1,3-GalT$^{-/-}$ organs in non-primate models of xenotransplantation. Thus, it is claimed that α1,3Gal-T is not the sole glycosyltransferase responsible for the formation of antigenic Galα1,3Gal epitopes. iGb3S enzyme raises to be the possible source of additional α-1,3Gal epitopes in GGTA1$^{-/-}$ KO pigs and any related animals. Therefore, the A3Gal-T2 enzyme of iGb3S gene is of interest in pig-

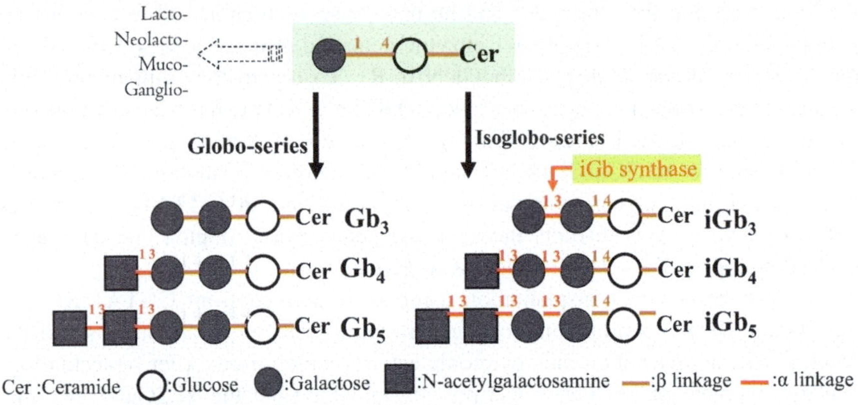

Fig. 11.3 *I*llustration of the major mammalian glycosphingolipid (GSL) pathways of globo-series, isoglobo-series, lacto-, neolacto-, muco-, and ganglio-series. Lactosylceramide (Lac-Cer) is a starting precursor of GSL synthesis. iGb synthase enzyme catalyzes Galα1,3linkage. iGb3 structure can be compared to α1,3Gal-transferase-generating α1,3Gal structure. iGb3 synthase enzyme action can be found in (https://sites.google.com/site/abobloodtype/37-α-1-3-gal-nac-transferase-family)

to-primate or pig-to-human immunological response. Currently, it was considered that the generation of Galα1,3Gal found in GGTA1$^{-/-}$ KO pigs can be catalyzed by iGb3 synthase (iGb3S) or glycosyltransferase family responsible for the ABO blood group of humans.

Except for iGb3, with regard to any possible synthesis of α1,3Gal antigen similar and α1,3Gal-mimicry antigens in the α1,3Gal-T KO pigs, non-α1,3Gal antigens have been found to induce a different type of new immunological responses in humans. However, structurally similar antigen structures to the α1,3-Gal are not evidenced yet. For example, plant lectins such as MAL and ECA isolated from *Maackia amurensis* and *Erythrina cristagalli*, respectively, recognize the glycan epitopes such as Neu5Acα2,3Galβ1,4GlcNAc and Galβ1,4GlcNAc as the precursor lactosamine structure for the Galα1,3Gal synthesis [54]. The structure is structurally different from Lew^{x} [Galβ1,4(Fucα1,3)Gal-NAcβ-] or sialyl-Lewx, but slightly similar. Similarly, several carbohydrate structures such as P1 (Galα4nLc4), X2 (synthesized by β1,3GalNAc-T), non-capped LacNAc precursor and fucosyl-H type-2 have been detected from the glycolipid antigens of pigs, although they are not immunogenic in humans due to their expression in the human cells [51]. Therefore, it seems that Gal-T KO does not express some α1,3-Gal-similar antigens in the pigs.

11.6.2 iGb3 Expression in Mice, Rats, and in GalT KO Mice

The glycosphingolipid (GSL) iGb3 is an additional epitope of the α1,3Gal glycans, as shown in rats and mice [55]. The iGbs is formed by the iGb3S enzyme, starting from precursor LacCer [56]. From the residual expression of α1,3Gal in α1,3Gal-T KO mice and α1,3Gal-T KO pigs as well as iGb3S as second enzyme to synthesize α1,3Gal in rats [55], it has been assumed that different enzymes from the known α1,3Gal-T may generate carbohydrate epitopes mimic to the known Galα1,3Gal antigens to be bound by xenoantigen-reactive Abs. The iGb3 is the possibly alternative epitope for the Galα1,3Gal antigen. The discovered iGb3S enzyme demonstrated the presence of terminal Galα1,3Gal antigenic epitopes attached to GSLs [53].

The iGb3 synthase enzyme is belonged to the ABO-blood group glycosyltransferase family and synthesizes isoglobo-series of GSLs. A big difference between α1,3Gal-T and iGb3S is in that iGb3S enzyme utilizes the substrate LacCer, which is a common precursor [53, 55]. So, there should be alternative pathways responsible for the α1,3Gal antigenic biosynthesis, raising a possibility of iGb3S expression even in α1,3Gal-T KO pigs. Puga et al. [57] investigated whether terminal residues of the Galα1,3Gal saccharides are absent or present in the α1,3Gal-T-lacking pigs as its KO phenotype. They used Galα1,3Gal specific Abs and specific plant lectins by immune-analytical tools using fluorescent microscopy and flow cytometry. To detect anti-α1,3Gal and H-type structures, human polyclonal α1,3Gal-specific Abs and α1,3Gal-specific IgM M86 as a mouse mAb, which Alexis Corporation (Switzerland) generated and sold the mAb, are specifically utilized [68]. *Bandeiraea simplicifolia* isolectin BS-IB4, which was FITC-conjugated, can also be used for its detection, as developed by Sigma-Aldrich (Darmstadt, Germany). α1,3Gal-T activity was enzymatically observed and α1,3Gal-T enzymatic assay is easily performed using the reaction mixture of a donor substrate UDP-[U-^{14}C]-Gal, UDP-Gal and acceptor substrate asialofetuin. Sugar structures were determined by mass spectroscopy (MS) such as ion trap MS using aortic endothelial neutral glycolipids. iGb3 synthase mRNA expression has been examined by RT-PCR using various organs and tissues including kidney, liver, lymph node, lung, spleen and thymus tissues of pig. For example, cell surfaces of α1,3Gal-T KO endothelium in aortic vessels of pig produced neither Galα1,3Gal epitopes nor iGb3 epitopes. Their α1,3Gal-T enzyme activities were also not present.

In lectin binding, levels of the sugar structures of H-type of ABO system were upregulated in α1,3Gal-T KO cells of pig. MS analysis was negative to reveal presence of the Galα1,3Gal epitopes in plasma membranes of α1,3Gal-T KO aortic endothelial cells (AECs) of pig and iGb3 product was completely not detected. In contrast, a fucosylated iGb3 structures were weakly present in extracts of both pig AECs and tissues. Expression of the iGb3S mRNA was found in all pig tissues of wild-type or α1,3Gal-T KO pigs. Although iGb3 could not be detected in the neutral glycolipids by MS, iGb3S mRNAs were expressed in all cells and tissues of WT or α1,3Gal-T KO pigs. The MS analysis excluded any possibility that iGb3 can be further elongated to isoglobo-series GSLs of iGb4, iGb5, and iGb6 GSLs [59]. The reason why the iGb3S protein is absent in the condition of its mRNA expression is not clear yet. However, one possibility is that the iGb3S enzyme needs enzymatic cofactors for its active enzyme activity. In addition, genetic differences are found between the examined animals. But neither α1,3Gal nor iGb3 were detected in α1,3Gal-T KO breed animals. Remarkably, polyclonal human anti-α1,3Gal Abs cannot recognize α1,3Gal antigens on α1,3Gal-T KO cells, suggesting that antibody-based DXR may not be related to residually expressed α1,3Gal glycans. Therefore, the DXR-responsible xenoantigenic epitopes are required to identify to overcome and prevent the rejection [11, 46, 60].

11.6.3 Positional Expression of iGb3 Synthase Enzyme in Pig

iGb3S-encoding iGb3S cDNA has been cloned from GGTA1$^{-/-}$ mouse thymus [61]. The mouse iGb3 synthase is predictably a type II transmembrane (TM) protein having 370 a.a. and glycosyltransferase catalytic domain is in a cytoplasmic region. Interestingly, rats have two different α1,3Gal-T enzymes of α1,3Gal-T and iGb3S. The pig α1,3Gal-T enzyme shows its high similarity with the known α1,3Gal-/GalNAc-T enzymes like the human A/B blood type transferases and Forssman synthase in its amino acid sequence. The rat iGb3S [53] can synthesize the Galα1,3Gal glycan structure; however, the amino acid similarity between the rat iGb3S and other α1,3Gal-T enzymes of pig, ox, and mouse is below 40%. The iGb3S is found only in rat. The rat α1,3Gal-T enzyme exhibited amino acid sequence similarities of 90% with the mouse α1,3Gal-T enzyme, 76% with the pig, and 75% with Ox [55]. The rat α1,3Gal-T enzyme had the lower amino acid sequence similarity of 42% with the rat iGb3 synthase. It has been considered that other species could also have iGb3S genes to produce the antigenic epitopes, hence bearing an alternative pathway for Galα1,3Gal antigenic biosynthesis, which is critically important for xenotransplantation. Although in genomic level, similar to the rats, pigs and mice have both two genes of iGb3S and α1,3Gal-T enzymes. In mice, like α1,3Gal-T, the iGb3S gene is comprised of 5 exons. It human homolog has been searched and found to be locate on chromosome 1, although the human homolog is existed as pseudogene, unprocessed form [1]. In the α1,3Gal-T (GGTA1) KO pigs, the iGb3 synthase seems to likely be expressed [62]. Like human α1,3Gal-T enzyme gene, the iGb3s gene (NM001080438) is present on human genome; however, the human iGb3s gene is not expressed due to mutations [63].

11.6.4 Expression Issue in Pig Tissues of iGb3 Synthase

iGb3 or the related isoglobo-series GSLs seem not to be expressed in the organs of pig, confirming the lacked iGb3 or isoglobo-series GSLs in pig organs. The present conclusion is that iGb3 is not an immunogenic epitope in pig-to-primate or pig-to-human xenotransplantation. Surprisingly, the iGb3 epitopes are frequently expressed in dendritic cells and thymus of human and mouse. In animals, iGb3 levels are expressed in species and tissue-specific manners. For example, rat thymus expresses a detectable and moderate level of iGb3, while thymus of human and mouse express undetectable iGb3 levels [64]. iGb3 expression level is not present in different tissues such as brain, eye, testis, kidney, liver, lung, spinal cord, spleen, stomach, thymus and intestine of mice as well as mouse plasma. In contrast, iGb3 expression is observed in dorsal root ganglia (DRG) of murine species. The differences in tissue- and species-specific expressions in animal iGb3 are not fully acceptable for the biological meaning. Rather giving some confusion on the significance to allow insights into the pig-to-human or to-primate xenotransplantation [64]. In addition, iGb3s gene expression is not correlated to iGb3 product levels, not matching with the epitope levels. For example, in mice tissues, mRNA expression of iGb3S gene is ubiquitous, but iGb3 product is expressed only in the DRG probably due to translational control or dynamic glycolipid biosynthesis. One possibility of the fact is that iGb3 is used as an iGb4 synthase substrate to convert iGb3 to iGb4 product. iGb4 is not the terminal Galα1,3Gal glycan. That explains a reason why results obtained from iGb3 and iGb3s are logically conflicted [65]. In primate-to-human or pig-to-human xenotransplantation, the above reason has been an important argument for the iGb3 role, and iGb3 product is not observed in pig tissues [51].

Actually, pigs seemed not to express iGb3 in meaningful levels to display cell damaged destruction, because at least minimal extent of α1,3Gal antigen amount is indeed needed for possible antibody-mediated xenograft rejection in a α1,3Gal-T KO mice [66]. It seems that deletion of the GGTA1 transferase (α1,3Gal-T) in pigs eliminates any Galα1,3 carbohydrate antigens, regardless of the existence of iGb3 epitopes in α1,3Gal-T-KO pig tissues, without antibody-mediated cell damaged destruction. Aortic endothelial cells (AEC) of α1,3Gal-T KO pigs were negative for detection in Gal α1,3Gal and iGb3, and α1,3Gal-T enzyme activity is also not found. Lectin analysis exhibited a largely increased glycan structures of blood group H-type in α1,3Gal-T KO pig cells, although AEC of wild-type pigs are negative. From the mass spectroscopic analysis, Galα1,3Gal and iGb3 were not observed in α1,3Gal-T KO pig AEC; however, small levels of a fucosylated iGb3 were observed from both wild-type and KO pig AEC. In the neutral GSLs, aGal epitope structure of Galα1,3Galβ1,4GlcβNAcβ1,3-LacCer was detected in wild-type pigs, but not in Gal-T KO pig AEC [59]. iGb3-fucosylated structure was observed in both cell types. Fuc-iGb3 is the GSL Fucα1,2Galα1,3Galβ1,4Glcβ1,1-Cer, as detected from pig intestines of both wild-type and α1,3Gal-T KO pigs [67].

The terminal fucosylated structures in pig AEC is recognized by specific lectin isolated from plant *Ulex europaeus* (UEA-I), which is a product of Sigma-Aldrich. The glycan structures of blood group H-type were increased, as detected by specific lectin using UEA-I of α1,3Gal-T KO pig AEC [68]. The branching pathway of fucose epitope, where the Fuc residue is linked to the internal Gal residue of Galα1,3 (Fucα1,2)Galβ1,3Glc-Cer, seems to be independently present from the conventional synthetic iGb3 pathway. In the fucosyl-iGb3 structure, fucosyltransferases of FuT-7 (EC 2.4.1.65), FuT-1, or FuT-2 (EC 2.4.1.69) are candidates. Although iGb3 is not found in the glycolipid neutral fractions, iGb3S mRNA expression is found in cells and tissues of both of WT or α1,3Gal-T KO animals. A possibility is that the iGb3S protein is absent, but its mRNA is present.

11.6.5 Historical Array of iGb3 Synthase and iGb3 Product

Sandrin's group hypothesized a central dogma, suggesting that pig organs and endothelial cells present iGb3 and consequently elicits immune interaction with human natural xenoantibodies [63]. For example, binding capacity of lectin BS-IB4 to iGb3 epitope is low, compared to α1,3Gal epitope, suggesting that the binding was considered not to be originated from the iGb3-binding. However, because IB4 has no capacity to bind iGb3, the comparative reduction in IB4 binding is based on the defected iGb3s. Nevertheless, the iGb3 levels are not the reflection of IB4 staining and iGb3s gene silencing does not reduce iGb3 staining level. If this is true, iGb3 synthase and α1,3Gal-T genes of pigs should be deleted for pig-to-human xenotransplantation. In addition, the Galα13Gal antigens formed by iGb3 synthase is observed by immunohistochemistry using a specific antibody named 15.401 in the kidneys and pancreas of α1,3Gal-T KO mice. Endothelial cells also express Galα13Gal epitopes produced by iGb3 synthase. Using the GGTA1$^{-/-}$ KO cells of pig, pig-baboon xenotransplantation has proved that there is still formation of a residual α1,3Gal-recognizing antibodies with chronic cytotoxicity. This suggests that additional sources may be present in the pigs for synthesis of α1,3Gal epitopes. Hence, it was possible to conclude that the iGb3s may alternatively biosynthesize α1,3Gal epitopes. iGb3S may synthesize a residual α1,3Gal epitope in GGTA1$^{-/-}$ pig, while other possibilities are controversial to synthesis of α1,3Gal epitopes [69]. However, the iGb3 or terminal Galα1,3Gal epitope-containing isoglobo-series GSL is not observed in α1,3Gal-T KO pigs, as reported by Diswall et al. [51]. α1,3Gal-containing GSLs were not detected even in neutral and acidic GSL fractions. iGb3 is an alternative source of Galα1,3Gal antigenic epitopes [63, 70].

However, iGb3s is assumed to have any role in acute rejection, because anti α1,3Gal Abs can distinguish the LacNAc (GGTA1)-originated α1,3-Gal carbohydrate epitopes and LacCer (iGb3) core structures. Human naturally occurred Abs bind to LacNAc form of α1,3Gal, compared to the LacCer species of ceramides, indicating the antibodies distinguish overall structure among the identical terminal disaccharide. Thus, iGb3-reactive Abs from GGTA1$^{-/-}$ mice raise a question for iGb3s role to produce the antibody-recognizing sugar. However, the specificities to antibodies has been questioned from the iGb3S production of pol-GAL GSLs that structured Galα1,3Galα1,3Galα1,3Galα1,3Galβ1,4Glcβ1,1-Cer. It is considered that iGb3s may synthesize various products rather than just iGb3. In addition, GGTA1$^{-/-}$ KO mice still produce low amounts of naturally formed anti-α1,3Gal Abs, compared to Old World primates and humans. Hence, the mice can produce iGb3 mimetic epitopes-specific antibodies, such as poly-Gal-L glycosphingolipids.

But, the current evidences obtained from pig organs including kidneys, heart, intestine, and pancreas clearly say "it is lackingly deficient." iGb3-lacking evidences in pig endothelial cells are obtained from immunologic, genetic, and biochemical aspects [51, 70]. If any animal expresses iGb3, the Galα1,3Gal-reactive antibodies should not be produced [63], or iGb3 must be absent in Galα1,3Gal-reactive antibody-bearing animal. Therefore, the above results still raise fundamental questions, regarding the specificity and accuracy of the produced Abs to be justified [59]. From MS analysis, NMR spectrometry analysis and antibody reactiveness, iGb3 and iGb3-based GSL derivatives are considered not to be present in pigs [65]. Several opposite results describing that cytotoxic anti-Galα1,3Gal antibodies are still produced even low levels in α1,3Gal-T KO mice and pigs without immune rejection responses against self-organs or self-blood vessels [71]. In more paradoxical results, Christiansen et al. [63] suggested that anti-iGb3 monoclonal antibodies were produced in α1,3Gal-T KO mice with the active iGb3 enzyme.

If the insistence and paradox of the co-presence of iGb3S expression and anti-α 1,3Gal Abs in animals are accepted, the following three cases are explained: (1) the α1,3Gal epitope-induced anti-α1,3Gal antibodies in α1,3Gal-T KO animals and human organs are not reactive to iGb3 antigen with unknown reasons; (2) the iGb3 product in α1,3Gal-T1 KO animals is further changed to non-reactive iGb4 and iGb5 series by endogenous enzymes; and (3) the iGb3S in α1,3Gal-T1 KO animals is not active to produce iGb3 in organs without reason. The above possibility (1), describing the non-reactivity of α1,3Gal epitope-induced anti-α1,3Gal antibodies with iGb3 in α1,3Gal-T KO animal and human, is supported [59]. The iGb3 conversion to other series of the possibility (2) is excluded because the iGb3 antigenic epitope in α1,3Gal-T1 KO animals is enzymatically further not changed to other iGb series [51, 65]. For the most acceptable point was originated from the fact that the iGb3 production is not linked to the transcriptional level of iGb3S gene [65, 70]. If mouse, pig and rat iGb3S genes are introduced into Chinese hamster ovarian (CHO) cells, iGb3 and iGb3-based GSLs such as the B4, B5, and B6 series are highly expressed [59, 65].

11.6.6 Explanation of Inconsistency Between Enzyme Activity of iGb3 Synthase and iGb3 Formation

The human iGb3 synthase cDNA also produce iGb3 at a trace amount in CHO cells; however, the human iGb3 synthase does not allow the formation of the additional iGb3-derived poly α1,3-Gal-GSLs, probably due to the weak iGb3 synthase activity [59, 63, 65]. To explain the above phenotype in human enzyme activity, a chaperone theory has been suggested [65], as explained by a lactose production-regulatory protein, α-lactalbumin, as a part of lactose synthetase (LS) in the milk of mammalian species. α-Lactalbumin forms the regulatory subunit of LS heterodimer and β1,4-galactosyltransferase (β4Gal-T1) forms the catalytic part. These proteins together produce lactose by LS via transferring Gal moieties to Glc residue. Multimeric α-lactalbumin binds Ca^{2+} and Zn^{2+} ions. A complex with β4Gal-T1, α-lactalbumin enhances the enzyme's affinity for Glu and inhibits the Gal polymerization. This pathway forms lactose by β4Gal-TI to LS conversion [72].

The similar case is previously explained in the core-1 β1,3galactosyltransferase cosmic genes [73]. Inhibitory chaperone proteins can be also explained for iGb3 synthase in a fashion like such iGb3 synthase trafficking to the trans Golgi complex to synthesize GSLs [65, 74]. Then, it is not surprising that the iGb3 synthesis is critically regulated in the mice, pigs, and humans. Considering the NKT cell-stimulating ligand iGb3, a large production of iGb3 in thymocytes should be avoid and to sustain the level of NKT cells. If iGb3 is largely expressed, NKT cells would be deleted during their development. Actually, a largely produced iGb3 in peripheral immune organs induced hyperreactive NKT cells in a mouse Fabry disease model, where the mouse expresses high levels of iGb3 product due to the defected α-galactosidase gene [75]. Then, the conflicts in the iGb3 expression in the pig is somewhat solved because such glycosyltransferases of iGb3 synthase to produce NKT cell ligands should strictly be controlled [65, 76].

11.6.7 Significance of iGb3 Synthase in Xenoantigen Synthesis

In rats, iGb3 synthesizes Galα1,3Gal when Gb3 is used as the acceptor substrate, indicating rat iGb3S can form iGb3 from LacCer and it is further acted to synthesize poly-α-1,3Gal epitopes. The globotrihexosylceramide Gb3 (CD77) and isoglobotrihexosylceramide iGb3 (isogloboside 3) as GSLs are structural isomers, which only one glycosidic bond is different, implicating the innate and adaptive immune system [77]. iGb3 is a hurdle in xenotransplantation, although the iGb3 antigenic structures are not yet clarified in pigs. Transfection of mouse iGb3S cDNA formed Galα1,3Gal structure of the isoglobo series pathway on cell surfaces. Hence, murine iGb3S gene has been recognized as another source for the Galα1,3Gal synthesis as the xenoantigen. The iGb3S-derived Galα1,3Gal is not affected by α1,2Fuc-T, whereas α1,3Gal-T is affected by α1,2Fuc-T. However, rat iGb3 synthase also synthesized Galα1,3Gal epitopes on the iGb3-bearing glycolipids.

The iGb3S can only form the iGb3 glycolipid using UDP-Gal and lactosyl-ceramide as donor and acceptor substrates, respectively and is classified as the isoglobo-series of GSLs because it acts as the precursor of iso-Forssman antigens and iGb4 isogloboside. Expression of the α1,3Gal-T gene in rats is found in various tissues and organs such as brain, kidney, liver and spleen. Genomic intron/exon organization of the α1,3Gal-T of rats is similar to that of the mouse. Only rat α1,3Gal-T synthesizes Galα1,3Gal epitopes on glycoproteins but not the iGb3 glycolipid. However, iGb3 synthase synthesizes Galα1,3Gal epitopes on GSL glycolipids to generate iGb3. It is, therefore, concluded that the Galα1,3Gal linkage antigens are strictly distinguished by two distinct Galα1,3Gal transferring enzymes of α1,3Gal-T and iGb3 synthase. The iGb3 synthase specifically synthesize poly-α-Gal structure on glycolipids. Because several poly-α1,3Gal antigens such as Galα1,3Galα1,4LacCer, Galα1,3Galα1,3Galα1,3Galα1,4-LacCer, and relatively large structures are found in rat small intestinal glycolipids, these are regarded as core Gb3 glycolipid structures. In the iGb3S transfected cells, poly-α1,3Gal antigens larger than Galα1,3-nLc4 structures were found; however, Gb3S cannot synthesize the poly-α1,3Gal structures. Compared to Gb3S, only iGb3S synthesizes poly-α 1,3Gal epitopes. Therefore, it is noted that Galα1,3Gal epitopes can be generated by the α1,3Gal-T and iGb3 synthase.

To bind the glycan structure, the isolectins IA and IB produced by *Griffonia (Bandeiraea) simplicifolia* (GS-IA4 and GS-IB4) are known to have about 89% homology. The GS-IA and GS-IB have different specificities for terminal α1,3GalNAc epitope, which is a known antigen for blood group A antigen and Forssman antigen, and terminal α1,3Gal epitope as a blood type B antigen and the Galα1,3Gal xenoantigenic epitope, respectively [78]. Although the lectin GS-IB4 binds only to terminal Galα1,3Gal epitope, the GS-IB4 also prefers to recognize GalNAcα1,3Gal. However, the lectin GS-IA4 also recognizes Galα1,3Gal epitopes. The specificity difference between GS-IA and GS-IB is caused by a single a.a. difference in the recognition site, which is located at amino acids 106 [79].

The iGb3S cannot glycosylate glycoproteins, as evidenced by the experiments that used GS-IB4 lectin for GS-IB4 lectin binding capacity to Galα1,3Gal glycan antigens on glycoprotein. GS-IB4 could not detect iGb3S-synthesized cell surface Galα1,3Gal structures, but effectively detect the α1,3Gal-T-synthesized determinant in rats. Thus, the iGb3 glycolipid specific mouse monoclonal antibody has been used to detect the Galα1,3Gal on iGb3 on target cells. In the rats, two different glycolipid Galα1,3Gal structures of the iGb3 and Galα1,3Galβ1,4GlcNAcα1,3Galβ1,4Glcβ1-Cer, which is called Galα1,3-nLc4, were reported [80]. Galα1,3-nLc4 as a Galα1,3Gal glycolipid structure is formed by the α1,3Gal-T of many different species [81]. The two α1,3Gal-T enzymes are important for the production of α1,3Gal-T KO and Galα1,3Gal KO pigs. In pigs, the fucosylated iGb3 (Fucα1,2Galα1,3-LacCer was found in stomach mucosa [67]. Therefore, this issue of pig iGb3 structure may causes a problem in xenotransplantation to induce HR by the iGb3 synthase-produced Galα(1,3)Gal glycolipids. To reach the Galα1,3Gal removal in pigs, α1,3Gal-T and iGb3 synthase should be considered.

11.6.8 Innate Immunological Role of iGb3 in iNKT Cells

In the biological roles, Gb3 has initially been known as the verotoxin-binding receptor of host cells in hemolytic syndrome occurred in the small blood vessels of damaged and inflamed kidneys, which lead to kidney failure, as in representatively hemolytic uremic syndrome (HUS). HUS is common in children, as caused by infection with verotoxin-producing *Escherichia coli* strains. Natural killer T (NKT) cells are attractively popular to immunologists and NKT cells fascinate basic immunologic researchers for innate immune cells and conventional adaptive T cells. NKT cells influence autoimmune responses to tumors and infections in mice [82]. NKT cells show a limited T-cell receptor (TCR)-α chain diversity and therefore NKT cells called termed invariant NKT cells (iNKT). They bind to certain bacterial glycolipids as antigens presented by CD1d, a nonpolymorphic antigen-presenting molecule of innate APCs.

iNKT cells are not tolerant of self, although lymphocytes are absolutely tolerant of self. iNKT cells are due to self-antigen-contacted activation in the thymus. A single GSL, iGb3, is used for thymus iNKT cell differentiation and peripheral iNKT cell activation. iGb3 is the self-antigen required for iNKT cells. From the two results that iGb3 is not detected in the thymus or in DCs from mice and humans and that iGb3S-deficient mice have normal number and action of iNKT cells, other self-antigens may be present for iNKT cells [82]. The semi-invariant receptor of NKT cells binds to GSL glycan-antigens expressed on the CD1d antigen as a monomorphic molecule. In the natural iNKT cell-selecting ligand, iGb3 has been known as a first glycolipid antigen in the thymus [83]. T-cell receptors recognize CD1d-presented iGb3 [84]. iGb3 as an immune-modulating GSL in mice, CD1d-bound iGb3 is presented to the TCR of iNKT cells [85]. Dendritic cell iGb3 activates NKT function in peripheral tissues, and iGb3 thus is a stimulating ligand in pig organs. Then it acts to reject α1,3-Gal-T deleted xenotransplanted organs in hosts [63]. In tumor regression, iGb3-acted bone marrow-derived DCs can display an iNKT cells-driven anti-tumor action [86].

In the case of iGb3 function in the positive selection of iNKT cells, iGb3 seems not to be a natural iNKT cell ligand. The iGb3S has an acceptor substrate specificity specific for a terminal Gal residue [87]. TCRs of NKT cells "flatten" and bind the iGb3 saccharide presented on CD1d. The terminal α1,3Gal recognizes the CD1d "pocket" region to optimize TCR recognition. iGb3-primed bone marrow-derived DCs induce the iNKT cell-mediated cytotoxicity against tumor. Historically, because NKT cell differentiation and function are mediated by endogenous antigens, the identification of those ligand antigens is important. iGb3 is the key endogenously natural ligand in human and mice. The unique population of NKT cells expresses a human invariant Vα14Jα18 TCR-Vα24Jα18 to recognize CD1d-presented glycolipid antigens. During activation, NKT cells regulate immune responses by production of excess amounts of IL-4 and IFN-γ [88]. Thus, NKT cell deficiency indicates disease onset such as tumor progression and development, autoimmune diseases and inflammation in mice and humans [88].

Historically, a GSL α-galactosyl-ceramide (α-Gal-Cer) obtained from a marine sponge [89] was suggested to be an NKT cells agonist for a CD1d recognition in both humans and mice [90]. Since α-1,3Gal-Cer is produced from the mammal, the β-hexaminidase A and B deficiency in lysosome inhibits NKT cell differentiation [91]. Then, as the candidate to stimulate NKT cells, iGb3 was considered. Using NKT cells from mouse and human as well as hybrodoma NKT cells, responding to iGb3, as well as isolectin B4 (GS-IB4) inhibition of iGb3, iGb3 was considered to be self-antigen for human NKT cells [91]. iGb3 is the endogenous natural and major ligand for NKT cell differentiation and self-recognition. iGb3 as human and mouse NKT cell ligand or agonist [92], iGb3 is required for the mice NKT cell TCR presentation. The iGb3 is regarded as a factor for mice and human NKT cell differentiation and self-recognition. Although it has been hypothesized that iGb3 acts as a ligand for development of mice NKT cells; however recently, the hypothesis has been controversial and ignored [93], because iGb3 detection was failed in mouse or human thymus. Using iGb3S KO mice, it was reported that NKT cell differentiation was still normal. Thus, it is concluded that mice NKT cell differentiation does not need iGb3. NKT cell differentiation defect as in Hex-b–lacking mouse seems to be the mutated lysosome storage disease, cause a defect in glycolipid presentation [94]. The iGb3 has no capacity to activate human NKT cells, with still question of the iGb3 role in NKT cells of human. However, the iGb3 role in NKT cell development is controversial from the fact that mice NKT cell differentiation is normally occurred even without the enzyme iGb3 synthase (iGb3S).

11.6.9 Controversial Aspect on iGb3 Function in Controversial

At the present time, this lacking α1,3Gal-containing GSLs such as iGb3 is one of the most controversial issue in the glycobiology. Moreover, two xenogenic genes of GGTA-1 and A3Gal-T-2 (iGb3S) are inactive for their function in baboons and humans. To solve the conflict and argue as well as any ambiguity on immune cross reactivity in pig-to-primate or pig-to-human xenotransplantation for the iGb3S role and iGb3 product, Butler et al. [95] created GGTA1 KO and iGb3s gene (A3Gal-T2) double KO pigs using the CRISPR/Cas system. The immune reactivity of iGb3s products has then been examined in GGTA-1$^{-/-}$ KO pigs as well as GGTA-1$^{-/-}$ and A3Gal-T2$^{-/-}$ double-KO pigs. Antibody-dependent complement-dependent cytotoxicity (CDC) and antibody-binding profiles were examined to compare between GGTA-1$^{-/-}$ and GGTA1^{-}//A3Gal-T2$^{-/-}$ double KO pigs, as two genes of GGTA-1 and A3Gal-T2 are not active in baboons and humans. The CRISPR/Cas 9 technology targeted the two genes of pig GGTA-1 and pig A3Gal-T2. These targeted cells were directly applied in a single pig pregnancy to generate somatic cell nuclear transfer-created single KO and double KO piglets.

The targeted pigs were grown to adult maturity and were examined for, the GSL profiles to check iGb3 production. GSLs were also analyzed from renal tissues, because the defected iGb3s gene may modulate the GSL profile in pigs. Using peripheral blood mononuclear cells (PBMC), cytotoxicity against human and baboon sera was tested for comparative cytotoxicity. IB4 lectin was also examined through sugar-specific presence for α1,3Gal and iGb3 staining. The iGb3S KO pig showed the altered the GSL profile of renal tissue and iGb3 content was absent. Serum PBMC cytotoxicities of the baboon and human and α-1,3Gal and iGb3 contents were not changed by the genetic iGb3S inactivation. Thus, from the non-alteration in serum PBMC cytotoxicities of baboon and human as well as the α-1,3Gal and iGb3 immune undetection, it is concluded that iGb3S does not affect HR in pig-to-primate and pig-to-human xenotransplantation. Therefore, it is suggested that iGb3S enzyme does not synthesize sugar antigen to exert antibody-driven immune rejection in pig-to-human or pig-to-primate xenotransplantation.

However, a very small level to weaker α1,3Gal generation is observed in the iGb3S-deficient animal cells, although antibody binding levels and cytotoxic levels are not changed by inactivation of iGb3S gene. In addition, defect in iGb3S gene showed a significant change in the GSL profile of kidney, the cytotoxic profiles, Galα1,3Gal alterations and antibody bindings observed on pig PBMCs of human and baboon sera were not changed.

In the view of the antigenicity of iGb3 and iGb3s pig-to-human xenotransplantation, the above facts provoke a conflict and controversy [65]. For example, as raised even in the pig-to-primate xenotransplantation, iGb3 role and importance are still arguable, because iGb3 detection is succeeded in pig organs and tissues [51]. That is a reason why iGb3s gene (A3Gal-T2) KO pigs in a GGTA-1$^{-/-}$ background is created and this is interested in comparison with expression of α1,3Gal epitopes, antibody binding profiles, and antibody-dependent CDC between GGTA-1$^{-/-}$ and GGTA-1$^{-/-}$ /A3Gal-T2$^{-/-}$ double KO pigs [95].

From the previous studies, deduction and translation of murine-based results to pig-to-human xenotransplantation are non-logic, causing misunderstandings, because GGTA1 gene KO mice still contain even low contents of naturally formed anti α1,3Gal Abs, compared to old world primates and humans [59]. Whereas, because baboons or humans are not immunized with α1,3Gal zeno antigens, pig PBMN cells isolated from GGTA-1$^{-/-}$ and GGTA-1$^{-/-}$/A2Gal-T2$^{-/-}$ pigs is minor. Then, for the role of iGb3 contents in HR, wild-type pig iGb3 was extremely low. iGb3 is not observed in heart, kidney, liver and pancreas of pig [65], although the possibility that iGb3 conversion to iGb4 is catalyzed by iGb4 synthase. From the above conclusion, iGb3S gene KO pigs show the changed profile of the renal GSLs; however, any effects of Galα1,3Gal contents, antibody bindings, and cytotoxicities of human and baboon sera on pig PBMC were ignorable. Recent triple KO pigs of xenoantigens showed the reduced antibody binding level for xenoantigens, indicating potential possibility of non-α1,3Gal xenoantigens and xenopeptides [22, 96].

From doubt issue and history of GGTA2 and iGb3, it was concluded that iGb3 is not related to antibody-mediated controversial [97]. Then, the GGTA2 gene function in pigs is still controversial, although GGTA2 is not a key factor in xenograft rejection. However, as a residual weak content of α-1,3Gal antigenic epitopes, isoglobotrihexosylceramide is raised. The isoglobotrihexosylceramide synthase (iGb3S or A3Gal-T-2) is an another member of the α-1,3-galactosyltransferase family and synthesizes the isoglobo-series GSLs of an α1,3-Gal-terminal disaccharide known as iGb3. iGb3 is converted to iGb4 because it is used as an iGb4 substrate. However, iGb4 is negative for usage as a terminal Galα1,3Gal antigen. Thus, iGb3s is suggested as Galα1,3Gal xenoantigen in immune responses of the pig-to-human or pig-to-primate xenotransplantation. iGb3 synthesis is species-dependent and specific, as iGb3 amounts are regularly present in rat thymus. In contrast, iGb3 amounts are completely negative in human and mouse thymus [64].

iGb3 also is species- and tissue-specific in its production, as iGb3 expression in mice is not observed in brain, intestine, kidney, liver, plasma spinal cord, spleen, stomach, thymus and testis. However, murine dorsal ganglia express iGb3. iGb3 binds to TCR and CD1d of iNKT cells in mice [98]. Function of iGb3 expression is still unknown in the pig-to-primate or pig-to-human xenotransplantation [64]. The iGb3 and iGb3s are not so important in the pig-to-human xenotransplantation [65], but the importance of iGb3 should be considered. iGb3S enzyme can replace a residual α1,3Gal antigenic content in GGTA-1 KO animals, although such a role is argued. Still question is why iGb3 is not detected in pigs [51]. Using iGb3S gene (A3Gal-T-2)-KO pigs in a α1,3Gal-$T^{-/-}$ KO pig, antibody bindings and antibody-mediated CDC have been reported [99]. iGb3 is α1,3Gal terminal saccharide. From iGb3 as a substrate, various iGb3 derivatives of α1,3Gal-terminated GSLs including B4, B5, and B6 are convertedly (or derivatively) produced [59]. This conversional possibility of iGb3 indicates that iGb3S may generate the Galα1,3Gal epitopes in α1,3Gal-$T1^{-/-}$ KO animals. The gene is silent due to mutations and consequent non-synthesis of iGb3.

11.7 Non-Gal Glycan Xenoantigen, *N*-Glycolylneuraminic Acid (NeuGc), or Hanganutziu–Deicher (HD) Antigen in Xenotransplantation

The most recent results justified that the serum contents of α1,3Gal-reactive Abs and non-α1,3Gal-reactive Abs as isotype IgM and IgG are different, depending on age, blood group ABO types, diet, gender, childhood living place and vaccination history in a cohort study of 75 healthy humans [100]. The different levels of α1,3Gal-reactive IgM and IgG type antibodies and also non-α1,3Gal-reactive IgM and IgG type antibodies in humans reflect its importance of Neu5Gc as the non-α1,3Gal antigen. Apart from the Galα1,3Gal antigenic epitope, the second swine sugar xenoantigens, not α1,3-Gal-T antigen-related, have been searched and identified to removal of the antigens. Humans also carry anti-Neu5Gc Abs of IgM and IgG types to recognize AECs of pigs and anti-Neu5Gc Abs also recognize RBCs of pigs.

11.7.1 General Aspect of NeuGc

Sia is a generic term of NeuAc and derivatives of NeuAc. Sias are a nine-carbon acidic sugar family and N-acetylneuraminic acid is mostly common in vertebrates. Gunnar Blix discovered and isolated sialic acids from salivary mucins in 1936 and he named it "sialic acid" as a meaning of the Greek word, saliva (σίαλον) [101]. Five years later, Ernst Klenk isolated the sialic acids from brain glycolipids in 1941 and he named it "neuraminic acid" as a meaning of neuronal specificity. NeuAc or 5-*N*-acetylneuraminic acid is the most general from of Sia saccharide in CMAH-negative animals such as humans. The family of Sia varies depending on the C-5 carbon and easily modified to its derivatives such as NeuGc or 5-*N*-glycolylneuraminic acid and KDN or 2-keto-deoxynonulosonic acid [102]. CMAH enzyme modifies activated nucleotide sugar CMP-Neu5Ac to CMP-Neu5Gc in the cytosolic space. Chemical diversity of Sia structures is caused from enzyme action to modify Neu5Ac, Neu5Gc, and Kdn structures in the lumen of the Golgi apparatus.

Particularly distinct saccharide residue in the humoral rejection response for xenografts is NeuGc or *N*-glycolyl-neuraminic acid, the termed Hanganutziu–Deicher (HD) xenoantigen, as a Sia family member. The HD term has been originated from the facts that Hanganutziu (1924) and, 2 years later, Deicher (1926) described the patient sera obtained from anti-tetanus horse serum-injected patients for therapeutic immunization expressed blood serum diseases. The symptoms were caused by factors like serum heterophilic antibodies that define Hanganutziu–Deicher Abs. The HD-reactive Abs bind to their specific antigens and this binding property were separately evidenced by Higashi group [103] and Merrick group [104]. The HD antigens were characterized as NeuGc-based glycolipids and glycoproteins. The Sias are characteristic of 9-C saccharides in structural backbone attached to ganglioside and glycoproteins. The Neu5Gc is hydroxylated derivative of Neu5Ac.

For NeuGc distribution in mammals, NeuGc-terminated saccharides on gangliosides and glycoproteins are generally present in monkeys and pigs, however, not present in chickens or humans. In molecular aspects, the number of surfaced gangliosides on plasma membranes is largely expressed rather than that of glycoproteins. In addition, gangliosides are largely expressed in neuronal tissue and moderately in other cells, tissues and organs in mammals. Although human gangliosides are physiologically and biochemically crucial in mammals, gangliosides-specific antibodies produced in mammals such as human. The anti-ganglioside antibodies are specific to NeuAc, although which carbohydrate structures and antigen epitopes of gangliosides are not known. HD antigen as recognized to NeuGc is found in animals including lower invertebrates including echinodermata species and higher vertebrates including mammals such as mice, cows, apes and monkeys. However, they are not distributed in avians such as birds and humans. *N*-glycans of plasma membrane glycoproteins of pig kidneys have the α1,3Gal antigenic epitopes as the most abundant glycan structure compound. However, the most of *N*-glycan structures in pig were Neu5Gc form epitope as non-α1,3Gal carbohydrate antigens [42]. After α1,3Gal identification for the main xenoantigenic target for pig-specific Abs of human, NeuGc was discovered on the vascular endothelium [105]. The human anti-NeuGc Abs developed would cause immune failure to survival of xenografted pig organs xenotransplanted to a human recipient. Moreover, islets of adult pigs produce sulfated and high mannosyl *N*-glycan types, and they are not observed in humans, indicating possibility of new non-α1,3Gal specificities [106]. Approximately more than 80% of healthy and normal humans generate NeuGc-recognizing specific Abs [107], although there is a discrepancy between the results, such discrepancies are presumably derived by the antigenic glycan structures present on ECs and RBCs of pigs [108].

11.7.2 Defect in NeuGc Synthesis and Anti-NeuGc Antibody in Human

Humans have a single-exon deletion, leading to genetic mutation of the CMAH gene, which is an essential enzyme in biosynthesis of NeuGc. Therefore, humans are such NeuGc negative animal in mammals. Mechanistically, the repeating Alu family inactivated the CMAH gene and this event enhances the generation of NeuAc residue as the NeuGc precursor. For the downstream events, several sialylation is followed by increased α2,6-linked sialylation by ST6Gal-I and substrate-dependent enforced changes in their Siglec receptor genes. The CMAH mutation elicits the non-genetic phenomenon of metabolic NeuGc incorporation during human dietary food take with defensive front line of anti-Neu5Gc Abs of "autogenic xenoantigen" production [109]. Therefore, the human CMAH is not basically observed due to the simple deletion of exon 6 in the full length genome [110].

The second xenoantigen NeuGc is absent in humans. However, Neu5Gc expression is commonly detected in mammals even in human's closest like chimpanzees or Old World monkeys which was separated from humans 12–25 million years ago. Therefore, all the known mammals including baboons, monkeys, pigs, and rats produce the NeuGc saccharide as distinct antigen. A NeuGc as xenoantigen has been targeted, because the NeuGc is produced by the specific hydroxylase enzyme, named the CMAH. Thus, in NeuGc structure, an extra linked hydroxyl (-OH) group is present on the acetyl substituent position of neuraminic acid. The enzyme CMAH catalyzes the reaction of the Neu5Gc antigen synthesis from the NeuAc as substrate, because NeuGc structure is highly similar to that of NeuAc as a human type. Mutation of the human CMAH gene disrupts the expression of the human type CMAH enzyme, blocking extra transferring of the hydroxyl group to the original NeuAc residue. Recently, Varki et al. revisited the function of Sas species [102]. Neu5Gc conversion from Neu5Ac is occurred via addition of one oxygen atom at the sugar nucleotide level by the above enzyme CMAH.

However, anti-HD antigen-specific human natural antibody is produced in human sera, and the human anti-HD-specific antibody occupies most anti-non-α1,3Gal Abs which are generated in humans [108]. This suggests that the HD antigen is the second major xenoantigen, which follows the α1,3Gal-antigen, in humans. For example, in the experiment that human patient extracorporeally received by perfusion of pig kidney organ, the staining of α1,3-Gal xenoantigens and HD reactive gangliosides were largely detected, where gangliosides of NeuGc-GM3 and NeuGc-GD3 are analyzed [111]. Human sera contain naturally occurred anti-HD IgM and IgG antibodies [112]. The anti-HD antibody may also induce HR when it encounters the HD antigen and the HD antigen can also induce humoral responses-based delayed type of rejection [113]. This is one of the differences between the NeuGc and α1,3-Gal xenoantigen during xenotransplantation. The antibodies are likely to be produced only in some human diseases, as reported for Guillain-Barre syndrome having a neurological phenotype of the acute axonal motor neuropathy and ganglioside *N*-glycolyl GM3 in retinoblastoma cancers [114]. Furthermore, anti-

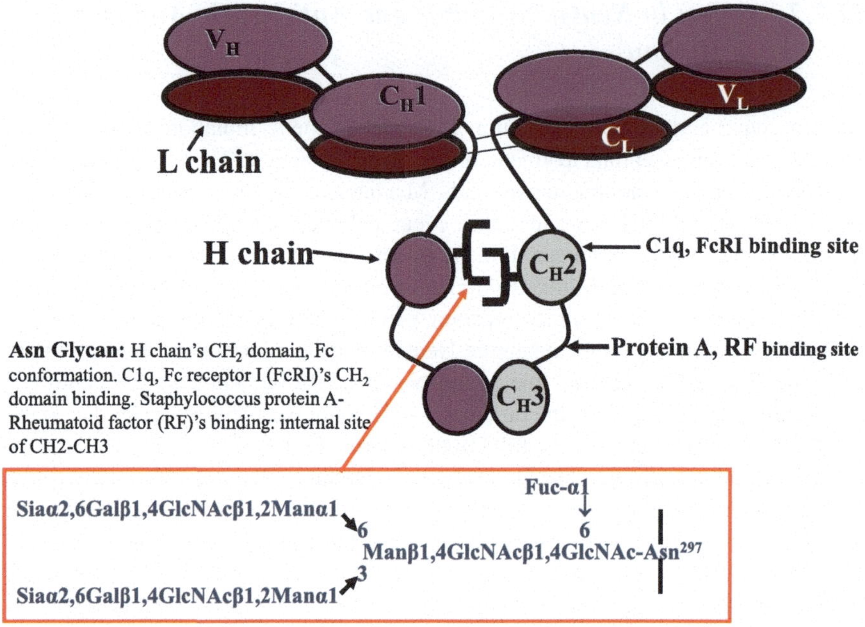

Fig. 11.4 Asn297-attached N-glycan structure, which carries a core Fuc-α1,6 residue, has been engineered to remove the Fuc-α1,6 residue to enhance FcγR of NK cells or macrophages

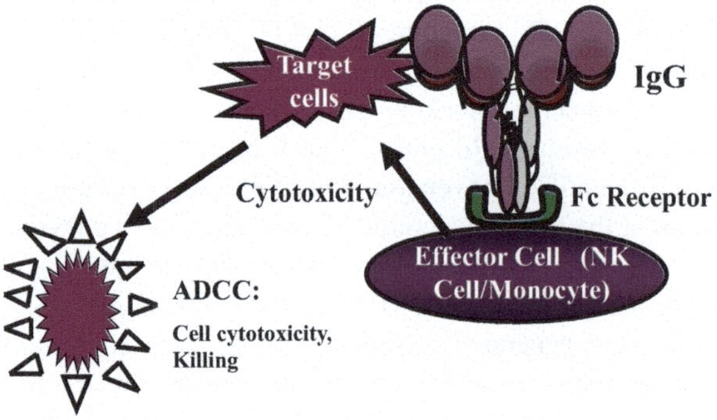

Fig. 11.5 Induction of ADCC, antibody-dependent cell cytotoxicity by specific IgG antibody and its Fc receptor, which kills target cells by effector cells such of NK cells and innate immune cells of macrophages, DCs, leukocytes, etc.

gangliosides antibodies have many biological functions of complement activation, apoptosis, angiogenesis and growth [115].

The major anti-non-α1,3Gal Abs recognize the NeuGc antigenic epitopes. Neu5Gc-specific Abs raise antibody-dependent cell cytotoxicity (ADCC) or CDC. In ADCC progression, the Asn297-attached N-glycan structure, which lacks a core Fuc-α1,6-linked residue, has been engineered to remove the Fuc-α1,6-linked residue (Figs. 11.4 and 11.5). As illustrated in Fig. 11.6, When the concentration of NeuGc epitopes expressed on cultured pig AEC was calculated, 6.3×10^7 NeuGc epitopes are estimated [116], as compared to the 2×10^7 Gal epitopes reported by Galili group

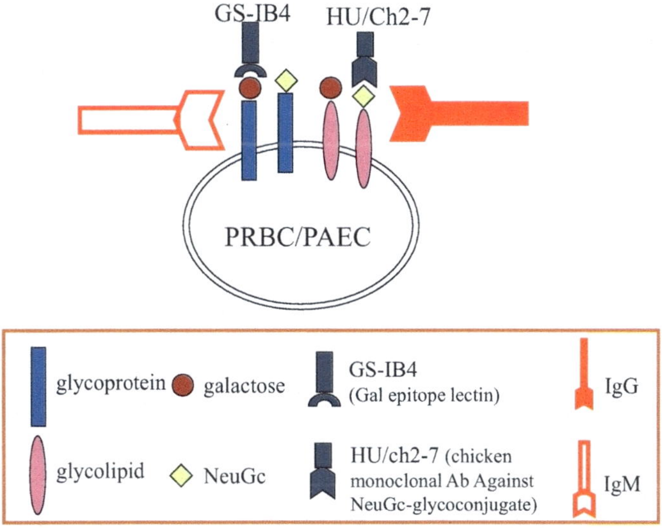

Fig. 11.6 Induction of CDC, complement-dependent cytotoxicity by Neu5Gc-specific antibody. NeuGc-specific IgG or IgM class Abs are produced by human. Experimentally, HU/ch2-7 (chicken mAbs specific for NeuGc-glycoconjugates) has been raised. Pig RBC or pig AEC expresses α1,3Gal antigenic epitopes and NeuGc saccharides on their cells and tissues. The α1,3Gal antigenic epitopes are recognized by GS-IB4 lectin

[117]. The expressed NeuGc level is more than that of Gal, indicating potent immunogenicity of NeuGc. In addition, the cohort analysis for the levels of Neu5Gc-specific Abs and non-α1,3Gal/non-Neu5Gc-specific Abs have been done from humans. From the cohort analysis, their binding capacities of human anti-nonα1,3Gal Abs and anti-non-α1,3Gal/non-Neu5Gc Abs as IgM and IgG types to RBCs and AECs of pig, isolated from pig α1,3Gal-T KO/CD46 and pig α1,3Gal-T KO/CD46/Neu5Gc KO were examined [100].

An interesting finding is that the mens level of anti-Neu5Gc-specific Abs as IgG type was larger, compared to that of women in humans. In the α1,3Gal-T KO/CD46/Neu5Gc KO pigs, the binding capacity of pig RBCs and pig AECs with human antibody is decreased rather than those from α1,3Gal-T KO/CD46 cells. Almost 100% of human sera contain IgM and IgG Abs as anti-Neu5Gc antibody, which recognize AECs of pigs and 50% contain anti-Neu5Gc Abs to recognize pig RBCs. The level of antibodies to Neu5Gc and non-α1,3Gal/non-Neu5Gc pig antigens reflects the status of pig RBCs for clinical transfusion [118]. This is based on the facts that serum IgM and IgG Abs of humans recognize pig RBCs, as pig RBCs produce pig-specific antigenic carbohydrates but not immune receptor antigens of MHC C-I or MHC C-II such as pig-corresponded SLA. However, pig AEC express pig-specific carbohydrates as well as immune receptor antigens of pig-specific MHC C- I and MHC C-II. Thus, interaction of human serum with pig RBCs reflects the Abs binding to carbohydrate antigenic epitopes.

11.7.3 Mechanistic Explanation of Production of Human Abs Specific for NeuGc Saccharide

For the loss of the NeuGc saccharides among neuraminic acids in the human lineage, infectious pathogenic theory is suggested because pathogens use such Neu5Gc-based Sias as their receptors to recognize, bind and infect host cells of mammals. The loss of Neu5Gc results in Neu5Ac accumulation. The cellular environments reflect the changes in relative interactions of pathogens with humans [109]. Under such circumstance, humans are stressed to defense to NeuGc-specific pathogens [119] and become susceptible to NeuAc-specific pathogens. For example, erythrocyte Sia-binding preference of malaria parasites is different from humans and African Non-Human Hominids (NHHs). Earlier, Varki group reported that genetic elimination of Neu5Gc gene and NeuGc production enabled to escape the prevalence of NHH malaria in ancestral hominins but instead a malaria species of *Plasmodium falciparum*, which is a human type of "malignant malaria," is invasive to humans, because an NHH malaria strain evolved to recognize human erythrocyte surfaced Neu5Ac saccharides [45]. For example, it was hypothesized that a malaria-causing protozoa recognize Neu5Gc and select the candidate of the CMAH-negative strains.

Under natural adaptation, some CMAH-negative females produced anti-Neu5Gc antibodies and the antibodies denied the sperm-oocyte interaction, preventing fertilization even with CMAH positive males or negative males [102]. For naturally preformed anti-carbohydrate antibodies, the cross-reactive immune responses with bacterial exopolysaccharides have been explained as a consequent immune reaction. Furthermore, the newly appeared synthesis of NeuGc sugar influences biological behaviors of bacterial endotoxins and enterointestinal parasites. Representatively, the Gram-negative bacterial endotoxin, lipopolysaccharide (LPS)-induced activation of mouse B-cells decreases in expression levels of human CMAH mRNA [120, 121]. For example, it is assumed that if a human does not express anti-NeuGc antibodies, the human would be not communicated with any NeuGc-lacking bacteria. Nevertheless, it has been suggested that Abs specific for NeuGc sugar are produced by a consequence of dietary consumed exogenous NeuGc, as they are incorporated into capsular LPSs of *Haemophilus influenzae* and induce immune responses [122]. It has been also discovered that NeuGc is found in endothelial cells of pigs as xenoantigenic epitopes.

The human antibodies against sialic acid were obtained when human was planted with pig tissues were found by Kobayashi et al. [116] from the study on pig islets and kidney-planted Swedish patient serum, detecting GM3-NeuGc gangliosides by ELISA and immuno-TLC [111]. When the patients, who received grafts of fetal pancreatic islets, were examined for Abs specific for carbohydrate antigens [13], NeuGc-specific antibody titers were found. It has also been evidenced that NeuGc expressed in pig islets together with some Sia-bearing antigenic molecules, reactive for naturally formed Abs of humans [45]. In baboon graft-planted with α1,3Gal-T-KO pig hearts, antibodies against pig heart gangliosides were observed [1]. However, from the fact that baboon is a NeuGc-producing mammal species, the produced anti-gangliosides antibodies should be against NeuAc-containing sialylated carbohydrate structures. Thus, it is thought that diverse sialyl carbohydrates having neuraminic acid, regardless NeuGc and NeuAc, can induced antibody production when human are exposed to pig organ/tissue/cells.

11.7.4 CMAH Gene-KO Pig and Disruption of NeuGc Production in Pig

Since the generation of pig α-1,3Gal-T KO mutant, non-α-1,3Gal antigens such as Neu5Gc induced acute humoral xenograft rejection with complement activation. Ubiquitous expression of NeuGc epitope antigens are well known in the endothelium of vascular vessels in the heart, kidney, liver, and pancreas of pig, and also on islet cell clusters, as well as on the cornea [123]. The NeuGc is also strongly expressed in aortic valves and also in pulmonary valves of pigs [124]. Patients temporarily treated with grafts of pig skin easily induced the development of high NeuGc-specific IgG Abs level and the production levels were sustained for long time [125]. This indicates that humans are easily sensitized to the non-α1,3Gal antigenic NeuGc saccharide. The immunological effects of NeuGc-deficient pig organs using the CMAH gene-KO pigs have not been approached to explore in vivo roles in transplantation yet due to absence of relevant animal models to apply. However, NeuGc saccharide is known not to express in some distinct species such as New World monkeys [126], which therefore can develop anti-NeuGc antibodies.

Introduction of pig NeuGc-deficient KO strains to the NeuGc-deficient species of New World monkeys may give immunological roles of the NeuGc species in xenotransplantation and how well the grafts are tolerated. In the CMAH gene-siRNA experiments, the HD antigen synthesis have been reduced or eliminated in pig cells such as PK15 kidney cells, giving a possibility of knocking down (KD) and KO of the NeuGc-synthesizing CMAH gene [127]. The siRNA knocking down of CMAH gene in kidney PK15 and endothelial cells of pig exhibited the decreased HD antigenicity. Thus, this provisional aspect triggered to start the accelerated studies on the double KO animals of α1,3-Gal-T KO and HD-KO pig in the many academic and commercial institutes to reduce the xenoantigenicity in pigs. The double KO pigs exhibited the immune-protection from human natural antibodies [128]. Hearts, kidneys, and livers of α1,3Gal-T KO pigs except for lungs [129] are reported to have increased Neu5Gc expression rather than wild-type pigs, and α1,3Gal-T-KO seems to affect the non-α1,3Gal antibody reaction, as found for α1,3Gal-T-KO neonatal pig islets.

Events of high mannose, truncated glycans, xylosylation and fucosylation were observed on glycoconjugates in the α1,3Gal-T/CMAH transgenic pig serum [130]. Fucose-attached glycans are belonged to blood group ABO antigens, indicating that the Fucose-attached glycans are possibly targeted by the human humoral immunity. In recent report [131], valves from pig α1,3Gal-T KO/CD46/NeuGc KO strains showed less IgM and IgG binding activities. Pig α1,3Gal-T KO/CD46/NeuGc KO strains are completely negative for the expression of the specific carbohydrate antigenic epitopes after relevant gene KO operation. Only pig heart valves of α1,3Gal-T KO/CD46/NeuGc KO KO strains produced minimal levels of human IgM/IgG recognizing Abs when they were incubated with Abs-containing sera of human. The antibody-recognizing level was similar to the binding level to a heart valve of human.

Seemingly, in the analysis of pig α1,3Gal-T KO and pig α1,3Gal-T KO/NeuGc KO strains, binding capacity of IgM and IgG Abs of human to aortic tissues and corneas isolated from pig α1,3Gal-T KO and pig α1,3Gal-T KO/NeuGc KO strains was decreased in comparison to binding levels of pig wild-type strains. Although binding levels of human Abs to pig α1,3Gal-T KO/NeuGc KO AECs was dramatically reduced, when compared to pig α1,3Gal-T KO AECs, but no significant difference of binding capacities was observed between α1,3Gal-T KO and α1,3Gal-T KO/NeuGc KO corneal ECs of pig strains. Thus, it is not yet proved that the NeuGc deficiency in pig corneas of α1,3Gal-T KO/NeuGc KO strains has a merit and advantage over α1,3Gal-T KO pig corneas [8]. Basically, because the deficiency of NeuGc sugars in pig aortic and AECs of pig α1,3Gal-T KO strains is related to the decreased level of specific binding of Abs in human, it has been considered that α1,3Gal-T KO/NeuGc KO pigs possibly allow better outcome in clinical xenotransplantation.

11.7.5 Acquisition and Presence of NeuGc and Anti-NeuGc Antibodies in Human

Interestingly, the human cancer cells have aberrant glycosylation with terminal glycans of Neu5Gc. Expression of Neu5Gc species is negative in human normal cells [132] because of an inactivating gene expression caused by genetic CMAH gene mutation of human. Hence, expression of NeuGc saccharide in cancer patients is possible by red meat dietary or dairy products Neu5Gc absorption [133]. The Neu5Gc expression is correlated with tumor invasiveness and metastasis. Why tumor cells are acquisitive for the Neu5Gc glycans? This is in part explained by the tumor microenvironmental condition of hypoxia. The hypoxia environments strongly induce the sialic acid transporter sialin expression, increasing the contents of sialic acids and Neu5Gc on the tumor cell surfaces [134]. NeuGc can metabolically be extrageneously incorporated into human cells during food ingestion of NeuGc-containing meats [135]. Therefore, NeuGc species introduced by milk products or meats can be transported to the cytosolic area by the known delivery mechanisms such as lysosomal transporting system and pinocytosis. Then those delivered NeuGc species acts as self-usable substrates by the glycan metabolic and catabolic enzymes of human to incorporate to human glycoconjugates.

Because the sialylated glycans upregulate cell–cell adhesion and cell mobility, the acquired NeuGc and sialic acids are surface sialylated to upregulate the progressive and metastatic behavior of tumor cells. Such cellular increasingly absorbed Sia species including Neu5Gc are reported in cancer cells of breast, colon, lung, ovarian and prostate cancers of human [132]. NeuGc incorporation to cancer cells refers to a tumor-specific phenotype. The tumor Neu5Gc expression also develop xenoautoantibodies of anti-Neu5Gc antibodies [136]. However, Neu5Gc is a foreign antigen in the immune response to elicit generation of polyclonal anti-Neu5Gc antibodies. During beginning of the NeuGc study, it has been recognized that production of NeuGc-reactive Abs of humans is a consequence of any side-effect during active immunization or tumor immune responses because NeuGc was also found in various tumor cells of human. In case of normal tissues of human, merely incorporation of NeuGc was from dietary sources. Such NeuGc-specific Abs are found in normal individuals as healthy humans [137]. Thus, NeuGc expression is a risk factor for carcinoma, cardiovascular and inflammation diseases. Naturally formed NeuHc-specific Abs are mainly IgG type with a minor content of IgM and IgA. Compared to anti-ABO blood group Abs and anti-α1,3Gal Abs, anti-Neu5Gc Abs are observed during 6–12 months after birth, as induced by dietary NeuGc-took normal microorganisms [122]. Approximately 0.1% of total immunoglobulins are specific for Neu5Gc epitopes, and the Neu5Gcα2,6Lac are comparably reactant to anti-α1,3Gal antibodies. Anti-Neu5Gc antibodies effectively elicit AHR of α1,3GalT-KO organs.

11.8 The Third Xenoreactive Antigen, SD^a Blood Group Antigen, GalNAcβ1,4[Neu5Acα2,3]Gal β1,4GlcNAcβ1,3Gal Terminal Glycan

11.8.1 Background of Sd^a β1,4N-Acetylgalactosaminyltransferase-2 (β1,4GalNAcT-II or B4GALNT2 or Previous GALG-T2)

The Sda antigenic molecule has the distinct roles of (1) its down-regulation is coupled with the upregulation of sialyl Lewis antigens and cancer metastatic potentials in colon cancer. (2) It induces the cytotoxic lysis activity of cytotoxic T lymphocytes of murines. (3) It protects the muscular dystrophy disease of mouse model. (4) It prolongs the vWf half-span to protect the mouse bleeding disorder. In pigs, pig AECs binding of human serum is critical in the clinical xenotransplantation using pig organs. The binding capacity of human Abs to pig AECs is generally weak; however, all human serum has completely IgM and IgG against pig AECs, which are not specific for α1,3-Gal antigen or Neu5Gc antigen, but these antibodies are likely directed to Sd^a [22] and somewhat minority are against unknown pig antigens [130]. From the pig AEC binding to anti-non-α1,3Gal IgG and IgM types as well as of anti-non-α1,3Gal/non-Neu5Gc IgG and IgM, the putative target xenoantigen is Sd^a, a candidate for human anti-pig antibodies. Thus, the Sd^a's cognate biosynthetic enzyme β1,4N-acetylgalactosaminyltransferase-2 corresponding to the β1,4GalNAcT-II or B4GALNT2 or previously GALG-T2 for Sd^a synthesis with the carbohydrate structure is suggested [22]. Thus, α1,3Gal or Neu5Gc or Sd^a-negative pigs most seem to mutate for optimized source development of cells and organs required for clinical application in xenotransplantation. In part, serum binding to pig AECs also imply the reflection of binding to SLA and this is important for the clinical pig organ xenotransplantation.

Although the blood group ABO is the most general system in human, there are still "minor" group systems in human blood group. They include Duffy, Lewis, and Kelly as the surfaced phenotypes expressed on human tissues and blood. The Sd^a group, a histo-blood group, has a high incidence. The antigens of histo-blood group systems are also glycan structures, which are expressed in body fluids and tissues. The histo-blood group carbohydrate Sd^a antigen has been discovered in 1967 as another antigen types of the "non-blood group ABO system" or histo-blood groups present in the erythrocytic surface and exudates of the Caucasians [130]. Sd^a antigen is also found in the Caucasian-related races. The immunodominant antigen of Sd^a is known to a β1,4-GalNAc residue. In humans, the Sd^a is present in human tissues in tissue-type and cell-type specificities [138]. Interestingly, the Sd^a antigen is particularly expressed in the large intestinal goblet cells and also in the gastrointestinal tract cells such as epithelial cells located on the epithelial brush borders. The blood group glycosyltransferase B4galnt2, that synthesizes a glycan antigen Sd^a similar to A blood type [139].

As the Sd^a antigen is similar to A blood group, B4GALN-T2 enzyme is highly expressed in normal colon. The β1,4GalN-T2 enzyme synthesizes the distinct carbohydrate of GalNAcβ1,4[Neu5Acα2,3]Galβ1,4GlcNAcβ1,3Gal structure, known as the blood group Sd^a carbohydrate, by the terminal attachment of a β1,4GalNAc residue to a Sia residue-attached lactosamine acceptor substrate. The plant *Dolichos biflorus* agglutinin (DBA) is a specific lectin to the terminal α-GalNAc residue and Sd^a glycan-attached β-GalNAc residue [140], as the DBA has previously been utilized the Sda pentasaccharide study [141]. The DBA lectin as a glycoprotein 111 kDa consists of 4 subunits and shows a glycan specificity toward α-linkage GalNAc. DBA is used to define secretor status of A blood group individual due to possible techniques for hemagglutination inhibition and to define blood group typing. The biotinyl DBA lectin as a marker is used for renal collecting ducts because biotinylated DBA is deal as examination intermediate for glycoconjugates. Recently, a novel β1,4GalN-T2 to synthesize a rare SD^a blood group antigen has been found to reject pig cardiac xenografts [142]. This implies that genetic modification may influence the xenoantigenic carbohydrate phenotypes of pigs and carefully select developed pig lines.

Pig α1,3Gal-T KO-planted organ xenograft has still been xenoantigenic in a mode of antibody mediated rejection. This indicates non-α1,3Gal protein and xenogeneic carbohydrate antigens in pigs. Another carbohydrate xenoantigen recognized as relevant target to xenotransplantation is a pentasaccharide oligosaccharide synthesized by the β1,4GalN-T2 gene. In the case of blood system, erythrocyte surface is covered by the blood group carbohydrates antigens [138]. The carbohydrate antigens of histo-blood group in fluids and tissues specify the individual immunodominant sugar phenotype. The erythrocyte surfaced carbohydrates are frequently expressed in other tissues and the antigens are alternatively termed "histo-blood group" antigenic substances since 1989 [143]. The carbohydrate antigens of histo-blood group show polymorphic phenotype with the "natural" Abs against the antigen [144]. The ABO blood group is representatively known and other non-blood group ABO systems are Duffy, Kelly, and Lewis, as specified in the immunophenotypes.

11.8.2 *β1,4GalNAcT-II Enzyme Specificity*

The Sd^a antigen, β1,4-linked GalNAc, is synthesized by glycosyltransferase, β1,4GalNT2. The enzyme name is referred to the HUGO (http://www.genenames.org/) to follow the officially termed symbols for the human genes. The current B4GALNT2 was previously called as the UDP-GalNAc:Neu5Acα2,3Galβ-R β1,4 N-GalNAc-T2 (abbreviated to β1,4GalNAc-T-II or GALG-T2). In Fig. 11.7, xenoantigenic modification of the precursor glycan NeuAcα2,3Galβ1,4GlcNAcβ1-R backbone has been illustrated to yield sLe^X and Sd^a antigen by fucosyl-T6 and β1-4 GalNAc-T2, respectively. Sd^a antigen is similar to blood group A antigen. Like α1,3-Gal, the pentasaccharide carbohydrate generated by the GALG-T2 enzyme is

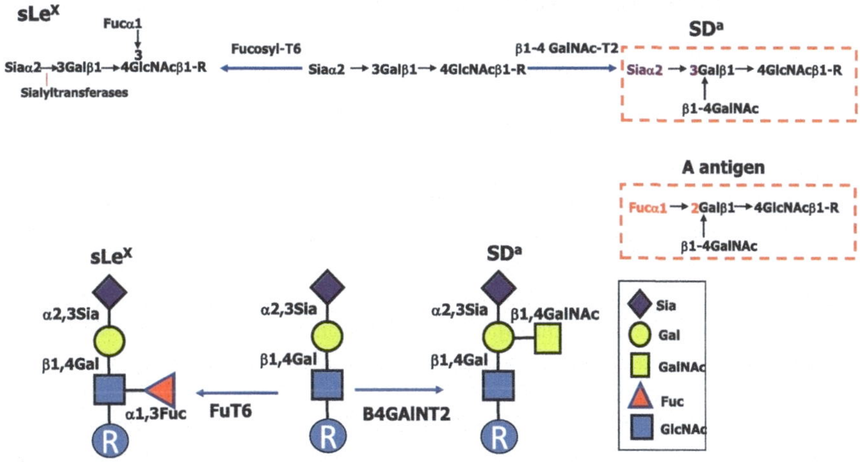

Fig. 11.7 Comparison of xenoantigenic modification. Precursor glycan NeuAcα2,3Galβ1,4GlcNAcβ1-R backbone is modified to sLe^X and Sd^a antigen by two different enzymes of fucosyltransferase-6 (FucT-6) and β1-4 GalNActransferase 2, respectively. Sd^a antigen is similar to blood group A antigen

absent in Old World monkey and human. Thus, the anti-β4GalN-T2 Abs in primates have been tested on transplanted grafts of pigs in the pig-to-NHP xenotransplantation. A candidate of non-β1,3Gal xenogeneic carbohydrate antigens was isolated in pigs using by expression screening approach using cDNA libraries [145], discovering cDNAs synthesized from endothelial cells of pigs. The cDNA-borne antigens of pigs bound to IgG antibody like such rejection antibodies obtained during pig-to-primate xenotransplantation. Among them, a pig β1,4GalNT2 gene homologous to the bovine β1,4GalNT2 was isolated. A β1,4GalNT2-catalyzed glycan has been regarded as the third xenoreactive antigen and the third xenoreactive glycan antigens are expressed on pig cell surfaces. However, humans and some NHPs have natural antibodies against the third xenoreactive antigen. The β1,4GalN-T2 gene expression is abundant in pig endothelial cells.

Transfection of pig β1,4GalN-T2 to human embryonic kidney (HEK) cells enhanced binding capacity with anti-Sd^a of the blood group SID producing a GalNAcβ1,4[Neu5Acα2,3]Galβ1,4GlcNAcβ1,3Gal terminal glycan and DBA as well as cytotoxic sensitivity to complement mediated lysis. Thus, pig β1,4GalNT2-produced new immunogenic non-β1,3Gal carbohydrates is regarded as the non-β1,3Gal xenoantigen in pig-to-primates xenotransplantation. Furthermore, β4GalN-T2-KO pigs negative for the Sd^a antigen expression are also now available [22], and construction of organ transplants obtained from double α1,3-Gal-T KO/β4GalNT2 KO pigs is under investigation in monkeys by a research group in the Emory University and University of Indiana, USA.

11.8.3 *Sd^{a+} Erythrocytic Agglutination by Anti-Sd^a Antibodies and Roles in Homing*

Sd^{a+} erythrocytes are agglutinated upon reaction with anti-Sd^a antibodies and the agglutinating capacity is the Sd^a activity [8]. The Sd^a antigens are expressed in milk, meconium, human saliva and urine. The Sd^{a-} antigen-negative individuals also express Sd^a substances in the urine or saliva [146]. The Sd^a antigens are also found in kidney and urine of animals such as guinea-pig. As a Sd^a antigen carrier, the urinary Tamm–Horsfall glycoprotein (THGP) is known [147]. In the structure aspect of the Sd^a/Cad antigens, the Sd^a antigen structure is GalNAcβ1,4 (Neu5Acα2,3)Galβ1,4GlcNAcβ1,3Gal structure. The Cad antigen was reported in 1968 in a Mauritian family [148]. GalNAc-recognizing lectins are found from *Dolichos biflorus* and *Helix pomatia*. These two agglutinins agglutinate the Cad erythrocytes by recognizing that Cad antigen is a GalNAc [149]. Because Cad antigens are reacted with anti-Sd^a antibodies [150], Cad antigen is regarded as a structurally similar carbohydrate antigen to Sd^a antigen. Then, consequently, Cad antigen was termed super Sid or Sd^{a++} antigen.

The glycophorin A-linked Cad structure belongs to a type 3 O-glycan structure of GalNAcβ1,4(Neu5Acα2,3)Galβ1,3(Neu5Acα2,6)GalNAc-Ser/Thr [151]. This is a potentially blocking inhibitor to inhibit the anti-Sd^a agglutination reaction, when Sd^a $^+$ erythrocyte cells are agglutinated by specific agglutinins. In Cad individuals carry the β1,4-linked GalNAc sialylparagloboside of GalNAcβ1,4(Neu5Acα2,3) Galβ1,4GlcNAcβ1,3Galβ1,4Glc-Cer structure [152]. Thus, the Sd^a/Cad-active structures carries the common structure of GalNAcβ1,4(Neu5Acα2,3)Gal-β-R. Interestingly, this trisaccharide is well known as a glycan chain of the GM2 of GalNAcβ1,4(Neu5Acα2,3)Galβ1,4Glc-Cer, while the Sd^a/Cad antigen-synthesizing GalNAc-transferase is not active to the GM2 synthesis from GM3 substrate [153]. Therefore, blood group Sd^a and Cad antigens carry the same antigen carbohydrate determinants on N-glycan or O-glycan glycoproteins and gangliosides such as long sialylparagloboside. In conclusion, Sd^a and Cad antigens are synthesized by the same enzymes. The B4GALNT2 as Sd^a synthase is a catalytic enzyme of transferring of GalNAc residue from donor UDP-GalNAc substrate to the acceptor *N*-glycoproteins substrate such as fetuin or THGP, but not to their asialoglycan substrates, suggesting that the α2,3-sialylated sugar chain is further attached with β1,4GalNAc]. GalNAc residue is transferred to a free α2,3-sialylglycans only, but not to α2,6-sialoglycan structure or even to the α2,3-sialylated gangliosides such as GM3. Sd^a synthase can use the O-linked structure Neu5Acα2,3Galβ1,3GalNAcol as substrate. The human kidney Sd^a synthase is not similar to the synthase of the guinea-pig Sd^a synthase as guinea-pig enzyme transfers GalNAc residue to acceptor glycophorin [154], while the Sd^a synthase enzyme of human uses a long sialylparagloboside, except for α2,3-sialylated GM3 [150].

Mouse cloned B4GalNT2 cDNA has a type II TM protein with 510 a.a, as commonly found as a type 2 Golgi-glycosyltransferase topologies, and a limited number of potential *N*-glycosylation site [155]. Like B4GALNT2 of other origins, the cloned recombinant mouse B4GalNT2 enzyme strictly required an α2,3-Sia linkage in the acceptor glycans. The B4GALNT2-encoding cDNA gene has been cloned from human in 2003 [47]. The B4GALNT2 gene located on chromosome 17q21.33 locus has 11 exons coded for the gene in human chromosome. The human B4GALNT2 gene has two different transmembrane peptides with a 566 aa and 506 aa cytoplasmic tails. Another trans-Golgi glycosyltransferase B4GALT1 acts toward α2,3-sialyl type 1 glycan of Galβ1,3GlcNAc, type 2 glycan of Galβ1,4GlcNAc and core 1 O-glycans of Galβ1,3GalNAc acceptor substrates [47]. The carbohydrate chains of the Cad/Sda glycan antigens or B4GALNT2 enzyme acceptor substrate are not simple. For example, a similar enzyme, B4GALNT1 (GM2 synthase) adds β1,4GalNAc residue to the Gal residue of GM3 ganglioside to form ganglioside GM2 [156]. Conclusively, B4GALNT2 needs the α2,3-sialylated glycan chains, while B4GALNT1 does not require the structure [156].

11.8.4 Mucin Sd^a Antigen of Gastrointestinal and Colon Cells

Expression of the Sd^a is found on core 3 *O*-glycoprotein glycans of mucin type in the descending colon from ileum to descending colon of human, but not found in colon mucins of human fetals. Sd^a antigenic expression is highly regulated during developmental stage and increasingly upregulated with age and life [157]. The mAbs reactive for Sd^a antigen and also ganglioside GM2 detect Sd^a antigen. The Sd^a roles and functions in healthy and normal colons are not elucidated yet for its biological meaning and importance. Sd^a level is decreased during oncodevelopment in human colon carcinoma cells and is completely negative in *O*-glycan structures of mucins present in gastrointestinal tract cancer tissues. In contrast, the level of the sialyl Lewis X or sLe^x of Neu5Aca2,3Galb1,4[Fucα1,3]GlcNAc structure known as a selectin ligand is increasingly expressed in the gastrointestinal track cancer tissues. The structures of the Sd^a and sLe^x glycans are commonly shared with together, and therefore, their *in vitro* synthesis of the two Sd^a and sLe^x glycan saccharide structures is known to be mutually exclusive. The cooperative action of Galactosyl-Transferases of B4GAL-T1 or B3GAL-T5 and *α*2,3-Sialyl-Transferases of ST3Gal-3, ST3Gal-4 or ST3Gal-6 produces a common glycan precursor as the enzyme substrate for of α1,3 and α1,4-fucosyl-Transferases (FUT-3, FUT-4, FUT-5, FUT-6, and FUT-7) and then produces the final product of sLe^x carbohydrate. If the carbohydrate precursor is used for B4GALN-T2 produces Sd^a. FUT-6 mainly generates sLe^x carbohydrate in colon. Expression of the B4GALN-T2 gene is regulated by DNA modification via DNA hypermethylation in tumor cells [158, 159] and influences to sLe^{x}/^{a} expression. Enforced B4GALN-T2 cDNA expression elicits Sd^a expression in gastrointestinal cells [42] with eliminated glycan ligands required for selectin binding and consequently decreases potential tumor metastasis [160].

B4GALNT1 and B4GALNT2 protein sequences share with together, indicating a high homology. Sd^a antigen is not found in colon mucins of human fetals, while the high Sd^a activity is observed in human newborn tissues including feces and saliva as well as urine. In the colonic crypt, the poorly differentiated cells highly express the B4GALN-T2 activity. In colonic crypts of human, the colon epithelial layer consists of a single columnar sheet of epithelial cells to form the functional unit of the intestinal crypt. The colon is made of numerous crypts and crypts have stem-cells made by the stem cells themselves in the stem-cell niche. Also, the crypts are surrounded with mesenchymal cells. Colon epithelial cells have four differentiated cell types, including colonocytes (or absorption cells), goblet cells for mucus secretion, endocrine cells for peptide hormone secretion, and Paneth cells. These all cells are originated from a single stem cell. In human colonic cancer, B4GALNT2 activity is down-regulated and Sd^a antigen is poorly expressed in colon cancer [161]. The monoclonal antibody KM531 known to recognize GM2 and Sd^a antigen [162].

B4GALNT2 acts to glycosylate sialylparagloboside [21] and not GM3, as B4GALNT1 (GM2 synthase) expression is inhibited in gastric cancer while is

Table 11.2 Sda glycan structures in the known carriers

Sd^a glycan structures	Sd^a carriers
1. GalNAcβ1,4(NeuAcα2,3) Galβ1,4GlcNAc β1,3Gal	Red blood cells glycophorin, Tamm-Horsfall glycoprotein (THGP)
2. GalNAcβ1,4(NeuAcα2,3)Galβ1,3 (NeuAcα2,6)	RBCs cold agglutinin disease (CAD)
GalNAc-Ser/Thr	
GalNAcβ1,4(NeuAcα2,3) Galβ1,4GlcNAcβ1,3	
Galβ1,4Glc-ceramide	
3. GalNAcβ1,4(NeuAcα2,3)Galβ1,4Glc-ceramide	Ganglioside GM2
4. GalNAcβ1,4(NeuAcα2,3)Galβ-R	SD^a

increased. The Sd^a antigen gene expression is controlled by many different regulators such as promoter methylation and two different coding proteins are synthesized. From the MUC2 and DBA lectin staining, it was suggested that B4galnt2 enzyme glycosylated the MUC2 protein [163]. B4galnt2 also glycosylates the Sd^a/Cad antigens in mucins of colon crypts, glycoproteins and glycolipids of colon crypts [164, 165]. Several Sda glycan structures have been known as the attached glycan chains to the known carriers (Table 11.2).

Functionally, such carbohydrate antigens upregulate the behaviors of differentiation and homing of small intestinal epithelial lymphocytic cells. This indicates that B4galnt2 expression induces formation of proper phenotype of endothelial cells [166]. The glycosylation of selectin receptors recruits CD3$^+$ cells and neutrophils [165]. In addition, the selectin receptor glycosylation also potentiates leukocyte infiltration [167]. The glycosylation of selectin receptors involves in synthesis of selectin-interacting carbohydrate ligands. In gastrointestinal cancers, B4galnt2 expression also eliminates or decreases metastatic dissemination in cell motility of the cancer cells, which caused by the roles of the Sd^a antigen [168].

Un colon cancers, the B4GALNT2 and Sd^a expression is rapidly reduced and instead, the selectin ligand sLex expression involved in metastasis is concomitantly elevated. The Sd^a antigen synthesis down-regulated in colon cancers consequently results in enhanced expression of sLe antigens. Reversely, sialyl Lewis x (sLe^x) expression is downregulated by B4GALNT2-expressed Sd^a production, because antigens of sialyl Lewis a (sLe^a) and sLe^x are fucosyl glycans. They are overexpressed in colon cancer due to the B4GALNT2, but not by cognate fucosyltransferases [169]. However, how B4GALNT2 expression is linked with the down regulation of sLex is not clearly concluded in tissues yet. The B4GALNT2 gene of human is alternatively transcribed into two transcripts of the long exon form known as an exon 253 nt-1L and the short exon form known as an exon 38 nt-1S, which are diverse in the first exon. Among the two transcript variants, normal and cancer tissues of colon as well as cell lines of colon exclusively express only the short exon form of exon 38 nt-1S, whereas the fibroblast CCD112CoN cell line of the embryonic colon express the long exon form of exon 253 nt-1L

[170]. Apart from the two transcripts, an additional variant form has been predicted by simulation by in silico as a third transcript. In tissues, B4GALNT2 protein isoforms induce Sd^a expression, but inhibit sLe^x expression.

Regarding the functional impact of B4galnt2 expression in endothelial cells, the B4galnt2-transferred GalNAc residues are implicated in immune responses of hosts with its biological potentials for immune cells' homing to the endothelium. In mice, Sd^a antigen regulates the life span of blood coagulation factor, von Willebrand factor (vWf) in circulatory system. Although B4galnt2 the expression is not broadly occurred, but tightly controlled by a tissue-specific expression in mice. If not controlled or ubiquitously expressed through the whole tissues, the mice are harmful and detrimental from the hemolysis. Theoretically, in mice, the tissue specific expression of B4galnt2 gene confer the differentially expressed switching role toward blood vessels from intestinal gut and the process is controlled by common allele [171]. This common allele is very important and called "modifier of von Willebrand factor-1 (MvWf-1) [172]. If vascular expression of B4galnt2 gene aberrantly glycosylate the vWf to form the β1,4GalNac-glycosylated glycosyl-vWF, this glycosyl-vWf protein will be rapidly cleared from blood circulation system [173]. MvWf-1 has been described for the first time in the RIIIS/J inbred mouse strain, and the RIIIS/J-like B4galnt2 alleles contributes to the gut B4galnt2 tissue-type switching expression, which is characteristic of epithelial phenotype, to blood vessel, which is characteristic of endothelial phenotype, as the commonly occurring genetic property found in population of wild-type mice [174]. The genetic variation in mice has been substantially maintained for long period more than million years, although this is detrimentally harmful to bleeding time, which is prolonged. But this is beneficial due to protection from host-pathogen interactions [175].

11.8.5 Sd^a Antigen Increases in the Cytotoxicity of Murine Cytotoxic T Lymphocytes

Since 1970, some mouse CTL populations of cytotoxic T cells produced a carbohydrate structure which is specifically recognized by GalNAc-binding lectins, where the GalNAc-binding lectin, VVL, is produced from plant *Vicia villosa* [176]. VVL binds preferentially to α- or β-linked terminal GalNAc carbohydrates such as α-GalNAc residue-Ser/Thr as the Tn antigen. The disaccharyl Galα1,3GalNAc is VVL inhibitor. The GalNAc-specific carbohydrate antigen is specific for CTL, but not other lymphocyte populations such as NK cells. In 1984, the mice CTL cell B6.1.SF.1 line named VV6 clone, which is resistant to VVL was isolated and the clone lacked β1,4-linked GalNAc on O-linked glycoprotein [177], due to GalNAc-transferase-deficiency [178]. Murin populations of CTL have the high GalNAc-transferring enzyme (GalNAc-T) activity. The MAbs CT1 and CT2 obtained from the CTL-reactive murine clone bind to antigenic glycan structures expressed in CTL and diminish cytotoxic lysis of target cells. CT1 and CT2 antibodies recognized Sia-bearing carbohydrate epitope [51, 69]. Both Abs of CT1 and CT2 recognize Tamm–Horsfall glycoprotein of Sd^a-positive individuals. In addition, the Abs of CT1 and CT2 also recognize erythrocytes of Cad individuals [179]. The CT antigen expression was associated with the CD-8 (Lyt-2) antigen-expressing cells, also showing its cytotoxicity.

11.8.6 B4GALNT2 and Sd^a Antigen in Prevention of the Muscle Pathology

Sd^a antigen biosynthetic enzyme, β1,4-GalNAc-T2 (GALGT2; B4GALNT2) overexpression improves muscle pathology in Duchenne muscular dystrophy (DMD). The overexpressed B4GALNT2 and Sd^a antigen prevents the muscle diseases and muscle pathologic pathies such as congenital muscular dystrophy 1A (MCD1A) raised by the laminin α2-chain gene (LAMA2) mutation, limb girdle MD 2D that is the SGCA gene's pathogenic variants-causing autosomal recessive disorder and DMD, as proved in in the disease models of mouse [65]. The mechanistic explanation is that muscle expression of GALGT2 induces the dystroglycan glycosylation with the glycan glycosylation of CTL. The enzyme activity enhances the dystrophin overexpression and surrogate laminin α2 chain overexpression, which are known as the disease inhibitors. It was reported that expression of GALGT2 gene dramatically decreases the muscle pathological level by mouse experiment of Thomas et al. [180]. Muscular overexpression of GALGT2 gene does not induce substantially dystroglycan glycosylation with the CTL glycans but increases dystrophin glycosylation and surrogate laminin α2 chain in matured skeletal myofibers.

Gene therapy using GALGT2 gene decreases uptake of serum IgG Abs into damaged and injured myofibers, protecting mutant muscles from injury. Muscular dystrophy (MD) is caused by mutations that impair the dystroglycan glycosylation with *O*-Man residue phosphate-linked carbohydrates. MD is congenitally diverse in its expression severity, caused by glycosylation gene mutation of a dystroglycan with *O*-Man phosphate-attached carbohydrates, which are essential for interaction between laminin and dystroglycan protein. For example, the known MD forms include muscle eye brain disease, Fukuyama congenital MD, congenital MD 1C and 1D to adult-onset limb girdle MDs (LGMDs) and Walker–Warburg syndrome (WWS). The MD-causing factor is in dystroglycan glycosylation problem, and thus, MD is called dystroglycanopathies. Dystroglycan of the dystrophin-associated glycoprotein in striated muscles is glycosylated into two α and β dystroglycan proteins to bind to one another. A TM β-dystroglycan protein is in the sarcolemmal membrane of muscle, whereas α-dystroglycan is not a TM protein but it is a peripherally-present membrane protein. Several proteins of agrin, laminin 211, and perlecan present on ECM interact with α dystroglycan, while the intracellular β-dystroglycan region interacts with cytoskeletal proteins such as plectin 1 and dystrophin utrophin, potentiating binding to filamentous actin and microtubules. Thus, dystroglycan is a linker of the membrane and the ECM, specifically known laminin 211 protein is a key basal lamina component surrounded by all skeletal myofiber via the membrane-actin cytoskeleton interaction. ECM proteins interact with the O-mannose (Man)-phosphate (O-Man-P)-linked oligosaccharides in its mucin-like domain of dystroglycan. Dystroglycan glycosyltransferase mutation causes for congenital MD or limb girdle MD form. The most severe MD includes muscle eye brain disease, Fukuyama congenital MD and WWS. Dystroglycanopathies influence developments of skeletal muscles and brain/eye in the cerebral cortex because of the glial limitans-pial basement membrane loss and retinal detachment. Fukutin-related protein (FKRP) gene mutation causes for dystroglycanopathies such as congenital MD 1C, LGMD2I, and WWS, due to α-dystroglycan and laminin binding problems. FKRP and FKRP-homologous fukutin (FKTN) mutations cause for congenital MD or LGMD. A dystroglycanopathy, named LGMD2I, is caused by FKRP mutations-derived abnormal dystroglycan glycosylation with the decreased O-Man-P-attached glycan level required for ECM interaction and decreased dystroglycan ribitol 5-phosphate (R-5-P) level synthesized by FKRP and fukutin. FKRP and fukutin bear R-5-P transferase activity, and a tandem R-5-P moiety is found in a laminin-binding glycan, dystroglycan; however, tandem R-5-P moiety is not present in the FKRP-domain, fukutin-domain, and isoprenoid synthase domain-null cells [180].

The overexpressed GALGT2 gene needs the native glycosylation of a dystroglycan because the native dystroglycan glycosylation is important for GALGT2 activity for its effective glycosylation. This is confirmed from the Chinese hamster ovary cells, which GALGT2 glycosylation only occurs when GlcNAc-T enzyme is highly expressed. Therefore, improper dystroglycan glycosylation results from FKRP deficiency causes for improper GALGT2-elicited molecular property in skeletal muscle. Glycolipid GALGT2 glycosylation also influence the MD extent. The glycolipid GALGT2 substrate is a sialylpentosylceramide. A dystroglycan exhibits multi-glycosylated forms in various tissues. Changes in dystroglycan glycosylation can be used for a therapeutic factor of GALGT2. Overexpressed GALGT2 enzyme suppresses the muscle pathological severity in the FKRP P448Lneo-mutant model for LGMD2I despite a FKRP-deficient glycosylation. GALGT2 gene therapy can decrease serum IgG capture to injured myofiber damages. This protects FKRP mutant muscles from muscle injured damages. However, GALGT2 gene expression and number of rAAVrh74.MCK.GALGT2 vgs in muscle are decreased within 6-month of the myofiber expression loss. GALGT2 overexpression can suppress disease pathology. GALGT2-edited FKRP P448Lneo muscles increase muscle regenerating myofiber number and muscle regeneration. GALGT2 expression can enhance muscle regeneration in dystrophic muscles. Endogenous GALGT2 expression in muscle is increased after acute injuries when embryonic myosin is elevated. GALGT2-defected muscles exhibit the decreased growth of muscle upon acute muscle injuries and damages. Dystroglycan glycosylation loss is found in embryonic muscle development and reduces laminin a2 expression. A therapy regenerates and protects mature myofibers. Knock-in model of mouse of FKRP mutation of human exhibits muscle disease pathology and dysfunction [181]. FKRP P448L mutation induces severe muscle pathology with a similar phenotype to LGMD2I, which exhibit adult muscle necrosis, inflammation, fibrosis, muscle damage, regeneration and muscle wasting. The mice model resembles human LGMD2I in terms of the decreased dystroglycan glycosylation in skeletal and heart muscles.

Recently, GALGT2 gene therapy has been recognized to treat the dystroglycan-defective diseases. For example, muscle pathology was decreased in FKRP P448Lneo-negative mice known as a limb girdle muscular dystrophy 2I model by GALGT2 gene therapy [180]. rAAVrh74.MCK.GALGT2-tested FKRP P448Lneo-negative muscles had nucleated myofiber reduction and myofiber IgG uptake with myofiber diameter enhancement. Dystroglycan glycosylation was not caused by overexpression of GALGT2 gene in FKRP P448Lneo-negative muscles in mature skeletal myofibers, giving the cytotoxic T-cell glycan or enhanced dystrophin and laminin a2 surrogate expression. The FKRP P448Lneo-mouse model [181] was tested for a surrogate gene therapy. rAAVrh74.MCK.GALGT2 inhibits muscle pathology in the dyW mouse model for congenital MD 1A, Duchenne MD model and the Sgca/mice model for LGMD2D. Overexpression of GALGT2 gene glycosylates a dystroglycan in skeletal muscle and because the GALGT2 needs the dystroglycan expression. GALGT2 overexpression suppresses the muscle pathology progression and development in FKRP. GALGT2 therapy of limb muscles improves the muscle degeneration extent and increases regeneration cycle as central nucleation reduction of myofibers and myofiber size improvement.

O-Man phosphate-linked carbohydrates in dystroglycan protein are crucial for binding of laminin molecules. The problem is in glycosylation of a dystroglycan and MD is therefore called dystroglycanopathies. Dystroglycans of two α- and β-dystroglycan proteins are a key molecule of the muscular dystrophin-associated glycoproteins [181]. They bind noncovalently to each other. β-Dystroglycan molecule is a typical transmembrane glycoprotein present in the sarcolemmal membrane of muscle. In contrast, α-dystroglycan molecule is a glycoprotein present in peripheral membranes. ECM proteins of plasma membrane recognize a dystroglycan molecule via the O-Man-phosphate-linked carbohydrates attached to its mucin-like domains, and genetic mutation in dystroglycan glycosylation leads to MD. Abnormal glycosylation reduces *O*-Man-phosphate-linked carbohydrate expression important for binding of ECM to dystroglycans [182]. Therefore, dystroglycan-glycosylase enzyme can prevent muscle pathology. Such B4GALNT2 expression in adult skeletal synthesizes the CTL glycan structure of the GalNAcβ1,4[Neu5Ac/Galα2,3]-Galβ1,4GlcNAcβ-glycans [183].

The CTL carbohydrate structure was firstly discovered in $CD8^+$ T cells as a CTL population and is now elucidated to express in various organs of human including colon tissue [138]. Overexpression of GALGT2 prevents the disease development of MD-related muscle disease of the Duchenne MD mouse model, the congenital MD 1A model of dyW mouse and the LGMD2D model of $Sgca^{-/-}$ mouse [184]. Overexpression of GALGT2 gene enhances the α-dystroglycan glycosylation with the normal level of CTL glycans, also and overexpression of GALGT2 gene enhances the ectopic overexpression of dystrophin and surrogate laminin α2 molecules, which normally recognize dystroglycan molecule in mice and in macaques, preventing muscle injury and muscle pathology for most the entire animal life span. In addition, it was known that GALGT2 gene-transgenic overexpression transiently prevents muscle disease in dystroglycan-deleted mice, suggesting the central role of GALGT2 glycosylation to dystroglycan. GALGT2 expression protects dystroglycan-dysfunction disease when regular glycosylation with O-Man-phosphate-linked carbohydrates is impaired.

11.8.7 Creation of CRISPR9/Cas-9-Based Triple α1,3Gal-T, CMAH, and β4GalN-T2 Triple KO Pigs

Using the recent CRISPR9 method, the triple α1,3Gal-T, CMAH, and β4GalN-T2 triple genes-deleting pig cells have been created. The binding levels of the pig-derived myeloid cells with naturally preformed Abs of human, containing IgG and IgM Abs, are largely decreased. The xenoreactivity of the triple α1,3Gal-T, CMAH, and β4GalN-T2-deleting pig cells was much lowered rather than the α1,3Gal-T and Neu5Gc-deleting cells. Thus, β4GalNT2-derived glycan antigen has also been suggested as the third xenoreactive antigen. A recent report elucidated binding level of Abs from waitlisted recipient patients of renal transplanted grafts to the above porcine KO cells of triple glycan deficiency using the CRISPR/Cas-9 technology to generate pigs with triple KO (TKO) and pig MHC-class I SLA [185]. PBMCs obtained from SLA-identical WT, GGTA1 KO pigs, GGTA1/CMAH double KO pigs, and GGTA1/CMAH/B4GalN-T2 TKO pigs were screened for binding levels of human Abs using 820 patients-obtained sera. The GGTA1/CMAH/B4GalN-T2 TKO cells were screened using WT pig RBCs-deleted sera to remove Abs reactive against cell surfaces but leave Abs reactive for SLA molecules. Moreover, Abs reactive for humans and pigs were also examined for tests to know the cross-species recognition level and cross-species reaction level to single-antigen HLA molecule. Such genetic modification or deletion of KO pigs dramatically decrease antibody recognition level of waitlisted recipients of human patients. Tests using the serum also reduced recognition level of SLA class I KO cells. Xenoantigen-binding Abs recognition of the triple gene TKO pigs was reduced. But certain HLA-specific Abs in sensitized patient recipients easily cross-react with MHC-class I SLA molecule. Thus, as a MHC class-I molecule, pig SLA class I should be a targeting candidate for CRISPR/Cas-9 genome editing technology in future xenotransplantation. The SLA class-I-deleted KO mutant of pigs can be created with the pig SLA I-deleted locus transgene with patient's gene of own HLA allele, generating patient HLA allele-inserted single class I HLA match. In fact, such engineered cells are generated and removed anti-MHC-class I-specific Abs recognition in pig cross-matches, but pigs with these cells were not yet created [97].

11.8.8 *Intestinal Mucosal B4galnt2-Synthesized Glycans and Microbial Resistance*

Host intestinal mucosal glycans lined with the luminal surfaces attach to some microorganisms. The surfaced mucus layer glycans exert function as the primary barrier of microbial habitats. B4galnt2-synthesized glycan structures are diverse. If B4galnt2 expression is lost, the interactive behaviors of host autogenic cells and parasitic microbes are altered [163]. B4galnt2-specific host glycosylation influences microbial resistance. The carbohydrate blood group antigens are co-evolved with hosts and their pathogens as α-1,2-fucosyltransferase (FUT2) synthesizes the H antigen in secreted fluids and tissues. FUT2 defection loses ABO and H type glycans, referring the "nonsecretor" phenotype. Nonsecretor status is found in inflammatory Crohn's disease and sclerosing cholangitis. Epithelial glycosylation protects immune cells from infection. Glycan mutations give benefits for pathogens, For example, the *Helicobacter* strains recognize the gastric mucosal Lewis antigens [186]. The blood group B4galnt2 is mice-specific. It synthesizes a Sd(a) antigen similar to A blood type and involved in intestinal microbiota. The glycan profile can be changed in B4galnt2-lacking animas or non-expressing animals. β1,4-GalNac residues/Sd(a) deficiency changes in animal glycans for bacteria or immune cells. Intestinal mucin blood group α-1,2 fucosylglycan receptors are used for the binding and attaching sites for bacterial Salmonella species, although Salmonella bacteria can not directly recognize and interact with B4galnt2-GalNAc residues. B4galnt2 expression alter pathogen infection susceptibility and commensal microbial habitation. B4galnt2 gene-engineered mice exhibited the specific bacterial composition and high Salmonella infection. The epithelial B4galnt2-dependent microbiota also increased intestinal inflammation and invasion of *S. typhimurium*. B4galnt2-expression-dependent intestinal microbiota are species and phylogenetic diverse. Intestinal mucus glycans and carbohydrates confer benefit for commensal gut bacteria. The mucosal B4galnt2 expression indicator species were *Prevotella* and *Bacteroide* for glycan degradation [187]. *Turicibacter sanguinis* species is used as B4galnt2-deficient mice indicator in the gut microbiomes of mice and human [188]. Similarly, *Barnesiella* are present in the absence of B4galnt2 glycans [189]. B4galnt2-deficient mice are less pathologic upon *S. typhimurium* infection.

References

1. Diswall M, et al. Structural characterization of alpha1,3-galactosyltransferase knockout pig heart and kidney glycolipids and their reactivity with human and baboon antibodies. Xenotransplantation. 2010;17:48–60.
2. Fischer K. Efficient production of multi-modified pigs for xenotransplantation by 'combineering', gene stacking and gene editing. Sci Rep. 2016;6:29081. https://doi.org/10.1038/srep29081.

3. Platt JL, et al. Immunopathology of hyperacute xenograft rejection in a swine-to-primate model. Transplantation. 1991;52:214–20.
4. Byrne GW, McGregor CG, Breimer ME. Recent investigations into pig antigen and anti-pig antibody expression. Int J Surg. 2015;23(Pt B):223–8.
5. Ashton-Chess J, et al. The effect of immunoglobulin immunadsorptions on delayed xenograft rejection of human CD55 transgenic pig kidneys in baboons. Xenotransplantation. 2003;10: 552–61.
6. Zhang J, Xie C, Lu Y, Zhou M, Qu Z, Yao D, Qiu C, Xu J, Pan D, Dai Y, Hara H, Cooper DKC, Ma S, Li M, Cai Z, Mou L. Potential antigens involved in delayed xenograft rejection in a Ggta1/Cmah Dko pig-to-monkey model. Sci Rep. 2017;7(1):10024. https://doi.org/10.1038/s41598-017-10805-0.
7. Ahrens HE, et al. siRNA mediated knockdown of tissue factor expression in pigs for xenotransplantation. Am J Transplant. 2015;15:1407–14. https://doi.org/10.1111/ajt.13120.
8. Oropeza M, et al. Transgenic expression of the human A20 gene in cloned pigs provides protection against apoptotic and inflammatory stimuli. Xenotransplantation. 2009;16:522–34.
9. Petersen B, et al. Transgenic expression of human heme oxygenase-1 in pigs confers resistance against xenograft rejection during ex vivo perfusion of porcine kidneys. Xenotransplantation. 2011;18:355–68.
10. Chen G, Sun H, Yang H, et al. The role of anti-non-Gal antibodies in the development of acute humoral xenograft rejection of hDAF transgenic porcine kidneys in baboons receiving anti-Gal antibody neutralization therapy. Transplantation. 2006;81:273–83.
11. Baumann BC, Stussi G, Huggel K, Rieben R, Seebach JD. Reactivity of human natural antibodies to endothelial cells from Galalpha(1,3)Gal-deficient pigs. Transplantation. 2007;83:193–201.
12. Ezzelarab M, Ayares D, Cooper DK. Carbohydrates in xenotransplantation. Immunol Cell Biol. 2005;83:396–404.
13. Blixt O, Kumagai-Braesch M, Tibell A, Groth CG, Holgersson J. Anticarbohydrate antibody repertoires in patients transplanted with fetal pig islets revealed by glycan arrays. Am J Transplant. 2009;9:83–90.
14. Yeh P, Ezzelarab M, Bovin N, et al. Investigation of potential carbohydrate antigen targets for human and baboon antibodies. Xenotransplantation. 2010;17:197–206.
15. Park SJ, et al. Production and characterization of soluble human TNFRI-Fc and human HO-1 (HMOX1) transgenic pigs by using the F2A peptide. Transgenic Res. 2014;23:407–19.
16. Kim GA. Generation of CMAHKO/GTKO/shTNFRI-Fc/HO-1quadruple gene modified pigs. Transgenic Res. 2017;26:435–45.
17. Vadori M, Cozzi E. The immunological barriers to xenotransplantation. Tissue Antigens. 2015;86(4):239–53.
18. Zhang Z, Gao B, Zhao C, Long C, Qi H, Cooper DKC, Hara H. The impact of serum incubation time on IgM/IgG binding to porcine aortic endothelial cells. Xenotransplantation. 2017;24:e12312.
19. Harnden I, Kiernan K, Kearns-Jonker M. The anti-nonGal xenoantibody response to alpha1,3-galactosyltransferase gene knockout pig xenografts. Curr Opin Organ Transplant. 2010;15(2): 207–11.
20. Knosalla C, et al. Renal and cardiac endothelial heterogeneity impact acute vascular rejection in pig-to-baboon xenotransplantation. Am J Transplant. 2009;9:1006–16. https://doi.org/10.1111/j.1600-6143.2009.02602.x.
21. Kuwaki K, et al. Heart transplantation in baboons using alpha1,3-galactosyltransferase gene-knockout pigs as donors: initial experience. Nat Med. 2005;11:29–31. https://doi.org/10.1038/nm1171.
22. Estrada JL, Martens G, Li P, et al. Evaluation of human and non-human primate antibody binding to pig cells lacking GGTA1/CMAH/β4GalNT2 genes. Xenotransplantation. 2015;22: 194–202.

23. Breimer ME. Gal/non-Gal antigens in pig tissues and human non-Gal antibodies in the GalT-KO era. Xenotransplantation. 2011;18:215–28.
24. Landsteiner K, Levine P. Further observations on individual differences of human blood. Exp Biol Med. 1927;24(9):941–2. https://doi.org/10.3181/00379727-24-3649.
25. Kaczmarek R, Buczkowska A, Mikolajewicz K, Krotkiewski H, Czerwinski M. P1PK, GLOB, and FORS blood group systems and GLOB collection: biochemical and clinical aspects. Do we understand it all yet? Transfus Med Rev. 2014;28(3):126–36. https://doi.org/10.1016/j.tmrv.2014.04.007.
26. Hellberg A. P1PK: a blood group system with an identity crisis. ISBT Sci Ser. 2020;15:40–5. https://doi.org/10.1111/voxs.12505.
27. Mikolajczyk K, Sikora M, Hanus C, Kaczmarek R, Czerwinski M. One of the two N-glycans on the human Gb3/CD77 synthase is essential for its activity and allosterically regulates its function. Biochem Biophys Res Commun. 2022;617(Pt 1):36–41. https://doi.org/10.1016/j.bbrc.2022.05.085.
28. Yeh CC, Chang CJ, Twu YC, Hung ST, Tsai YJ, Liao JC, Huang JT, Kao YH, Lin SW, Yu LC. The differential expression of the blood group P1-A4GALT and P2-A4GALT alleles is stimulated by the transcription factor early growth response 1. Transfusion. 2018;58(4):1054–64. https://doi.org/10.1111/trf.14515.
29. Thinley J, Nathalang O, Chidtrakoon S, Intharanut K. Blood group P1 prediction using multiplex PCR genotyping of A4GALT among Thai blood donors. Transfus Med. 2021;31(1):48–54. https://doi.org/10.1111/tme.12749.
30. Duk M, Kusnierz-Alejska G, Korchagina EY, Bovin NV, Bochenek S, Lisowska E. Anti-α-galactosyl antibodies recognizing epitopes terminating with α1,4-linked galactose: human natural and mouse monoclonal anti-NOR and anti-P1 antibodies. Glycobiology. 2005;15(2):109–18. https://doi.org/10.1093/oxfordjournals.glycob.a034964.
31. Yu LC, Twu YC, Chang CY, Lin M. Molecular basis of the adult I phenotype and the gene responsible for the expression of the human blood group I antigen. Blood. 2001;98(13):3840–5. https://doi.org/10.1182/blood.v98.13.3840.
32. Yu LC, Lin M. Molecular genetics of the blood group I system and the regulation of I antigen expression during erythropoiesis and granulopoiesis. Curr Opin Hematol. 2011;18(6):421–6.
33. Reid ME. The gene encoding the I blood group antigen: review of an I for an eye. Immunohematology. 2020;20(4):249–52.
34. Marsh WL. Anti-i: a cold antibody defining the Ii relationship in human red cells. Br J Haematol. 1961;7:200–9.
35. Feizi T. Demonstration by monoclonal antibodies that carbohydrate structures of glycoproteins and glycolipids are onco-developmental antigens. Nature. 1985;314(6006):53–7.
36. I antigens. Identification of a UDP-GlcNAc:GlcNAc beta 1-3Gal(−R) beta 1-6(GlcNAc to Gal) N-acetylglucosaminyltransferase in hog gastric mucosa. J Biol Chem. 1984;259(21):13385–90.
37. Van Poll D, Nahmias Y, Soto-Gutierrez A, et al. Human immune reactivity against liver sinusoidal endothelial cells from GalTalpha(1,3)GalT-deficient pigs. Cell Transplant. 2010;19:783–9.
38. Shimizu A, Hisashi Y, Kuwaki K, et al. Thrombotic microangiopathy associated with humoral rejection of cardiac xenografts from alpha1,3-galactosyltransferase gene-knockout pigs in baboons. Am J Pathol. 2008;172:1471–81.
39. Oriol R, Ye Y, Koren E, et al. Carbohydrate antigens of pig tissues reacting with human natural antibodies as potential targets for hyperacute vascular rejection in pig-to-man organ xenotransplantation. Transplantation. 1993;56:1433–42.
40. Bouhours D, Liaigre J, Naulet J, et al. A novel pentaglycosylceramide in ostrich liver, IV4-beta-Gal-nLc4Cer, with terminal Gal(beta1-4)Gal, a xenoepitope recognized by human natural antibodies. Glycobiology. 2000;10:857–64.

41. Hallberg EC, Holgersson J, Samuelsson BE. Glycosphingolipid expression in pig aorta: identification of possible target antigens for human natural antibodies. Glycobiology. 1998;8:637–49.
42. Kim YG, Gil GC, Harvey DJ, et al. Structural analysis of alpha-Gal and new non-Gal carbohydrate epitopes from specific pathogen-free miniature pig kidney. Proteomics. 2008;8:2596–610.
43. Zhu A. Binding of human natural antibodies to nonalphaGal xenoantigens on porcine erythrocytes. Transplantation. 2000;69:2422–8.
44. Buhler L, Xu Y, Li W, et al. An investigation of the specificity of induced antipig antibodies in baboons. Xenotransplantation. 2003;10:88–93.
45. Komoda H, Miyagawa S, Kubo T, et al. A study of the xenoantigenicity of adult pig islets cells. Xenotransplantation. 2004;11:237–46.
46. Byrne GW, Stalboerger PG, Davila E, et al. Proteomic identification of non-Gal antibody targets after pig-to-primate cardiac xenotransplantation. Xenotransplantation. 2008;15:268–76.
47. Montiel MD, Krzewinski-Recchi MA, Delannoy P, et al. Molecular cloning, gene organization and expression of the human UDP-GalNAc: Neu5Acalpha2-3Galbeta-R beta1,4-N-acetylgalactosaminyltransferase responsible for the biosynthesis of the blood group Sda/Cad antigen: evidence for an unusual extended cytoplasmic domain. Biochem J. 2003;373:369–79.
48. Miyagawa S, Ueno T, Nagashima H, Takama Y, Fukuzawa M. Carbohydrate antigens. Curr Opin Organ Transplant. 2012;17(2):174–9.
49. Komoda H, Miyagawa S, Omori T, Takahagi Y, Murakami H, Shigehisa T, Ito T, Matsuda H, Shirakura R. Survival of adult islet grafts from transgenic pigs with N-acetylglucosaminyltransferase-III (GnT-III) in cynomolgus monkeys. Xenotransplantation. 2005;12(3):209–16.
50. Chung TW, Kim KS, Kim CH. Reduction of the Gal-alpha1,3-Gal epitope of mouse endothelial cells by transfection with the N-acetylglucosaminyltransferase III gene. Mol Cells. 2003;16(3):368–76.
51. Diswall M, Gustafsson A, Holgersson J, et al. Antigen-binding specificity of anti-αGal reagents determined by solid-phase glycolipid-binding assays. A complete lack of αGal glycolipid reactivity in α1,3GalT-KO pig small intestine. Xenotransplantation. 2011;18:28–39.
52. Song KH, Kim CH. Sialo-Xenoantigenic glycobiology: molecular glycobiology of sialylglycan-xenoantigenic determinants in pig to human xenotransplantation. Heidelberg, New York, Dordrecht, London. USSN 2195-3546: Springer; 2012. https://doi.org/10.1007/978-34094-9.
53. Keusch JJ, Manzella SM, Nyame KA, et al. Expression cloning of a new member of the ABO blood group glycosyltransferases, iGb3 synthase, that directs the synthesis of isoglobo-glycosphingolipids. J Biol Chem. 2000;275:25308–14.
54. Gusstafsson A, Ayares DL, Cooper DKC, et al. Carbohydrate phenotyping of endothelial cell glycoproteins from alpha1,3galactosyl-transferase gene knockout and wild type pigs. Xenotransplantation. 2005;12:375.
55. Taylor SG, Mckenzie IF, Sandrin MS. Characterization of the rat alpha(1,3) galactosyltransferase: evidence for two independent genes encoding glycosyltransferases that synthesize Galalpha(1,3)Gal by two separate glycosylation pathways. Glycobiology. 2003;13:327–37.
56. Li Y, Thapa P, Hawke D, et al. Immunologic glycosphingolipidomics and NKT cell development in mouse thymus. J Proteome Res. 2009;8:2740–51.
57. Puga Yung GL, Li Y, Borsig L, Millard A-L, Karpova MB, Zhou D, Seebach JD. Complete absence of the aGal xenoantigen and isoglobotrihexosylceramide in a 1,3galactosyltransferase knock-out pigs. Xenotransplantation. 2012;19:196–206.

58. Baumann BC, Forte P, Hawley RJ, Rieben R, Schneider MK, Seebach JD. Lack of galactose-alpha-1,3-galactose expression on porcine endothelial cells prevents complement-induced lysis but not direct xenogeneic NK cytotoxicity. J Immunol. 2004;172:6460–7.
59. Zhou D, Levery SB. Response to Milland et al.: carbohydrate residues downstream of the terminal Galalpha(1,3)Gal epitope modulate the specificity of xenoreactive antibodies. Immunol Cell Biol. 2008;86:631–2. author reply 633-634.
60. Pierson RN III, Dorling A, Ayares D, et al. Current status of xenotransplantation and prospects for clinical application. Xenotransplantation. 2009;16:263–80.
61. Milland J, Christiansen D, Lazarus BD, Taylor SG, Xing PX, Sandrin MS. The molecular basis for galalpha(1,3)gal expression in animals with a deletion of the alpha1,3galactosyltransferase gene. J Immunol. 2006;176(4):2448–54.
62. Sandrin IF, Christiansen D, Milland J. The impact of the α1,3-galactosyltransferase gene knockout pig on xenotransplantation. Curr Opin Organ Transplant. 2007;2:154–7.
63. Christiansen D, Milland J, Mouhtouris E, et al. Humans lack iGb3 due to the absence of functional iGb3-synthase: implications for NKT cell development and transplantation. PLoS Biol. 2008;6:e172.
64. Speak AO, Salio M, Neville DC, Fontaine J, Priestman DA, Platt N, et al. Implications for invariant natural killer T cell ligands due to the restricted presence of isoglobotrihexosylceramide in mammals. Proc Natl Acad Sci U S A. 2007;104(14):5971–6.
65. Tahiri F, Li Y, Hawke D, Ganiko L, Almeida I, Levery S, et al. Lack of iGb3 and isoglobo-series glycosphingolipids in pig organs used for xenotransplantation: implications for natural killer T-cell biology. J Carbohydr Chem. 2013;32(1):44–67.
66. Murray-Segal L, Gock H, Cowan PJ, D'apice AJ. Anti-Gal antibody-mediated skin graft rejection requires a threshold level of Gal expression. Xenotransplantation. 2008;15:20–6.
67. Slomiany BL, Slomiany A, Horowitz MI. Characterization of blood-group-H-active ceramide tetrasaccharide from hog-stomach mucosa. Eur J Biochem. 1974;43:161–5.
68. Hara H, Long C, Lin YJ, et al. In vitro investigation of pig cells for resistance to human antibody-mediated rejection. Transpl Int. 2008;21:1163–74.
69. Christiansen D, Milland J, Mouhtouris E, Vaughan H, Pellicci DG. McConville of isoglobotrihexosylceramide in mammals. PNAS. 2007;104(14):5971–6.
70. Kiernan K, Harnden I, Gunthart M, Gregory C, Meisner J, Kearns-Jonker M. The anti-non-gal xenoantibody response to xenoantigens on gal knockout pig cells is encoded by a restricted number of germline progenitors. Am J Transplant. 2008;8(9):1829–39.
71. Dor FJ, Tseng YL, Cheng J, Moran K, Sanderson TM, Lancos CJ, Shimizu A, Yamada K, Awwad M, Sachs DH, Hawley RJ, Schuurman HJ, Cooper DK. alpha1,3-galactosyltransferase geneknockout miniature swine produce natural cytotoxic anti-Gal antibodies. Transplantation. 2004;78(1):15–20.
72. Brew K, Vanaman TC, Hill RL. The role of alpha-lactalbumin and the A protein in lactose synthetase: a unique mechanism for the control of a biological reaction. Proc Natl Acad Sci U S A. 1968;59(2):491–7.
73. Wang Y, Ju T, Ding X, Xia B, Wang W, Xia L, He M, Cummings RD. Cosmc is an essential chaperone for correct protein O-glycosylation. Proc Natl Acad Sci U S A. 2010;107(20):9228–33.
74. Yamaji T, Nishikawa K, Hanada K. Transmembrane BAX inhibitor motif containing (TMBIM) family proteins perturbs a trans-Golgi network enzyme, Gb3 synthase, and reduces Gb3 biosynthesis. J Biol Chem. 2010;285(46):35505–18.
75. Darmoise A, Teneberg S, Bouzonville L, Brady RO, Beck M, Kaufmann SH, Winau F. Lysosomal alpha-galactosidase controls the generation of self lipid antigens for natural killer T cells. Immunity. 2010;33(2):216–28.
76. Kleene R, Berger EG. The molecular and cell biology of glycosyltransferases. Biochim Biophys Acta. 1993;1154(3–4):283–325. Review
77. Porubsky S, Luckow B, Bonrouhi M, Speak A, Cerundolo V, Platt F, Gröne HJ. Glycosphingolipids Gb3 and iGb3. In vivo roles in hemolytic-uremic syndrome and

iNKT cell function. Pathologe. 2008;29 Suppl 2:297–302. https://doi.org/10.1007/s00292-008-1040-0. German.
78. Lescar J, Loris R, Mitchell E, Gautier C, Chazalet V, Cox V, et al. Isolectins I-A and I-B of Griffonia (Bandeiraea) simplicifolia. Crystal structure of metal-free GS I-B(4) and molecular basis for metal binding and monosaccharide specificity. J Biol Chem. 2002;277:6608–14.
79. Milland J, Yuriev E, Xing PX, McKenzie IF, Ramsland PA, Sandrin MS. Carbohydrate residues downstream of the terminal Galalpha(1,3)Gal epitope modulate the specificity of xenoreactive antibodies. Immunol Cell Biol. 2007;85(8):623–32.
80. Ogiso M, Nishiyama I, Saito N, Okinaga T, Hoshi M, Komoto M. Localization of neutral and acidic glycosphingolipids in rat lens. Glycobiology. 1995;5:187–94.
81. Blanken WM, Van den Eijnden DH. Biosynthesis of terminal Gal alpha 1–3Gal beta 1--4GlcNAc-R oligosaccharide sequences on glycoconjugates. Purification and acceptor specificity of a UDP-Gal:N-acetyllactosaminide alpha 1–3-galactosyltransferase from calf thymus. J Biol Chem. 1985;260:12927–34.
82. Kronenberg M, Gapin L. Natural killer T cells: know thyself. Proc Natl Acad Sci U S A. 2007;104(14):5713–4. https://doi.org/10.1073/pnas.0701493104.
83. Schumann J, Mycko MP, Dellabona P, Casorati G, MacDonald HR. Cutting edge: influence of the TCR Vbeta domain on the selection of semi-invariant NKT cells by endogenous ligands. J Immunol. 2006;176(4):2064–8.
84. Pellicci DG, Clarke AJ, Patel O, Mallevaey T, Beddoe T, Le Nours J, Uldrich AP, McCluskey J, Besra GS, Porcelli SA, Gapin L, Godfrey DI, Rossjohn J. Recognition of β-linked self glycolipids mediated by natural killer T cell antigen receptors. Nat Immunol. 2011;12(9):827–33.
85. Cheng JM, Dangerfield EM, Timmer MS, Stocker BL. A divergent approach to the synthesis of iGb3 sugar and lipid analogues via a lactosyl 2-azido-sphingosine intermediate. Org Biomol Chem. 2014;12(17):2729–36.
86. Dias BR, Rodrigues EG, Nimrichter L, Nakayasu ES, Almeida IC, Travassos LR. Identification of iGb3 and iGb4 in melanoma B16F10-Nex2 cells and the iNKT cell-mediated antitumor effect of dendritic cells primed with iGb3. Mol Cancer. 2009;8:116.
87. Godfrey DI, MacDonald HR, Kronenberg M, Smyth MJ, Van Kaer L. NKT cells: what's in a name. Nat Rev Immunol. 2004;4:231–7.
88. Godfrey DI, Kronenberg M. Going both ways: immune regulation via CD1d-dependent NKT cells. J Clin Invest. 2004;114:1379–88.
89. Morita M, Motoki K, Akimoto K, Natori T, Sakai T, et al. Structure-activity relationship of alpha-galactosylceramides against B16-bearing mice. J Med Chem. 1995;38:2176–87.
90. Brossay L, Chioda M, Burdin N, Koezuka Y, Casorati G, et al. CD1d-mediated recognition of an alpha-galactosylceramide by natural killer T cells is highly conserved through mammalian evolution. J Exp Med. 1998;188:1521–8.
91. Zhou D, Mattner J, Cantu C 3rd, Schrantz N, Yin N, et al. Lysosomal glycosphingolipid recognition by NKT cells. Science. 2004;306:1786–9.
92. Cheng L, Ueno A, Cho S, Im JS, Golby S, et al. Efficient activation of valpha14 invariant NKT cells by foreign lipid antigen is associated with concurrent dendritic cell-specific self recognition. J Immunol. 2007;178:2755–62.
93. Porubsky S, Speak AO, Luckow B, Cerundolo V, Platt FM, et al. Normal development and function of invariant natural killer T cells in mice with isoglobotrihexosylceramide (iGb3) deficiency. Proc Natl Acad Sci U S A. 2007;104:5977–82.
94. Schumann J, Facciotti F, Panza L, Michieletti M, Compostella F, et al. Differential alteration of lipid antigen presentation to NKT cells due to imbalances in lipid metabolism. Eur J Immunol. 2007;37:1431–41.
95. Butler JR, Skill NJ, Priestman DL, Platt FM, Li P, Estrada JL, et al. Silencing the porcine iGb3s gene does not affect Galα3Gal levels or measures of anticipated pig to human and pig to primate acute rejection. Xenotransplantation. 2016;23(2):106–16.

96. Lutz AJ, Li P, Estrada JL, Sidner RA, Chihara RK, Downey SM, Burlak C, Wang ZY, Reyes LM, Ivary B, Yin F, Blankenship RL, Paris LL, Tector AJ. Double knockout pigs deficient in N-glycolylneuraminic acid and galactose α-1,3-galactose reduce the humoral barrier to xenotransplantation. Xenotransplantation. 2013;20(1):27–35.
97. Yung GP, Schneider MK, Seebach JD. Immune responses to α1,3 galactosyltransferase knockout pigs. Curr Opin Organ Transplant. 2009;14:154–60.
98. Sanderson JP, Brennan PJ, Mansour S, et al. CD1d protein structure determines species-selective antigenicity of isoglobotrihexosylceramide (iGb3) to invariant NKT cells. Eur J Immunol. 2013;43:815–25.
99. Butler JR, Skill NJ, Priestman DL, Platt FM, Li P, Estrada JL, Martens GR, Ladowski JM, Tector M, Tector AJ. Silencing the porcine iGb3s gene does not affect Galα3Gal levels or measures of anticipated pig-to-human and pig-to-primate acute rejection. Xenotransplantation. 2016;23(2):106–16.
100. Gao B, Long C, Lee W, Zhang Z, Gao X, Landsittel D, Ezzelarab M, Ayares D, Huang Y, Cooper DKC, Wang Y, Hara H. Anti-Neu5Gc and anti-non-Neu5Gc antibodies in healthy humans. PLoS One. 2017;12(7):e0180768.
101. Blix FG, Gottschalk A, Klenk E. Proposed nomenclature in the field of neuraminic and sialic acids. Nature. 1957;179:1088.
102. Varki A, Schnaar RL, Schauer R. In: Varki A, Cummings RD, Esko JD, Stanley P, Hart GW, Aebi M, Darvill AG, Kinoshita T, Packer NH, Prestegard JH, Schnaar RL, Seeberger PH, editors. Sialic acids and other nonulosonic acids. Essentials of glycobiology [Internet]. 3rd ed. Cold Spring Harbor (NY): Cold Spring Harbor Laboratory Press; 2015-2017. Chapter 15.2017.
103. Higashi H, Naiki M, Matsuo S, Okouchi K. Antigen of 'serum sickness' type of heterophile antibodies in human sera: identification as gangliosides with N-glycolylneuraminic acid. Biochem Biophys Res Commun. 1977;79:388.
104. Merrick JM, Zadarlik K, Milgrom F. Characterization of the Hanganutziu–Deicher (serum-sickness) antigen as gangliosides containing N-glycolylneuraminic acid. Int Arch Allergy Appl Immunol. 1978;57:477.
105. Bouhours D, Pourcel C, Bouhours JE. Simultaneous expression by porcine aorta endothelial cells of glycosphingolipids bearing the major epitope for human xenoreactive antibodies (Gal alpha 1-3Gal), blood group H determinant and N-glycolylneuraminic acid. Glycoconj J. 1996;13:947–53.
106. Miyagawa S, Maeda A, Kawamura T, et al. A comparison of the main structures of N-glycans of porcine islets with those from humans. Glycobiology. 2014;24:125–38.
107. Zhu A, Hurst R. Anti-N-glycolylneuraminic acid antibodies identified in healthy human serum. Xenotransplantation. 2002;9:376–81.
108. Miwa Y, Kobayashi T, Nagasaka T, et al. Are N-glycolylneuraminic acid (Hanganutziu-Deicher) antigens important in pig-to-human xenotransplantation? Xenotransplantation. 2004;11:247–53.
109. Varki A. Colloquium paper: uniquely human evolution of sialic acid genetics and biology. Proc Natl Acad Sci U S A. 2010;107(Suppl 2):8939–46. https://doi.org/10.1073/pnas.0914634107. Epub 2010 May 5.
110. Irie A, Koyama S, Kozutsumi Y, et al. The molecular basis for the absence of N-glycolylneuraminic acid in humans. J Biol Chem. 1998;273:15866–71.
111. Magnusson S, Mansson JE, Strokan V, et al. Release of pig leukocytes during pig kidney perfusion and characterization of pig lymphocyte carbohydrate xenoantigens. Xenotransplantation. 2003;10:432–45.
112. Saethre M, Baumann BC, Fung M, et al. Characterization of natural human antinon-Gal antibodies and their effect on activation of porcine Gal-deficient endothelial cells. Transplantation. 2007;84:244–50.

113. Basnet NB, Ide K, Tahara H, et al. Deficiency of N-glycolylneuraminic acid and Gala1–3-Galb1–4GlcNAc epitopes in xenogeneic cells attenuates cytotoxicity of human natural antibodies. Xenotransplantation. 2010;17:440–8.
114. Torbidoni AV, Scursoni A, Camarero S, Segatori V, Gabri M, Alonso D, Chantada G, de Dávila MT. Immunoreactivity of the 14F7 Mab raised against N-glycolyl GM3 ganglioside in retinoblastoma tumours. Acta Ophthalmol. 2015;93(4):e294–300.
115. Zhang X, Kiechle FL. Review: glycosphingolipids inhealth and disease. Ann Clin Lab Sci. 2004;34:3–13.
116. Kobayashi T, Yokoyama I, Suzuki A, et al. Lack of antibody production against Hanganutziu–Deicher (HD) antigens with N-glycolylneuraminic acid in patients with porcine exposure history. Xenotransplantation. 2000;7:177.
117. Galili U. Anti-Gal antibody prevents xenotransplantation. Sci Med. 1998;5:28.
118. Wang ZY, Martens GR, Blankenship RL, Sidner RA, Li P, Estrada JL, et al. Eliminating xenoantigen expression on swine RBC. Transplantation. 2017;101:517–23.
119. Campanero-Rhodes MA, et al. N-glycolyl GM1 ganglioside as a receptor for simian virus 40. J Virol. 2007;81:12846–58.
120. Song KH, Kwak CH, Jin UH, Ha SH, Park JY, Abekura F, Chang YC, Cho SH, Lee K, Chung TW, Ha KT, Lee YC, Kim CH. Housekeeping promoter 5'pcmah-2 of pig CMP-N-acetylneuraminic acid hydroxylase gene for NeuGc expression. Glycoconj J. 2016;33(5): 779–88. https://doi.org/10.1007/s10719-016-9671-5.
121. Naito Y, Takematsu H, Koyama S, Miyake S, Yamamoto H, Fujinawa R, Sugai M, Okuno Y, Tsujimoto G, Yamaji T, Hashimoto Y, Itohara S, Kawasaki T, Suzuki A, Kozutsumi Y. Germinal center marker GL7 probes activation-dependent repression of N-glycolylneuraminic acid, a sialic acid species involved in the negative modulation of B-cell activation. Mol Cell Biol. 2007;27:3008–22.
122. Taylor RE, Gregg CJ, Padler-Karavani V, et al. Novel mechanism for the generation of human xeno-autoantibodies against the nonhuman sialic acid N-glycolylneuraminic acid. J Exp Med. 2010;207:1637–46.
123. Lee W, Miyagawa Y, Long C, et al. Expression of NeuGc on pig corneas and its potential significance in pig corneal xenotransplantation. Cornea. 2016;35:105–13.
124. Lee W, Hara H, Cooper DK, Manji RA. Expression of NeuGc on pig heart valves. Xenotransplantation. 2015;22:153–4.
125. Scobie L, Padler-Karavani V, Le Bas-Bernardet S, et al. Long-term IgG response to porcine Neu5Gc antigens without transmission of PERV in burn patients treated with porcine skin xenografts. J Immunol. 2013;191:2907–15.
126. Springer SA, Diaz SL, Gagneux P. Parallel evolution of a self-signal: humans and new world monkeys independently lost the cell surface sugar Neu5Gc. Immunogenetics. 2014;66:671–4.
127. Song KH, Kang YJ, Jin UH, et al. Cloning and functional characterization of pig CMP-N-acetylneuraminic acid hydroxylase for the synthesis of N-glycolylneuraminic acid as the xenoantigenic determinant in pig-human xenotransplantation. Biochem J. 2010;427:179–88.
128. Butler JR, Paris LL, Blankenship RL, Sidner RA, Martens GR, Ladowski JM, et al. Silencing porcine CMAH and GGTA1 genes significantly reduces xenogeneic consumption of human platelets by porcine livers. Transplantation. 2016;100(3):571–6.
129. Park JY, Park MR, Bui HT, et al. Alpha1,3-galactosyltransferase deficiency in germ-free miniature pigs increases N-glycolylneuraminic acids as the xenoantigenic determinant in pig-human xenotransplantation. Cell Reprogram. 2012;14:353–63.
130. Burlak C, Bern M, Brito AE, et al. N-linked glycan profiling of GGTA1/CMAH knockout pigs identifies new potential carbohydrate xenoantigens. Xenotransplantation. 2013;20:277–91.
131. Lee W, Long C, Ramsoondar J, Ayares D, Cooper DK, Manji RA, Hara H. Human antibody recognition of xenogeneic antigens (NeuGc and Gal) on porcine heart valves: could genetically modified pig heart valves reduce structural valve deterioration? Xenotransplantation. 2016;23(5):370–80. https://doi.org/10.1111/xen.12254.

132. Samraj AN, Laubli H, Varki N, Varki A. Involvement of a non-human sialic acid in human cancer. Front Oncol. 2014;4:33.
133. Lofling JC, Paton AW, Varki NM, Paton JC, Varki A. A dietary non-human sialic acid may facilitate hemolytic-uremic syndrome. Kidney Int. 2009;76:140–4. https://doi.org/10.1038/ki.2009.131.
134. Yin J, et al. Hypoxic culture induces expression of sialin, a sialic acid transporter, and cancer-associated gangliosides containing non-human sialic acid on human cancer cells. Cancer Res. 2006;66:2937–45. https://doi.org/10.1158/0008-5472.CAN-05-2615.
135. Varki A. Potential impact of the nonhuman sialic acid N-glycolylneuraminic acid on transplant rejection risk. Xenotransplantation. 2011;18:1–5.
136. Padler-Karavani V, et al. Human xeno-autoantibodies against a non-human sialic acid serve as novel serum biomarkers and immunotherapeutics in cancer. Cancer Res. 2011;71:3352–63. https://doi.org/10.1158/0008-5472.CAN-10-4102.
137. Padler-Karavani V, Yu H, Cao H, et al. Diversity in specificity, abundance, and composition of anti-Neu5Gc antibodies in normal humans: potential implications for disease. Glycobiology. 2008;18:818–30.
138. Dall'Olio F, Malagolini N, Chiricolo M, Trinchera M, Harduin-Lepers A. The expanding roles of the Sd(a)/cad carbohydrate antigen and its cognate glycosyltransferase B4GALNT2. Biochim Biophys Acta. 2014;1840(1):443–53.
139. Lo Presti L, Cabuy E, Chiricolo M, Dall'Olio F. Molecular cloning of the human beta1,4 N-acetylgalactosaminyltransferase responsible for the biosynthesis of the Sd(a) histo-blood group antigen: the sequence predicts a very long cytoplasmic domain. J Biochem. 2003;134 (5):675–82.
140. Piller V, Piller F, Cartron JP. Comparison of the carbohydrate-binding specificities of seven N-acetyl-D-galactosamine-recognizing lectins. Eur J Biochem. 1990;191:461–6.
141. Kamada Y, Muramatsu H, Arita Y, et al. Structural studies on a binding site for Dolichos biflorus agglutinin in the small intestine of the mouse. J Biochem. 1991;109:178–83.
142. Byrne GW, Du Z, Stalboerger P, Kogelberg H, McGregor CG. Cloning and expression of porcine beta1,4 N-acetylgalactosaminyl transferase encoding a new xenoreactive antigen. Xenotransplantation. 2014;21:543–54.
143. Clausen H, Hakomori S. ABH and related histo-blood group antigens; immunochemical differences in carrier isotypes and their distribution. Vox Sang. 1989;56:1–20.
144. Cartron JP, Colin Y. Structural and functional diversity of blood group antigens. Transfus Clin Biol. 2001;8:163–99.
145. Byrne GW, Du Z, Stalboerger P, Kogelberg H, McGregor CG. Cloning and expression of porcine β1,4N-acetylgalactosaminyl transferase encoding a new xenoreactive antigen. Xenotransplantation. 2014;21:543–54.
146. Conte R, Serafini-Cessi F. Comparison between the erythrocyte and urinary Sda antigen distribution in a large number of individuals from Emilia-Romagna, a region of Northern Italy. Transfus Med. 1991;1:47–9.
147. Byrne G, Ahmad-Villiers S, Du Z, McGregor C. B4GALNT2 and xenotransplantation: a newly appreciated xenogeneic antigen. Xenotransplantation. 2018;25(5):e12394.
148. Cazal P, Monis M, Caubel J, Brives J. Herediatry dominant polyagglutinability: private antigen (Cad) corresponding to a public antibody and a lectin of Dolichos biflorus. Rev Fr Transfus. 1968;11:209–21.
149. Tollefsen SE, Kornfeld R. The B4 lectin from Vicia villosa seeds interacts with N-acetylgalactosamine residues on erythrocytes with blood group Cad specificity. Biochem Biophys Res Commun. 1984;123:1099–106.
150. Sanger R, Gavin J, Tippett P, Teesdale P, Eldon K. Plant agglutinin for another human blood-group. Lancet. 1971;1:1130.
151. Herkt F, Parente JP, Leroy Y, Fournet B, Blanchard D, Cartron JP, Van Halbeek H, Vliegenthart JF. Structure determination of oligosaccharides isolated from Cad erythrocyte

membranes by permethylation analysis and 500-MHz 1H-NMR spectroscopy. Eur J Biochem. 1985;146:125–9.
152. Gillard BK, Blanchard D, Bouhours JF, Cartron JP, Van Kuik JA, Kamerling JP, Vliegenthart JF, Marcus DM. Structure of a ganglioside with Cad blood group antigen activity. Biochemistry. 1988;27:4601–6.
153. Blanchard D, Piller F, Gillard B, Marcus D, Cartron JP. Identification of a novel ganglioside on erythrocytes with blood group Cad specificity. J Biol Chem. 1985;260:7813–6.
154. Piller F, Blanchard D, Huet M, Cartron JP. Identification of a α-NeuAc-(2--3)-β-Dgalactopyranosyl N- acetyl-β-D-galactosaminyltransferase in human kidney. Carbohydr Res. 1986;149:171–84.
155. Smith PL, Lowe JB. Molecular cloning of a murine N-acetylgalactosamine transferase cDNA that determines expression of the T lymphocyte-specific CT oligosaccharide differentiation antigen. J Biol Chem. 1994;269:15162–71.
156. Hidari JK, Ichikawa S, Furukawa K, Yamasaki M, Hirabayashi Y. β1–4-Nacetylgalactosaminyltransferase can synthesize both asialoglycosphingolipid GM2 and glycosphingolipid GM2 in vitro and in vivo: isolation and characterization of a beta 1–4N-acetylgalactosaminyltransferase cDNA clone from rat ascites hepatoma cell line AH7974F. Biochem J. 1994;303(Pt 3):957–65.
157. Dall'Olio F, Malagolini N, Di Stefano G, Ciambella M, Serafini-Cessi F. Postnatal development of rat colon epithelial cells is associated with changes in the expression of the β 1,4-N-acetylgalactosaminyltransferase involved in the synthesis of Sda antigen and of α 2,6-sialyltransferase activity towards N-acetyllactosamine. Biochem J. 1990;270:519–24.
158. Kawamura YI, Toyota M, Kawashima R, Hagiwara T, Suzuki H, Imai K, et al. DNA hypermethylation contributes to incomplete synthesis of carbohydrate deter-minants in gastrointestinal cancer. Gastroenterology. 2008;135(142–51):e3.
159. Wang HR, Hsieh CY, Twu YC, Yu LC. Expression of the human Sd(a) beta-1,4-N-acetylgalactosaminyltransferase II gene is dependent on the promotermethylation status. Glycobiology. 2008;18:104–13.
160. Kawamura YI, Kawashima R, Fukunaga R, Hirai K, Toyama-Sorimachi N, Tokuhara M, et al. Introduction of Sd(a) carbohydrate antigen in gastrointestinal cancer cells eliminates selectin ligands and inhibits metastasis. Cancer Res. 2005;65(14):6220–7.
161. Robbe-Masselot C, Herrmann A, Maes E, Carlstedt I, Michalski JC, Capon C. Expression of a core 3 disialyl-Lex hexasaccharide in human colorectal cancers: a potential marker of malignant transformation in colon. J Proteome Res. 2009;8:702–11.
162. Dohi T, Ohta S, Hanai N, Yamaguchi K, Oshima M. Sialylpentaosylceramide detected with anti-GM2 monoclonal antibody. Structural characterization and complementary expression with GM2 in gastric cancer and normal gastricmucosa. J Biol Chem. 1990;265:7880–5.
163. Rausch P, Steck N, Suwandi A, Seidel JA, Künzel S, Bhullar K, Basic M, Bleich A, Johnsen JM, Vallance BA, Baines JF, Grassl GA. Expression of the blood-group-related gene B4galnt2 alters susceptibility to *Salmonella* infection. PLoS Pathog. 2015;11(7):e1005008.
164. Serafini-Cessi F, Monti A, Cavallone D. N-glycans carried by Tamm-Horsfall glycoprotein have a crucial role in the defense against urinary tract diseases. Glycoconj J. 2005;22(7–9): 383–94. https://doi.org/10.1007/s10719-005-2142-z.
165. Blanchard D, Capon C, Leroy Y, Cartron JP, Fournet B. Comparative study of glycophorin a derived O-glycans from human cad, Sd(a+) and Sd(a-) erythrocytes. Biochem J. 1985;232(3): 813–8. https://doi.org/10.1042/bj2320813.
166. Li YT, Li SC, Hasegawa A, Ishida H, Kiso M, Bernardi A, Brocca P, Raimondi L, Sonnino S. Structural basis for the resistance of Tay-Sachs ganglioside GM2 to enzymatic degradation. J Biol Chem. 1999;274(15):10014–8. https://doi.org/10.1074/jbc.274.15.10014.
167. Lowe JB. Glycan-dependent leukocyte adhesion and recruitment in inflammation. Curr Opin Cell Biol. 2003;15(5):531–8. Epub 2003/10/02
168. Kawamura YI, Adachi Y, Curiel DT, Kawashima R, Kannagi R, Nishimoto N, et al. Therapeutic adenoviral gene transfer of a glycosyltransferase for prevention of peritoneal

dissemination and metastasis of gastric cancer. Cancer Gene Ther. 2014;21(10):427–33. Epub 2014/09/13. https://doi.org/10.1038/cgt.2014.46.
169. Trinchera M, Malagolini N, Chiricolo M, Santini D, Minni F, Caretti A, Dall'Olio F. The biosynthesis of the selectin-ligand sialyl Lewis x in colorectal cancer tissues is regulated by fucosyltransferase VI and can be inhibited by an RNA interference based approach. Int J Biochem Cell Biol. 2011;43:130–9.
170. Montiel MD, Krzewinski-Recchi MA, Delannoy P, Harduin-Lepers A. Molecular cloning, gene organization and expression of the human UDP-GalNAc:Neu5Acalpha2-3Galbeta-R beta1,4-N-acetylgalactosaminyltransferaseresponsible for the biosynthesis of the blood group Sda/Cad antigen: evidence for an unusual extended cytoplasmic domain. Biochem J. 2003;373:369–79.
171. Johnsen JM, Levy GG, Westrick RJ, Tucker PK, Ginsburg D. The endothelial-specific regulatory mutation, Mvwf1, is a common mouse founder allele. Mamm Genome. 2008;19 (1):32–40. https://doi.org/10.1007/s00335-007-9079-4.
172. Mohlke KL, Nichols WC, Westrick RJ, Novak EK, Cooney KA, Swank RT, et al. A novel modifier gene for plasma von Willebrand factor level maps to distal mouse chromosome 11. Proc Natl Acad Sci. 1996;93(26):15352–7. Epub 1996/12/24.
173. Mohlke KL, Purkayastha AA, Westrick RJ, Smith PL, Petryniak B, Lowe JB, et al. Mvwf, a dominant modifier of murine von Willebrand factor, results from altered lineage-specific expression of a glycosyltransferase. Cell. 1999;96(1):111–20. Epub 1999/02/16. https://doi.org/10.1016/S0092-8674(00)80964-2.
174. Johnsen JM, Teschke M, Pavlidis P, McGee BM, Tautz D, Ginsburg D, et al. Selection on cis-regulatory variation at B4galnt2 and its influence on von Willebrand factor in house mice. Mol Biol Evol. 2009;26(3):567–78. Epub 2008/12/18. https://doi.org/10.1093/molbev/msn284.
175. Staubach F, Künzel S, Baines AC, Yee A, McGee BM, Bäckhed F, et al. Expression of the blood-group-related glycosyltransferase B4galnt2 influences the intestinal microbiota in mice. ISME J. 2012;6(7):1345–55. Epub 2012/01/27. http://www.nature.com/ismej/journal/vaop/ncurrent/suppinfo/ismej2011204s1.html
176. Kimura A, Orn A, Holmquist G, Wigzell H, Ersson B. Unique lectin-binding characteristics of cytotoxic T lymphocytes allowing their distinction from natural killer cells and "K" cells. Eur J Immunol. 1979;9:575–8.
177. Conzelmann A, Kornfeld S. β-Linked N-acetylgalactosamine residues present at the nonreducing termini of O-linked oligosaccharides of a cloned murine cytotoxic T lymphocyte line are absent in a Vicia villosa lectin-resistant mutant cell line. J Biol Chem. 1984;259: 12528–35.
178. Conzelmann A, Kornfeld S. Amurine cytotoxic T lymphocyte cell line resistant to Vicia villosa lectin is deficient in UDP-GalNAc: β-galactose β 1,4-Nacetylgalactosaminyltransferase. J Biol Chem. 1984;259:12536–42.
179. Conzelmann A, Lefrancois L. Monoclonal antibodies specific for T cell-associated carbohydrate determinants react with human blood group antigens CAD and SDA. J Exp Med. 1988;167:119–31.
180. Thomas PJ, Xu R, Martin PT. Mouse model of limb girdle muscular dystrophy 2IB4GALNT2 (GALGT2) gene therapy reduces skeletal muscle pathology in the FKRP P448L mouse model of limb girdle muscular dystrophy 2I. J Pathol. 2016;186:2429e2448. https://doi.org/10.1016/j.ajpath.2016.05.021.
181. Blaeser A, Keramaris E, Chan YM, Sparks S, Cowley D, Xiao X, Lu QL. Mouse models of fukutin-related protein mutations show a wide range of disease phenotypes. Hum Genet. 2013;132(8):923–34. https://doi.org/10.1007/s00439-013-1302-7.
182. Topaloglu H, Brockington M, Yuva Y, Talim B, Haliloglu G, Blake D, Torelli S, Brown SC, Muntoni F. FKRP gene mutations cause congenital muscular dystrophy, mental retardation, and cerebellar cysts. Neurology. 2003;60:988–92.

183. Hoyte K, Kang C, Martin PT. Definition of pre- and postsynaptic forms of the CT carbohydrate antigen at the neuromuscular junction: ubiquitous expression of the CT antigens and the CT GalNAc transferase in mouse tissues. Brain Res Mol Brain Res. 2002;109:146–60.
184. Xu R, DeVries S, Camboni M, Martin PT. Overexpression of Galgt2 reduces dystrophic pathology in the skeletal muscles of alpha sarcoglycan-deficient mice. Am J Pathol. 2009;175: 235–47.
185. Martens GR, Reyes LM, Butler JR, Ladowski JM, Estrada JL, Sidner RA, Eckhoff DE, Tector M, Tector AJ. Humoral reactivity of renal transplant-waitlisted patients to cells from GGTA1/CMAH/B4GalNT2, and SLA class I knockout pigs. Transplantation. 2017;101(4): e86–92.
186. Amorim I, Freitas DP, Magalhães A, Faria F, Lopes C, Faustino AM, et al. A comparison of helicobacter pylori and non-helicobacter pylori helicobacter spp. binding to canine gastric mucosa with defined gastric glycophenotype. Helicobacter. 2014;19(4):249–59.
187. Kaoutari AE, Armougom F, Gordon JI, Raoult D, Henrissat B. The abundance and variety of carbohydrate-active enzymes in the human gut microbiota. Nat Rev Micro. 2013;11(7): 497–504. https://doi.org/10.1038/nrmicro3050. Epub 2013/06/12 http://www.nature.com/nrmicro/journal/v11/n7/abs/nrmicro3050.html#supplementary-information.
188. Bosshard PP, Zbinden R, Altwegg M. Turicibacter sanguinis gen. nov., sp. nov., a novel anaerobic, gram-positive bacterium. Int J Syst Evol Microbiol. 2002;52(4):1263–6. Epub 2002/08/01. https://doi.org/10.1099/00207713-52-4-1263.
189. Weiss GA, Chassard C, Hennet T. Selective proliferation of intestinal Barnesiella under fucosyllactose supplementation in mice. Br J Nutr. 2014;111:1–9. Epub 2014/01/15. https://doi.org/10.1017/S0007114513004200.

Chapter 12
Other Non-α1,3Gal Antigens

12.1 Minor and Additional Pig Non-α1,3Gal Carbohydrate Antigens

Since understanding of the impact in xenoantigenic determinant Galα1,3Gal-R and anti-α1,3Gal Abs during xenoantigenic damages and injuries, other non-α1,3Gal glycan xenoantigens, the non-α1,3Gal epitope-reactive immune responses, and their related inflammation reactions have not well been explained. The α1,3Gal-T enzyme required for the biosynthesis of α1,3Gal-carrying antigenic epitopes [1] and α1,3Gal glycan-terminating GSLs is also discovered from bovine cells. Barone et al. [2] found and elucidated both structures of non-acidic neutral GSLs and acidic GSLs from pulmonary and aortic endothelium of naïve pig aortic vessel using α1,3Gal-specific *Griffonia simplicifolia* IB4 lectin, GS-IB4. To characterize the antibody reactivity of patient tissues with bioprosthetic valves of pig valves, acidic GSLs and non-acidic neural GSLs were isolated from pulmonary and aortic cusps of pig valves. The isolated GSLs have been analyzed for their characterization by MS analysis and recognition properties and patterns of glycan-binding ligands. The non-acidic neutral GSLs have been specified as globotetraosyl-Ceramide, H-type 2 pentaosyl-ceramide, Fuc-gangliotetraosyl-Ceramide, and Galα1,3neolactotetraosyl-Ceramide. The acidic GSLs contained gangliosides including GM1, GM2, GM3, Fuc-GM1, GD1a and GD3, and sulfatides. All gangliosides consisted of NeuAc residues. Significantly, the main component, NeuGc of NeuAc variant, was present in various organs of pig. To humans, it can induce development of Abs. Such glycan structures bound with complex lipids in pigs can possibly be target candidates for the immune system of human.

As mentioned, potential carbohydrate antigens included Neu5Gc, Forssman, β-LacNAc, and α-LacNAc. The Forssman carbohydrate antigen is considered as a potential pig non-α1,3Gal epitope targeted for natural Abs of primates. The Forssman carbohydrate has a complex structure of glycolipids with the similar structural motifs of the blood group type ABO antigens and α1,3Gal-glycan

C.-H. Kim, *Glycoimmunology in Xenotransplantation*,
https://doi.org/10.1007/978-981-99-7691-1_12

antigens. Expression of Forssman carbohydrate antigen is found in chickens, guinea pigs, hamsters, horses, sheep, chickens, hamsters and mice. However, the Forssman carbohydrate antigen is not present in the ox, pigeon, rabbit and rats [3]. Forssman carbohydrate antigen-specific Abs are known to be heterophilic. Forssman carbohydrate antigen is not produced in humans, instead, but humans express the Forssman precursor globoseries named Gb4 globoside. Another possible non-α1,3Gal carbohydrate is β-LacNAc. The β-LacNAc is exposed in the removal condition of the α1,3Gal epitope, as in α1,3Gal-T KO pigs. Hence, primates can produce Abs against the β-LacNAc antigen. However, β-LacNAc-binding IgG and IgM Abs are minorly found in baboon and human; thus, this β-LacNAc carbohydrate seems not to be related in the graft rejection response of α1,3Gal-T KO pigs.

Previously, terminal Galα2-recognizing human Abs are also known [4]. A cross-reactivity between Galα1,3Gal epitopes and Galα2Gal epitopes was found to decrease the human serum cytotoxic activity to PK15 cells of pig kidney-derived cell line. In addition, although Galα1,4Gal-terminating globotriaosylceramide glycolipids are also main carbohydrate expressed in animal and human tissues, any cross-reaction activity of anti-α1,3Gal Abs with this carbohydrate is ignorable. The GalGalα1,4Gal glycolipid carbohydrates are normally expressed as ubiquitous constituents in humans. Humans are also known to have a large amount of α-lactosamine (Galα1,4GlcNAc-)-reacting antibodies. Therefore, these carbohydrate antigens are not the issues in pig-to-human xenograft transplantation.

From the extensive MS analysis of *N*-glycan carbohydrate structures of corneal ECs and kidney ECs of pig and pig keratocytes, over 100 sialylated and neutral *N*-glycoproteins including non-Gal carbohydrate antigens were found. Some potentially immunogenic carbohydrate antigens are suggested in the pig-to-human xenotransplantation, depending on theoretical aspect, but not experimental results. They include the Forssman glycolipid. The Forssman antigen is not produced in human, but rarely detected in some human cancer [5] and human kidney [6]. Both baboons and humans produce a high amounts of Abs reactive for the Forssman disaccharyl carbohydrate. Moreover, antibody levels of anti-Forssman-specific Abs are significantly compared to that of anti-α1,3Gal antibodies.

Abs against Forssman carbohydrate are independent of anti-α1,3Gal Abs because Abs against Forssman carbohydrate are not diminished even after serum adsorption of α1,3Gal antigens. Human and baboons are negative for Forssman expression. Mice is also Forssman-positive. From the facts that anti-Forssman carbohydrate Abs are present in high level in pigs but Forssman antigens are not stained on pig tissues, it has been suggested that the Forssman antigen is not expressed and negative in pigs. Thus, the Forssman glycan antigen does reject pig grafts in primates. WT and α1,3Gal-T KO pigs contained high amounts of anti-Forssman glycan Abs, repeatedly indicating that pig is negative for Forssman antigen synthesis [7]. Pig tissues are negative for immunohistochemical staining with the Forssman antigen expression [5]. Forssman glycolipids is not biochemically characterized in pig tissues [5]. Abs specific for Forssman glycan have been detected in WT pig and α1,3Gal-T KO pig as well as in humans [5].

12.2 Blood Group AO Antigen in Pig

The blood group ABO(H) antigenic glycans are not unique antigens to only humans and the antigens are widely expressed over animal species in nature [5]. The pig blood groups in transplantation are also acted as antigens because minor histocompatibility antigens and blood group AO carbohydrate antigens are also crucial target determinant for the immune responses of rejection. Thus, transfusion aspect should be care of transplantation in pigs. In the surfaced carbohydrates on erythrocytes, the representative non-α1,3Gal glycan antigens should be the ABH(O) blood group glycans due to their similarities in linkage mode. The pig A blood group has been reported from the pig erythrocyte surfaces and further characterized later [8–10]. Pigs have at least one blood group antigen among the four blood group determinants of O or "–" (blank), A or Aw (weak A) phenotype. The distinct difference of the "–" (blank) and O phenotypes is based on reaction between cattle antisera to the J-s antigens as the equivalently corresponded blood group. Among the J-s antigens, the anti-s antiserum is the anti-H substance equivalent in human, and thus, this means the O type. If anti-A or anti-s serum does not react, it is the "–" phenotype. Pig blood group A is typed by commercial mAbs, which are being used in typing of human blood groups. Therefore, the pig erythrocytes are agglutinated with anti-A antigen mAbs.

Anti-H mAbs specific for H antigen is not distinguishable for the "–" and O phenotypes. Interestingly, the pig blood A and H antigens are not found in the organs, tissues and cells of pig at their birth [10]. The pig H blood group antigen is found in goblet cells and plasma membranes of intestines. Young pigs do not express the type AO phenotypes. Pigs have A and H [11], but do not have a B antigen. Interestingly, pigs also carry an AO system likely to the human blood ABO antigenic system. The H and A blood group antigenic glycans of pigs are well characterized [12]. The pig A blood group antigenic glycan is present largely in the tubular cells, but not in the vascular system. The pig A blood group antigenic glycan is recognized by anti-A blood group Abs of human. Pig A blood group antigenic glycans expressed on pig erythrocytes are thus used in blood group typing. Pig blood group A type antigenic glycans are detected in the blood plasma of pigs and also adsorbed to erythrocytic cells, indicating that hemolysis death of pig erythrocytes is not taken place due to the low antigenic levels on the pig erythrocytes. However, anti-A blood group Abs present in the blood plasma are responsive to blood group A antigens in the plasma of donor pigs, depending on the plasma volume transfused rather than the erythrocyte itself. Hence, RBCs with a minimum volume of A–O compatible blood or blood plasma is recommended for blood transfusion. Transfusion hemolysis is not raised because pigs were not generally transfused, except for the use of research. But in pigs, utilization of A–O incompatible transfusion applications raises adverse reactions. In fact, the A–O-incompatible transfusion-received pigs display decreased fibrinogen levels caused by increased fibrinolysis with fibrin degradation and pulmonary hypertension.

The two pig genetic loci responsible for the synthesis of erythrocyte antigen S (AO inhibitor) and erythrocyte antigen A (EAA) makes the inheritable blood group A phenotype [13]. The EAAA gene allele encodes a UDP-GalNAc transferase. UDP-GalNAc transferase transfers the A antigenic glycan to the H antigenic glycan. Type of the recessive EAAO allele carries the gene deletion, producing the inactivated enzyme, which is not functional [14]. The blood group H antigens are expressed on glycolipids or glycoproteins by the S or H loci gene product of α1,2-fucosyltransferase (Fuc-T) with difference on precursor substrates. The S locus-encoding enzyme Fuc-T2 is known to prefer the type I chain substrates, whereas the H locus-encoding enzyme Fuc-T1 is known to prefer the type II chain substrates, which the two type I and type II are different from the terminal Gal structure and subterminal GlcNAc structures of Galβ1,3GlcNAc and Galβ1,4GlcNAc. The S gene is known as a strict human homolog of the Lewis antigenic and human ABO blood group antigenic genes [8]. The recessive Ss allele encodes an inactivated enzyme gene, which is enzymatic not functional. Seemingly, the pig blood group H gene is also a strictly homolog of the H blood type antigenic gene (Fuc-T1) of human. Chromosomal locations of the S locus and EAH locus are nearby closed on the chromosome 6 of pigs, while the chromosomal EAA locus is located on the chromosome 1 of pigs [15].

Homozygous pigs for a recessive FucT1 gene (FUT1 M307AA) allele exhibit lowered FucT1 enzymatic activity and result in resistance to F18 fimbrilla *Escherichia coli* infection as a diarrhea-causing agent in pig [16]. In addition, the H antigen in the type II chains is known as the *E. coli* adhesion receptor. *E. coli* cholera and heat labile toxins bind effectively to glycolipids expressed in the intestinal mucosa of the pig blood group A pig rather than the glycolipids in non-A pigs [17]. *E. coli*, causing urinary tract infection, easily adhere to the blood group P antigens. The P antigen-deficient patients exhibit their adhesion resistance to the urinary tract-infectious *E. coli* [18]. On the other hand, the known blood group Lewis b antigens are used as the adhesion receptor for gastrointestinal bacteria *Helicobacter pylori* required for the peptic ulcer disease [19].

In the pig genome map, pig chromosomal blocks are the regions equivalent to certain human chromosomes (synteny). The syntenic region is described at http://www.toulouse.inra.fr/lgc/pig/compare/compare.htm. In pig, chromosome 6 location is the corresponding synteny to the regions of Secretor and H genes of human chromosome 19 region [20]. The same region of chromosome 6 of pigs is also syntenic to the Rh antigen genes of human chromosome 1, as the RH gene location of pigs is present in the same chromosome 6 region. Similarly, the ABO genes of human are located on the chromosome 9 of human as the synteny to the EAA region of the chromosome 1 of pigs.

Pig A antigenic substances present in erythrocytes are the substances adsorbed from the pig plasma in order to express them in pig, rather than directly expressed on glycoproteins [21]. This is strictly in contrast to the human erythrocytes, which the human erythrocytes directly express the A antigenic substances. However, this is also similar to the case of human erythrocytic Lewis antigens, which are adsorbed to human erythrocytes [22]. In addition, in human, A antigen expression in tissues is found in both the chain type I and type II [23]. But the recessive Sss genotyped pigs, which pig erythrocytes are consequently negative for the A group antigen, can produce A antigen on type II chain in pig tissues. The A antigens are also present on collecting ducts, distal convoluted tubules, Henle loop, and urothelium of pig kidneys. However, pig endotheliums are negative for the expression of A and H antigenic substances [24]. The A antigenic substance is present in pig endothelial cells of heart tissues, particularly in the endocardium of pigs.

The A blood type antigen of pigs can be reactive for immune response to human recipients of blood group O or B type [25]. Regarding to the pig antigens for blood groups, human A and B blood type antigens are glycan antigenic epitopes reactive for natural Abs of human. Although the blood group A antigen is the complete barrier graft in pig-to-human xenotransplantation, the pig blood group A is not problematic for clinical xenotransplantation if we test the pig A or O blood group. Structurally, the A/H carbohydrate structures are linked to the two type 1 core glycan chain and 2 core glycan chain. Although the type 1 core glycan-linked Le^a and Le^b antigenic epitopes are absent in pigs, the core type 2 glycan chain isomers of Le^X and Le^Y antigens are found [12]. The blood group systems of type 1-Le^a and -Le^b as well as the type 2 core-Le^X and -Le^Y antigenic glycans are also not problematic in pig-to-human xenotransplantation, too. Additionally, although other blood type antigenic glycans were reported, those antigenic epitopes were not characterized even in the respect of the protein or carbohydrate epitopes. In the pig kidney extracorporeal xenotransplantation, the B group recipient and the pig RBC donor was A group [26].

Since endothelial cells of pigs are positive for α1,3Gal antigens and negative for A blood group, reserving that anti-α1,3Gal Abs are expected to induce the graft-specific thrombotic death. Also, in the case of tubular cells, which are positive for blood group A antigens and negative for α1,3Gal antigens, the tubular cells were destroyed by complement cascade and IgM Abs soon after perfusion time 65 min, due to damage from non-α1,3Gal Abs. Mismatched types of the blood group AO antigens of skin allografts in recipients shows a substantially survival effect [27]. Other grafts of solid organs show the similar effect shown in the above skin allografts in pigs. Genotype and phenotypes are crucial in breeding pigs because the two non-A parents mating can make A offspring in the condition of the Ss allele homozygous in the parents.

On the other hand, in the pig erythrocyte analysis using lectin, which specifically bind to carbohydrates and are previously frequently used in human blood typing, such blood group specific lectins agglutinate pig erythrocytes [28]. For example, *Ulex europaeus* lectin specific for [alpha]-fucosyl glycans recognizes the H antigenic molecule of human, but does not agglutinate AO phenotype in erythrocytes of pigs. *Ulex europaeus* lectin and anti-H monoclonal antibody equally recognized the H substance produced by the pig FucT2 gene-transfected COS cells [29]. Regardless of AO phenotype, human erythrocyte A1 subtype-binding *Dolichos biflorus* lectin and human A erythrocyte-binding *Phaseolus lunatus* lectin agglutinate 95% and 83% of pig erythrocytes, respectively. *Dolichos biflorus* lectin also strongly bind pig endothelial cells [30]. Although the pig blood group H substance and pig blood group A substance are similar, the unexpectedly observed differences in lectin binding reactions on human and pig substances presumably reflect their substantial differences.

12.3 Double Phenotypic Modification of Gal-T-KO/Fuc-T TG Pigs

The double phenotypically engineered Gal-T-KO/FucT-TG pigs are not beneficial for the GalT-KO pig in reaction with human Abs with no merit in xenotransplantation. α1,3Gal-T KO pigs completely eliminated α-1,3Gal-glycosphingolipids (GSLs) and instead, slightly increased LacNAc precursor levels. GSL glycolipids also compete with α1,3Gal-T for LacNAc acceptor substrates including H type 2 blood group (Fucα1,2Galβ1,4GlcNAc-R), P1 antigen (Galα1,4Galβ1,4GlcNAc-R) and x2 (GalNAcβ1,3Galβ1,4GlcNAc-R) [31]. The human blood group H-transferase (α1,2Fuc-T) transgenic α1,3Gal-negative pigs generate α1,3Gal-antigen-free organs of pigs with the uncapped LacNAc precursors because the α1,3Gal-T KO pigs carry non-xenoantigenic H blood group. This prevents potential synthesis of xenoantigenic non-α1,3Gal antigens. Thus, the α1,3Gal-T KO/α1,2Fuc-T TG pigs show more effective results compared to the pig Fuc-T TG or pig α1,3Gal-T KO strains in avoiding xenoantigenic rejection of xenografts. The α1,2Fuc-T-transgene also reduces the NeuGc antigen levels in the transgenic cells of pigs. Sialyltransferases (Sia-Ts) are also competitive with Fuc-Ts for LacNAc acceptor precursors to synthesize sialyl-glycolipids. The reaction property of Abs in human blood sera against α1,3Gal-T KO/Fuc-T TG GSL glycolipids and α1,3Gal-T KO GSL glycolipids is similar [31].

The human Ab-reactive neutral glycolipids were also detected. H blood group substances were detected to high levels in α1,3Gal-T KO AEC and PBMC of pigs. α1,3Gal-T-KO-induced compensatory glycosylation consequently take places to newly produce immunogenic non-α1,3Gal antigens. Enforced expression of human α1,2 Fuc-T gene in pigs competes with α1,3Gal-T enzyme for LacNAc acceptor precursors. Transgenic α1,2Fuc-T, consequently inhibit α1,3Gal-antigen synthesis in the pig cells and reduce the production of xenoantigen-specific natural Abs, and this increase in the pig resistant capacity to serum-mediated cytotoxicity such as ADCC or CDC of human bloods [32]. Cooperatively double expression of α1,2Fuc-T and CD59 genes of human further increase the resistant capacity of pigs to serum-mediated cytotoxicity of human bloods [33]. For example, chondrocyte cells obtained from the α1,2Fuc-T TG pig cells-grafted α1,3Gal-T KO mice are protected from cellular, delayed and humoral rejections when accounted with human serum [34]. Pig kidneys from multiply engineered Gal-TKO/FucT/hCD55/hCD59/hCD39TGs were not protected from host Abs and complementation when transplanted into baboons [35]. Glycolipid antigenic substances isolated from the α1,3Gal-T KO/α1,2Fuc-T TG intestine of pig exhibited minor or no different rejective responses from α1,3Gal-T KO pigs. Thus, the additional α1,2Fuc-T gene expression in α1,3Gal-T KO pig is not effective to reduce reactivity of human Abs.

12.4 Absence of Lewis Le^a and Le^b Antigens in Pig

The pig blood AO type is different from the human blood ABO group. For example, Pig B-antigen is absent [36]. Several H-antigenic glycan structures and only A-antigenic structure are found on the *O*-glycans. Mucin-type *O*-glycans are made of Core 1, Core 2 and Core 4 chains that are sialyl, sulfayl, and α1,2 fucosyl glycans. Lewis antigens (α1,3Fuc- and α1,4Fuc-) in pigs are not present due to the absence of the Lewis gene in pigs. The blood group A-antigenic structures are present in all tissues and all pigs represent blood A group although A/O polymorphic diversity is not present in pigs]. Only one A-antigen is present on the intestinal *O*-glycans only in the young aged pigs since pig intestinal mucus A-antigens are not present until aged 5 weeks after birth.

Pigs lack the Lewis antigen, Le gene, while they bear the Lewis X gene. Hence, the pig organs have fucosyl-Le^x type GSL glycolipids. However, pig does not have the antigens of Le^a and Le^b [14]. In addition, because only pig type-I glycans are the adsorbed substances from the blood serum, pig erythrocytes are consequently absent for all type of Lewis antigenic substances. Therefore, the GSL glycolipids with Lewis types are not functioned as pig blood type antigens. In contrast, the Lewis Le^a and Le^b antigens of human are also biosynthesized on the same type-I glycan. In addition, the human A, B and H blood groups are consequently released or secreted to plasma and tissues. Another blood group-synthetic enzyme, fucosyltransferase (Fuc-T3 or Le) catalyzes a Fuc residue transferring to the subterminally present GlcNAc. If the "Secretor" gene (Fuc-T2) is held, next, a second Fuc residue is additionally transferred to the terminal Gal residue to form the Lewis Le^b antigens. Non-secretors have only the subterminal Fuc residue, calling for the antigen Le^a. Type-II glycan chains are fucosylated by another type of fucosyl-transferase (X gene-encoded) to form Le^x antigen, as humans carry it, and hence, it is not a blood group antigen.

Terminal histo-blood group structures are often involved in bacterial adhesion. For example, GSL glycolipids with blood group H/A antigens function as F18 fimbrial enterotoxic *E. coli* receptor (ETEC F18) [37], the pig gastrointestinal infectious factor. In the human ABO antigens, two genetic loci of the H and SE required for α1,2Fuc-Ts synthesize the antigen H of the Fucα1,2Galβ1 precursor which is required for antigens of A and B. α1,2Fuc-T1 adds the Fuc residue to a terminal Gal residue in an α1,2-glycosidic linkage to form the H blood group antigens. The pig α1,2Fuc-T1 gene is orthologous to the human H gene, and pig α1,2Fuc-T2 is orthologous to the SE gene of human, which is present in the small intestines [38].

12.5 Other Blood Group Antigen Systems

The 16 pig blood groups are known and S gene regulates the A-O blood group expression [39]. The 16 pig blood groups are called as each single letter with respective indications of A to P blood groups. Letter of EA derived from "erythrocyte antigen" is used. The antigenic blood groups distributed are present in erythrocytes, and however, the exceptional cases of blood group antigens of A, E, G, L, and N are found in leukocytes or general tissues [40]. The blood groups are also decided by pig sera obtained from the pigs sensitized to pregnant blood and injected by allogeneic erythrocytes or skin transplants. Other named EAE and EAM groups are not clear whether they are epitopes. Five EAC, EAH, EAJ, EAK, and EAP groups are not clear whether they are detectable, hence calling a "blank" phenotype and thus expressed by the "–" symbol like allele name of (EAC−).

Apart from the AO blood group antigen and carbohydrate antigen $Le^{x}/^{y}$, pig erythrocyte blood group antigen systems are known [41]. The pig blood group antigen system is mainly restricted to pig erythrocytes; however, other blood group systems of E, G, L, and N are classified to the histo-blood group antigen systems due to location on cells/tissues as well [42]. Histo-blood group antigens are crucial in complement-dependent hemolysis and allotransplantation of pig skins [42]. Therefore, the pig histo-blood group antigens function as immune-rejection antigenic targets in the pig-to-human xenotransplantation.

12.6 Minor Blood Group Antigens Expressed in Pig Tissues or Cells

The blood group E, G, L and N systems are known. Among them, the histo-blood E group shows a relatively complex type with 18 epitope factors of Ea to Et, which are bound by antisera antibodies. Each Allele is named for each factor as such "Eaeg" or as a numerical designation. In the E histo-blood group alleles, 17 alleles of the E1 to E17 are included [14]. The histo-blood E group is highly immunogenic and expressed on erythrocytes, granulocytes and lymphocytes [43]. Thirty percent of sows bear high titer Abs against the blood groups. In addition, 46% of the 30% are due to the histo-blood group E-specific Abs [44]. Littermate-derived skin grafts, which SLA are matched, show the prolonged and longer survival than non-littermates if the skin grafts are matched for the histo-blood group E antigen. In fact, pigs received the Ea antigen-mismatched skin grafts highly produce anti-Ea antigen-specific Abs. Therefore, this indicates that the histo-blood antigens E belongs to a blood group antigen as well as a minor histocompatibility antigen [47]. In the pig chromosome location of the EAE gene, the pig chromosome 9 long arm is the EAE gene location, as this is syntenic to the chromosome 1 of humans, where the complement receptor CR1 is located on the chromosome 1. In the human CR1, the protein CR1 is the blood group Knops antigen. In erythrocytes, the

CR1 protein is produced in T/B cells, NK cells, granulocytes and also on macrophages and follicular DCs, indicating that the blood group E gene location is in the CR1 gene region.

On the other hand, the blood group G bears two different alleles named EAGa and EAGb. Anti- α1,3Gal antisera is absorbed with lymphocytes, lung and spleen tissue extracts. Skin grafts-received pigs from the G antigen-mismatched donors produces blood group G-specific Abs [46]. The pig EAG gene is known to present on pig chromosome 15. The L type has six alleles with 12 factors. Skin graft-received pigs from the L antigen-mismatched donors produce blood group L-specific Abs [46]. The pig EAL gene is known to locate on the pig chromosome 4. In addition, the pig blood group N bears three independent gene alleles, which comprised of EANa, EANb, and EANbc. Anti-blood group N Abs are easily induced by injection of skin grafts or erythrocytes. Similarly, anti-blood group Na Abs can be absorbed out with lung tissue extract, spleen extracts and lymphocytes. Pig EAN gene location is found in pig chromosome 9.

12.7 Blood Group Antigens Expressed in Erythrocyte Membranes, Not for Tissues and Rh Antigen

The pig blood groups H, I and M antigens-reactive antisera are known not to be absorbed by tissue extracts or lymphocytes. The pig blood group B is present on glycophorin but found limitedly in erythrocytes. Also, pig skin grafts mismatched cannot induce the blood group K-reactive Abs in sera and also to the blood group C, D, F, J and O. As well known, the blood groups of B and F bear two different alleles of EABa and EABb linked to glycophorin protein [41]. Location of the pig EAB gene is found on pig chromosome 8, as the similar location of chromosome 4q31.2 of humans, which is the location for the A and B glycophorin genes [47]. The blood group F also bears two different alleles of EAFa and EAFb, which both EAFa and EAFb are located on the chromosome 8 of pigs. They are compared with the human M/N and S/s blood groups linked to the A and B glycophorins. The blood groups of B and F are known to be extremely immunogenic, causing for the newborn hemolytic diseases. Monoclonal Abs are made for the antigens of Ba and Fa in experimental approach [47].

On the other hand, for the blood group C and J of pigs, the blood groups of C and J are related to the pig MHC genes, where the gen location is ion pig chromosome 7 [48]. The order of the C and J genes is located in the order of SLA–EAJ–EAC. Among them, the antigen J gene bears two different co-dominant alleles named EAJa and EAJb. In contrast, another antigen C gene bears only one allele named EACa and the C antigen-deficient pigs termed EAC−strain. For blood group D antigen, the D gene carries two different alleles named EADa and EADb. The EAD blood group gene is located on the pig chromosome 12 [16, 49].

For the I antigenic gene, the gene bears two different alleles named EAIa and EAIb, where location of the blood group EAI antigen gene is present on pig chromosome 18. For the K antigen, the K gene bears diversely six alleles, characteristic of seven different factors named EAKacef, EAKacf, EAKade, EAKae, EAKbf, and EAK- (negative) [50]. EAK gene location is in pig chromosome 9. On the other hand, blood group M antigen is known to be highly complex, which bears 13 different factors from Ma-to-Mm factor [51]. In addition, the blood group M antigen complex bears 20 alleles named EAMab(e), EAMaem, EAMaejm, EAMade(m), EAMb, EAMbcd, EAMbcdi, EAMbd, EAMbdg, EAMbd(f)m, EAMcd, EAMcdi, EAMcdk, EAMd, EAMdjk, EAMdk, EAMef, EAMefm, EAMh, and EAM– (negative). Pig chromosome 11 shows the EAM gene location. For the O antigen gene, it bears two different alleles named EAOa and EAOb. Gene location of EAO is in pig chromosome 6, and it encodes for the separate genes of EAH (FUT1) locus and S (FUT2) locus. Interestingly, location of the pig EAO gene is closely near on the end of the chromosome 6 telomere of pigs and this location is syntenic to the chromosome 1 of humans [47]. A new blood group P antigen is also proposed [51] with one allele designated EAPa. Based on previous results obtained from the transfused pigs, the transfused pigs produce blood groups-reactive Abs and thus quickly clear transfused erythrocytes. After injection of Ea blood, pigs develop anti-Ea antibodies with quick clearance of Ea erythrocytes. Blood group Abs cause newborn pig hemolysis disease. Since the immunoglobulins are not transport to pig placenta, piglets receive their maternal Abs from the mother's colostrum after birth. The Ba, Ea, Fa, and Kb-responsive Abs of pigs cause hemolytic diseases, and therefore, the cross-match compatible blood is required when pigs receive transfusions.

With regard to Rh antigen, pigs are known to bear an only single RH gene, but no blood group antigen in pigs. RH gene is polymorphic in fourth intron region and the RH gene location is on pig chromosome 6, which is closely near to the blood group antigens of EAH and S loci of pig genome [52], although polymorphism is not found in the coding region. For the human case, the blood group Rh D, C, c, E and e antigens of humans show extremely immunogenic property, as they are causing factors during the transfusion responses and the newborn pigs hemolytic disease. The known human Rh antigenic genes include the RHD and RHCE, produced by a gene duplication [53].

12.8 Swine Leukocyte Antigens (SLA) Antigens

In human allotransplantation, the antigens of human major histocompatibility complex (MHC) are known as a substantial target molecules of humoral immune rejection, which has been done for more than 60 years [54]. Although humans produce SLA class II-reactive Abs, the actual approaches are less well performed and established. However, human humoral immune reaction against pig class-II SLA antigens has been elucidated by their recognizing reaction to the HLA C-I positive and HLA C-II negative platelets of human, which are experimentally pooled. Generally, the immune system of human responds directly to the antigenic SLA, if SLA-reactive Abs are produced, consequently to induce xenograft damage [55].

Humans have antibodies to target epitopes on the HLA. Since the single HLA beads are experimentally developed, the target-binding spectrum of HLA-reactive Abs has been simply analyzed during clinically applied allotransplantation and binding properties of donor specific Abs (DSA) to HLA C-I and C-II antigens are easily facilitated for their detections [56]. The developed single HLA beads easily analyze HLA-specific antibodies in allotransplantation and donor HLA-I and HLA–II-specific antibodies. For example, specifically-reactive Abs are substantial barriers in kidney transplantation regardless any type of xenotransplantation or allotransplantation, as histocompatibility test indicates a direction of allotransplantation [57]. Anti-HLA Abs are known to be produced by pregnancies, previous blood transfusions in patient recipients or failed renal allograft-transplantation [58]. Anti-HLA class-I-reactive Abs, which are specific to donors, are believed as decision parameter of survival of renal allograft transplantation, like antibody-dependent rejection [59]. Cross reactivity of human HLA and SLA might imply for successful indication of xenotransplantation that a xenograft-received patient is followed by an allotransplantation [60].

The swine SLA complex or swine MHC is the important subject of the pig-to-human xenotransplantation, because anti-HLA Abs of human are easily cross-react with SLA in the pig-to-human xenotransplantation. Cross reaction of swine SLA and human HLA is also a problem in successful xenotransplantation [61]. The sensitive reactivity of pooled single antigen beads enabled understanding on class-II Abs for the long term graft survival, because human HLA-reactive Abs cross-react with the homologously identical SLA C-I and C-II. Such HLA-reactive Abs can cross-react with pig SLA C-I and C-II [62], although cross-reactivity of human HLA class II is not clear. Certain human HLA C-I-reactive Abs also shows cross-reactivity with pig SLA C-I antigens, explaining the actual cross-match reaction. Thus, people can continuously try to react to the triple xenoantigen-deleted KO pigs. Some human HLA C-I-specific Abs can cross-react with pig SLA C-I antigens and can be used to screen the cross-matching of the pig [63]. On other point, for the existence of anti-SLA class-II Abs, pig SLA class-II like human HLA generates DR and DQ, except for DP [64]. The anti-class-II Abs are well documented as a key mediator to transplant glomerulopathy diseases and loss of transplanted grafts in allotransplantation [65]. Both pig SLA class-II-DR and class-II–DQ seem to function as xenoantigens, because anti-DQ specific Abs are known as the major HLA Abs in renal allografted recipient patients. Hence, SLA DQ might be a strong xenoantigen because anti-pig SLA class-II Abs in human patients easily recognize the pig SLA class-II DR and/or class-II DQ.

In preclinically-operating pig-to-primate xenotransplantation, recipient hosts with lower xenoreactive Abs are possibly survived for longer and prolonged length of time under the usage condition of immunosuppressive conditions, which T-cells are suppressed by relevant drugs [66]. When recognition of human Abs with xenoantigenic glycan-deleted KO cells of pigs was compared with SLA C-I KO cells of pigs and also the human-reactive and pig-reactive IgGs were examined, the results showed that HLA and SLA epitope molecules are similar but SLA class-I molecules are xenoantigens [67]. SLA and HLA proteins are extensively homologous with sequence and structural identities. Therefore, the SLA as pig homologs of HLA is also considered to be xenoantigen in humans. In the amino acid levels and protein structures, the SLA and HLA proteins are 75–80% homologous to HLA protein. Thus, they share cross-reactivity with their humoral reaction epitopes. Therefore, human HLA-reactive Abs in the previous allograft-sensitized patient recipients caused for cross-reaction with pig SLA antigens.

The C-I SLA KO pigs were established [68]. The HLA C-I-related epitope proteins present on pig cells are recognized as similar SLA molecules due to high sequence and structural similarities. HLA-specific monoclonal antibodies against SLA molecules were isolated from humans [69]. Some HLA class-II-reactive Abs confer the cross-reaction with the SLA class-II molecules expressed on those KO pigs, indicating that human- and pig-reactive Abs can bind to common sites of human HLA alleles [67]. From the designed experiment that peritoneal blood mononuclear cells (PBMCs) obtained from SLA C-I KO pigs were used, SLA class-I has been evidenced to be a powerful xenoantigen because the HLA class-I and SLA class-I epitopes are shared together [63]. These MHC antigens are directly associated with binding of the HLA-specific Abs to the SLA class-I antigens.

HLA-reactive Abs in sensitized humans undergo the cross-reaction with SLA antigens to elicit cytotoxic cell death [70]. For example, a SLA C-I protein having a 45-kDa molecular weight is reacted with anti-HLA IgG. For human who is not HLA-sensitized, xenoantigen-reactive Abs capable of recognition of pig lymphocytes can be eliminated by experimental pig RBC absorption technique [71]. If human is sensitized with SLA, pig lymphocytes will still cross-match even after removal of xenoreactive antibodies, the reason is why the anti-HLA class-II Abs are mainly IgG isotype. In human cytotoxicity against pig, human IgM binding induced the cytotoxicity to pig WT PBMCs, but not human IgG type. However, interestingly, both human IgG and IgM recognitions induced the cytotoxic cell death to PBMCs of α1,3Gal-T KO pig [63]. Activity of anti-non-α1,3Gal Abs is displayed by IgM type Abs.

Human monoclonal HLA class-I Abs also show the cross-reaction with pig SLA-positive mononuclear cells in CDC assays. HLA-reactive monoclonal antibodies are mainly IgM type and the HLA epitopic antigens show the similar conservation in the amino acid sequence of SLA molecules. Anti-HLA class-I antibody-carrying patients can therefore easily discriminate the pig organs lacking their antibodies-reactive SLA antigens [70]. In the recipient patients immunosuppressed who received islet xenografts of fetal pigs, specific xenoantibodies were generated without direct evidence of reactivity [72], and the pig MHC class I antigen-specific antibodies were reacted with the α1,3Gal antigenic epitopes present on the side chains of pig MHC-I carbohydrate. The reaction role of SLA-specific Abs in patient recipients who are highly HLA-sensitized, is not explained to date. In the fetal pig islets-planted human patients, the immune response to the pig SLA molecules is reported [72].

Highly reactive patients to pig antigens might need additional genome editing from the donor pigs before clinical renal xenotransplant trials. If an additional antigen is discovered, a relevant genome editing technology can be applied to remove and eliminate the antigens in xenotransplantation. Therefore, this genome editing strategy can create a xenotransplantable donor pigs with a more suitable cross-matches [73]. Because elimination of pig class I SLA xenoantigens is suggested, genome editing approaches to deal with the SLA gene area can be used to generate donor pigs with the non-cross-matches with humans. Those SLA-eliminated pigs are suitable for prevention of SLA class-I-recognizing Abs in cross-matched reaction. Recent CRISPR/Cas genome editing relevantly eliminates the SLA C-II molecule or its epitope, by deletion of the pig DQ locus using CRISPR/Cas or replacement of the pig DQ epitope gene sequences.

12.9 T, Tn, and Sialyl-Tn Antigens

Thomsen–Friedenreich antigen or T antigen is the Galβ1,3GalNAc-α-O-Thr/Ser antigen. Tn antigen is the GalNAc-α-O-Ser/Thr and sialyl-Tn antigen is the NeuAcα2,6GalNAc-α-O-Ser/Thr. These antigen epitopes are well studied in the tumor vaccination aspects of the cellular and humoral immunities in malignant cancers [74]. In humans having those antigen-specific preformed antibodies, the mechanistic explanation of the antibody production has been attributed to a consequence of specific bacterial flora habit and colonization, like ABO blood group- and α1,3-Gal-specific Abs. T antigen and Tn antigen are also existed in RBCs [75]. T antigens are also present in endothelium of pig kidney and tubular cells of pig tissues. T antigens are expressed in pig fibroblast and pig GalT-KO endothelial cells. However, the levels of the T and Tn antigens are decreased upon α1,3-Gal antigenic disruption [29]. Using the PNA to detect T-antigen, the T-antigen is known to express in pig fibroblasts. However, the Lewis antigen structures of Le^x antigen or sialyl-Le^x antigen are not present in any the pig ECs. It was also known that pig α1,3-Gal-T KO donor does not generate new type glycolipids to recognize human serum Abs [76]. Binding levels of *Lotus tetragonolobus* lectin, LTL [77], and fungus *Aleuria aurantia* (Pezizaceae) lectin (AAL) [78] specific for Le^x and sialyl-Le^x were not changed in α1,3Gal-T KO pigs, but binding levels of plant *Maackia amurensis* MAL, known as a Sia and sulfate binding lectin [79] and seed *Erythrina cristagalli* lectin ECA [80], which binds to a common structure of LacNAc (type 2 chain) residue in N-glycans, were increased in pig α1,3-Gal-T KO donor in comparison to the pig wild-type strains [81].

For Fuc residue binding, LTL lectin as a Fuc-binding lectin is isolated from lotus seeds of *Tetragonolobus purpureus*, which is also termed *Lotus tetragonolobus* with the common terms of winged pea or asparagus Pea, binds to fucose residues and specifically recognizes its major motif, Le^x, calling Le^x-specific lectin. LTL recognizes α1,3-Fuc residues in Le^x, Le^y, 6′sulfo-SLe^x, but not α1,2-Fuc glycans. The l-Fuc-binding lectin termed AAL was isolated from the orange peel fungus *Aleuria aurantia* fruiting bodies by Yazawa [78]. Unlike LTL lectin specific for the Fuc α1,2-linked residues, AAL recognizes preferentially Fucα1,6-linked GlcNAc or to Fucα1,3-linked LacNAc structures. AAL also binds Fuc-linked nucleic acids. AAL recognizes the α-l-Fuc residue with a strong affinity to GlcNAcβ1,4(Fucα1,6) GlcNAc core in *N*-glycans and weak affinity to Fucosyl-glycans. The AAL agglutinates human ABO erythrocytes.

12.10 Lectin Analysis to Detect Glycoantigens in Pigs

In xenotransplantation, glycan-based carbohydrate antigens are still representatively significant hurdles to overcome, which are caused by non-α1,3Gal antigenic epitopes, although the α1,3Gal-T KO pigs were created and non-α1,3Gal antigenic epitopes are being studied. Therefore, to search putative glycoantigens, lectin blot analysis using fibroblast cells and endothelial cells from pig α1,3Gal-T KO strains and α1,3Gal-T KO-GlcNAc-transferase-III (GnT-III) was previously performed by Miyagawa et al. [76]. The pig EC exhibited several binding capacities with the lectins from European spindle plant, *Euonymus europaeus* (EEL) (or agglutinin EEA), which binds to blood group B and H carbohydrates, with a strong affinity for terminal Fuc glycans and, to some extents, Man residues [76], and GSI-B4 to bind α1,3Gal epitopes.

Lectins of *Glycine max* (SBA), *Helix pomatia* (HPA), *Wisteria floribunda* (WFA), and GSI-A4 recognizable to the GalNAc residues in the Tn antigen (on known as Thomsen–Friedenreich precursor) were bounded with the pig ECs. Moreover, the human ECs were strongly bound to ECA, MAL, *Trichosanthes japonica* I (TJA-I), and *Ulex europaeus* I (UEA-I) lectins. However, the α1,3Gal-T KO pig ECs were very weakly bound with the lectins of EEL and GSI-B4. In addition, the binding to lectins of *Bauhinia purpurea* alba (BPL), SBA, HPA, and GSI-A4 have been apparently lost. In contrast, the α1,3Gal-T KO pig ECs were increasingly bound with the lectins of *Sambucus sieboldiana* (SSA), *Sambucus nigra* (SNA), and TJA-I, what are specific for α2,6-Sia, when compared to the wild-type ECs of pig. For pig fibroblast cells, *Maclura pomifera* (MPA) and *Arachis hypogaea* (PNA) bind Galβ1,2GalNAc residues bound to pig fibroblasts. Thus, pig fibroblasts contain the Thomsen–Friedenreich (T) antigens. Thus, GalNAc residue-specific T-antigen and Tn-antigens as non-α1,3Gal antigens were also decreased with α1,3Gal in the KO-pig [76], instead, α2,3 and α2,6 Sia expression was increased compared with human ECs. Apart from the Forssman antigen, the terminally linked GalNAc residue level was highly present in cells from wild-type pigs, but reduced in the pig α1,3Gal-T KO cells, due to cross talk of pig α1,3Gal-T enzyme and GalNAc-transferase enzyme [76]. In contrast, terminally attached α1,3-Gal or β-Gal residues and α-GalNAc residues are not present on pig cells [82]. In a pig-to-baboon transplantation, it was shown that common α- or β-Gal residues and α-GalNAc residues are not target residues for non-α1,3Gal Abs. However, the terminally linked GalNAc residue can be recognized as a non-α1,3Gal antigen. In the LacNAc, LacNAc levels was rather increased in GalT-KO mice, as the LacNAc epitope binds natural antibodies [83]. However, intensity levels of ECA and RCA120 to bind lactose and LacNAc were not changed [76].

12.11 Protein Antigens Detection Using Non-α1,3Gal Antibodies

Apart from the xenoantigenic carbohydrate residues, the protein sequence difference of some homologous proteins between human, primates and pigs raises an immune antixenograft reaction [84]. The species-specific protein epitopes in xenoantigenic glycoproteins induce IgG production against carbohydrate residues like α1,3Gal. Many membrane proteins, which are immunoreactive, in liver endothelial cells of α1,3Gal-T KO pig can also be bound by natural IgG and IgM Abs of human. IgG antibody production is induced by α1,3Gal-recognizing B-cells and APCs when the T-helper cells were exposed with immunogenic xenopeptides. The activated T-cells induce growth, proliferation, isotype switching and somatic mutation of B lymphocytic cells [85]. Membrane glycoprotein fibronectin-specific natural antibodies are known in human sera. The glycoprotein expression is significantly high on the ECs surfaces of kidney and liver tissues of pig wild-type and α1,3Gal-T KO strains. Fibronectin is the second-largely expressed membrane protein, where collagen is the most largely expressed. Human IgG and IgM Abs equally bind to human and pig fibronectins. Fibronectin-specific antibodies cross-react with CD98 and megalin [86]. Fibronectin also elicit the antibody production in pig-to-primate xenotransplantation, as other proteins of inflammation regulator like annexin A2, hemostasis like CD9, and protein C receptor of ECs or the complements (CD46 and CD59) also elicit the reaction [87]. They induce AHR. In addition, the anti-HLA antibodies recognize the swine MHC or SLA complex, presumably due to recognition of MHC epitopes shared with the two species [88]. The antibodies are cytotoxic to pig cells, requiring better selection of the matched donor pig haplotype.

Until now, the non-α1,3Gal protein epitopes bound by human non-α1,3Gal Abs have been searched. For example, Zhu group found an RBC 45-kD protein expressed of pigs, which is reacted with non-α1,3Gal Abs and NeuGc determinants-carrying proteins [70] When the pig embryonic ventral mesencephalon membrane proteins were blotted with Galα1,3Gal Abs-depleted serum of humans, three differently banded proteins were observed to 210 kDa, 105 kDa, and 50 kDa proteins [89]. Patients were graft-planted with pig α1,3Gal epitope-depleting glutaraldehyde-treated patellar tendon in order to replace anterior cruciate ligament failed [90]. Form anti-α1,3Gal Abs-removed sera of the xenograft-transplanted patient recipients, several proteins were found as candidates in the α1,3Gal-deleted ligament and kidney of pigs. In a kidney of α1,3Gal-T KO pigs transplanted to baboon as a recipient, a 47-kD protein to induce non-α1,3Gal Abs was also identified from pig AECs. Byrne et al. [91] found a non-α1,3Gal antigen of pig fibronectin, an extracellular matrix protein, which is a highly glycosylated, from the induced IgG Abs reacted with α1,3Gal-T KO PAEC proteins, when they have searched non-α1,3Gal Abs-targeted proteins using the pig to baboon transplantation as a heart xenotransplantation model. Technically, the recipient baboon was splenectomized and the α1,3Gal KO baboon recipient was immune-suppressed using anti-CD20 mAb and anti-thymocyte globulin mAb by combined treatment with immune suppressors of sirolimus and tacrolimus. Then, from the induced IgG Abs reacted with α1,3Gal-T KO pig AEC proteins.

12.12 Immune Responses of Non-α1,3Gal Antibody Production in Trials of Pig-to-Human Clinical Xenotransplantation

In clinical trials, patient recipients of humans can be subjected to pig organ xenograft transplantation or perfused for clinical trials, as previously established in thoracic surgery for heart valve replacements [92]. Extracorporeal perfusion of pig liver in patient recipients having fulminant dysfunction of hepatic organs and two extracorporeal pig kidney perfusion were tried [93]. Encapsulated pig islets [94] were also used on to overcome non-α1,3Gal antibodies. During the 1980s and 1990s, three independent trials were clinically performed in human subjects to avoid the xenoantigenic barriers in such xenografting operation of pig fetal islets into immunosuppressed renal patients. α1,3Gal-deficient cruciate ligaments of pig were also xenografted in orthopedic surgery [95].

References

1. Joziasse DH, Shaper JH, Van den Eijnden DH, Van Tunen AJ, Shaper NL. Bovine alpha 1–3-galactosyltransferase: isolation and characterization of a cDNA clone. Identification of homologous sequences in human genomic DNA. J Biol Chem. 1989;264:14290–7.
2. Barone A, Benktander J, Teneberg S, Breimer ME. Characterization of acid and non-acid glycosphingolipids of porcine heart valve cusps as potential immune targets in biological heart valve grafts. Xenotransplantation. 2014;21:510–22.
3. Wu GD, Fujii G, Johnson E, Swensson J, Oakley O, Cramer DV. Failure of anti-Forssman antibodies to induce rejection of mouse heart xenografts. Xenotransplantation. 1999;6:90–7.
4. Wieslander J, Mansson O, Kallin E, et al. Specificity of human-antibodies against Gal-alpha-1-3gal carbohydrate epitope and distinction from natural antibodies reacting with Gal-alpha-1-2gal or Gal-alpha-1-4 Gal. Glycoconj J. 1990;7:85–100.
5. Yeh P, Ezzelarab M, Bovin N, et al. Investigation of potential carbohydrate antigen targets for human and baboon antibodies. Xenotransplantation. 2010;17:197–206.

6. Young WW Jr, Hakomori SI, Levine P. Characterization of anti-Forssman (anti-Fs) antibodies in human sera: their specificity and possible changes in patients with cancer. J Immunol. 1979;123:92–6.
7. Breimer M. Chemical and immunological identification of the forssman pentaglycosylceramide in human kidney. Glycoconj J. 1985;2:375–85.
8. Yamamoto F, Yamamoto M. Molecular genetic basis of porcine histo-blood group AO system. Blood. 2001;97(10):3308–10.
9. Sprague L. On the recognition and inheritance of the soluble blood group property "Oc" of cattle. Genetics. 1958;43:913.
10. Rasmusen BA. Gene interaction and the A–O blood-group system in pigs. Genetics. 1964;50:191.
11. King TP, Kelly D. Ontogenic expression of histo-blood group antigens in the intestines of suckling pigs: lectin histochemical and immunohistochemical analysis. Histochem J. 1991;23:43.
12. Aminoff D, Morgan WTJ, Watkins WM. Specific serological characters of the mucoids of hog gastric mucin. Nature. 1946;158:879.
13. Diswall M, Angstrom J, Karlsson H, et al. Structural characterization of alpha1,3-galactosyltransferase knockout pig heart and kidney glycolipids and their reactivity with human and baboon antibodies. Xenotransplantation. 2010;17:48–60.
14. Smith DM, Newhouse M, Naziruddin B, Kresie L. Blood groups and transfusions in pigs. Xenotransplantation. 2006;13(3):186–94.
15. Thurin J, Blaszczyk-Thurin M. Porcine submaxillary gland GDP-L-fucose: beta-D-galactoside alpha-2-l-fucosyltransferase is likely a counterpart of the human secretor gene-encoded blood group transferase. J Biol Chem. 1995;270:26577.
16. Rohrer GA, Vogeli P, Stranzinger G, Alexander LJ, Beattie CW. Mapping 28 erythrocyte antigen, plasma protein and enzyme polymorphisms using an efficient genomic scan of the porcine genome. Anim Genet. 1997;28:323.
17. Meijerink E, Neuenschwander S, Fries R, et al. A DNA polymorphism influencing alpha(1,2) fucosyltransferase activity of the pig FUT1 enzyme determines susceptibility of small intestinal epithelium to Escherichia coli F18 adhesion. Immunogenetics. 2000;52:129.
18. Barra JL, Monferran CG, Balanzino LE, Cumar FA. Escherichia coli heat-labile enterotoxin preferentially interacts with blood group A-active glycolipids from pig intestinal mucosa and A- and B-active glycolipids from human red cells compared to H-active glycolipids. Mol Cell Biochem. 1992;115:63.
19. Johnson JR, Roberts PL, Stamm WE. P fimbriae and other virulence factors in Escherichia coli urosepsis: association with patients' characteristics. J Infect Dis. 1987;156:225.
20. Boren T, Falk P, Roth KA, Larson G, Normark S. Attachment of Helicobacter pylori to human gastric epithelium mediated by blood group antigens. Science. 1993;262:1892.
21. Reguigne-Arnould I, Couillin P, Mollicone R, et al. Relative positions of two clusters of human alpha-l-fucosyltransferases in 19q (FUT1-FUT2) and 19p (FUT6-FUT3-FUT5) within the microsatellite genetic map of chromosome 19. Cytogenet Cell Genet. 1995;71:158.
22. Hanagata G, Gasa S, Sako F, Makita A. Human blood group A and H glycolipids in porcine plasma. Evidence for acquisition of the erythrocyte antigens from plasma. FEBS Lett. 1990;261:312.
23. Hirsch HF, Graham HA. Adsorption of Lec and Led from plasma onto red blood cells. Transfusion. 1980;20:474.
24. Slomiany A, Slomiany BL, Horowitz MI. Structural study of the blood group A active glycolipids of hog gastric mucosa. J Biol Chem. 1974;249:1225.
25. Rydberg L, Molne J, Strokan V, Svalander CT, Breimer ME. Histo-blood group A antigen expression in pig kidneys – implication for ABO incompatible pig-to-human xenotransplantation. Scand J Urol Nephrol. 2001;35:54.

26. Oriol R, Ye Y, Koren E, Cooper DK. Carbohydrate antigens of pig tissues reacting with human natural antibodies as potential targets for hyperacute vascular rejection in pig-to-man organ xenotransplantation. Transplantation. 1993;56:1433–42.
27. Rydberg L, Molne J, Strokan V, et al. Histo-blood group A antigen expression in pig kidneys–implication for ABO incompatible pig-to-human xenotransplantation. Scand J Urol Nephrol. 2001;35:54–62.
28. Leight GS, Kirkman R, Rasmusen BA, et al. Transplantation in miniature swine. III: effects of MSLA and A-O blood group matching on skin allograft survival. Tissue Antigens. 1978;12:65.
29. Swanson JL, Cooling L. Porcine red blood cells express a polyagglutinable red blood cell phenotype. Transfusion. 2005;45:1035. author reply 1036.
30. Cohney S, Mouhtouris E, McKenzie IF, Sandrin MS. Molecular cloning and characterization of the pig secretor type alpha 1,2fucosyltransferase (FUT2). Int J Mol Med. 1999;3:199.
31. Darr D, McCormack KM, Manning T, et al. Comparison of Dolichos biflorus lectin and other lectin-horseradish peroxidase conjugates in staining of cutaneous blood vessels in the hairless mini-pig. J Cutan Pathol. 1990;17:9.
32. Diswall M, Benktander J, Ångström J, Teneberg S, Breimer ME. The alpha1,3GalT knockout/alpha1,2FucT transgenic pig does not appear to have an advantage over the alpha1,3GalT knockout pig with respect to glycolipid reactivity with human serum antibodies. Xenotransplantation. 2014;21(1):57–71.
33. Costa C, Zhao L, Burton WV, et al. Expression of the human alpha1,2-fucosyltransferase in transgenic pigs modifies the cell surface carbohydrate phenotype and confers resistance to human serum-mediated cytolysis. FASEB J. 1999;13:1762–73.
34. Costa C, Zhao L, Burton WV, et al. Transgenic pigs designed to express human CD59 and H-transferase to avoid humoral xenograft rejection. Xenotransplantation. 2002;9:45–57.
35. Costa C, Brokaw JL, Fodor WL. Characterization of cartilage from H-transferase transgenic pigs. Transplant Proc. 2008;40:554–6.
36. Le Bas-Bernardet S, Tillou X, Poirier N, et al. Xenotransplantation of galactosyl-transferase knockout, CD55, CD59, CD39, and fucosyl-transferase transgenic pig kidneys into baboons. Transplant Proc. 2011;43:3426–30.
37. Holgersson J, Jovall PA, Samuelsson BE, Breimer ME. Structural characterization of non-acid glycosphingolipids in kidneys of single blood group O and A pigs. J Biochem (Tokyo). 1990;108:766.
38. Coddens A, Diswall M, Angstrom J, Breimer ME, Goddeeris B, Cox E, Teneberg S. Recognition of blood group ABH type 1 determinants by the FedF adhesin of F18-fimbriated Escherichia coli. J Biol Chem. 2009;284:9713–26.
39. Bao WB, Ye L, Pan ZY, Zhu J, Du ZD, Zhu GQ, Huang XG, Wu SL. The effect of mutation at M307 in FUT1 gene on susceptibility of Escherichia coli F18 and gene expression in Sutai piglets. Mol Biol Rep. 2012;39:3131–6.
40. Hojny J, Stratil A. Report on the pig and sheep blood group and polymorphic protein workshops (Libechov, 9 to 11 August 1978). Anim Blood Groups Biochem Genet. 1978;9:245.
41. Saison R, Ingram DG. Production of specific haemagglutinins in pigs after receiving skin homografts. Nature. 1963;197:296. A: Academic Press, 1998: 389.
42. Schmid DO, Buschmann HG, Hammer. Blood groups in animals. Lengerich: Pabst Science Publishers; 2003. p. 186–227.
43. Hojny J, Nielsen PB. Allele Ebdgjmr (E17) in the pig E blood group system. Anim Genet. 1992;23:523.
44. Klaudy J, Hruban V, Hradecky J, Pazdera J, Pech V. The presence of blood group and lymphocyte antigens on porcine granulocytes. Anim Blood Groups Biochem Genet. 1981;12:67.
45. Abe T, Mogi K, Oishi T, Himeno K, Hosoda T. A subclinical case of hemolytic disease of newborn pigs caused by anti-Ea. Nippon Juigaku Zasshi. 1970;32:139.
46. Hojny J, Hradecky J, Pazdera J. The blood group factor Kf and allele Kae in the pig. Anim Blood Groups Biochem Genet. 1979;10:175.

47. Gonzalez A, Friend M, Moreno A, Pintado CO, Vogeli P, Llanes D. A monoclonal antibody to swine erythrocytes recognizes the B blood group on the major glycophorin. Anim Genet. 1995;26:351.
48. Onda M, Kudo S, Fukuda M. Genomic organization of glycophorin A gene family revealed by yeast artificial chromosomes containing human genomic DNA. J Biol Chem. 1994;269:13013.
49. Hradecky J, Hruban V, Pazdera J, Klaudy J. Map arrangement of the SLA chromosomal region and the J and C blood group loci in the pig. Anim Blood Groups Biochem Genet. 1982;13:223.
50. Cepica S, Moser G, Schroffel J Jr, et al. Chromosomal assignment of porcine EAD, EAO, LPR and P3 genes by linkage analysis. Anim Genet. 1996;27:109.
51. Nielsen PB, Vogeli P. A new Kd subgroup designated Kg in the porcine K blood group system. Anim Blood Groups Biochem Genet. 1982;13:65.
52. Hojny J, Schroffel J Jr, Geldermann H, Cepica S. The porcine M blood group system: evidence to suggest assignment of its M1 factor to a new system (P). Anim Genet. 1994;25(Suppl 1):99.
53. Omi T, Vogeli P, Hagger C, et al. cDNA cloning, mapping and polymorphism of the porcine Rhesus (RH) gene. Anim Genet. 2003;34:176.
54. Avent ND, Reid ME. The Rh blood group system: a review. Blood. 2000;95:375.
55. Zachary AA, Leffell MS. Desensitization for solid organ and hematopoietic stem cell transplantation. Immunol Rev. 2014;258(1):183–207.
56. Breimer ME. Gal/non-Gal antigens in pig tissues and human non-Gal antibodies in the GalT-KO era. Xenotransplantation. 2011;18:215–28.
57. Pei R, Lee JH, Shih NJ, Chen M, Terasaki PI. Single human leukocyte antigen flow cytometry beads for accurate identification of human leukocyte antigen antibody specificities. Transplantation. 2003;75(1):43–9.
58. Tambur A, Claas F. HLA epitopes as viewed by antibodies: what is it all about? Am J Transplant. 2015;15:1148–54.
59. Patel R, Terasaki PI. Significance of the positive crossmatch test in kidney transplantation. N Engl J Med. 1969;280:735–9.
60. Lefaucheur C, Loupy A, Hill GS, et al. Preexisting donor-specific HLA antibodies predict outcome in kidney transplantation. J Am Soc Nephrol. 2010;21:1398–406.
61. Cooper DK, Dou KF, Tao KS, Yang ZX, Tector AJ, Ekser B. Pig liver xenotransplantation: a review of progress toward the clinic. Transplantation. 2016;100(10):2039–47.
62. Ladowski JM, Reyes LM, Martens GR, Butler JR, Wang ZY, Eckhoff DE, Tector M, Tector AJ. Swine leukocyte antigen (SLA) class II is a xenoantigen. Transplantation. 2017;102(2):249–54. https://doi.org/10.1097/TP.0000000000001924.
63. Diaz Varela I, Sanchez Mozo P, Centeno Cortes A, Alonso Blanco C, Valdes CF. Cross-reactivity between swine leukocyte antigen and human anti-HLA-specific antibodies in sensitized patients awaiting renal transplantation. J Am Soc Nephrol. 2003;14(10):2677–83.
64. Martens GR, Reyes LM, Butler JR, et al. Humoral reactivity of renal transplant-waitlisted patients to cells from GGTA1/CMAH/B4GalNT2, and SLA class I knockout pigs. Transplantation. 2017;101(4):e86–92.
65. Smith DM, Lunney JK, Ho CS, et al. Nomenclature for factors of the swine leukocyte antigen class II system, 2005. Tissue Antigens. 2005;66(6):623–39.
66. Wiebe C, Pochinco D, Blydt-Hansen TD, et al. Class II HLA epitope matching-A strategy to minimize de novo donor-specific antibody development and improve outcomes. Am J Transplant. 2013;13(12):3114–22.
67. Higginbotham L, Mathews D, Breeden CA, et al. Pre-transplant antibody screening and anti-CD154 costimulation blockade promote long-term xenograft survival in a pig-to-primate kidney transplant model. Xenotransplantation. 2015;22:221–30.
68. Martens GR, Reyes LM, Butler JR, Ladowski JM, Estrada JL, Sidner RA, Eckhoff DE, Tector M, Tector AJ. Humoral reactivity of renal transplant-waitlisted patients to cells from GGTA1/CMAH/B4GalNT2, and SLA class I knockout pigs. Transplantation. 2017;101(4):e86–92.
69. Reyes LM, Estrada JL, Wang ZY, et al. Creating class I MHC-null pigs using guide RNA and the Cas 9 endonuclease. J Immunol. 2014;193:5751–7.

70. Mulder A, Kardol MJ, Arn JS, et al. Human monoclonal HLA antibodies reveal interspecies crossreactive swine MHC class I epitopes relevant for xenotransplantation. Mol Immunol. 2010;47:809–15.
71. Barreau N, Godfrin Y, Bouhours JF, et al. Interaction of anti-HLA antibodies with pig xenoantigens. Transplantation. 2000;69:148–56.
72. Hara H, Ezzelarab M, Rood PP, et al. Allosensitized humans are at no greater risk of humoral rejection of GT-KO pig organs than other humans. Xenotransplantation. 2006;13:357–65.
73. Satake M, Kawagishi N, Rydberg L, et al. Limited specificity of xenoantibodies in diabetic patients transplanted with fetal porcine islet cell clusters. Main antibody reactivity against alpha-linked galactose-containing epitopes. Xenotransplantation. 1994;1:89–101.
74. Li P, Estrada JL, Burlak C, et al. Efficient generation of genetically distinct pigs in a single pregnancy usingmultiplexed single-guide RNA and carbohydrate selection. Xenotransplantation. 2015;22:20–31.
75. Springer GF. Immunoreactive T and Tn epitopes in cancer diagnosis, prognosis, and immunotherapy. J Mol Med. 1997;75:594–602.
76. Miyagawa S, Takeishi S, Yamamoto A, Ikeda K, Matsunari H, Yamada M, Okabe M, Miyoshi E, Fukuzawa M, Nagashima H. Survey of glycoantigens in cells from alpha1-3galactosyltransferase knockout pig using a lectin microarray. Xenotransplantation. 2010;17:61–70.
77. Diswall M, Schuurman HJ, Dor F, Rydberg L, Breimer ME. Biochemical studies of Gal antigens in small intestine and pancreas from alpha1,3-galactosyltransferase knock-out miniture swine. Xenotransplantation. 2005;12:407.
78. Bojar D, Meche L, Meng G, Eng W, Smith DF, Cummings RD, Mahal LK. A useful guide to lectin binding: machine-learning directed annotation of 57 unique lectin specificities. ACS Chem Biol. 2022;17(11):2993–3012. https://doi.org/10.1021/acschembio.1c00689.
79. Yazawa S, Furukawa K, Kochibe N. Isolation of fucosyl glycoproteins from human erythrocyte membranes by affinity chromatography using aleuria aurantia lectin. J Biochem. 1984;96 (6):1737–42. https://doi.org/10.1093/oxfordjournals.jbchem.a135006.
80. Geisler C, Jarvis DL. Effective glycoanalysis with Maackia amurensis lectins requires a clear understanding of their binding specificities. Glycobiology. 2011;21(8):988–93.
81. Berman E, Brown JH, Lis H, Sharon N. Binding of [1-13C]galactose-labeled N-acetyllactosamine to Erythrina cristagalli agglutinin as studied by 13C-NMR. Eur J Biochem. 1985;152(2):447–51. https://doi.org/10.1111/j.1432-1033.1985.tb09217.x.
82. Fouquaert E, Peumans WJ, Smith DF, Proost P, Savvides SN, Van Damme EJM. The "old" Euonymus europaeus agglutinin represents a novel family of ubiquitous plant proteins. Plant Physiol. 2008;147:(pg. 1316-1324).
83. Zhu A. Binding of human natural antibodies to nonalphaGal xenoantigens on porcine erythrocytes. Transplantation. 2000;69:2422–8.
84. Milland J, Christiansen D, Sandrin MS. Alpha1,3-galactosyltransferase knockout pigs are available for xenotransplantation: are glycosyltransferases still relevant? Immunol Cell Biol. 2005;3:687–93.
85. Galili U. Xenotransplantation and ABO incompatible transplantation: the similarities they share. Transfus Apher Sci. 2006;35:45–58.
86. Tanemura M, Yin D, Chong AS, Galili U. Differential immune responses to alpha-gal epitopes on xenografts and allografts: implications for accommodation in xenotransplantation. J Clin Invest. 2000;105:301–10.
87. Chihara RK, Lutz AJ, Paris LL, et al. Fibronectin from alpha 1,3-galactosyltransferase knockout pigs is a xenoantigen. J Surg Res. 2013;184:1123–33.
88. Byrne GW, Du Z, Sun Z, Asmann YW, McGregor CG. Changes in cardiac gene expression after pig-to-primate orthotopic xenotransplantation. Xenotransplantation. 2011;18:14–27.
89. Zhu A, Hurst R. Anti-N-glycolylneuraminic acid antibodies identified in healthy human serum. Xenotransplantation. 2002;9:376–81.

90. Sumitran S, Liu J, Czech KA, et al. Human natural antibodies cytotoxic to pig embryonic brain cells recognize novel non-Galalpha1,3Gal-based xenoantigens. Exp Neurol. 1999;159:347–61.
91. Stone KR, Abdel-Motal UM, Walgenbach AW, et al. Replacement of human anterior cruciate ligaments with pig ligaments: a model for anti-non-gal antibody response in long-term xenotransplantation. Transplantation. 2007;83:211–9.
92. Byrne GW, Stalboerger PG, Davila E, et al. Proteomic identification of non-Gal antibody targets after pig-to-primate cardiac xenotransplantation. Xenotransplantation. 2008;15:268–76.
93. Mcgregor CG, Carpentier A, Lila N, et al. Cardiac xenotransplantation technology provides materials for improved bioprosthetic heart valves. J Thorac Cardiovasc Surg. 2011;141:269–75.
94. Breimer ME, Bjorck S, Svalander CT, et al. Extracorporeal ("ex vivo") connection of pig kidneys to humans 1. Clinical data and studies of platelet destruction. Xenotransplantation. 1996;3:328–39.
95. Valdes-Gonzalez RA, Dorantes LM, Garibay GN, et al. Xenotransplantation of porcine neonatal islets of Langerhans and Sertoli cells: a 4-year study. Eur J Endocrinol. 2005;153:419–27.

Chapter 13
Blood-Mediated Inflammatory Reaction (IBMIR) and Prevention of IBMIR

In transplantation, function loss of grafts is primarily observed in the early time of transplantation, mainly by two known responses of the instant blood-mediated inflammatory reaction (IBMIR) and hyperacute rejection (HR) [1]. It was reported that the first few hours have been reported to display the IBMIR after cell infusion of xenoantigenic islets [2]. IBMIR is characteristic of innate inflammatory responses and thrombotic pathway displayed by a serial reaction complement activation, immune cell infiltration, platelet adhesion, and coagulation [3]. IBMIR is frequently coupled with thrombosis and releases inflammatory mediators such as monocyte chemoattractant protein-1 (MCP-1), macrophage migration inhibitory factor (MIF), and IL-8. Thrombin fundamentally activates platelets. In addition, thrombin can activate innate immune cells of monocytes and neutrophils. Islets-producing tissue factors mainly cause the IBMIR as a causing factor [2, 4]. Therefore, the thrombosis reaction is prevented by blocking of tissue factors in vitro [3]. The IBMIR is also reported in three different transplantations of allo-, auto-, and xenotransplantation for islets [5], as also caused by incompatible species between thrombotic factors of humans and regulatory molecules of pigs. IBMIR is explained to be not only a response occurred in transplantation of allogeneic and xenogeneic islets, but also a responsive phenomenon to the islets and the highly immunogenic cells surrounded with acinar tissues [5]. However, such hurdles are now resolved by α1,3Gal-T KO animals and transgenic gene manipulation of human genes of CD46/CD55/CD59 complement regulators [6]. Using the genetic engineering technology in the immunological-challenged baboon model, HR- and IBMIR-based phenomena have been prevented by the transgenic islets of pigs. Like the HR, IBMIR can be mitigated by genetically disrupting the α1,3Gal-T gene and by transducing human CD46, CD55, and CD59 complement regulators as transgenic genes, respectively, as the blood mediates inflammatory rejection and the regulators protect the inflammation.

C.-H. Kim, *Glycoimmunology in Xenotransplantation*,
https://doi.org/10.1007/978-981-99-7691-1_13

References

1. Bennet W, Sundberg B, Groth CG, et al. Incompatibility between human blood and isolated islets of Langerhans: a finding with implications for clinical intraportal islet transplantation? Diabetes. 1999;48:1907–14.
2. Moberg L, Johansson H, Lukinius A, et al. Production of tissue factor by pancreatic islet cells as a trigger of detrimental thrombotic reactions in clinical islet transplantation. Lancet (London, England). 2002;360:2039–45.
3. Moberg L, Korsgren O, Nilsson B. Neutrophilic granulocytes are the predominant cell type infiltrating pancreatic islets in contact with ABO-compatible blood. Clin Exp Immunol. 2005;142:125–31.
4. Ji M, Yi S, Smith-Hurst H, et al. The importance of tissue factor expression by porcine NICC in triggering IBMIR in the xenograft setting. Transplantation. 2011;91:841–6.
5. Naziruddin B, Iwahashi S, Kanak MA, et al. Evidence for instant bloodmediated inflammatory reaction in clinical autologous islet transplantation. Am J Transplant. 2014;14:428–37.
6. Hawthorne WJ, Salvaris EJ, Phillips P, et al. Control of IBMIR in neonatal porcine islet xenotransplantation in baboons. Am J Transplant. 2014;14:1300–9.

Chapter 14
Protection of Cellular Antigens from Xenoreactive Responses as Overcoming Strategies

14.1 Introduction

β-D-Mannoside β1,4*N*-acetylglucosaminyl-transferase-III (GnT-III) as a catalytically branching enzyme of glycoprotein *N*-glycans generates a bisecting GlcNAc residue in N-linked oligosaccharides. GnT-III as a bisecting enzyme glycosylates to add a GlcNAc residue to the core Man residue of the complex type *N*-glycans. The GnT-III enzyme therefore inhibits further enzymatic glycosylation of glycans by competitively acting glycosyltransferases of α1,3-D-mannoside β1,4-*N*-acetylglucosaminyltransferase-IV (GnT-IV) and α-1,6-D-mannoside β1,6-*N*-acetylglucosaminyltransferase V (GnT-V), in the Golgi apparatus [1]. The bisected GlcNAc residue itself blocks the additional glycosylation activities of complex type *N*-glycans by other competitive glycosyltransferases [2]. Non-α1,3Gal antigenicity is mainly found in the terminal residues in the N-linked oligosaccharides of glycoproteins. Therefore, the GnT-III action reduces the antigenic levels of non-α1,3-Gal antigen-bearing N-linked oligosaccharides [3, 4].

In pig-to-human xenotransplantation, downregulation of the Galα1,3Galβ1,4GlcNAc-R as α-1,3Gal antigenic epitope is crucial in mouse and swine tissues to prevent HR. As one model to downregulate the Galα1,3Galβ1,4GlcNAc-R, GnT-III-Tg pigs have been created to use as organ donors. In the transgenic GnT-III-Tg pigs, both the complement- and NK cell-mediated lyses of pig cells were also reduced. In the GnT-III-Tg pigs, the α1,3Gal-T, GnT-IV, and GnT-V enzymes are not changed in their activities, indicating no cross-talking of the three enzymes (α1,3Gal-T, GnT-IV, and GnT-V) and GnT-III enzyme. Moreover, GnT-III enzyme reduces the xenoantigen levels in heart grafts of pigs during pig-to-cynomolgus monkey (crab-eating macaque) transplantation model. In another report, when the GnT-III gene was transduced into aorta ECs of mouse model, the level of normal human serum (NHS)-caused CMC (complement-mediated cell death or cytolysis) was decreased [5]. GnT-III expression also showed the antigenicity suppression of mouse aortic ECs to natural Abs of

C.-H. Kim, *Glycoimmunology in Xenotransplantation*,
https://doi.org/10.1007/978-981-99-7691-1_14

human, as confirmed by the reduced recognition of isolectin GSIB4-1 against the α-1,3Gal epitopes. The reaction levels of the transfectant glycoproteins to GSIB4-1 isolectin and NHS was also decreased. Thus, the GnT-III-based competitive strategy is to control the synthesis of target carbohydrate chains [6] and consequently can remodel the surface antigenicity of target cells [7]. GnT-III-mediated antigenic elimination represents the branching blocking of addition of the glycan oligosaccharide chains or defected maturation, by bisected GlcNAc residue in the Man structure of core *N*-glycans. Consequently, competing GnT-IV and GnT-V enzymes are disturbed in their glycosylations to substrate tri-structures in the Golgi stack [8]. The *N*-glycan structures in the GnT-III-transfected pig ECs were determined to have the bi-, tri-, and tetra-antennary structures of *N*-glycan as complex type. These structures, lacking α1,3Gal antigenic structure, were mainly bisected structures without α1,3Gal residues [8].

14.2 Masking of α1,3Gal Antigen by βD-Mannosideβ1,4*N*-Acetylglucosaminyl-Transferase III (GnT-III)

In pig islet cells, the major antigenicity of the cells is present in the *N*-glycan carbohydrates of glycoproteins and GnT-III-generated antigenicity reduction prolongs the survival length of pig islets. For example, during the experiment to examine whether GnT-III gene-transgenic cells of pig islets prolong their survival length in pig to cynomolgus monkey transplantation model [1], the GnT-III-mediated antigenicity reduction prolongs the survival length of GnT-III gene-transgenic cells of pig islets in cynomolgus monkey recipients. Although GnT-III gene is weakly expressed in islets of wild-type pigs, the GnT-III-enforced expression in pig GnT-III-transgenic islets reduced xenoantigenic reactivity. GnT-III-transgenic pig islets prolonged survival length rather than wild-type islets of pigs in cynomolgus monkey model. Levels of anti-pig islet Abs was reduced, too. Xenoantigenicity reduction caused by artificial GnT-III expression gives the prolonged survival length of pig islets, emphasizing the crucial roles of non-H-D antigens and non-α1,3Gal antigens because the two antigens are constituents of glycoprotein *N*-glycans in the early immune rejections of pig islets in the cynomolgus monkey model. When the wild-type pig islets transplanted into cynomolgus monkeys were compared with GnT-III-Tg pigs transplanted into cynomolgus monkeys, the wild-type pig islets were fast lysed via host Abs-mediated destruction in the rejection process [9].

Since the α1,3Gal-T KO pigs were produced, the known glycoantigens are expected and explored from α1,3Gal-T KO pigs. From the study on the pig glycoprotein glycoantigens from pig fibroblasts and ECs, the α1,3Gal-T KO and wild-type pig fibroblasts and ECs were compared using lectins [10]. The lectins *Euonymus europaeus* (EEL) isolated from European spindle tree and GSI-B4 known as

Bandeirea simplicifolia agglutinin BS-I, strongly bind to α-1,3Gal epitopes. Seemingly, other lectins such as *Glycine max* (SBA) as soy bean agglutinin, *Griffonia simplicifolia* I-A4 (GSI-A4), roman snail *Helix pomatia* (HPA) and *Wisteria Floribunda* (WFA) isolated from Japanese Wisteria seeds bind to GalNAc residues as in the Tn antigen (or known as Thomsen-Friedenreich precursor antigen).

In contrast, in the human ECs, *Maackia amurensis* (MAL), gorse seed *Ulex europaeus* agglutinin I (UEA-I), *Erythrina cristagalli* (ECA) isolated from coral tree seeds, and *Trichosanthes japonica* agglutinin I (TJA-I) bind strongly the glycoantigens. In the ECs from the α1,3Gal-T KO pigs, EEL and GSI-B4 exhibits non-binding activities and *Bauhinia purpurea* alba lectin (BPL), HPA, GSI-A4 and SBA completely lost the binding capacities, whereas lectins of *Sambucus nigra* (SNA), *Sambucus sieboldiana* (SSA), and TJA-I increasingly recognize α2,6Sia residue. For the wild-type fibroblasts, the HPA, GSI-A4 and SBA lectins bind α1,3Gal antigens for the strongest levels. The homozygous-α1,3Gal-T KO pig fibroblasts showed lower binding capacity. In addition, *Maclura pomifera* (MPA) and *Arachis hypogaea* (PNA) lectins, specifically bound for Galβ1,2GalNAc residues on the T antigen (known as Thomsen-Friedenreich antigen), bound the glycoantigens from homozygous-α1,3Gal-T KO pig fibroblasts, indicating the existence of the T antigen or Thomsen-Friedenreich antigen in fibroblasts of pig. From the above results, the pig cells were shown to have the terminal GalNAc residue, not for the Forssman, indicating the presence of non-α1,3Gal antigens in the wild-type pig, as the non-α1,3Gal antigens were reduced the α1,3Gal-T KO pig cells. The results reflect the cross talking between α1,3Gal-T of pigs and GalNAc-Transferases of pigs to form the terminal GalNAc residues a non-α1,3Gal antigen in host cells.

Another strategy to downregulate the α1,3Gal antigens by catalytic competition in glycan biosynthesis of terminal oligosaccharides of the enzymes of α1,3Gal-T and terminal glycosyltransferases, which they use common substrates as their acceptors, has been suggested, as they enzymes commonly localize in the trans-Golgi compartment stacks and networks. For example, α1,2-Fuc-T [11], α1,3-Fuc-T [11], α2,3-Sia-T [12], and α2,6-Sia-T are indeed candidates [13]. To prevent formation of the Galα1,3Galβ1,4GlcNAc-R (α1,3Gal antigenic epitopes), except for the bisecting GnT-III, two glycosyltransferases including α2,3/α2,6-Sia-T and α1,2-Fuc-T have also been considered in a fashion of competition (Fig. 14.1).

14.3 Pig ST3Gal III, ST6Gal I, and α1,2Fuc-T Competition with α1,3Gal-T for the Acceptor Substrates in the Trans-Golgi Network

In the trans-Golgi network, group II glycosyltransferases involves in enzymatic competition for terminal glycosylation with α1,3Gal-T in order to use the common acceptor substrate. The known group II glycosyltransferases of ST3Gal III (α2,3-sialyltransferase), ST6Gal I (α2,6-sialyltransferase), and α1,2Fuc-T

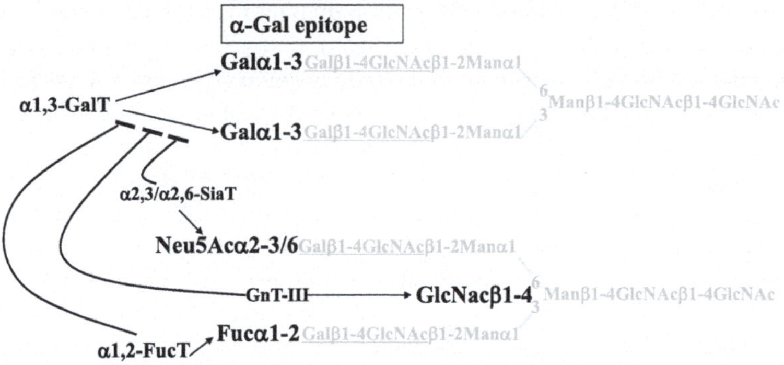

Fig. 14.1 Competitive inhibition of Galα1,3Galβ1,4GlcNAc-R (α-1,3Gal antigenic epitope) formation by glycosyltransferases of bisecting GnT-III, α2,3-Sia-T/α2,6-Sia-T, and α1,2-Fuc-T

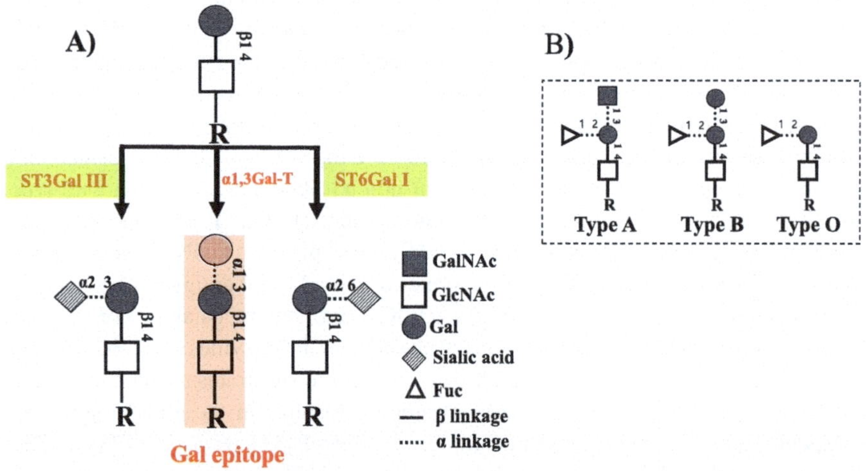

Fig. 14.2 ST3Gal III and ST6Gal I compete with α1,3Gal-T for the common acceptor substrate in the trans-Golgi network of pigs. (**a**) α1,3Gal epitope structure and competition. (**b**) AB(O)H blood type structure

(α1,2-fucosyltransferase) exhibit moderate competition with α1,3Gal-T [12] The overexpressed GnT-III and the competitive ST6Gal I, ST3Gal III and α1,2Fuc-T enzymes attenuate the pig EC antigenic properties (Fig. 14.2). The GSL α1,3-Gal epitope levels were not reduced by the GnT-III transfection. However, the levels are reduced by the ST6Gal I, ST3Gal III, and α1,2FT transfections [13]. Therefore, ST6Gal I, ST3Gal III, and α1,2FT enzymes are competitive with α1,3Gal-T to reduce the α1,3Gal epitope in GSLs and glycoproteins, whereas GnT-III enzyme is effective for only glycoproteins.

With regard to fucosylation and fucosyltransferases (FUTs), FUTs transfer Fuc residue from GDP-fucose to glycoconjugates. FUTs prefer to glycan O-glycan, N-glycan and GSLs types. FUT 1 and FUT 2 are specific for α1,2-Fuc linkage.

FUT 3–7 and FUT 9–11 are specific for the α1,3-Fuc linkage. FUT 3 and FUT 5 catalyze the α1,4-Fuc formation. FUT 8 forms the α1,6-Fuc linkage. FUT8 is a functional regulator of cancer cells. In the physiological FUT 8 role, FUT8 transgenic mouse bears an excess core fucosylation and contributes to abnormal lipid metabolism. FUT8 KO mouse showed the postnatal death, and thus, FUT8 is crucial for normal development of lung and brain with homeostatic maintenance. FUT8 is upregulated in several malignant tumors like non-small cell lung cancer (NSCLC) and enhances tumor progression and recurrence. On the other hand, two different FUT enzymes of POFUT1 and POFUT2 are known. POFUT1 fucosylates Ser/Thr residue of EGF-like attached to N-, O-, and lipid-linked glycans through FUTs. POFUT2 fucosylates directly thrombospondin repeats.

14.4 Complements System

Xenotransplantation is understood by the infectious, physiologic and immunologic hurdles to pursue overcoming the complex and combine hurdles. Several events including natural antibody recognition to the vascular endothelial cells, antibody-based complementation and consequent coagulation are requiring for intervention. The next barrier is a delayed xenograft/acute vascular rejection by its multiple pathways driven by immunologic recognition and activation of endothelial cells (ECs). The final hurdle is the cellular immune response like allograft rejection. Thus, immunologic tolerance expressed by unresponsiveness is desired. Glycan epitopes present on pig vascular ECs react with naturally-formed antibodies of human. Because coagulation pathways reject grafts, methods of creating available organs are important. Human complement regulatory protein (such as hDAF) transgenic pigs and immunosuppression induce the decreased xenoreactive α1,3Gal antibodies and leads to the decreased acute vascular rejection (AVR). The Ethics Committee of the International Xenotransplantation Association has thus recommended to keep ethical standards and ensure xenotransplantation success come from immunosuppression to the innate immune system and donor specific tolerance. Rejection and infectious agent transmission are future issues.

Inflammation and immunity are involved in endothelial cell (EC) injury and thrombus formation. Humoral immune response directs against the endothelium to induce coagulation. Complement system removes pathogens from bloods and tissues in collaborative synergies between innate immunity and adaptive immunity. Its functional roles include opsonization, inflammatory reaction and directly destroying pathogens. Various components (C1–C9) and adaptive factors (Factor B, D, H, I, and P) are composed [14]. Complement is a multi-proteolytic cascade upon microbial invaders and is an immune regulation component and cellular physiological regulator. To regulate complementation, complement inhibitor (CI), complement factor (CF), and complement receptor (CR) are present in various cell types [15]. CFs and CRs have the classical pathway components such as C1q (C1qA, C1qL2, C1qL3, C1qL4), C2, and C4B. The lectin pathway components have ficolin-1 (FCN1), ficolin-2 (FCN2), MBL [mannose (Man)-binding lectin], and MASP (MBL-associated serine proteinase). The alternative pathway components have C3, CF118B (CFB), CFD, and CF properdin (CFP). The various convertase enzymes are known, including C3 and C5. The terminal pathway components are C6, C7, C8α, C8β, C8γ, and C9. CR receptors have C5aR1 and C5aR2. CIs include C1 inhibitor (C1INH), C4BP (C4-binding protein), CD35, CD46, CD55, CD59, CFH, CFI, CLU (clusterin), COMP (cartilage oligomeric matrix protein), CPB2 (carboxypeptidase B2), CPN1, CPN2, Cub and CSMD-1 (sushi multiple domain protein-1), PLG (plasminogen), PTX-3 (pentraxin-3), SMAP-1 (small MBL-associated protein-1), SMAP2, SUSD-4 (sushi-domain containing protein-4), VTN (vitronectin), and vWF.

When complement activation occurs by either antibody binding (classical pathway), lectin-bacterial oligosaccharide binding (lectin pathway), or C3b protein binding (alternative pathway), two major complement convertases named C3 and C5 convertases cleave the C3 and C5. All three pathways generate the C3 and C5 convertases. The resulted cleavage products C3b and C5b opsonize the cells or microbes, named phagocytosis. The phagocytosed debris recruits to exert innate immune cell activation and induce anti-microbial killing and form the membrane attack complex (MAC). Therefore, if complement activation is dysregulated, the host cells are damaged and hurt. To protect host cells, they generate complement inhibitors to prevent cell damage and inflammation. Several cells such as renal cells, hepatocytes and immune cells express many complement cascade components (Table 14.1). Hepatocytes greatly generate various complement cascade components, only except for C1q and C7 [16]. Macrophages and T/B cells also generate complement components to induce immune responses [17, 18]. Complement regulates T-cell function during APC interaction. Anaphylatoxin (C5a,C3a) binding to their receptors C3aR and C5aR1 stimulates T cells by inflammatory cytokine IL-1 and local C3 expression as well as the suppressive TGF-β and IL-10 suppression. However, T cells are suppressed in the absence of anaphylatoxin, and C5aR2 sequesters C5a binding to inhibit the pro-inflammatory factor expression and to express TGF-β and IL-10.

Complement products such as C3, C5, and their cleaved fragments are used for antigen-presenting cells (APCs) and T-cell differentiation into Th1 and Th17 cell types [18]. Blocking C3aR and C5aR1 generates immune suppressive Treg cells, enhancing graft survival and protection [19]. CR type I (CR1), which is called CD35 and C3b/C4b receptor, has polymorphic 160, 190, 220, 250, and 285 kDa membrane glycoproteins present on human peripheral blood cells of B/T-cell subsets, monocytes, some NK cells, erythrocytes, leukocytes, and macrophages, glomerular podocytes, and splenic follicular DCs, but not on basophils. CD35 binds to C3b, C4b, and iC3b. The CD35 recognizes complement-coated cells to kill by neutrophil and monocyte phagocytosis. In addition, CD35 also inhibits two complement pathways of the classic and alternative pathways. The CD35 contains Knops blood group system. Each leukocyte expresses more largely 10,000–30,000 molecules than erythrocyte having 300–800 CR1 molecules. CR1 binds immune complexes to deliver to the liver or spleen macrophagic internalization and elimination. After

Table 14.1 Complement inhibitor (CI), complement factor (CF), and complement receptor (CR)

Common CIs: CD55, CLU, PTX3, SMAP,
Hepatocyte: CIs (C1INH, C4BP, CD46, CD59, CFH, CFHR1-5, CFI, PLG, VTN), CFs (C1r, C1s, C2-C9, CFB, FCN, MBL, MASP), CRs (C5aR2)
Renal cells: CIs (CD46, CD55, CD59, PLG, VTN, VWF), CFs (C1q, C2-C4, CFB, CFD, C9), CRs (CD35)
Neutrophils: CIs (C4BP, CD46, CD55, CD59), CFs (C3, C6, C7, CFB, CFP, FCN), CRs (CD35, CR3, CR4, C3aR, C5aR1)
Monocyte: Cis (C1INH, C4BP, CFH, CFI), CFs (C1q, C1r, C1s, C2-C8, C9, CFB, CFD, CFP), CRs (CD35, CR3, CR4, C3aR, C5aR1)
Macrophages: Cis (C1INH, CFH, CFI), CFs (C1q, C1r, C1s, C2-C5, CFB, CFD, CFP), CRs (CD35, CR3, CR4, CRIg, C3aR, C5aR1, C5aR2)
DCs: CIs (C4BP, CFH, CFI), CFs (C1q, C1r, C1s, C2-C5, C7-C9, CFB, CFD, CFP), CRs (CD35, CR3, CR4, CRIg, C3aR, C5aR1)
B cells: Cis (CD46, CD55, CD59), CF (C5), CRs (CD35, CD21, CR4, C5aR1)
T cells: CIs (C1QBP, CD46, CD55, CD59, CFH, SUSD4), CFs (C3, C5, CFB, CFD, CFP), CRs (CD35, C1qBP, C3aR, C5aR1, C5aR2)
Neonatal pig Sertoli cells (NPSC): CIs (C1INH, C4BP, CD35, CD46, CD55, CD59, CFH, CFI, CLU, COMP, CBP2, CSMD1, PLG, PTX3, SMAP, SUSD4, VTN, VWF), CFs (C1q, C1s, C2, C3, C4, C5, C6, C7, C8, C9, CFB, CFD, CFP, FCN1, FCN2, MASP, MBL), CRs (CD35, C5aR1, C5aR2)
*Human testicular tissue cells: CIs (*C1INH, CD35, CD46, CD55, CD59, CRIg, CFH, CFHR1, CFI, CLU, COMP, CPB1, CPN1, CSMD1, PTX3, SMAP2, SUSD4, VTN, VWF), CFs (C1q, C1s, C2, C3, C4, C5, C6, C7, C8, C9, CFD, CFP, FCN1, MASP1), CRs (CD21, CD35, C3aR, C5aR1, C5aR2, CRIg).

delivery of captured immune complex to liver/spleen/macrophage, free erythrocytes are circulated to the blood to repeatedly clear those immune complexes. The CR1 gene locates on chromosome 1 (1q32) and consists of 39 exons over 133 kb of DNA [20]. The exons encode approximately 60 amino acid length's SCRs in the CR1 protein. Seven different SCRs form large domains of long homologous repeats (LHRs). Among the CR1s, CR1-1 protein is the most common form with four LHRs (A, B, C and D), a TM region, and a cytoplasmic tail region (CYT). The C3b and C4b recognition sites are present in SCRs 8–9 and 15–16 (LHRs B and C) and in SCRs 1–2 (LHR-A), respectively. The CR1 gene has two more polymorphic regions besides the Knops of human blood group system. Four polymorphic different genes encode 190 kDa (CR1*3), 220 kDa (CR1*1), 250 kDa (CR1*2), and 280 kDa (CR1*4). The third common polymorphism is found in E-CR1 by quantitative differences.

14.4.1 Roles of Complement Inhibitors of Pig Sertoli Cells in Xenograft Survival

Neonatal pig Sertoli cells suppress human natural antibody-mediated lysis. Testicular Sertoli cells nourish the germ cells to mature via spermatogenesis and protect germ cells via immunoregulation and anti-autoimmune responses [21]. Testicular Sertoli cells are known to distinctly modulate the complement system of hosts' recipients. Allografted and xenografted testicular Sertoli cells can survive for prolonged period longer than 90 days. Sertoli cell complement system can provide the germ cell protective effects by testis immune priority and environmental graft protection for xenotransplantation. Neonatal pig Sertoli cells (NPSC) are tolerant to human complement attack allowing its survival and attributing to the expressed CIs of CD46, CD55 and CD59 as a potential mechanism for immune evasion and xenograft survival. The identified carious complement inhibitors are differentially expressed by neonatal pig Sertoli cells (NPSC) compared to islet cells and EC cells. NPSC express various complement inhibitors of membrane-bound CIs of CD46, CD55, and CD59. From a protective role in male reproduction and fertility, Sertoli complement system may open use basis of xenografts and allografts survivals. The prolonged survival of testicular Sertoli cells is based on their immunoregulatory molecules to protect from the host recipient's immune rejection. Xenogeneic Sertoli cells attenuate the complement system to prevent MAC assembly via expression of complement inhibitors that prolong their xenogeneic graft survival period. Allografted and xenografted Sertoli cells are known to prolong their long-term survival without immunosuppressants and protect co-grafted cells [22]. The Sertoli cell complement inhibitory proteins enable for the pig Sertoli cell survival against human complement. Treg cell number is increased in Testicular Sertoli cells compared to other cell grafts [15, 23]. Upregulated C5aR2 by Testicular Sertoli cells reduces C5a levels, generating Treg cells. In contrast, the upregulated C5aR1 and C5aR2 induce Testicular Sertoli cells to respond to complement activation to modulate effector immune cell responses [24].

Pig Sertoli cells and human testicular tissue cells express their similar complement-associated proteins [25, 26]. For example, in pig Sertoli cells, for CIs, C1INH and C4BP are distributed in plasma. C1INH targets C1r and C1s to combine C1r/s from C1q. C4BP (540–590 kDa) is also in plasma to target C4b and prevent combination of the assembled CP-C3 convertase. CD35, CD46, CD55 and CD59 are associated with membrane. Plasma CFH (115 kDa) targets C3b and Bb to cleave C3b and dissociate Bb from C3b. Plasma CFI (88 kDa) targets C4b and C3b to inactivate C4b/C3b through α chain cleavage. Plasma COMP (524 kDa) targets C1q, MBL and C3 to prevent C1q/MBL interaction and stabilize C3. Plasma CPB2/N1/N2 targets C3a and C5a to remove carboxy terminus to form desArg. Plasma PTX3 targets C1q to prevent C1q activation and interaction. Plasma SMAP1/2 targets MASP1-3 to bind MASP1-3 and prevent its activity. Plasma VTN targets C7 for C8 recruitment prevention. Plasma vWF targets C3b to cleave C3b. CD35 (190–250 kDa) on membrane targets C1q, C4b, C3b and MBL to confer their competitive interaction with C1q/MBL/C4b/C3b. Membrane CD46 (51–68 kDa) targets C4b and C3b to cleave the C3b/C4b. Membrane CD55 (60–75 kDa) targets C3b to decay convertases and inhibit C3 cleavage during complement function. Membrane CD59 (18–23 kDa) targets C8 for C8 suicide inhibitory interaction. Membrane CSMD1 targets C3b and C7 to cleave C3b and prevent C7 binding to C6. Membrane SUSD4 targets C1q and MBL to prevent C1q/MBL interaction and C2 cleavage. CLU and PLG are both plasma and membrane forms. Among them, CLU targets C6, C7, C8 and C9 to inhibit the assembled MAC formation and blocks C9 insertion, while PLG targets C3, C3b, C3d, and C5 to decay convertases and inhibit C3.

14.4.2 Complement Regulators (C Regs)

Human complement-regulatory proteins (CRPs) are involved in avoidance of xenoreactive responses, as CD55 (DAF). They are representatively known in the cases of DAF (decay accelerating factor), CD46 (or MCP) as MCP (membrane cofactor protein) and MIRL (membrane inhibitor of reactive lysis) (CD59).

Glycosylphosphatidylinositol (GPI) is phosphoglyceride or phosphatidylinositol (PI) derivative linked to glucosamine (GlcN) and Man residue linker bound to the inositol residue and an ethanolamine phosphate or phosphoethanolamine (EtNP). GPI anchor refers to the protein C-terminus-linked GPI occurred during posttranslational modification (PTM) pathway. The anchored proteins have carboxyl-terminal signal peptides required for GPI-structural modification. The C-terminal amino acids of matured proteins bind to EtNP followed by three Man and glucosamine (GlcN) residues and the glycan core further attach to PI. Functional and structural aspect of GPI localization is in positional direction of its anchored protein to plasma membrane localization as caused by its own structural fate. GPI anchors bring its anchored proteins to membranes to present them to lipid rafts or apical surfaces. The known GPI-anchor functionality diversely ranges from enzyme activities to surfaced antigen motifs and cell–cell interacting adhesion capacities. In addition, GPI-anchored proteins involve in various signal transductions in receptor–ligand interaction events. For example, MCP, DAF, and CD59 inhibit the membrane attack complex formation. It is known that combinations of MCP + CD59 and MCP + DAF are efficient rather than other combination DAF + CD59 from cells injury induced by complement cytotoxicity. It is also known that DAF or MCP alone more potently acts rather than CD59 alone to suppress membrane attack complex (MAC) formation during complement pathway.

Biosynthetic pathway is summarized as follows: (1) Polypeptide chain synthesis via ER membrane-attached ribosomes polypeptide chain is subjected to N-terminal leader peptide splitting by signal peptidases. (2) carboxyl-terminal hydrophobic peptide of the target polypeptide is linked to lumen side of ER-membrane. (3) the protein in lumen side is delivered to the transamidation reaction site, where transamidase complex and precursor GPI are located with the presence of ATP,

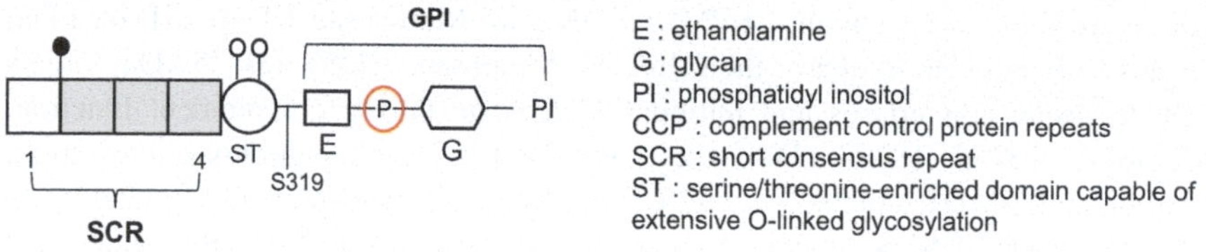

Fig. 14.3 Schematic structure of GPI-anchored proteins

Fig. 14.4 Basic GPI-anchored protein structure

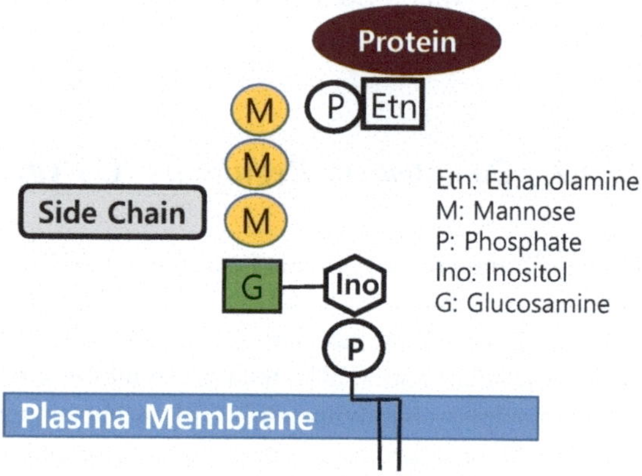

GTP and BiP chaperon, by translocation pathway. (4) C-terminal amino acid of protein, surrounded by hydrophobic region, is enzymatically split off by transamidase complex. (5) newly appeared carboxyl-terminal peptide region (w-site) is linked to precursor GPI ethanolamine head.

GPI-anchored proteins are structured with ethanolamine, glycans, phosphatidylinositol, complement control protein repeats, short consensus repeat (SCR), and Ser/Thr-region for heavy O-glycosylation (Figs. 14.3, 14.4, and 14.5). In human, three complement regulatory proteins (hCRP) are known: CD46 termed MCP, CD55 antigen termed DAF, which constitutes the Cromer blood group system (CROM), and CD59 antigen termed MIRL (Table 14.2). CD55 (DAF) and CD59 are GPI anchors attached to plasma membrane, while CD46 (MCP) and CR1 are adhered to plasma membrane through TM domain regions (Fig. 14.6).

CD55 (or DAF) inhibits the C3 and C5 convertase formation, dissociating C3 convertases at human cell surfaces. For molecular action of CD55 (DAF), CD55 protein interacts with C3b fragment and C4b fragment that condense with hydroxyl group or amino groups in cell surfaces during local generation of native C3b and C4b proteins in C3 and C4 activation event. Binging of DAF to cell-associated C4b and C3b proteins interferes with their catalysis capacity of the C2 and B factor conversion to active C2a and Bb enzymes, this event consequently prevents the C4b2a and C3bBb formation in the complement cascade [27]. Therefore, CD55 destabilizes and prevents the C3 and C5 convertase formation to inhibit complement activation, preventing complement damages [27]. The CD55 glycoprotein involves

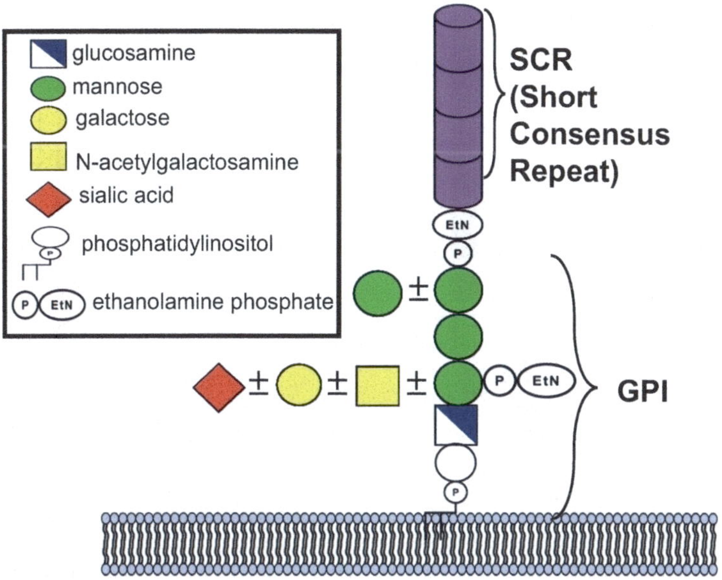

Fig. 14.5 Systemic GPI-anchored protein structure

Table 14.2 Three major types of hCRPs

C Reg types	Function roles	Featured property
CD46 (MCP)	Membrane cofactor protein, regulation of C3b and C4b	
CD55 (DAF)	C3 convertase inactivation	Correlation with D/P Cre
CD59 (MAC-IP/ MIRL)	C9 inactivation	Attaches to host cells via a GPI anchor

in the complement cascade regulation via binding to complement proteins that accelerates their decay, leading to disruption of the cascade and prevention of damage to host cells. For the CD55 gene expression, alternatively spliced transcript variants are formed. The major spliced variant encodes a membrane-bound protein, but alternative transcripts generate soluble proteins.

CD46 (or MCP) recognizes C3b and C4b proteins known as complement-activated products to enhance complement factor (CF)-I-derived inactivation of C3b and C4b. Therefore, CD46 is a cofactor of a serine protease CF-I. Complement factor I cleaves C3b and C4b proteins deposited on host tissues and this protects host cells from complement-mediated cell injuries. CD59 (MIRL) induces inhibition of the MAC formation by its interaction with complement C8 and/or C9 during the complex assembly. Hence, CD59 (MIRL) inhibits the incorporation of multiple C9 copies into the complement MAC. The incorporation of multiple C9 copies is essential for formation of osmolytic pore. Consequently, CD59 (MIRL) glycoprotein

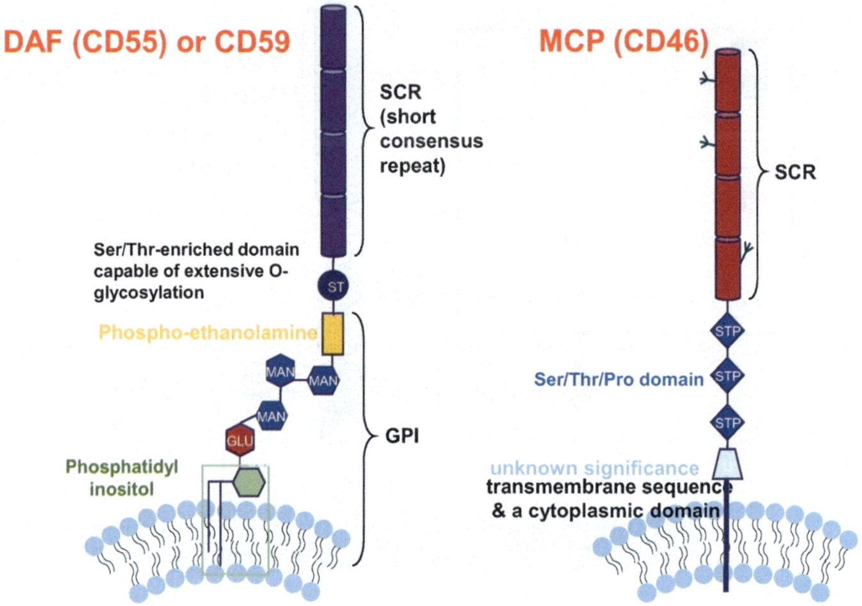

Fig. 14.6 *DAF (CD55) VS MCP (CD46).* DAF (CD55) and MIRL (CD59) are GPI anchors attached to plasma membrane, while CD46 (MCP) and CR1 are attached to plasma membrane through transmembrane domains

regulates complement-mediated cell lysis. CD59 (MIRL) deficiency results in hemolytic anemia and thrombosis.

CD59 (MIRL), DAF (CD55) and MCP (CD46) collaboratively act to inhibit the MAC formation. DAF (CD55) or MCP (CD46) alone acts more potently compared to CD59 (MIRL) alone to inhibit MAC formation. However, combinatory actions of DAF (CD55) + MCP (CD46) and CD59 (MIRL) + MCP (CD46) exhibit higher activities compared to single action in each alone.

After the use of human CRPs and removal of α1,3Gal epitope for the HR reduction, the delayed antibody-mediated xenograft rejection characteristic of the recipient consumptive coagulopathy (CC) and graft thrombotic microangiopathy (TM) is still caused. Then, the issue of Neu5GC has been resolved by CMAH KO. However, issues of the interspecies molecular incompatibilities and dysregulating coagulation have been appeared during the late xenograft rejection. The TM and CC are mediated by molecular incompatibilities between intrinsic interspecies and by immune injuries caused by innate/adaptive immunities. The innate complement system as a cascade reactive process in innate immune system protects the hosts against bacterial attacks. The complement system kills invaded bacteria by opsonization and phagocytosis. Complex formation between the Abs and xenoantigens contributes to the complement system activation and HR. Pig cells obtained from transgenic pigs with multiple genes displayed protective phenotypes from human CDC or complement-dependent lysis. Defection of endothelial activation reduced inflammation-responding gene expressions related with leukocytes homing, cytokine-induced MHC class-II function and apoptotic pathway. Because xenoreactive antibodies are the causing factor for HR, downstream immune reaction covering complement system and ADCC are crucial for the rejection. Except for ADCC, complement issue has been considered for long time with xenotransplantation. Control of complement activation of pig cells is achieved by transgenic human CRP expression.

MCP (CD46), DAF (CD55), and MIRL (CD59) act as the crucial complement regulators in cellular membranes of host cells. In human complement system, the expressed CIs such as MCP (CD46), DAF (CD55), and MIRL (CD59) complement inhibitors are abundant in human tissues. Three MCP (CD46), DAF (CD55), and MIRL (CD59) CIs inhibit complementation reaction at each C3 convertase, C5 convertase, and MAC level. The combined inhibition of triple proteins shows the cross-activation prevention of the complement cascade at later steps rather than action stage of C3 convertase, by the complement-binding coagulation factors [28]. Complement-protecting cellular antigens of MCP (CD46), DAF (CD55), and MIRL (CD59) are well studied. Among them, MCP (CD46) and DAF (CD55) combination remarkably reduces complement-dependent lysis of host cells. In addition, CD46 (MCP), CD55 (DAF), and CD59 (MIRL) together completely show the protective effects in vitro [29].

14.4.3 CD46 (MCP, Membrane Co-Protein)

CD46 is also another form of complement regulatory protein or membrane cofactor protein as an inhibitory complement receptor. Although α1,3Gal epitopes induce IgM, IgG, and C5b-9 depositions in xeno-perfusion tissues in humans, hCD46/HLA-E-double transgenic expression on vascular endothelium of pigs reduces complement activation and prevents the pro-coagulative and anti-fibrinolytic formation of the pig endothelial tissues in human-to-pig limb xeno-perfusion with human bloods. hCD46 expression prevents the complement activation at the terminal pathway by inhibiting the C3b and C4b, which are central proteins in complementation system. CD46-expressing cells are protected from the attack of C3 convertase complex (Fig. 14.7). Recent transgenic technology allowed the artificial production of hCRPs and thrombomodulin molecules of human in baboons with a longer survival for more than 900 days [30].

14.4.4 CD55 (Decay-Accelerating Factor, DAF)

A DAF glycoprotein CD55 is encoded by the *CD55* gene and discovered in 1969. Three CD55 gene variants were found in human edema patients, with a frameshift, causing abnormal translation termination of *CD55* mRNA [31]. CD55 is a negative

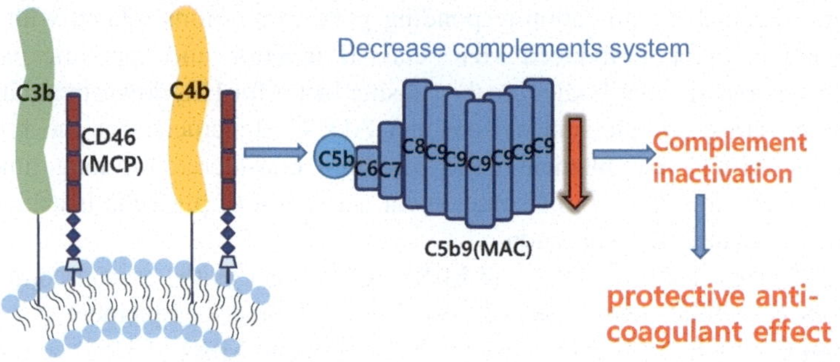

Fig. 14.7 *MCP (CD46).* CD46 (MCP) attached to plasma membrane through transmembrane domains blocks C3b and C4b known as the central proteins of complementation system via inhibiting the complement activation terminal pathway

Fig. 14.8 Protein and glycan structure of *CD55*

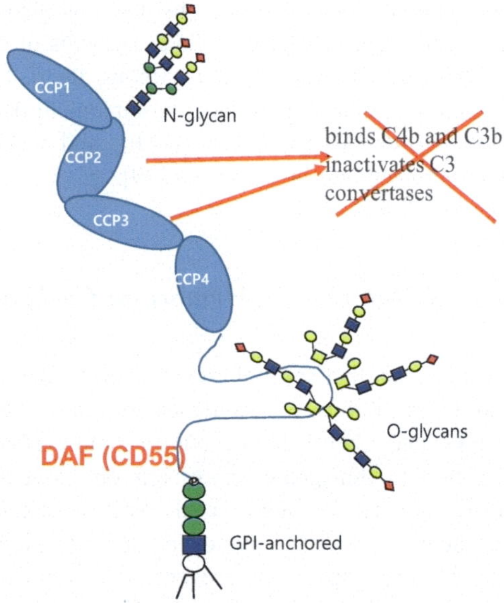

complement regulator and regulates the complement system on the erythrocyte surface. CD55 as a GPI-anchored protein consists of protein domains of complement control protein (CCP), one N-glycan, and multiple O-glycan structures. N-glycan structure between CCP1 and CCP2 is located. Stalk region bridges between CCP and GPI. The stalk region has a variety of O-glycans. N-glycan stabilizes protein structure and O-glycan protects the protein from protease attack. CCP2 and CCP3 stimulate C3 and C5 convertase of complement pathway and alternative pathway (Fig. 14.8). CD55 (DAF) expression is also found in malignant cancer cells, implicating in angiogenesis, metastasis and tumorigenesis. CD55 reduces complement mediated lysis of tumor cells in hosts to prevent apoptotic cells death of tumor cells. To do such role in cancer cells, CD55 binds to the CD97 seven-span transmembrane protein. Three known CD97 ligands are CD55, chondroitin sulfate and integrin α5β1. Among them, CD55 binds to CD97 on APCs to exhibit immune tolerance.

CD97 expression has widely been found in the multiple immune cell surfaces, hematopoietic progenitor and stem cells, muscle cells, epithelial cells and their malignancies, including DCs, granulocytes, lymphocytes, macrophages, monocytes and SMCs (smooth muscle cells). *N*-glycans in the epidermal growth factor (EGF) domains of CD97 is the site for CD55 binding [32]. CD97 (known as BL-Ac) as a leukocyte adhesion marker implicated in tumor invasion is the EGF-TM7 family of adhesive G-protein coupled receptors (GPCRs) present in the leukocyte surfaces and epithelial malignancy subsets. Among CD97 ligands, DAF/CD55, which binds to the first and second EGF-like domains of CD97 [33]. Chondroitin sulfate B (CSB) also interacts with the fourth EGF-like domain of CD97 [34]. Integrin α5β1 and αvβ3 bind to an RGD motif in the EGF-like domains of CD97. CD90 (Thy-1) also binds to the GAIN domain of CD97 [35]. *N*-terminal domains of CD97 consist of alternative-splicing EGF-like domains. Three alternatively spliced variants are known. The human CD97 *N*-terminal fragment contains three to five EGF-like domains and mouse CD97 has three to four EGF-like domains [36]. CD97-ligand CD55 binding activates T-cell function to enhance cytokine production. CD97 expression involves in auto-inflammatory diseases such as rheumatoid arthritis (RA). CD97 expression on macrophage with its ligand CD55 on fibroblasts potentiates to interact between CD97 and CD55, and hence to recruit macrophages into the RA synovial tissues [37]. Deficient mice CD97 or CD55 reduce synovial inflammatory response and joint damaged injury in RA. Because CD55 expression is restricted to the endothelium, ECs CD55 and robust CD97 in infiltrating leukocytes promote immune cell migration. Interestingly, soluble N-terminal CD97 fragment forms are observed in the patient sera with RA [37].

DAF (CD55) destabilizes to prevent the C3 and C5 convertase formation, preventing complement damages and inhibiting complement activation. Its expression is increased during lymphocyte activation to function in adhesion and migration of cells. Normally, DAF (CD55) interacts with C3b and C4b to inactivate the C3 convertases. CD55 deficiency activates complement cascade due to the active C3 and C5 convertase. DAF (CD55) protected the CD55-expressing cells from complement attack, suppressing the C3 and C5 convertase formations. CD55 deficiency cannot inhibit C3 and C5 convertase, resulting in a self-membrane attack complex (MAC) formation, as known for Paroxysmal nocturnal hemoglobinuria (PNH). The MAC formation promotes inflammation. For example, TNF and INF-γ levels are improved and IL-10 level is decreased. Thrombomodulin production is decreased and thrombosis occurs (Fig. 14.9) [27]. In neuroblastoma cells, CD55 elevates growth and invasiveness. CD55 is upregulated in an HIF-2α-positive neuroblastoma cells, which are invasive but has poor adhesiveness to ECM proteins of collagen and fibronectin. Expression of CD55 is highly related to poor neuroblastoma patient prognosis. CD55 proteins which show molecular weights of 75–100 kDa on SDS-PAGE analysis followed by Western immunoblots are detected in several tumor cell lysates such as A431 cells (human epidermoid squamous carcinoma

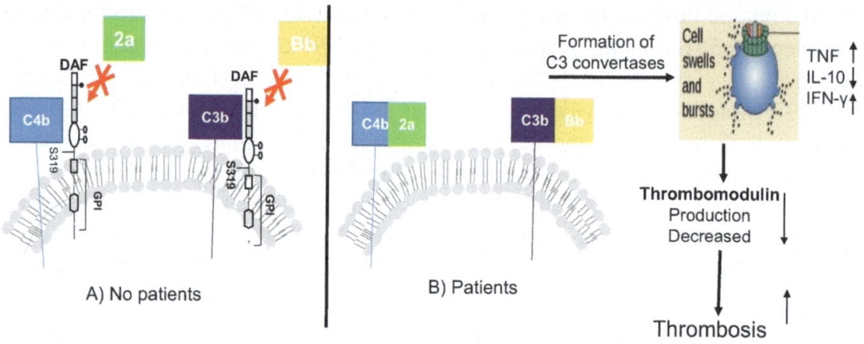

Fig. 14.9 DAF (CD55) deficiency and thrombosis. (**a**) Healthy CD55 expression (no patients). (**b**) CD55-deficient patients' thrombomodulin production is decreased and thrombosis occurs (patients)

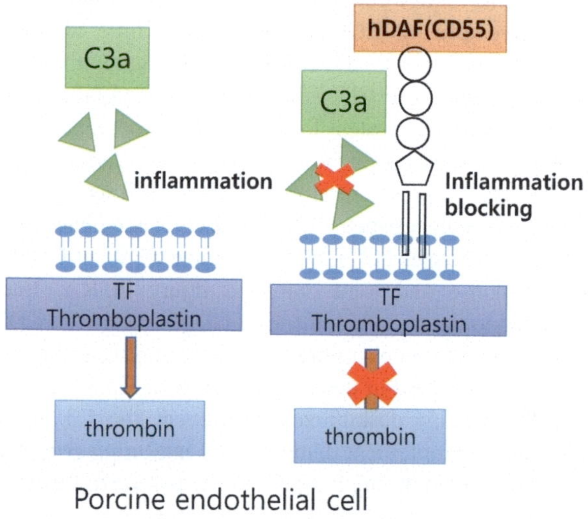

Fig. 14.10 hDAF expression in porcine endothelial cells inhibits the blood coagulation system via blocking complement C3a activation and thromboplastin

cells), HeLa cells, MCF7 cells, and MDA-MB-231 cells (the latter two cells are human breast cancer cells) [38]. CD55 extracellular region has been detected by Mab.

Complement regulatory proteins (hCRPs) of human can be introduced into the donor pigs [38]. hDAF in porcine endothelial cells (ECs) inhibits the blood coagulation system, which is a problem in pig-human transplantation due to complement activation (Fig. 14.10). DAF reduces immune rejection of xeno-antigen, α1,3Gal without effect on IgG dependent immune rejection [39]. Transgenic hDAF pigs, which lack Galα1,3Gal antigenic epitope-specific Abs exhibit the reduced AXR. Thus, to eliminate HR, inactivation of GGTA1 gene, complement system regulation and carbohydrate structure modification of surfaced glycoproteins are required. Transgenic pig expression of a human CRP, CD55 (hCD55) of human, protects the host recipients from early stage of xenograft failure of α1,3Gal-T KO organs of pig in baboon recipients. Expression of CD55 blocks HR and induces the restricted complement activation in α1,3Gal-T KO xenografts of cardiac organ in pig-to-primate xenotransplantation [40]. Similarly, using the above genes, pig renal xenotransplantation pig-to-macaque gave longer survival for more than 227 days with α1,3Gal-T KO/CD55 (DAF) TG pig than the expected [28]. This consequently accelerates their decay, and therefore, they are name to the DAF.

DAF induces the tolerant phenotypes of $CD103^+$ DCs and Tregs. The DAF regulates cytokine and SHP expression. Soluble complement 5b-9 (sC5b-9) activates the alternative complement pathway [41]. Defects in C3ar1/C5ar1 signaling enable autocrine production of TGF-β cytokine in $CD4^+$ cells and Treg induction in pigs. DAF induces Treg differentiation, suppressing T_{eff} proliferation. DAF expression decreases DC MHC II expression and increases PD-L1/ICOS-L expression with upregulations of $CD103^+$ DC migration, PD1 expression, Treg TGF-b1, and transition of $CD4^+$ cells. For example, Ocular tolerance induces TGF-b1 expression in the Anterior chamber (a.c.). CD40 decreation in the cornea, IL-10 decrease in the spleen, pTregs decrease in the a.c. and increase in natural killer T-cell activity have been known. DAF is indispensable in the pTreg induction as well as expression of $CD103^+$ and $CD4^+$ cellular markers. Defects in C3ar1/C5ar1 signaling enable autocrine cytokine production of TGF-β in $CD4^+$ cells and program Treg induction in pigs. DAF induces the tolerogenic $CD103^+$ DCs and Tregs functions. For example, DAF induces ocular tolerogenic response. Oral tolerance is mediated by DAF-dependent Tregs immunosuppressive function [38]. Protein-losing enteropathy (PLE)-caused painful and diarrhea patients, who are associated with intestinal primary lymphangiectasia, display edema caused by hypoproteinemia symptom, malabsorption disease, and angiopathic thromboembolic disease. Protein-losing enteropathy (PLE) exhibits protein loss through intestinal mucosa and is characterized by synthesis<loss of albumin, leading to hypoalbuminemia. Its symptoms display weight loss, abdominal pain, diarrhea, fever, etc.

On the other hand, CR proteins of CD46, CD55 and CD59 in membrane are correlated with complement (C) activation products in human diseases such as peritoneal dialysis (PD)-receiving patients. Complement (C) activating products are crucial and PD treatment process corrects the CD55 level in the peritoneal mesothelial cells. In the peritoneal dialysis (PD) patients, which is apart from the xenotransplantation, the peritoneal mesothelial cells of PD patients express CD55, not CD46 and CD59. CD55 expression is in high correlation with dialysate-to-plasma creatinine concentration ratio, which is known as a factor for peritoneal function. Human CD55 level in peritoneal mesothelial cells is also in high correlation with sC5b-9 level. In contrast, C3 and C4 levels are not correlated with sC5b-9 level in the PD fluids (Fig. 14.11). In contrast, diabetes patients are not correlated with CD55 level. PD-based therapeutic approaches thus modify the CD55 protein level in peritoneal mesothelial cells and leads to activation of complementation system, requiring for restricted activation of the patient complement system, who receive therapeutic actions such as PD.

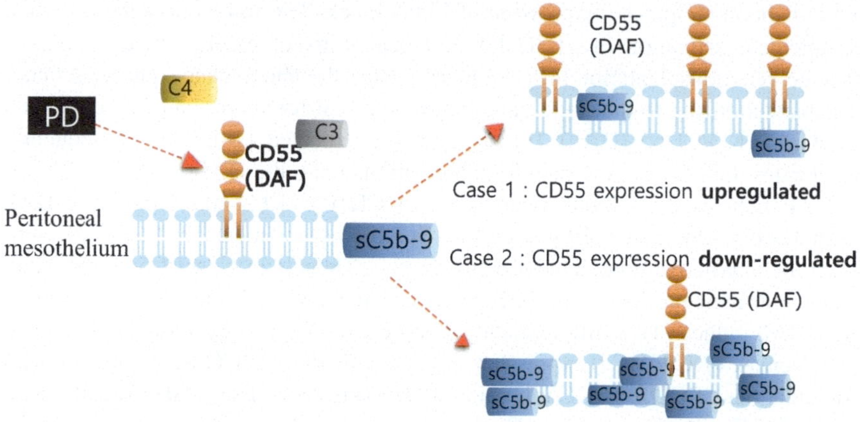

Fig. 14.11 Changes in CD55 expression level of CD55 in mesothelial cells of peritoneal dialysis-receiving patients

14.4.5 CD59 (MAC-IP/MIRL)

As the last example, CD59 is called membrane inhibitor of reactive lysis because it binds C8 and C9 complements to inhibit C9 polymerization and the final MAC formation. Three complementation pathways upon complementation activation are classified to the classical, alternative, and lectin pathways [42]. During the complementation activation, each pathway includes each cascade component converted from the serine proteinases leads to inflammatory, humoral and cellular immune responses [43]. Among the complementation protein fragments, the complement component (C3) is essential in the whole complementation events and also regulates the acquired immune response [44].

C3 component is predominantly biosynthesized in hepatocytes, while some cells including macrophages, DCs, activated T cells, fibroblasts, pneumonocytes, polymorphonuclear leukocytes, proximal tubular ECs, and uterine epithelial cells can synthesize C3 component to some levels [45]. Consequently, CD59 protects cells from lysis. From the above fact, genetically enforced human CD59 expression in pigs protects the tissues and organs of pigs from human complement attack [46]. On the other hand, human HT and CD59-co-expressing murine splenocytes are resistant against the human serum [47]. Similarly, CD59 expression in α1,3Gal-T KO splenocytes of mice are also resistant [48]. In addition, the co-expressed factors of CD55 and CD59 show greater protective activity than the co-expressed CD55 and CD46 [49]. The enforced anticoagulant expression or its induction or procoagulant elimination on the vascular endothelium xenografts shows the prolonged organ survival time. Other candidate factors are such CD62P inhibitor, CD39 known as ectonucleoside triphosphate diphosphohydrolase-1, CD73, hirudin, GPIIbIIIa known as the platelet fibrinogen receptor antagonist, TFPI and TM, or fibrinogen-like protein (FLP)-2, tissue factor KO genes, and protease activated receptor (PAR)-3 and PAR-4.

14.4.6 Human CRPs and Thrombomodulin in Anti-Coagulation

Human transduced CRPs and removed α1,3Gal epitope reduce the HR in pig-to-primate xenotransplantation. With α1,3Gal-T KO pigs, if CRPs such as human CD46 (or MCP), CD55 [50], and CD59 [51] simultaneously inhibit human complement activation, the xenotransplantation efficiency will be much improved. MCP (or CD46) stimulates the catalytic reaction of the C3b and C4b degradations by Factor I. CD46 subsequently inhibit downstream components involved in complement pathway and decreases the injury levels of α1,3Gal-T KO organs, which the complement system is directly associated. hCD46 expressing α1,3Gal-T KO lungs, which are expressing hCD46, lower C3a formation during the perfusion [52]. Transgenic co-expression of hCD46 additionally and effectively inhibits lung injury, indicating additionally enhanced complement regulation. For example, hCD46 expression in α1,3Gal-T KO cells of pig lungs regulates activation of platelets with the deposited CD41 in lung and the expressed CD62 in platelets as well as the decreased genesis of thrombin. Among platelet markers, it is known that CD41 known as integrin subunit α IIb and CD62 (or P-selectin) known as platelet activation marker are highly expressed on activating platelet surfaces. CD41 expression is high on both resting and activating platelet surfaces, the CD62 expression is low on resting platelet surface but is high on activating platelet surfaces [53]. The CD41 associates with another CD61 known as β3 chain to generate CD41/CD61-complexed form that acts as a coagulation receptor for fibrinogen, fibronectin, thrombospondin, vitronectin and vWf.

Because human CRPs have no a directly inhibitory activity of clotting cascade or the platelet functions, the effects are attributed to indirect effects. For example, the diminished complement function leads to lower endothelial vWF expression and secretion in human platelets. The hCRP expression in α1,3Gal-T KO pig organs contributes to anti-inflammation, anti-coagulation, and anti-thrombosis. The transgenic introduction of human CRPs such as hMCP (hCD46) or hDAF (hCD55) is therefore capable of protection of the human recipients against human complement-mediated rejection [54]. hCD46 transgenic pig hearts gave cardiac xenograft survival [55]. Recently, it was proved that α1,3Gal-T KO pig lungs that multiple-modified transgenes such as hCD55 and hCD59 are expressed, exhibited highly-improved functions of lungs in a xeno-transgenic model of perfusions [56].

With unknown reasons, hCD46 expression suppresses the human T-cell proliferation and also downregulates aggregation potentials of human platelets [57]. The enforced hCD46 expression in the pig organs also protected transgenic pigs' organs from the recipient's complement system [58]. In 2012, baboons transplanted with α1,3Gal-T KO/hCD46 Tg pig-derived hearts exhibited the prolonged survival length of xenografts in the recipients [59]. At present, transgenic pigs, which express all three CD46/CD55/CD59 transgenes, have been produced [60]. Produced organs of the multiple transgenic pigs are resistant against transplantation immune rejection. Therefore, triple genetic-engineered pigs having hCD46 and hTBM (human thrombomodulin) in α1,3Gal-T KO pigs (GTKO/hCD46/hTBM) was established and the triple Tg pig cardiac organs are reproducible with a long-term survival period for 1 year for pig cardiac xenografts in pig-to-baboon xenotransplantation [30]. The triple GTKO/hCD46/hTBM Tg pigs, which are combined with an immune modulatory antibody of αCD40 (2C10R4), prolonged the survival period of xenografts in baboons. Immune-suppression driven by αCD20 antibody-treated B cells depletion protected the transplanted xenografts and improved survival period of recipients. However, the xenografts ultimately were rejected by delayed rejection [59]. Expression of cell surfaced glycoprotein CD40 is wide on APCs of B cells, DCs, and macrophages, and in inflamed ECs and fibroblasts [61]. Inhibition of the T-cell co-stimulating pathways increases grafts survival, delays or prevents Abs elaboration, and suppresses chronic rejection. αCD40/CD154 mAbs can delay antibody-mediated rejection (AMR), as the CD40 (2C10R4)/CD154 (IDEC-131, 5C8H1) axis blockade activates peripheral regulatory tolerance and αCD40 mAb and αCD154 mAb treatment suppress B cell sensitization or maturation [62].

Functional blocking of Abs-treated B cells and suppressive downregulation of B-cell function are caused by the treatment with αCD20-specific Abs. Because the harmonized interplay of self-defensive systems including coagulation, complementation, anti-pig Abs and T lymphocytes leads to immune rejection responses against transplanted xenografts, it is considered to be likely that the triple modification of the GTKO/hCD46Tg/hTBM Tg genes prevents the complication of rejection process. In addition, another triple gene modification of the GTKO/hCD46Tg/NeuGcKO pigs showed the reduced response of human Abs after clinical transplantation of cornea and islets grafts [63]. By local transgenic introduction and expression of human genes encoding for thromboregulatory molecules in a pig GTKO/hCD46Tg xenografts, xenograft injuries related with coagulation response are prevented. A recent study [30] has described that heterotopic GTKO.hCD46 pig cardiac with the additional human thromboregulatory protein, thrombomodulin (hTBM) [64], xenotransplantation to baboon [65] reduced thrombodysregulation. GTKO/hCD46/hTBM pigs primarily target the anti-CD154-CD40 co-stimulation pathway, because they express human CRP, CD46 and human thrombomodulin. When baboon received the genetically modified GTKO.hCD46.hTBM pig hearts exhibited long-term survival for up to 1–3 years, preventing humoral rejection and coagulation dysregulation, consequently leading to long-term cardiac xenograft survival.

14.4.7 CRISPR/Cas9 Technology in Pig Gene Targeting

CRISPR/Cas9 gene editing technology has been utilized to target pig genes together with the somatic cell nuclear transfer to construct the complement C3-deleted piglets [66]. The generated complement pig C3 KO strain was used as a pig animal model for C3 function in xenografts. To disrupt the pig complement C3 gene function, the exon 26 site-targeting two sgRNAs have been designed, as described in public skill (http://crispr.mit.edu/) and used. The complement C3 KO cells were then constructed by the TD-tomato plasmid co-transfected with Cas9sgRNA1 targeting vector in primary PFFs isolated from a Bama mini fetus. Biallelic mutations with homozygous and heterozygous types in the target region of C3 gene was obtained, yielding the C3-deficient piglets. The complement C3 protein was completely destroyed, as lysis assay for the CDC confirmed that the C3 deletion completely eliminated the complement pathway activation in the pig C3 KO strains. However, it is considered that the complement C3 disruption may raise unexpected immune deficiency phenotype in the pig C3 KO strains, as the founder strains of C3 KO pig cannot sexually produce F1 progeny. As another strategy to avoid phagocytosis, the α1,3Gal-T KO/hCD47 Tg pigs has been suggested to be used [67]. Signal-regulatory protein alpha (SIRPα) and CD47 interaction has been known to decrease host phagocytosis capacity occurred by macrophages, although some interspecies incompatibility between the CD47–SIRPα interaction of the different organisms [68]. Integrin-associated protein (IAP) as a TM protein, human CD47, interacts with the thrombospondin-1 (TSP-1) and SIRPα as ligands. SIPRα-CD47 binding bi-directionally induces the innate alloresponse of recipients against bone marrow grafts by donor SIRPα interaction with recipient CD47. In addition, the SIRP-α-CD47 interaction induces the primary innate response of allorecognition and also generates monocyte memory necessary for SIRPα-CD47 pathway in chronic rejection of allografts [69]. In addition, the SIRPα-CD47 interaction induces inhibitory signals by SIRPα to prevent macrophage phagocytosis of host cells, suppress neutrophil migration and the inflammatory TNF cytokine production. SIPRα-CD47 binding bi-directionally induces the innate alloresponse of recipients against bone marrow grafts by donor SIRPα interaction with recipient CD47. Therefore, SIRP-α-CD47 binding modulates the alloimmune response [70].

14.5 α1,2-Fucosyltransferase (α1,2-FucT) and Lysosomal α-Galactosidase Enzymes Reduce the Galα(1,3)Gal Epitope Expression and Xenoreaction

Human induction of HR response against pig organs involves in the interaction between Galα1,3Gal antigenic epitopes with specific Abs and complements. To decrease the xenoantigenic levels of Galα1,3Gal epitopes, transgenic expression of α-Galactosidase in human lysosome to pigs can be considered [71]. The cultured

COS cells and pig ECs, which overexpress the human α-Galactosidase gene, showed the reduced level of cell surfaced production of Galα1,3Gal antigens and also cellular reactivity with human natural Abs. On the other hand, α1,2-fucosyltransferase (α1,2Fuc-T) is known to decreases the Galα1,3Gal antigenic epitope levels [72]. As an alternative strategy to decrease the epitope level and the HR, human α1,2Fuc-T (or known as H-transferase, HT) gene was integrated into the pig genome. This alternative method is to reduce the Galα1,3Gal epitope number due to transducing incorporation of the human α1,2Fuc-T (or HT) gene into the pig genome. However, the α1,2Fuc-T gene expression does not completely eliminate the Galα1,3Gal antigenic epitope. The enzyme α1,2Fuc-T specifically transfer the Fuc residue to the acceptor substrate LacNAc and α1,3Gal-T enzyme uses the same acceptor substrate, indicating that the two enzymes are competitive to utilize the same acceptor substrate [73].

Although pigs have an endogenous α1,2Fuc-T gene, the α1,2Fuc-T-producing H epitope is not present on the ECs surfaces in blood vessels [74]. Despite the expression of α1,2Fuc-T, the α1,3Gal epitope is not completely eliminated from the cell surfaces. Thus, *in vitro* α-Galactosidase and α1,2Fuc-T co-expression effectively remove the terminal epitope residues from the surfaced cells of pigs [75]. As a transgene, the human α1,2Fuc-T gene-overexpressing mice have almost 90% more-decreased levels of Galα1,3Gal antigenic epitopes because of the fucosylation-driven xenoantigenic masking. In gene dosage effect, transgenic pigs with both the human CD59 and the α1,2-FuT genes showed the reduced complement-dependent lysis in the aortic EC and peripheral blood mononuclear cells (PBMC) [72]. As a transgene, human α1,2Fuc-T gene-overexpressing mice also have almost 90% more reduction level of the Galα1,3Gal antigens because of the xenoantigenic masking by fucosylation [76]. In porcine xenogeneic cartilage, high expression of α1,2Fuc-T (HT) protects from Galα1,3Gal-dependent responses of antibody production [77].

In another α1,2Fuc-T-transgenic α1,3Gal-T KO mice, the α1,2Fuc-T gene-expressing α1,3Gal-T KO mice in pig cartilage remarkably decreased the host immune responses [78]. The number of Galα1,3Gal-epitope-recognizing B cells was dramatically increased in the grafted spleen of the control mice. However, in contrast, the number of Galα1,3Gal-epitope-recognizing B cells was decreased in the α1,2Fuc-T grafted α1,3Gal-T KO mice. Moreover, the Galα1,3Gal-epitope-recognizing B cells in the mice spleen grafted with α1,3Gal-T KO spleen produce anti-Galα1,3Gal epitope-specific IgM Abs, indicating the antibody producer. However, production response of the anti-Galα1,3Gal epitope-specific Abs of even either IgG or IgM subtypes was not observed in the α1,3Gal-T KO mice grafted with α1,2Fuc-T gene because the production response was blocked by absence of Galα1,3Gal antigen-induced activation of B-cells. With regard to the production response of anti-pig Abs, production of the anti-pig Abs was apparently decreased in the α1,2Fuc-T-transgenic α1,3Gal-T KO as analyzed by cohort implants.

Apart from xenoantigenic Galα1,3Gal epitopes, other glycan structures were responded to elicit the IgM production in the model animal. The low production of anti-pig Abs of IgG isotype in the α1,2Fuc-T-transgenic mice α1,3Gal-T KO grafts is explained due to diminished presentation level of xenoantigens to the host T cells. The Galα1,3Gal epitope-recognizing B cells function as APCs, consequently to elicit the T-cell activation. From the survived evidences of the α1,2Fuc-T transgenic grafts in the α1,3Gal-T KO mice, it has been suggested that the reactivity of naturally formed Abs is reduced and the response of anti-Galα1,3Gal-specific Abs is eliminated with the reduced production of the anti-pig Abs. Complement activation and ADCC were concomitantly reduced. Specific recognition of α1,2Fuc-T grafts with NK cells and macrophages is also reduced [79]. Human NK cells can directly and antibody-independently adhere to the carbohydrates such as Galα1,3Gal epitope on xenogeneic cells [79]. Adhesion and activation of human monocytes on α1,2Fuc-T-expressed porcine ECs were reduced. Monocyte cell adhesion and human NK cell-dependent cytotoxicity of fibroblasts were decreased in the α1,2Fuc-T transgenic pigs [80]. Although α1,2Fuc-T expression is not final destination in xenotransplantation, α1,2Fuc-T expression by pig genetic modification is a rationale of xenograft acceptance in both clinical applications in tissue- and cell-leveled transplantations.

14.6 MicroRNA (miRNA) to Reduce the Xenoreaction

14.6.1 Functional and Regulation of miRNAs in Pig miRNAome in Xenotransplantation

MicroRNAs (miRNAs) are defined as non-coding RNAs shorter than 25 nucleotides in length. Therefore, the known miRNAs have approximately 21–25 nucleotides length-carrying non-coding RNA classes, which is now called the microRNA (miRNA). Current data-based information of miRNAs indicated that miRNAs are conserved among animal species [81]. When sizes of pig and human genomes are compared, the similar numbers of protein-coding genes have been estimated [82]. Therefore, the number of pig-derived miRNAs has also been considered to have similar number to humans. Currently, only 411 pig miRNAs in the number have been known in the public miRBase (v21). The pig miRNA number is quite low when compared to mouse (1982) and human (2588) [83], probably due to limited genome information of pig.

For the functions and regulating mechanisms of miRNAs in pig miRNAome, pig miRNAs can be applicable for understanding in regulatory roles of human miRNAs. Using bioinformatic approaches of pig genomes, pig genome-expressed miRNAs and their targeting genes are obtained in order to utilize as miRNA biomarkers in pig-to-human xenotransplantation. miRNAs act as gene expression and signaling pathway regulators that involve in normal cellular system and pathogenic diseases.

They are regulatory molecules of various phenotype diseases. miRNAs regulate expression of target genes by blocking of the protein translation and degradation of the target mRNAs [83]. The functional roles of miRNAs how to regulate the hematopoietic and nervous systems have been suggested, as well as how to modulate the innate and adaptive immunities in inflammation and diseases of cancers have well been understood. For example, changes in miRNA expression regulate the immune and inflammatory responses. LPS induction of the human mononuclear THP-1 cells upregulates the three miRNA expressions, which include miR-132, miR-146a/b, and miR-155 [84]. The known inflammatory cytokines of TNF-α and IL-1β as well as toll-like receptor (TLR)-specific ligands also elicit miR-146 expression by activation of transcription factors such as NF-κB in the THP-1 cells [84]. The known two molecules of TNF receptor (TNFR)-associated factor 6 (TRAF6) and IL-1 receptor (IL1R)-associated kinase (IRAK)-1 directly enhance the miR-146 target in the TLR/IL-1β signaling pathway of THP-1 cells [85]. Furthermore, stimulation of THP-1 cells with LPS, interferon (IFN)-γ, IFN-β and polyinosinic acid highly elevate the expression level of miR-155 gene. In addition, cytokine induction of THP-1 cells with IFN-γ or IFN-β significantly increases the miR-155 expression level in the innate immune cells [86]. Furthermore, miR-155 activates the translational synthesis of TNF-α protein, suggesting the complexed functionality of miR-155 in immune response modulation [87]. However, in the transplantation field, the miRNA role has not well been understood yet in xenotransplantation.

Recently, FK506, a macrolide immunosuppressive antibiotic prolonged the survival period in xenotransplantation of rat cardiac grafts, elucidating that miR-146a and miR-155 protect the response in mouse to rat xenotransplantation model [87]. FK506 inhibited the rejection level of cardiac xenografts in mouse-to-rat xenotransplantation [88], as similarly in other xenotransplantation types [89]. miR-146a is expressed in an NFκB-dependent manner [84] and gene expression of miR-146a/b is depended on expression levels of proinflammatory cytokines [90]. FK506 inhibited the rejection level of the xenografts. However, phenotypes of miR-146a upregulation and RelA downregulation known as NF κB family were specifically detected [87]. Hence, it has been concluded that miR-146 has a protective effect on the xenograft rejection through the NF-κB downregulation known as a core factor for the AVR. miR-155 inhibited and suppressed behavioral nTreg cells strengthened by conventional $CD4^+$ Th cells [91]. In addition, miR-155 deficiency induced the impaired both innate and adaptive immunities and also decreased the level of IgG1 production in hosts [92].

In xenotransplantation, the lentiviral-delivered miRNA targeted therapy was applied and the expression levels of different miR-146a and miR-155 were observed during xenograft rejection [87]. miRNAs are biomarkers for checking of xenograft status because miRNA controls the immune cell homeostasis. Using miRNA technology, relevant strategies toward RNA interference to modulate the immune repertory in transplantation can be designed. If one miRNA is manipulated, multiple mRNAs are regulated and multiple miRNAs can also regulate one mRNA, indicating that miRNA may provide novel therapeutic tools in xenotransplantation. The level of miRNA expression is increased in cardiac xenograft rejection in mouse-to-rat xenotransplantation. The levels of miR-146a and miR-155 expressions were increased in cardiac xenografts of xenotransplantation rather than cardiac allografts of allotransplantation. Therefore, miRNAs have been considered as target therapeutics in immune responses of transplantation, especially in xenotransplantation. In xenotransplantation, the analysis of miR-146a and miR-155 is of our future interests to understand their roles. The miR-146a and miR-155 expressions levels are also analyzed in a FK506-treated xenograft model of cardiac organs. FK506 easily inhibited the immune rejection of cardiac xenografts in a mouse-to-rat xenotransplantation rats. FK506 treatment of the recipients showed the altered expression levels of miR-146a and miR-155, suggesting that the miRNAs regulate the immune rejection to confer the protective effect.

14.6.2 Possibility of Pig miRNA Profiles as Monitoring Biomarkers in Xenotransplantation

The miRNAs have been suggested as ideal biomarkers of diseases because biomarkers are important for monitoring of early diagnosis, treatment prediction and response. Also, in organ transplantation, biomarkers are important finally to gain the successful xenotransplantation with injury prevention of grafts. miRNA profiles as biomarkers of quality of organ grafts/ischemia reperfusion injury [93] can show the indication between miRNAs and graft function or acute rejection. Previously miRNAs have been suggested to be available biomarkers of allograft transplantation in human [92]. For example, three miRNAs of miR-142-5p, miR-155, and miR-223 were used as sensitive and specific prediction marker of acute rejection in renal allografts of human [94]. The stable expression of miRNAs confers to noninvasive biomarkers in functional monitoring of xenografts. High similarities are known in morphology and physiology in pig and human, indicating possibility of pig miRNAs as biomarkers in pig-to-human xenotransplantation. Moreover, due to high similarity in the miRNA nucleotide sequences between animal species, the circulating miRNAs of pigs in human body may be utilized as markers monitoring for pig-to-human xenotransplantation. Circulating miRNA profiles as biomarkers can also be used for the diagnostic, therapeutic, and prognostic parameters involved in immune suppression, graft toxicity and rejection.

References

1. Komoda H, Miyagawa S, Omori T, Takahagi Y, Murakami H, Shigehisa T, Ito T, Matsuda H, Shirakura R. Survival of adult islet grafts from transgenic pigs with N-acetylglucosaminyltransferase-III (GnT-III) in cynomolgus monkeys. Xenotransplantation. 2005;12(3):209–16.
2. Miyagawa S, Murakami H, Takahagi Y, et al. Remodeling of the major pig xenoantigen by N-acetylglucosaminyltransferase III in transgenic pig. J Biol Chem. 2001;276:39310–9.
3. Koyota S, Ikeda Y, Miyagawa S, et al. Down-regulation of the alpha-Gal epitope expression in N-glycans of swine endothelial cells by transfection with the N-acetylglucosaminyltransferase III gene: modulation of the biosynthesis of terminal structures by a bisecting GlcNAc. J Biol Chem. 2001;276:32867.

4. Chung TW, Kim KS, Kim CH. Reduction of the Gal-alpha1,3-Gal epitope of mouse endothelial cells by transfection with the N-acetylglucosaminyltransferase III gene. Mol Cells. 2003;16 (3):368–76.
5. Tanemura M, Miyagawa S, Ihara Y, Matsuda H, Shirakura R, Taniguchi N. Significant downregulation of the major swine xenoantigen by N-acetylglucosaminyltransferase III gene transfection. Biochem Biophys Res Commun. 1997;235(2):359–64. Erratum in: Biochem Biophys Res Commun 1997; 238(2): 686.
6. Ihara Y, Nishikawa A, Tohma T, Soejima H, Niikawa N, Taniguchi N. J Biochem. 1993;113:692–8.
7. Tanemura M, Miyagawa S, Ihara Y, Matsuda H, Shirakura R, Taniguchi N. Biochem Biophys Res Commun. 1997;235:359–64.
8. Taniguchi N, Yoshimura M, Miyoshi E, Ihara Y, Nishikawa A, Fujii S. Glycobiology. 1996;6:691–4.
9. Komoda H, Miyagawa S, Kubo T, et al. A study of the xenoantigenicity of adult pig islet cells. Xenotransplantation. 2004;11:237.
10. Miyagawa S, Takeishi S, Yamamoto A, Ikeda K, Matsunari H, Yamada M, Okabe M, Miyoshi E, Fukuzawa M, Nagashima H. Survey of glycoantigens in cells from alpha1-3galactosyltransferase knockout pig using a lectin microarray. Xenotransplantation. 2010;17 (1):61–70.
11. Sandrin MS, Fodor WL, Mouhtouris E, Osman N, Cohney S, Rollins SA, Guilmette ER, Setter E, Squinto SP, McKenzie IFC. Nat Med. 1995;1:1261–7.
12. Tanemura M, Miyagawa S, Koyota S, Koma M, Matsuda H, Tsuji S, Shirakura R, Taniguchi N. J Biol Chem. 1998;273:16421–5.
13. Koma M, Miyagawa S, Honke K, Ikeda Y, Koyota S, Miyoshi S, Matsuda H, Tsuji S, Shirakura R, Taniguchi N. Reduction of the major xenoantigen on glycosphingolipids of swine endothelial cells by various glycosyltransferases. Glycobiology. 2000;10(7):745–51.
14. Liszewski MK, Atkinson JP. Membrane cofactor protein (MCP; CD46): deficiency states and pathogen connections. Curr Opin Immunol. 2021;72:126–34.
15. Thorgersen EB, Barratt-Due A, Haugaa H, Harboe M, Pischke SE, Nilsson PH, Mollnes TE. The role of complement in liver injury, regeneration, and transplantation. Hepatology. 2019;70:725–36. https://doi.org/10.1002/hep.30508.
16. Zhou Z, Xu MJ, Gao B. Hepatocytes: a key cell type for innate immunity. Cell Mol Immunol. 2016;13:301-546 315. https://doi.org/10.1038/cmi.2015.97.
17. Lubbers R, van Essen MF, van Kooten C, Trouw LA. Production of complement components by cells of the immune system. Clin Exp Immunol. 2017;188:183–94. https://doi.org/10.1111/cei.12952.
18. Ghannam A, Fauquert JL, Thomas C, Kemper C, Drouet C. Human complement C3 deficiency: Th1 induction requires T cell-derived complement C3a and CD46 activation. Mol Immunol. 2014;58:98-107, 556. https://doi.org/10.1016/j.molimm.2013.11.010.
19. Kwan WH, van der Touw W, Paz-Artal E, Li MO, Heeger PS. Signaling through C5a receptor and C3a receptor diminishes function of murine natural regulatory T cells. J Exp Med. 2013;210:257–68. https://doi.org/10.1084/jem.20121525.
20. Vik DP, Wong WW. Structure of the gene for the F allele of complement receptor type 1 and sequence of the coding region unique to the S allele. J Immunol. 1993;151:6214–24.
21. Kaur G, Wright K, Verma S, Haynes A, Dufour JM. The good, the bad and the ugly of testicular immune regulation: a delicate balance between immune function and immune privilege. Adv Exp Med Biol. 2021;1288:21–47. https://doi.org/10.1007/978-3-030-77779-1_2.
22. Dufour JM, Lord SJ, Kin T, Rayat GR, Dixon DE, Bleackley RC, Korbutt GS, Rajotte RV. Comparison of successful and unsuccessful islet/Sertoli cell Cotransplant grafts in streptozotocin-induced diabetic mice. Cell Transplant. 2007;16:1029–38. https://doi.org/10.3727/000000007783472417.
23. Kaur G, Wright K, Mital P, Hibler T, Miranda JM, Thompson LA, Halley K, Dufour JM. Neonatal pig Sertoli cells survive xenotransplantation by creating an immune modulatory

environment involving CD4 and CD8 regulatory T cells. Cell Transplant. 2020;29:963689720947102. https://doi.org/10.1177/0963689720947102.

24. West EE, Kolev M, Kemper C. Complement and the regulation of T cell responses. Annu Rev Immunol. 2018;538(36):309–38. https://doi.org/10.1146/annurev-immunol-042617-053245.
25. Washburn RL, Martinez-Marin D, Korać K, Sniegowski T, Rodriguez AR, Chilton BS, Hibler T, Pruitt K, Bhutia YD, Dufour JM. The Sertoli cell complement signature: a suspected mechanism in xenograft survival. Int J Mol Sci. 2023;24(3):1890. https://doi.org/10.3390/ijms24031890.
26. Guo J, Grow EJ, Mlcochova H, Maher GJ, Lindskog C, Nie X, Guo Y, Takei Y, Yun J, Cai L, et al. The adult 569 human testis transcriptional cell atlas. Cell Res. 2018;28:1141–57. https://doi.org/10.1038/s41422-018-0099-2. 570.
27. Ozen A, Comrie WA, Ardy RC, Domínguez Conde C, et al. CD55 deficiency, early-onset protein-losing enteropathy, and thrombosis. N Engl J Med. 2017;377(1):52–61. https://doi.org/10.1056/NEJMoa1615887.
28. Amara U, et al. Molecular intercommunication between the complement and coagulation system. J Immunol. 2010;185:5628–36.
29. Sei Y, Mizuno M, Suzuki Y, Imai M, Higashide K, Harris CL, Sakata F, Iguchi D, Fujiwara M, Kodera Y, Maruyama S, Matsuo S, Ito Y. Expression of membrane complement regulators, CD46, CD55 and CD59, in mesothelial cells of patients on peritoneal dialysis therapy. Mol Immunol. 2015;65(2):302–9.
30. Mohiuddin MM, Singh AK, Corcoran PC, Thomas ML 3rd, Clark T, Lewis BG, Hoyt RF, Eckhaus M, Pierson RN 3rd, Belli AJ, Wolf E, Klymiuk N, Phelps C, Reimann KA, Ayares D, Horvath KA. Chimeric 2C10R4 anti-CD40 antibody therapy is critical for long-term survival of GTKO.hCD46.hTBM pig-to-primate cardiac xenograft. Nat Commun. 2016;7:11138.
31. Strainic MG, Liu J, An F, Bailey E, Esposito A, Hamann J, Heeger PS, Medof ME. CD55 is essential for CD103+ dendritic cell tolerogenic responses that protect against autoimmunity. Am J Pathol. 2019;189(7):1386–401.
32. Wobus M, Vogel B, Schmücking E, Hamann J, Aust G. N-glycosylation of CD97 within the EGF domains is crucial for epitope accessibility in normal and malignant cells as well as CD55 ligand binding. Int J Cancer. 2004;112(5):815–22. https://doi.org/10.1002/ijc.20483.
33. Hamann J, Stortelers C, Kiss-Toth E, Vogel B, Eichler W, van Lier RA. Characterization of the CD55 (DAF)-binding site on the seven-span transmembrane receptor CD97. Eur J Immunol. 1998;28(5):1701–7. https://doi.org/10.1002/(SICI)1521-4141(199805)28:05.
34. Wandel E, Saalbach A, Sittig D, Gebhardt C, Aust G. Thy-1 (CD90) is an interacting partner for CD97 on activated endothelial cells. J Immunol. 2012;188(3):1442–50. https://doi.org/10.1002/(SICI)1521-4141(199805)28:05<1701::AID-IMMU1701>3.0.CO;2-2.
35. Wang T, Ward Y, Tian L, Lake R, Guedez L, Stetler-Stevenson WG, Kelly K. CD97, an adhesion receptor on inflammatory cells, stimulates angiogenesis through binding integrin counter receptors on endothelial cells. Blood. 2005;105(7):2836–44. https://doi.org/10.1182/blood-2004-07-2878.
36. Gordon S, Hamann J, Lin HH, Stacey M. F4/80 and the related adhesion-GPCRs. Eur J Immunol. 2011;41(9):2472–6. https://doi.org/10.1002/eji.201141715.
37. Hoek RM, de Launay D, Kop EN, Yilmaz-Elis AS, Lin F, Reedquist KA, Verbeek JS, Medof ME, Tak PP, Hamann J. Deletion of either CD55 or CD97 ameliorates arthritis in mouse models. Arthritis Rheum. 2010;62(4):1036–42. https://doi.org/10.1002/art.27347.
38. Shimizu I, Smith NR, Zhao G, Medof E, Sykes M. Decay-accelerating factor prevents acute humoral rejection induced by low levels of anti-alphaGal natural antibodies. Transplantation. 2006;81(1):95–100.
39. Cascalho M, et al. Xenotransplantation and the future of renal replacement. JASN. 2004;15 (5):1106–12. 15; 81(1): 95-100.
40. Azimzadeh AM, Kelishadi SS, Ezzelarab MB, Singh AK, Stoddard T, Iwase H, Zhang T, Burdorf L, Sievert E, Avon C, Cheng X, Ayares D, Horvath KA, Corcoran PC, Mohiuddin MM, Barth RN, Cooper DK, Pierson RN 3rd. Early graft failure of GalTKO pig organs in

baboons is reduced by expression of a human complement pathway-regulatory protein. Xenotransplantation. 2015;22(4):310–6.
41. MAHM M, RJF M, van der Velden TJAM, van de Kar NCAJ, van den Heuvel LPWJ, Volokhina EB, COMBAT Consortium. The role of properdin in C5 convertase activity and C5b-9 formation in the complement alternative pathway. J Immunol. 2021;207(10):2465–72.
42. Dunkelberger JR, Song WC. Complement and its role in innate and adaptive immune responses. Cell Res. 2010;20:34–50.
43. Ricklin D, Lambris JD. Complement in immune and inflammatory disorders: pathophysiological mechanisms. J Immunol. 2013;190:3831–8.
44. Singer L, Colten HR, Wetsel RA. Complement C3 deficiency: human, animal, and experimental models. Pathobiology. 1994;62:14–28.
45. Peng Q, Li K, Patel H, Sacks SH, Zhou W. Dendritic cell synthesis of C3 is required for full T cell activation and development of a Th1 phenotype. J Immunol. 2006;176:3330–41.
46. Niemann H, Verhoeyen E, Wonigeit K, Lorenz R, Hecker J, Schwinzer R, et al. Cytomegalovirus early promoter induced expression of hCD59 in porcine organs provides protection against hyperacute rejection. Transplantation. 2001;72:1898–906.
47. Costa C, Zhao L, Decesare S, Fodor WL. Comparative analysis of three genetic modifications designed to inhibit human serummediated cytolysis. Xenotransplantation. 1999;6:6–16.
48. Tange MJ, Salvaris E, Romanella M, et al. Additive effects of CD56 expression in Gal knockout mice in vitro but in an ex vivo model. Xenotransplantation. 1997;4:25–33.
49. Huang J, Gou D, Zhen C, et al. Protection of xenogeneic cells from human complement-mediated lysis by the expression of human DAF, CD59 and MCP. FEMS Immunol Med Microbiol. 2001;31:203–9.
50. Lee HJ, Lee BC, Kim YH, Paik NW, Rho HM. Characterization of transgenic pigs that express human decay accelerating factor and cell membrane-tethered human tissue factor pathway inhibitor. Reprod Domest Anim. 2011;46(2):325–32.
51. McGregor CG, Teotia SS, Byrne GW, Michaels MG, Risdahl JM, Schirmer JM, Tazelaar HD, Walker RC, Logan JS. Cardiac xenotransplantation: progress toward the clinic. Transplantation. 2004;78(11):1569–75.
52. Burdorf L, Stoddard T, Zhang T, Rybak E, Riner A, Avon C, Laaris A, Cheng X, Sievert E, Braileanu G, Newton A, Phelps CJ, Ayares D, Azimzadeh AM, Pierson RN 3rd. Expression of human CD46 modulates inflammation associated with GalTKO lung xenograft injury. Am J Transplant. 2014;14(5):1084–95.
53. Simak J, Holada K, Janota J, Stranak Z. Surface expression of major membrane glycoproteins on resting and TRAP-activated neonatal platelets. Pediatr Res. 1999;46:445–9.
54. He S, Feng C, Long C, Pan H, Wang N, Shi N, Wang F, Zhou Y, Chen H, Chu M, et al. Production of Wuzhishan miniature pigs expressing high-level human CD46. J Anim Vet Adv. 2014;13:616–22.
55. Byrne GW, Davies WR, Oi K, et al. Increased immunosuppression, not anticoagulation, extends cardiac xenograft survival. Transplantation. 2006;82:1787–91.
56. Westall GP, Levvey BJ, Salvaris E, Gooi J, Marasco S, Rosenfeldt F, et al. Sustained function of genetically modified porcine lungs in an ex vivo model of pulmonary xenotransplantation. J Heart Lung Transplant. 2013;2498(13):01341–7.
57. Ezzelarab MB, Ayares D, Cooper DK. Transgenic expression of human CD46: does it reduce the primate T-cell response to pig endothelial cells? Xenotransplantation. 2015;22(6):487–9.
58. McGregor CG, Davies WR, Oi K, Teotia SS, Schirmer JM, Risdahl JM, et al. Cardiac xenotransplantation: recent preclinical progress with 3-month median survival. J Thorac Cardiovasc Surg. 2005;130(844–851):37.
59. Mohiuddin MM, Corcoran PC, Singh AK, Azimzadeh A, Hoyt RF Jr, Thomas ML, et al. B-cell depletion extends the survival of GTKO.hCD46Tg pig heart xenografts in baboons for up to 8 months. Am J Transplant. 2012;12:763–71.

60. Zhou CY, McInnes E, Copeman L, Langford G, Parsons N, Lancaster R, et al. Transgenic pigs expressing human CD59, in combination with human membrane cofactor protein and human decay-accelerating factor. Xenotransplantation. 2005;12:142–8.
61. Zhang T, Pierson RN 3rd, Azimzadeh AM. Update on CD40 and CD154 blockade in trasplant models. Immunotherapy. 2015;7(8):899–911.
62. O'Neill NA, Zhang T, Braileanu G, Sun W, Cheng X, Hershfeld A, Laird CT, Kronfli A, Hock LA, Dahi S, Kubicki N, Sievert E, Hassanein W, Cimeno A, Pierson RN 3rd, Azimzadeh AM. Comparative evaluation of αCD40 (2C10R4) and αCD154 (5C8H1 and IDEC-131) in a nonhuman primate cardiac allotransplant model. Transplantation. 2017;101(9):2038–47. https://doi.org/10.1097/TP.0000000000001836.
63. Lee W, Hara H, Ezzelarab MB, Iwase H, Bottino R, Long C, Ramsoondar J, Ayares D, Cooper DK. Initial in vitro studies on tissues and cells from GTKO/CD46/NeuGcKO pigs. Xenotransplantation. 2016;23(2):137–50.
64. Wuensch A, et al. Regulatory sequences of the porcine THBD gene facilitate endothelial-specific expression of bioactive human thrombomodulin in single- and multitransgenic pigs. Transplantation. 2014;97:138–47.
65. Michler RE, McManus RP, Smith CR, Sadeghi AN, Rose EA. Technique for primate heterotopic cardiac xenotransplantation. J Med Primatol. 1985;14:357–62.
66. Zhang W, Wang G, Wang Y, Jin Y, Zhao L, Xiong Q, Zhang L, Mou L, Li R, Yang H, Dai Y. Generation of complement protein C3 deficient pigs by CRISPR/Cas9-mediated gene targeting. Sci Rep. 2017;7(1):5009.
67. Tena A, Kurtz J, Leonard DA, et al. Transgenic expression of human CD47 markedly increases engraftment in a murine model of pig-to-human hematopoietic cell transplantation. Am J Transplant. 2014;14:2713–22.
68. Ide K, Wang H, Tahara H, et al. Role for CD47-SIRPalpha signaling in xenograft rejection by macrophages. Proc Natl Acad Sci U S A. 2007;104:5062–6.
69. Garcia-Sanchez C, Casillas-Abundis MA, Pinelli DF, Tambur AR, Hod-Dvorai R. Impact of SIRPα polymorphism on transplant outcomes in HLA-identical living donor kidney transplantation. Clin Transpl. 2021;35(9):e14406. https://doi.org/10.1111/ctr.14406.
70. Per-Arne O, Alex Z, Yi-Fu F, Carl FL, Hattie DG, Frederik PL. Role of CD47 as a marker of self on red blood cells. Science. 2000;288(5473):2051–4.
71. Zeyland J, Woźniak A, Gawrońska B, Juzwa W, Jura J, Nowak A, Słomski R, Smorąg Z, Szalata M, Mazurek U, Lipiński D. Double transgenic pigs with combined expression of human α1,2-fucosyltransferase and α-galactosidase designed to avoid hyperacute xenograft rejection. Arch Immunol Ther Exp. 2014;62(5):411–22.
72. Costa C, Zhao L, Burton WV, Rosas C, Bondioli KR, Williams BL, et al. Transgenic pigs designed to express human CD59 and H-transferase to avoid humoral xenograft rejection. Xenotransplantation. 2002;9:45–57.
73. Larsen RD, Ernst LK, Nair RP, Lowe JB. Molecular cloning, sequence, and expression of a human GDP-L-fucose:beta-D-galactoside 2-alpha-L-fucosyltransferase cDNA that can form the H blood group antigen. Proc Natl Acad Sci U S A. 1990;87(17):6674–8.
74. Sharma VK, Bologa RM, Li B, Xu GP, Lagman M, Hiscock W, Mouradian J, Wang J, Serur D, Rao VK, Suthanthiran M. Molecular executors of cell death–differential intrarenal expression of Fas ligand, Fas, granzyme B, and perforin during acute and/or chronic rejection of human renal allografts. Transplantation. 1996;62(12):1860–6.
75. Luo Y, Wen J, Luo C, Cummings RD, Cooper DK. Pig xenogeneic antigen modification with green coffee bean alpha-galactosidase. Xenotransplantation. 1999;6(4):238–48.
76. Osman N, IF MK, Ostenried K, Ioannou YA, Desnick RJ, Sandrin MS. Combined transgenic expression of alpha-galactosidase and alpha1,2-fucosyltransferase leads to optimal reduction in the major xenoepitope Galalpha(1,3)Gal. Proc Natl Acad Sci U S A. 1997;94(26):14677–82.
77. Sommaggio R, Bello-Gil D, Pérez-Cruz M, Brokaw JL, Máñez R, Costa C. Genetic engineering strategies to prevent the effects of antibody and complement on xenogeneic chondrocytes. Eur Cell Mater. 2015;30:258–70.

78. Costa C, Brokaw JL, Wang Y, Fodor WL. Delayed rejection of porcine cartilage is averted by transgenic expression of alpha1,2-fucosyltransferase. FASEB J. 2003;17(1):109–11.
79. Inverardi L, Clissi B, Stolzer AL, Bender JR, Sandrin MS, Pardi R. Human natural killer lymphocytes directly recognize evolutionary conserved oligosaccharide ligands expressed by xenogeneic tissues. Transplantation. 1997;63(1318):1330.
80. Costa C, Barber DF, Fodor WL. Human NK cell-mediated cytotoxicity triggered by CD86 and Galα1,3-Gal is inhibited in genetically modified porcine cells. J Immunol. 2002;168:3808–16.
81. Song Z, Cooper DKC, Cai Z, Mou L. Expression and regulation profile of mature MicroRNA in the pig: relevance to xenotransplantation. Biomed Res Int. 2018;2018:2983908. https://doi.org/10.1155/2018/2983908.
82. Groenen MAM, Archibald AL, Uenishi H, Tuggle CK, Takeuchi Y, Rothschild MF, et al. Analyses of pig genomes provide insight into porcine demography and evolution. Nature. 2012;491(7424):393–8.
83. Bartel DP. MicroRNAs: target recognition and regulatory functions. Cell. 2009;136:215 233.
84. Taganov KD, Boldin MP, Chang KJ, Baltimore D. NF kappaB dependent induction of microRNA miR 146, an inhibitor targeted to signaling proteins of innate immune responses. Proc Natl Acad Sci U S A. 2006;103:12481 12486.
85. Hou J, Wang P, Lin L, Liu X, Ma F, An H, Wang Z, Cao X. MicroRNA 146a feedback inhibits RIG I dependent type I IFN production in macrophages by targeting TRAF6, IRAK1, and IRAK2. J Immunol. 2009;183:2150 2158.
86. O'Connell RM, Taganov KD, Boldin MP, Cheng G, Baltimore D. MicroRNA 155 is induced during the macrophage inflammatory response. Proc Natl Acad Sci U S A. 2007;104:1604 1609.
87. Zhao Z, Qi F, Liu T, Fu W. Effect of miR-146a and miR-155 on cardiac xenotransplantation. Exp Ther Med. 2016;12:3972–8.
88. Qi F, Zhu LW, He XH, et al. Research on function of combined drugs of immune rejection in mouse to rat cardiac xenotransplantation. Shangdong Med J. 2010;50:12 14.
89. Zou XM, Li XL, Li F, et al. Effect of immunosuppressant FK506 on syrian hamster to rat small bowel xenotransplantation. Zhong Hua Wei Chang Wai Ke Za Zhi. 2003;6:251 254.
90. Bhaumik D, Scott GK, Schokrpur S, Patil CK, Orjalo AV, Rodier F, Lithgow GJ, Campisi J. MicroRNA miR 146a/b negatively modulate the senescence associated inflammatory mediators IL 6 and IL 8. Aging (Albany NY). 2009;1:402–11.
91. Stahl HF, Fauti T, Ullch N, Bopp T, Kubach J, Rust W, Labhart P, Alexiadis V, Becker C, Hafner M, et al. MiRNA 155 inhibition sensitizes CD4+ Th cells for TREG mediated suppression. PLoS One. 2009;4:e7158.
92. Vigorito E, Perks KL, Abreu Goodger C, Bunting S, Xiang Z, Kohlhaas S, Das PP, Miska EA, Rodriguez A, Bradley A, et al. MicroRNA 155 regulates the generation of immunoglobulin class switched plasma cells. Immunity. 2007;27:847–59.
93. Shan J, Feng L, Luo L, et al. MicroRNAs: potentiazl biomarker in organ transplantation. Transpl Immunol. 2011;24(4):210–5.
94. Tao J, Yang X, Han Z, et al. Serum MicroRNA-99a helps detect acute rejection in renal transplantation. Transplant Proc. 2015;47(6):1683–7.

Chapter 15
Delayed Rejection of Xenograft (DRX)

Delayed rejection in organ transplantation is observed in a delayed type of immune rejection within a few hours to days after transplantation [1]. The delayed rejection is similar to the HR in its course and process. Human NK cells and CTLs are the major actors of delayed xenograft rejection (DRX) by cell mediation, but the DRX event is still in question [2]. DC is a regulator in anti-pig T-cell responses of human [2, 3]. The plasticity of DCs produces regulatory DCs, termed tolerant DCs, from undifferentiated progenitor cells such as $CD14^+$ monocytes through differentiation induction by cytokines of IL-4 and GM-CSF. Regulatory DCs weakly express costimulatory proteins with the high ratio of the secreted IL-10/IL-12. They induce the $CD4^+$ T-cell anergy state and the CD4-T regulatory cells (Tregs) expansion state that is antigen-specific [4, 5]. Human regulatory DCs can promote donor-specific tolerance in transplantation without any effect on the entire immune system [6]. Regulatory DC-based immune-regulatory therapies are recommended as adjunct immunosuppressive therapy treatments in transplant recipients. DCs mediate interaction between cells and release of cytokines as well as IFN-γ-acted CTL and NK cell activation and cytotoxicity. DCs' interplay between innate and acquired immune responses initiates or modulates the immune responses. In xenoantigenic rejection, DCs activate NK cell and CTL to cell-mediated inflammatory and xenoreactive responses. Thus, differentiated regulatory DCs can be utilized to protect pig EC from human cell-mediated immune reaction. Regulatory DCs of human can decrease xenoantigenic responses of NK cells and CTLs to ECs of pig [7].

Because both human NK cells and CTL cells respond to pig-derived xenoantigens for anti-pig immune cell responses, the following immune tolerance is induced by various parameters including mixed chimerism, ITIM-based negative modulation, thymic transplantation, transgenic expression of NK cell-inhibitory ligands (HLA), or Treg cells [2]. Regulatory DCs of human are an alternative tool in xenotransplantation [6–8]. Regulatory DC interaction with NK cells and CTL is important. However, delayed rejection is not involved in complement; however, it is mediated by IgG immunoglobulins. The progress of delayed rejection is slow, starting from the arterial lumen, but not started from arterioles or capillaries. DRX

C.-H. Kim, *Glycoimmunology in Xenotransplantation*,
https://doi.org/10.1007/978-981-99-7691-1_15

of xenotransplants is performed by ECs for type II activation with proinflammatory gene upregulation, chemokine secretion, and blood platelet activation. Difference in coagulation systems between the humans and pigs causes coagulative disorder. Vascular endothelial and blood cells express the known ectoenzyme termed ectonucleoside triphosphate diphosphohydrolase-1 (ENTPDase-1) known as CD39. The ectoenzyme CD39 regulates blood clotting and inflammation. CD39 inhibits ATP and ADP-based platelets. CD39 degrades extracellularly generated ADP and ATP to adenosine and adenosine monophosphate (AMP). Adenosine is an aggregation inhibitor of platelets. CD73 (ecto-5′-nucleotidase) degrades ATP into adenosine. Human endothelial cells (ECs) have low CD73 expression, but porcine ECs express the high level of CD73 [9]. Therefore, the transgenic hCD39 and hCD73 in the pig increase in production of adenosine. Human CD39 expression in pigs protects against myocardial damage [10].

References

1. Bach FH, Winkler H, Ferran C, Hancock WW, Robson SC. Delayed xenograft rejection. Immunol Today. 1996;17:379–84.
2. Griesemer A, Yamada K, Sykes M. Xenotransplantation: immunological hurdles and progress toward tolerance. Immunol Rev. 2014;258:241–58.
3. Yamada K, Sachs DH, Dersimonian H. Human anti-porcine xenogeneic T cell response. Evidence for allelic specificity of mixed leukocyte reaction and for both direct and indirect pathways of recognition. J Immunol. 1995;155:5249–56.
4. Gregori S, Tomasoni D, Pacciani V, et al. Differentiation of type 1 T regulatory cells (Tr1) by tolerogenic Dc-10 requires the IL-10-dependent Ilt4/Hla-GPATHWAY. Blood. 2010;116:935–44.
5. Torres-Aguilar H, Aguilar-Ruiz SR, Gonzalez-Perez G, et al. Tolerogenic dendritic cells generated with different immunosuppressive cytokines induce antigen-specific anergy and regulatory properties in memory Cd4(+) T cells. J Immunol. 2010;184:1765–75.
6. Van Kooten C, Lombardi G, Gelderman KA, et al. Dendritic cells as a tool to induce transplantation tolerance: obstacles and opportunities. Transplantation. 2011;91:2–7.
7. Madelon N, Puga Yung GL, Seebach JD. Human anti-pig NK cell and CD8+ T-cell responses in the presence of regulatory dendritic cells. Xenotransplantation. 2016;23(6):479–89.
8. Morelli AE, Thomson AW. Orchestration of transplantation tolerance by regulatory dendritic cell therapy or in-situ targeting of dendritic cells. Curr Opin Organ Transplant. 2014;19:348–56.
9. Khalpey Z, Yuen AH, Kalsi KK, Kochan Z, Karbowska J, Slominska EM, et al. Loss of ecto-50-nucleotidase from porcine endothelial cells after exposure to human blood: implications for xenotransplantation. Biochim Biophys Acta. 2005;1741:191–8.
10. Wheeler DG, Joseph ME, Mahamud SD, Aurand WL, Mohler PJ, Pompili VJ, et al. Transgenic swine: expression of human CD39 protects against myocardial injury. J Mol Cell Cardiol. 2012;52:958–61.

Chapter 16
Blood Coagulation as Coagulation Dysregulation

16.1 EPCR (Endothelial Protein C Receptor), CD36, and the α1,3Gal-T KO/hCD46/hEPCR Pig

Prior to human transplantation application, clarification and elucidation of the pig organ compatibility with the human host thromboregulation system are prerequisite in the pig-to-human or pig-to-primate xenotransplantation. Pig xenografts promote intensive thrombotic response of human blood, although pharmacological interventions and genetic targeting approaches have been made. To reduce acute thrombosis, human thromboregulatory genes have been transgenically introduced into pig donors, including endothelial protein C receptor (EPCR) and human CD36 (hCD36), and the α1,3Gal-T KO/hCD46/hEPCR pig has been generated. In addition, a triple KO pig donor also showed the decreased binding levels of human natural Abs in the acute antibody-dependent rejection.

In pig and human, differences in the coagulation systems at the molecular level are recognized as additional hurdles of pig-to-human xenotransplantation. Different from the direct xenoantigen, additional issue of vascular thrombotic microangiopathy still remained, and they include platelet activation, adherence, and clumping during xenograft transplantation to human and other primates [1]. The thrombotic issue raises a barrier of long-term survival of xenoorgans. Resultant blood coagulation concern is a minor risk due to the incompatibility of the two systems between blood coagulation and anti-coagulation regulation [2]. A thrombotic disease named thrombocytopenia is known to be associated with severely occurred complication of blood breeding and thrombotic microangiopathy of grafts. To overcome thrombotic microangiopathy and thrombocytopenia, the α1,3Gal-T KO pigs have further been genetic modified toward human CRP gene expression of CD39, EPCR, tissue factor pathway inhibitor (TFPI), and TM. Pig vWF replacement with human vWF is a powerful direction together with prevention of the thrombotic microangiopathy.

C.-H. Kim, *Glycoimmunology in Xenotransplantation*,
https://doi.org/10.1007/978-981-99-7691-1_16

16.2 CD39, CD73, and α1,3Gal-T KO/hCD73 Pigs

CD39 and purinergic signaling pathways protect from blood thrombosis, tolerance induction, and vascular inflammation in pig-to-human xenotransplantation. Vascular ATP and ADP regulate vascular thrombosis and inflammation capacity, converting to AMP form by CD39 (or ENTPDase-1) and then further converted to adenosine form by ectoenzyme CD73 known as ecto-5′-nucleotidase (e5NT) [3]. As a key molecule, CD39 (or ENTPDase-1) in vascular endothelium and Tregs hydrolyzes extracellular ATP form and ADP form to AMP form [4]. Therefore, ubiquitous and stable expression of CD73 prevents blood homeostasis [5]. CD39-associated molecules include CD39-dependent e5NT and ENTPDase1/CD39, and they function as thromboregulatory agents [6]. Actually, it was reported that pig heart ECs display a low level of CD73 expression [7]. Cellular suppressive CD39 molecule shows the specific distinguishment in resting and activating $FoxP3^{+}CD4^{+}CD25^{+}$ Tregs among the T-cell populations present in mouse, human, and primate [8] as well as T-cell populations of pigs. Hence, transgenic CD39 expression [9] promotes immune-suppressive pathways. The mice CD39 transgenic cardiac xenografts show their strong resistance to antibody-mediated rejection as a humoral-type rejection induced by elicited α1,3Gal epitope-specific Abs. The enforcedly expressed CD39 molecule in Treg cells and vasculature suppresses both ATP-activating and adenosine-elicited suppression of purinergic responses in the planted xenografts. Therefore, to prevent and protect the vascular inflammation and blood coagulation in the xenotransplanted organs of α1,3Gal-T KO pig, two human CD39 (hCD39) and human CD73 (hCD73) molecules have been suggested to artificially express in pigs [10]. If hCD73 is overexpressed in the ECs of pig, coagulopathy level is decreased during pig-to-NHP xenotransplantation. To avoid such risk, the hCD73 transgenic cells were established as donor cells [11]. By transferring the reconstructed embryos, the hCD73 TG pigs were established to use in the generation of multiple-manipulated strains like α1,3Gal-T KO/hCD46 pigs [11].

16.3 CD46, Thrombomodulin (TM), and α1,3Gal-T KO/hCD46/hTM Pigs

The pig TFPI is not capable of neutralization of the blood coagulation factor Xa of human and also cannot suppress the direct prothrombin activation to active thrombin form in human. Thrombomodulin is known as integral membrane protein present on the EC cell surfaces, and thrombomodulin inhibits coagulation. Pig thrombomodulin recognizes human thrombin to form thrombomodulin–thrombin complex, but the thrombomodulin–thrombin complex does not effectively activate protein C, which stimulates blood clotting reaction. When thrombomodulin binds thrombin, thrombomodulin activates Protein C activity to block coagulation. Protein C as a protease uses its cofactor to block FVa and FVIIIa, leading to anti-coagulation.

Because porcine thrombomodulin does not bind human thrombin, protein C is not activated [12]. Fibroblasts of human thrombomodulin transgenic pigs increased activated protein C production [13]. Human thrombomodulin expressing pig EC inhibits prothrombinase activity and consequently coagulation [14]. α1,3Gal-T KO/hCD46/human thrombomodulin pigs allowed the increased survival (945 days) of xenografts in pig-to-NHP xenotransplantation animal [15]. The vascular ECs generate the natural extrinsic coagulation pathway inhibitor, where its anticoagulant activity inhibits (1) factor Xa and (2) the Xa/inhibitor complex formation. Thus, subsequently, the issue factor (TF)–VIIa complex is inhibited. It is an inhibitor to prevent the formation of thrombin. Natural extrinsic coagulation pathway inhibitor controls the TF-dependent coagulation in xenotransplantation. Aggregation level of human platelets by aortic ECs of pig with the human natural extrinsic coagulation pathway inhibitor was reduced [16].

The vWF is a coagulating glycoprotein, which acts at the vascular damaged and injured sites. The vWF interacts with platelet receptors of human, contributing to enhance procoagulant capacity. It binds to FVIII in the blood. The vWF acts as a xenograft rejection actor [17]. Thus, vWF gene of pigs should be inactivated using the known gene engineering technology such as recently developed CRISPR/Cas9 techniques and others [18]. Alternative approaches known to avoid DXR are to transgene relevant genes engaged in anti-inflammation and anti-apoptosis responses to protect ECs of pig. The anti-inflammation and anti-apoptosis genes function as protectors of vascular endothelium. For example, human A20 gene known as TNF-α-induced protein-3 and HO-1. HO-1 is an anti-apoptotic and anti-inflammatory protein and exerts anti-apoptotic and anti-inflammatory activities. In addition, HO-1 shows a wide expression in human tissues. Therefore, HO-1 exhibits a broad range of anti-apoptosis and anti-inflammation activities. HO-1 [19] has also been used in xenotransplantation. HO-1 expression is known to suppress NK cell activity and generation of pro-inflammatory molecules including ICAM-1 (intercellular adhesion molecule-1), VCAM-1 (vascular CAM-1), C-C chemokine receptor type 5 (CCR5), and E-selectin (CD62, antigen-like family member-E) [20, 21]. HO-1 expression further inhibits aggregation of platelets to facilitate angiogenesis and to maintain microcirculation [22]. In xenografts of rodents, HO-1 expression decreases the level of thrombus formation and deposition of antibody IgM type and also prolongs survival period of organ xenografts [23, 24]. In pigs, HO-1 expression exerts the protection of ECs from TNF–αdriven apoptotic cell death, extension of *ex vivo* kidney perfusion efficiency with human bloods [25], protection from H2O2-induced fibroblast damages, and inhibition of cycloheximide- and TNF-α-mediated apoptotic cell deaths [26]. hHO1 and hA20 co-expression reduces the increased expression of cytokine-induced MHC C-II expression in aortic ECs of pigs [27]. Pig hHO-1 gene carrying ECs protects the EC cells from TNF-α-dependent apoptosis [25].

NF-κB-inhibitory A20 Zinc finger protein (or known as TNF-α-induced protein-3) also acts in an anti-inflammation [28]. Human A20 expression in pigs inhibited apoptosis Fas (CD59) ligands [29]. A20 protein suppresses activation of NF-κB transcriptional activity [30], and therefore, A20 protein inhibits upregulation of

IL-1β and TNF-α cytokines and pro-apoptotic molecules [31]. A20 protein protects pig aortic ECs from TNF-α-caused apoptotic cell death and exerts in part protective effect on ischemia/reperfusion-mediated tissue injuries [32]. It was known that A20 protein expression is ubiquitous in all tissues, and A20 significantly reduces E-selectin expression known an NF-κB activating indicator (lower than 65% of wild-type) in aortic ECs [33], through dual co-expression of human A20 and human HO-1 genes. The A20 and HO-1 genes are anti-apoptotic via inhibition of caspase 8 induction. Pig expression of various human regulatory molecules such as A20, CD46, CD55, CD59, and HO-1 had anti-apoptosis activities. Therefore, through various approaches of transgenic and KO technologies, molecular hurdles of pig-to-human xenotransplantation are currently overcome by genetically modified pigs, which are not distinguishable for the immune responses of human as the humoral immune type, and the risk of coagulation is also overcome.

References

1. Rocha PN, Plumb TJ, Crowley SD, Coffman TM. Effector mechanisms in transplant rejection. Immunol Rev. 2003;196:51–64.
2. Lin CC, Cooper DK, Dorling A. Coagulation dysregulation as a barrier to xenotransplantation in the primate. Transpl Immunol. 2009;21(2):75–80.
3. Crikis S, Cowan PJ, d'Apice AJ. Intravascular thrombosis in discordant xenotransplantation. Transplantation. 2006;82(9):1119–23.
4. Shalev I, Schmelzle M, Robson SC, Levy G. Making sense of regulatory T cell suppressive function. Semin Immunol. 2011;23:282–92.
5. Delikouras A, Fairbanks LD, Simmonds AH, Lechler RI, Dorling A. Endothelial cell cytoprotection induced in vitro by Allo- or xenoreactive antibodies is mediated by signaling through adenosine A2 receptors. Eur J Immunol. 2003;33(11):3127–35.
6. Enjyoji K, Sevigny J, Lin Y, et al. Targeted disruption of cd39/ATP diphosphohydrolase results in disordered hemostasis and thromboregulation. Nat Med. 1999;5:1010–7.
7. Khalpey Z, Yuen AH, Kalsi KK, Kochan Z, Karbowska J, Slominska EM, Forni M, Macherini M, Bacci ML, Batten P, Lavitrano M, Yacoub MH, Smolenski RT. Loss of ecto-5'nucleotidase from porcine endothelial cells after exposure to human blood: implications for xenotransplantation. Biochim Biophys Acta. 2005;1741(1-2):191–8.
8. Borsellino G, Kleinewietfeld M, Di MD, et al. Expression of ectonucleotidase CD39 by Foxp3+ Treg cells: hydrolysis of extracellular ATP and immune suppression. Blood. 2007;110:1225–32.
9. Fisicaro N, Londrigan SL, Brady JL, et al. Versatile co-expression of graft-protective proteins using 2A-linked cassettes. Xenotransplantation. 2011;18:121–30.
10. Cooper DK, Ekser B, Burlak C, Ezzelarab M, Hara H, Paris L, Tector AJ, Phelps C, Azimzadeh AM, Ayares D, Robson SC, Pierson RN 3rd. Clinical lung xenotransplantation–what donor genetic modifications may be necessary? Xenotransplantation. 2012;19(3):144–58.
11. Lee SC, Lee H, Oh KB, Hwang IS, Yang H, Park MR, Ock SA, Woo JS, Im GS, Hwang S. Production and breeding of transgenic cloned pigs expressing human CD73. Dev Reprod. 2017;21(2):157–65.
12. Van't Veer C, Golden NJ, Kalafatis M, Mann KG. Inhibitory mechanism of the protein C pathway on tissue factor-induced thrombin generation. Synergistic effect in combination with tissue factor pathway inhibitor. J Biol Chem. 1997;272:7983–94.

13. Petersen B, Ramackers W, Tiede A, Lucas-Hahn A, Herrmann D, Barg-Kues B, et al. Pigs transgenic for human thrombomodulin have elevated production of activated protein C. Xenotransplantation. 2009;16(6):486–95.
14. Miwa Y, Yamamoto K, Onishi A, Iwamoto M, Yazaki S, Haneda M, et al. Potential value of human thrombomodulin and DAF expression for coagulation control in pig-to-human xenotransplantation. Xenotransplantation. 2010;17:26–37.
15. Mohiuddin MM, Reichart B, Byrne GW, McGregor CG. Current status of pig heart xenotransplantation. Int J Surg. 2015;23:234–9.
16. Iwase H, Ekser B, Hara H, Phelps C, Ayares D, Cooper DK, et al. Regulation of human platelet aggregation by genetically modified pig endothelial cells and thrombin inhibition. Xenotransplantation. 2014;21(1):72–83.
17. Holzknecht ZE, Coombes S, Blocher BA, Li W, Zhou Q, et al. Immune complex formation after xenotransplantation: evidence of type III as well as type II immune reactions provide clues to pathophysiology. Am J Pathol. 2001;158:627–37.
18. Hai T, Teng F, Guo R, Li W, Zhou Q. One-step generation of knockout pigs by zygote injection of CRISPR/Cas system. Cell Res. 2014;24:372–5.
19. Loboda A, Jazwa A, Grochot-Przeczek A, Rutkowski AJ, Cisowski J, Agarwal A, et al. Heme oxygenase-1 and the vascular bed: from molecular mechanisms to therapeutic opportunities. Antioxid Redox Signal. 2008;10:1767–812.
20. Shen Z, et al. Immunoregulation effect by overexpression of heme oxygenase-1 on cardiac xenotransplantation. Transplant Proc. 2011;43:1994–7.
21. Wagener FA, Feldman E, de Witte T, Abraham NG. Heme induces the expression of adhesion molecules ICAM-1, VCAM-1, and E selectin in vascular endothelial cells. Proc Soc Exp Biol Med. 1997;216:456–63.
22. Bhang SH, et al. Combined gene therapy with hypoxia-inducible factor-1 alpha and heme oxygenase-1 for therapeutic angiogenesis. Tissue Eng Part A. 2011;17:915–26.
23. Zhen-Wei X, et al. Hemeoxygenase 1 improves the survival of discordant cardiac xenograft through its anti-inflammatory and antiapoptotic effects. Pediatr Transplant. 2007;11:850–9.
24. Tabata T, et al. Accommodation after lung xenografting from hamster to rat. Transplantation. 2003;75:607–12.
25. Petersen B, et al. Transgenic expression of human heme oxygenase-1 in pigs confers resistance against xenograft rejection during ex vivo perfusion of porcine kidneys. Xenotransplantation. 2011;18:355–68.
26. Yeom HJ, et al. Generation and characterization of human hemeoxygenase-1 transgenic pigs. PLoS One. 2012;7:e46646.
27. Fischer K. Efficient production of multi-modified pigs for xenotransplantation by'combineering', gene stacking and gene editing. Sci Rep. 6:29081. https://doi.org/10.1038/srep29081.
28. Lee EG, Boone DL, Chai S, Libby SL, Chien M, Lodolce JP, et al. Failure to regulate TNF-induced NFkappaB and cell death responses in A20-deficient mice. Science. 2000;289:2350–4.
29. Oropeza M, Petersen B, Carnwath JW, Lucas-Hahn A, Lemme E, Hassel P, et al. Transgenic expression of the human A20 gene in cloned pigs provides protection against apoptotic and inflammatory stimuli. Xenotransplantation. 2009;16:522–534.D.
30. Krikos A, Laherty CD, Dixit VM. Transcriptional activation of the tumor necrosis factor α-inducible zinc finger protein, A20, is mediated by κ B elements. J Biol Chem. 1992;267:17971–6.
31. Dixit V, Mak TW. NF-kappa B signaling. Many roads lead to Madrid. Cell. 2002;111:615–9.
32. Oropeza M, et al. Transgenic expression of the human A20 gene in cloned pigs provides protection against apoptotic and inflammatory stimuli. Xenotransplantation. 2009;16:522–34.
33. Cooper JT, Stroka DM, Brostjan C, Palmetshofer A, Bach FH, Ferran C. A20 blocks endothelial cell activation through a NF-kappaB-dependent mechanism. J Biol Chem. 1996;271(30):18068–73. https://doi.org/10.1074/jbc.271.30.18068.

Chapter 17
Xenogeneic and Allogenic Cellular Rejection (CR)

17.1 Introduction

In solid organ xenotransplantation, the humoral response of human immunity is a direct barrier due to its Abs binding to non-self-antigens of carbohydrates and proteins present in the pig vasculatures [1, 2]. During the last three decades, for the long-term solid organ survivals in xenotransplantation, vascular endothelial cells have been studied as a target for xenograft recipient immune system and grafted microvasculature has been received a great attention [3]. When antigenic deletion of α1,3Gal epitopes pigs prevents HR in recipients, instead DXR or cellular rejection (CR) becomes a secondary and major immune response as a huge barrier in xenotransplantation. To date in the xenotransplantation, the experimental and clinical evidence and process that systemic responses of host immune system cause the failure of xenotransplantation of pig xenografts in primates have largely been understood and explained. The mostly important barrier and hurdle in pig-to-human clinical xenotransplantation still remain immune-rejective response despite the animal development with the disrupted carbohydrate antigens.

In the cellular immunity, adaptive and innate immune cells cooperatively reject a xenograft. For example, NK cells, T cells, and macrophage cells mediate the cellular rejection events that are currently hot issue in clinical xenotransplantation. From the multiple modifications of antigens genes of a donor animal and the mechanism of xenograft rejection, long-term xenograft survival without systemic immunosuppression is the present interest and goal in xenotransplantation. For a long time, the antibody-mediated immune responses dealt currently as a main issue are importantly recognized in a rejected organ xenotransplantation of solid xenografts. However, cell-mediated immune responses and eventual rejection events become the secondarily frequent event in the cellular xenograft rejection. Currently, innate and adaptive immunities are accepted to collaboratively mediate the rejection mode in the xenotransplantation. Then, the innate immunity has largely been elucidated in the cellular xenograft rejection. Consequently, resolutions to overcome the xenografts rejection

C.-H. Kim, *Glycoimmunology in Xenotransplantation*,
https://doi.org/10.1007/978-981-99-7691-1_17

are under development in the different forms of xenotransplantation performances. For example, the immunosuppressive agents including immune blockades to eliminate and block such cellular xenograft rejection are recently under development, because such immunosuppression potentiates long-term xenograft survival in combined transgenic pigs.

In general, activating and inhibitory signaling pathways regulate the immune cells to immunologically activate and restrict. In the xenotransplantation, molecularly incompatible property between ligands and receptors in interspecies raise xenograft rejection. In the allotransplantation, the innate and primary immune cells of phagocytic cells of monocytes/macrophages, neutrophils, and NK cells exchange their signals to reject xenografts. In the molecular differences between interspecies, receptor/ligand interactions and glycosylation patterns regulate the migration, recognition of non-self and natural cell-mediated immune response. The CR is specialized with macrophage- and NK-mediated rejection. Pig organ or tissue transplanted to human is targeted by monocyte/macrophage, NK cells, and T cells. For xenotransplantation, immunomodulatory issue is a key to suppress the xenoimmune response. The microvascular circulation vessels include arterioles, capillaries, and venules [4] to supply nutrition and oxygen. The graft failure and chronic rejection are predominantly caused by the dysfunction of microvascular circulation vessels, suggesting that the microvascular system is essentially important for the transplantation of solid organs and preventive and strategic therapies of chronic rejection [5]. Substantially, the lined endothelial cells in the graft vasculatures are recognized as major targets for the host's immune responses mediated by thrombotic microangiopathy and antibody-mediated rejection (AMR) [6, 7]. Naturally preformed Abs elicited by xenoantigens of the endothelial vascular system primarily develop DXR that results from chronic activation or injury caused by Abs and activation of complement system [8].

CR takes place within several days after transplantation with the characteristic infiltrations of the innate immune cells, as frequently observed in mononuclear cells. Also, chronic rejection takes place within months or more after transplantation. However, no detailed information is available for chronic xenotransplant rejection. Damaged endothelium of vascular system is based on specific Abs and cytotoxic T cells, although the levels of such responses are weaker than other rejections. For the long-tern survival of patients in organ allotransplantation, the major limitation is the chronic allograft vasculopathy in large vascular vessels [9]. In xenotransplantation, if new target antigens are discovered, new genetic-engineered pigs can importantly be developed to lead to the reduction of chronic Abs-induced vascular endothelial activation and antigen-specific tolerance [10, 11]. A recent report hopefully activates the possibility to remove such CR or DXR, which searched the DXR or CR-involved xenoantigens [12]. They discovered many genes as candidates differentially expressed in the primary cells of renal microvascular ECs and aortic ECs from a GGTA1/CMAH double KO (DKO) pigs and cynomolgus monkeys immunized using GGTA1/CMAH DKO pig transcriptome sequencing. The expressed 101 genes as candidates from only primary renal microvascular ECs of pigs, but not expressed in aortic ECs of pigs or in human or monkey renal microvascular ECs.

Then, each candidate gene as separately subjected to knocking out in renal microvascular ECs of GGTA1/CMAH double-KO pigs. Among 101 gene candidates, 32 genes shoed reduction in antibody recognition, suggesting that the 32 gene candidates are immune-responding target genes involved in DXR and CR. The antibody recognition with the ECs of pig in the non-HR-responsive condition indicates that the EC cells were injured in vivo and develop fibrin-platelet thrombosis. Such conditions result in ischemic graft injuries. This indicates the consumptive coagulopathy and thrombotic microangiopathy development, which was previously known after transplantation of pig kidneys than transplantation of hearts [13]. Therefore, such important candidates as not the known carbohydrate antigens can light to the CR studies.

17.2 Systemic Inflammation in Transplantation

Since development of multiple genetically engineered pig lines, significant progress to understand immune responses is experimentally developed in xenotransplantation. Despite the development of available organs from genetic-engineered pigs, dysfunctional coagulation-triggered fibrin-platelet thrombosis and ischemic graft injuries as well as the graft-specific inflammatory responses are still considerable barriers [14]. Xenografts induce inflammatory response and systemic inflammation in the organ and cell xenotransplantation as a result of antigen recognition with the pig vasculature. The immune inflammation is not well defined, although migration of leukocytes such as blood granulocytes into tissue injuries is relatively well understood as crucial [15]. Also, C-reactive protein (CRP) level as an inflammatory biomarker has been increased [16]. Inflammatory process activates the blood coagulation system and pro-coagulant proteins and pro-inflammatory mediators such as thrombin propagate the process. Xenotransplantation induces the coagulation system and acquired response of immune system, where systemic inflammation in xenotransplantation crucially develops dysregulated coagulation and resistance induction is developed during immunosuppressive treatment [17]. In general, inflammation as natural, complex and essential responses responds to invasive agents, pathogenic infection and tissue damages performed by multiple interactions between proteins, low molecular mediators and secondary prostanoids. Inflammatory mediators of proteinous cytokines and chemokine are also important. For example, chemokines recruit innate immune cells to the sites of inflammation, while pro-inflammatory molecules such as cytokines regulate various immune cells to eliminate invasive pathogens and dead cells as well as repair tissues.

17.2.1 Inflammatory Coagulation System in Transplantation

Inflammation process eventually prevents the tolerance development of T cells during transplantation. Systemic inflammation in recipient is known to proceed coagulation in pig-to-primate xenotransplantation. Innate immune cell activation of recipients that express tissue factors increases in their cell numbers due to migrative accumulation after xenotransplantation. With immunosuppression, levels of IL-6, CRP and fibrinogen are upregulated in artery of pig recipients. For example, enhanced levels of CRP are detected in organs of pig recipients prior to the consumptive coagulopathy development. How the coagulation, inflammation and innate immune response are interacted to display the inflammation outcome in xenograft recipients is not clear until now [18]. However, the fact that coagulation proteins are highly pro-inflammatory and cytokines activate such coagulation is well defined. Therefore, it is emphasized that dysregulation of the coagulation system reflects the failed plantation of xenografted organs. Grafts and consumptive coagulopathy-based thrombotic microangiopathy (TM) also indicate xenograft rejection, because of these biochemical processes as main barriers to survived xenografts in NHPs [19]. In the allotransplantation, inflammatory signals are clearly known to activate T cells, and the signaling molecules are regarded as key factors to loss T-cell tolerance and prevent long-term survival of allografts, as well as the signals mostly activate antibody production of B cells [20]. In xenotransplantation, systemic responses of inflammation occurred in recipient xenograft should be revisited because it activates coagulation toward dysregulation, as well as increase resistance to immunosuppression after xenotransplantation [21].

Among the immune cell system, innate immune cell populations initially provoke inflammatory response and coagulation activation. Survival factor of xenografts is the innate immune cells, which initially act as a cellular barrier, in NHPs [22]. Pro-inflammatory cytokines are actual mediators after xenotransplantation. In the inflammation, innate immune cells like DCs and their produced pro-inflammatory molecules such as cytokines promote coagulation process. Pro-inflammatory cytokines prevent infection of pathogens. However, excessive-produced cytokines rather stimulate inflammatory response, as defined in cytokine storm. In addition, such pro-inflammatory cytokines activate coagulation. In fact, representatively important IL-6 and TNF-α known as pro-inflammatory cytokines stimulate tissue factor production to activate coagulation. Pig-to-primate xenotransplantation develops innate inflammatory response with local synthesis of inflammatory IL-6 cytokine [15].

17.2.2 Inflammatory C-Reactive Protein (CRP) Expression

In organ graft transplantation, inflammation responses precede the coagulation pathway activation in xenotransplantation. There is a reason that early and fast expression of CRP and IL-6 amplifies coagulation stimulation through upregulation of tissue factor (TF) expressed in innate immune cells. For example, the increased IL-6 expression levels reflect the increased CRP expression in human population [23]. Although in the coagulation process, the functional roles of chemokines and pro-inflammatory cytokines are not clearly explained yet in xenograft recipients [24], active IL-6 blockade is beneficially effective in xenotransplantation. Several cytokines such as IL-12, IL-8 and IFN-γ are expressed in baboons during pig-to-baboon xenotransplantation of organ xenografts with the enhanced levels of IL-6, monocyte chemotactic protein 1 (MCP1) and TNF-α expression [21]. IL-6 induces CRP production in hepatocytes and smooth muscle cells (SMCs) [24]. CRP and fibrinogen belong to reactive acute-phase proteins expressed during acute phase of inflammation. But CRP levels are increased in recipients of organ xenografts and level of fibrinogens gradually declines during coagulation activation due to gradual consumption of coagulative factors. Activated monocytes and DCs express tissue factor to activate coagulation [23]. Activated DCs additionally express functional tissue factors and thrombin receptors, and reversely such expressed thrombin potentiates DCs functions [25]. Activated pig ECs-produced tissue factor further contribute to TM and activate coagulation. Pig aortic ECs stimulate human $CD14^+$ monocytes to produce pro-inflammatory cytokines and upregulate tissue factor expression [26]. Inflammatory $CD11c^+$ DCs are monocyte-derived [27]. Macrophages and PBMCs express tissue factor after pig organ xenotransplantation [26, 28]. The transgenic human CRPs in pig xenografts protect the development of TM.

As an acute-phase protein CRP is also a potential marker of xenograft recipients. CRP is produced in humans and NHPs during infection and is also a signal for grafts-associated complications during allotransplantation of human organ grafts [29]. After transplantation, coagulation dysfunction is expressed as a type of thrombocytopenia, reduced fibrinogen and increased D-dimer, a fibrin degradation product, while inflammation develops as the increased CRP level. CRP promotes tissue factor expression [30] and level of CRP expression is upregulated in xenograft recipients of pig organs only within several days after xenotransplantation [21]. High CRP levels induce dysregulated coagulation and early xenograft failure of kidney organs [21]. The systemic dysfunction in organ xenograft recipients is based on heterogeneity of organ xenografts and organ-specific expression of vascular genes [31]. CRP deposition and CRP-positive immune cells are observed in the xonogenic organ recipients of heart, kidney, and lungs xenografts. Fibrin deposition and activated coagulation event reflect organ failure of pig xenografts in NHPs. Pig ECs produce their pro-coagulant proteins and develop TM in organ xenografts. Pig ECs production of pro-coagulant proteins is based on the recognition and binding of anti-pig Abs, which are such natural or elicited Abs [19]. Preventive induction of Abs prevents activation of pig ECs and pro-coagulant protein expressions. Although dysregulated coagulation system contributes to xenograft failure in non-human primates, anticoagulation is comparatively not efficient in prolonged survival of xenografts of pig xenografts in monkeys of pig-to-monkey xenotransplantation. Furthermore, anticoagulation is not efficient in enhancement of xenograft survival of pig heart in baboons of pig-to-baboon xenotransplantation [32]. In the hCRPs (CD47 or CD55/DAF)-expressing wild-type pigs, anti-coagulants ameliorate dysregulated coagulation during xenotransplantation, although the IL-6, fibrinogen, CRP, and systemic inflammation dysregulate coagulation [33].

The increased CRP reflects a prolonged inflammatory response. The known blockade of IL-6R clearly abrogated the increased CRP production and this indicates that the inflammatory response has been suppressed after xenotransplantation of pig organs [34]. In plasma, the D-dimer and CRP levels are involved in inflammation progress in xenotransplantation. Locally formed fibrin and its lysis are evidence partially involved in the inflammatory responses. Degradation products of fibrin expressed as D-dimer bear diverse functions in inflammation responses and, particularly in acute phase responses, which are mediated by diverse events such as activation of monocytes and neutrophils, IL6 secretion, and CRP and fibrinogen biosynthesis as acute-phase proteins in the surrounded cells such as hepatocytes [35]. Coagulation dysfunction includes several factors and develops MP in the consumptive coagulopathy and transplanted grafts in the recipients [35]. Fibrin thrombi, which are platelet-enriched, are accumulated in the vasculature of the grafts.

Consumptive coagulopathy and thrombomodulin™ are resulted by biochemical activation of coagulation pathway, as detected in organ xenotransplantation of pig xenografts to baboons [36]. Decreased levels of fibrinogens and platelet counts are observed immediately after xenotransplantation. D-dimer levels are reported to be dramatically increased after pig-to-baboon xenotransplantation of heterotopic hearts of pig [37]. Such changes in D-dimer, fibrinogens and platelet counts are characteristic of consumptive coagulopathy. Previously, it was reported in pig-to-baboons transplantation that an antagonist of IL-6R prevents a responding reaction of CRP to artery of pigs and is beneficial in organ xenotransplantation due to reduced inflammatory response. In addition, normal CRP levels are remained to keep when IL-6R antagonist, tocilizumab, is administered, even in the septic conditions, [38]. Therefore, although the artery patch grafts are a smaller size than others, the organ xenotransplantation model of pig-to-baboon artery xenotransplantation has some merits to obtain the regulated systemic inflammatory response after xenotransplantation. Genetic modifications of coagulation-regulatory and inflammation-regulatory proteins in donor pigs seem to need to preventively control systemic inflammatory responses observed in xenografts-transplanted recipients [15]. Further IL-6R antagonist therapy is considered necessary. To reduce the prolonged inflammation and immune responses, genetically modified pigs as donor animal protects humoral Abs responses and the resulted TM development [39]. Therefore, future therapeutic prevention is required for prevention of systemic and dysregulated coagulation develops in xenotransplantation recipients of xenografts without any usage of immunosuppressive agents to induce immunosuppression.

17.2.3 Allorecognition and Allograft Rejection Pathways

Allograft immune responses induce graft rejection by three independent manners. In allotransplantation, because allograft immune response causes graft rejection by three independent manners, different allograft-rejection mechanisms are explained by direct pathway, semi-direct pathway and indirect pathway during allogenic recognition-based allograft rejection. First, the direct pathway represents the allogenic molecule presentation of donor APCs, in a way that the MHC proteins of donor organs are exposed to lymphocytes of the recipients. This direct pathway elicits ACR process by the recipient T cells of $CD4^+$ and $CD8^+$ phonotypes. The donor APCs have the limited lifespan and thus are restricted to maintain alloresponse. However, the APCs present in the graft for the given period threaten the allograft survival. Thus, the direct rejection can be eliminated by strong immunosuppression. Second, indirect pathway represents the allogenic donor peptide presentation by recipient APCs using the typical antigen presentation for a graft lifespan by donor allogenic molecules. This pathway involves $CD4^+$ cell activation by the MHC-II-allopeptide. The activated $CD4^+$ T cells induce allopeptide-reactive B cells to produce donor-specific antibodies (DSA) [40]. Third, semi-direct pathway represents the immune response by extracellular vesicles formed by donor membrane-bound MHC and the vesicles interact with recipient APCs [41]. Thereafter, the recipient APCs are ready to present donor MHC allopeptides for the recipient $CD8^+$ and $CD4^+$ T-cell activation. The semi-direct pathway is indefinite for ACR for a few years after allotransplantation. Fourth, inverted direct pathway represents early *de novo* DSA genesis, which is based on $CD4^+$ T cells of donor organ binding to B cells of the recipients after presentation of the recipient MHC-II peptides. In the inverted direct pathway, donor T cells become rather actors as direct alloactivators and this is quite similar to the direct pathway, only except for the inverted subjects between donor vs. recipient cells. Donor T-cell activation is based on Graft vs. Host response. In the inverted direct pathway, however, the activated donor T-cell-generated cytokines lead to the Host vs. Graft B-cell responded class switching and expansion because recipient B cells act as APCs.

17.2.4 Role of Resident Memory T Cells (Trm) in Allogeneic Rejection

In solid organ transplantation, donor effector T-cell expansion is characteristic of GvH clone and a recipient effector T-cell expansion also causes "host versus graft (HvG) clones. The GvH and HvG magnitudes are decided by the relative donor T-cell level in the grafted organs [42]. De novo antibody production reactive for the grafted HLA, the donor-specific HLA antibody (DSA) indicates a high ACR risk in organ transplants [43]. Resident memory T cells (Trm) as non-circulating and memory cell subsets differ from central memory and circulating effector T cells.

Trm is present long period in barrier organs of gut, lung, skin and reproduction tract, but not for the large amounted T-cell memory subset. Trm is important for infectious immunity and the graft rejection [44]. Trm activates the graft loss upon persistent antigenic exposure. Among the Trm subsets, CD^{+} Trm is the largest group and present in circulation and spleen. In recipient animals, heart transplantation shows that skin alloreactive Trm reject the heterotopic grafted heart by alloreactive Trm. $CD4^{+}$ Trm which are alloreactive are Th17-like phenotype cells. Th17 cells are central players in thymoglobulin-resistant graft rejection. Graft donor lymphocytes replaced by recipients clinically induce graft rejection with specific HvG subset expansion. These cell groups acquire Trm phenotype and alloreactive Trm are central actors of the post-transplantation rejection. Immunosuppressive drugs limitedly regulate memory cells rather than naïve cells. Graft bears Trm subsets for GvH activity and the inverted direct allorecognition pathway.

17.2.5 Role and Importance of Lymphoid and Trm in the Intestinal Allotransplantation

Intestinal transplantation (IT) is for irreversibly failed intestine patients by the allograft transplantation with immune-suppression agents [45]. Alloantigenic activation occurs in the HvG response. The organized graft lymphoid tissue induces the allograft-specific mucosa and lymphoid immune responses by four different pathways. Historically, in the 1990s, Newell group reported $CD4^{+}$ and $CD8^{+}$ T-cell-associated ACR event in intestine grafts. Later, he described chronic graft rejection (CGR) in mice. The grafted intestine characteristically contains the gut-associated or gut-derived lymphoid tissue enriched with abundant epithelial cells. IT-driven immune response resembles that of inflammatory bowel disease (IBD), a representative autoimmune disease. Fundamentally intrinsic property of the intestinal immune cells along the gastrointestinal (GI) tract to protect the GI tissues against the persistently-accounted foreign antigens and resident microbiota is a fundamental factor of immune response induction to grafted intestine. In the intestinal transplantation, APCs subsets are diversely present in lamina propria or lymphoid structures.

IT represents distinct lymphoid structures with the mesenteric lymph nodes as the portal draining system and also the multi-lymphoid mucosal tissue present in the grafted organ. The lymphoid tissue-organized graft easily develops rejection responses. Recipient lymphocytes are present in mesenteric lymph nodes and graft Peyer patches. Recipient lymphocytes generate IFN-γ for ACR induction. Specialized intestinal structure with lymphoid follicles and Peyer's Patches helps T-cell and B-cell cooperation, not affected by immunosuppression [46]. These specialized sites can expand donor-specific B cells for DSA production. Due to the graft and intestinal grafted lymph donor B cells, B cells migrate to grafted lymphoid structures and donor T cells emigrate [47]. In the intestinal transplantation, the graft donor T cells determine the GvH response magnitude, but not in other solid organ transplantation

[48]. The intestinal grafts contain lymphoid precursor cells [49] capable of initial GvH cell expansion. Intestinal graft donor lymphocytes are replaced by recipients and specific HvG clones are expanded. These cells progressively obtain resident memory T-cell (Trm) phenotype and alloreactive Trm function as post-transplant central actors required for the graft rejection. Intestinal graft bears Trm subsets for GvH response and the inverted direct allorecognition pathway, and thus, immunosuppressive drugs are not effective for memory cells.

The alloresponse is mainly driven by intestinal mucosa. MHC C-II present on intestinal epithelial cells (IECs) is non-conventional APCs that interact with Th cells, but its role is not known. IECs secrete extracellular vesicles containing MHC-II molecules acquired by mononuclear phagocytes that induce adaptive immunity to microbial antigens [50]. This circuit involves a semi-direct pathway of intestinal allotransplantation. Among non-conventional lymphoid subsets present in the intestinal mucosa, innate lymphocyte type 3 cells (ILC3) have antigen presenting MHC C-II and present to conventional $CD4^{+}$ T cells and induce Tregs for commensal microbiota [51]. MHC-II deletion in ILC3 dysregulates CD^{+} T-cell responses for intestinal inflammation. ILC3 protects intestinal rejection via their generation of IL-22 [52]. Some non-conventional T-cell subsets of MAIT, CD8aa, or gdT lymphocytes contribute to alloresponse.

17.2.6 Serum Marker Expression in the Intestinal Allotransplantation

Some potential rejection markers were known as in plasma citrulline or fecal calprotectin. During ACR, tissue injured damage and immune activation, inflammatory response or metabolic changes are induced in the rejected intestinal grafts. Three pathological pathways include the acute rejection, direct graft damages or injuries and local and systemic inflammatory responses induced by the allogenic immune responses and the local metabolic changes by the intestinal graft damage. Intestinal mucosa is mainly targeted for the rejection and enterocyte death. Citrulline as a non-protein a.a. generated mainly by enterocytic cells, and it is the glutamine metabolic end product and an indicator of the functional enterocyte measurement. In IT recipients, serum, citrulline levels are decreased in ACR and correlated with rejection severe status. Thus, citrulline is an acute rejection biomarker [53]. The individually present variations cause for an overlapping parameter between citrulline levels of patient plasma with normal allografted organ histology and the levels in ACR histology. Plasma citrulline levels indicate the extent of mucosal injuries and damages. But the plasma citrulline level is not diagnostic biomarker for graft rejection or virus-derived enteritis. In villus enterocytic cells, intestinal fatty acid binding protein (I-FABP) can be used as a marker in the enterocyte metabolism in rodents [54]. Histamine-digesting enzymes are known for histamine *N*-methyltransferase and diamine oxidase. These enzyme activities are decreased in the rejecting allograft mucosa [55]. Enterocytic cell apoptosis indicates a central tissue injury mechanism in intestinal ACR process by cytotoxic T cells and the granule exocytosis, which contains serine protease granzyme and pore-forming perforin. Upon contact between cytotoxic T cells and enterocytic cells, perforin allows to enter for granzymes to the targeting cells to cleave the pro-apoptosis BH3 protein-binding domain death agonist, translocating to the mitochondrial membrane and enhance its permeability. Granzyme B and perforin in the cytoplasm are co-expressed in the rejecting intestinal allograft mucosa. Serum tumorigenicity-2 suppressor is a regulatory decoy receptor of IL-33 present in IBD [56]. Intestinal ACR injures the graft tissue by mediators such as granzymes, perforin, neutrophil protein, heat shock protein (HSP), IFN-γ, IL-1β, and TNF-α and also innate immune NK cells. Fecal calprotectin levels (FCL) are used for the initial differentiation between irritable bowel syndrome (IBS) and IBD and also to induce mucosal healing or inflammatory recurrence in IBD patients. Calcium-binding and zinc-binding Calprotectin is an S-100 protein family expressed in neutrophils and fecal calprotectin is by the neutrophil infiltration into the GI track tissue for infection or inflammation. Fecal stool calprotectin exhibits in vivo and in vitro degradation resistance against pancreatic secreted agents and intestinal proteases as well as bacterial degradation.

17.3 B-Cell Role in CR

B cells are essentially required to respond to humoral expression in immune system, and therefore, they are representatively an adaptive immune-responding populations during antigen (Ag) presentation and Abs biosynthesis and extracellular production. The B-cell action is followed by a differentiation event known as complex process when Ag receptor of B cells encounters their specific Ags, and they present the encountered Ags to their collaborator cells named T cells. Activation of B cells occurs by two different modes after either T-cell-independent or T-cell-dependent manners. Upon B-cell activation, B cells enter to cellular differentiation into antibody-secreting B cells named plasma cells and/or storage cells named memory B cells. In B cells, the changed B-cell phenotypes are causing factors for production of the autoimmune autoantibodies, autoantigen presentation, grafts rejection, allo-antigen presentation, primed alloreactive T cells and auto-reactive T cells. Likely to macrophages, DCs, T cells, and NK cells, B cells specifically express surface receptors, which bear Tyr-based inhibitory motif and immunoreceptor tyrosine-based inhibitory motif (ITIM) as inhibitory receptors. In fact, representatively, the Ig-like transcript 2 (ILT2)/leukocyte IgG-like receptor subfamily B member 1 (LILRB1), alternatively termed CD85j [57]. ILT2 recognizes the HLA C-I third domain, which is involved in β2-microglobulin having the strongest binding affinity to HLA-G, known as the non-classical HLA C-I. Interaction between ILT2 and HLA-G blocks cytocytic lysis of NK cells and $CD8^+$ T cells and also $CD4^+$ T-cell proliferation, suppressor T-cell production, and DC maturation [58]. HLA-G therefore suppresses B-cell capacities including proliferation, differentiation, and Ig production in both T-cell-independent and T-cell-dependent manners. In in a xenograft mouse model, HLA-G suppress Ab secretion of B-cell function. Such suppression mechanism(s) of B cells is derived from the interaction between ILT2 and HLA-G. Thus, B cells targeting therapeutic strategy can reduce inappropriately responded immune system during allo- and xenotransplantations.

To selectively eliminate xenoantibody-producing B lymphocytes, several approaches impact the xenotransplantation. B cells, which are resident on the peritoneal cavity and express anti-α1,3Gal antigen receptors, are classified as the anti-α1,3Gal-producing B-cell precursors in the spleen and are those natural antibody-producing B-cell types of Mac-1-negative B-1b phenotype. More specifically, they are B1b-like $CD21^{-}$/low IgM high $CD5^{-}$ $CD43^{+}$ Mac-1^{+} in phenotype [59]. These B cells, which are resident in peritoneal cavity, are not positive for secretion of anti-α1,3Gal-specific Abs. The cells also are negative for expression of Mac-1 (known as CD11b) but are positive for the $CD21^{-}$/low IgM high $CD5^{-}CD43^{+}$ phenotype. B cells, which express anti-α1,3Gal receptors, are phenotypically belong to mouse B-1b cells, where mouse B-1 cells known as the major B-cell subset are resident in the peritoneal cavity with the IgM^{hi} IgD^{low} $CD11b^{+}$ phenotype [60]. They are distinguished from the B-2B cell population that are resident in secondary lymphoid organ responsible for adaptive immune responses. The B-1 cell subset is further classified to CD-5-positive B-1a cell and B-1b cell subsets.

In peritoneal cavity, the α1,3Gal-xenoantigen-specific natural antibody-producing B-cell population is classified to $CD21^{-}CD23^{-}CD5^{-}CD43^{+}CD11b^{+}CD138^{-}$, in their phenotype, while the similar B-cell population of the spleens is phenotypically $CD11b^{-}$ [60]. The natural Abs-producing B-cell population resident in peritoneal cavities in α1,3Gal-T KO mice is therefore assumed to be the precursor B-cell types of splenic natural IgM Abs-secreting B-cell population [59]. Although the xenoantibody-producing B cells in NHPs and humans are not yet defined, B1b cells are postulated to produce xenoantibodies [61]. In baboons and humans, anti-α1,3Gal xenoantibody-secreting B-cell population is found in the spleen organs, as based on their recognition capacities of surfaced receptors such as CD20, CD138, and anti-Ig proteins [61]. The anti-idiotypic antibodies can determine these live cells. Xenoantibody-producing B cells are suggested to be the B1b cells, as demonstrated by 2G10 antibody known as anti-idiotypic antibody in humans and NHPs [62].

The anti-idiotype antibody (2G10) recognizes the Abs encoded by germline genes for a limited number of germline VH3 and VH4 regions [63]. The anti-idiotypic antibody (2G10) specifically detects VH3 gene-encoded Abs-secreting B-cell population in humans and NHPs. The anti-idiotypic antibody (2G10) also identifies xenoantibodies-producing B cells, as xenoantibody-secreting B cells of peripheral blood are determined as $2G10^{+}CD11b^{+}CD5^{-}$. The $CD11b^{+}2G10^{+}$ cells and antibody-positive germline progenitor cell swhen pig xenografts with Gal^{+} cells were exposed were changed in immunoglobulin heavy chain gene (IGHV). [64]. Germline VH genes are important in determination of B-cell antibody specificity and potentiate the pre-formed Abs to respond to antigenic glycan oligosaccharides via a use of VH gene and CDR3 restriction, while antigen-encountered B-2 cell responses involve in somatic mutation to give specificity [62].

Experimental ways of elimination of xenoantibody-producing B cells include anti-μ mAB-based depletion and the anti-idiotypic Abs development specific for xenoantibody-secreting B-cell populations [65]. The identification of α1,3-Gal and non-α1,3-Gal xenoantibody-producing B cells would be beneficial [66]. Administrative injection in baboons of polyclonal anti-idiotype Abs reactive to human α1,3-Gal-specific Abs reduces baboon serum cytotoxic capacity to pig donor cells by in vivo elicitation of anti-α1,3Gal IgG Abs, which are not cytotoxic [65]. Anti-idiotypic mAbs also neutralize the serum cytotoxicity of baboons to donor cells of pigs in vitro and also reduce the serum cytotoxicity of baboon. The efficient depletion of such B cells can increase the survival time of transplanted xenografts of pigs.

17.4 Macrophage Role in CR

In xenoantigenic rejection responses, roles of macrophages participated in the rejection have been known as a major hurdle to organ xenotransplantation. Macrophages are well known as a surveillant in CR, but CR mediated by monocyte/macrophage cells remains unclear yet in xenotransplantation. Innate immune cells of monocytes, macrophages and neutrophils induce host cellular immunity by displaying xenograft rejection by receptor (Abs)-mediated xenoantigen recognition.

Because inflammatory macrophages produce the pro-inflammatory cytokines, suppression of their production is an importantly desired factor to enhance xenotransplantation efficacy of islets. Macrophage cells actively function at the early stage of xenograft rejection in solid organ xenotransplantation, indicating the predictive failure of xenograft [67]. During the xenografts-response, the monocytes and differentiated macrophages are well studied in the cases of non-vascularized and vascularized xenografts. Like NK-cell immunity to xenoantigens, the xenograft rejection of macrophages is cooperatively regulated between non-self-derived activation signals and self-derived inhibitory signals. Macrophages initially phagocyte the xenoantigenic targets and provoke adaptive immunity through recruitment and antigen presentation. Upon encountering xenografts, macrophages recruit both $CD4^+$ T cells and $CD8^+$ T cells.

For islet xenografts, $CD4^+$ T-cell-dependent macrophage activation leads to destruction of islet xenografts, where macrophages infiltrate at the early stage of islet transplantation. In NOD-SCID animals, $CD4^+$-dependent macrophage activation destroys islet xenograft within early 8 days [68]. Macrophages also reject the peritoneally microencapsulated pig islets in a T-cell independent manner. For example, upon transplantation of encapsulated islets, recipient host macrophages and myofibroblasts produce chemotactic factors such as heat shock protein (HSP)-90, IL-8, high mobility group B1 (HMGB1) and macrophage inflammatory protein-1α

(MIP-1α), and then, consequently the cells infiltrated into the encapsulated islets, allowing capsular fibrotic hyperproliferation [69]. Relationship between macrophage infiltration and α1,3Gal antigen is not known yet, as observed in pig α1,3Gal-T KO islet xenografts [70]. In the studied macrophages in xenotransplantation, the FcγR signal pathway and galectin receptor activation are known to be activated by anti-pig antigen antibodies of naturally-formed α1,3Gal-reactive antibodies and by pig α1,3Gal antigen, respectively [71]. Also, Other macrophage activates through the IFN-γ produced from Th1 and NK cells, lectin pathway, and pattern-recognition proteins (PRPs).

Macrophages mediate the pro-inflammatory responses during xenotransplantation to damage tissues by leukocyte recruitment and activation of neutrophils by IL-8, G-CSF, and GM-CSF [72]. The macrophages present xenoantigens to T cells in order to generate Th1 and Th17 cells. Thus, the stage-specific macrophage function involves in reducing the risk of xenotransplantation. In the xenoantigenic extracellular matrices, macrophages degrade and remodel the designed scaffold, where such matrices are infiltrated by anti-inflammatory M2 macrophages, or chronic inflammatory M1 macrophages. At the earlier stage of the xenoantigenic extracellular matrices, mononuclear cells also infiltrate. Among them, M2-CD206$^+$ macrophages infiltrate tissues, whereas M1-CCR7$^+$ macrophage infiltration accumulates collagen, but not tissue remodeling. Pig small intestinal tissues and urinary bladder tissues induce the M2 macrophage population but suppress M1 and T-cell responses to give better outcomes, recruiting muscle-progenitor cells.

Xenotransplantation recruits and activates macrophages, where xenograft pro-inflammatory actors initially attract host monocytes to migrate and infiltrate into the xenograft by helps of β2-integrins and CD49d/CD29 known as cellular adhesion molecules [73]. Macrophage recruitment in xenotransplantation is much stronger than in allotransplantation and xenotransplantation seems to be T-cell-independent macrophage recruitment. The macrophage phagocytosis activity is tightly regulated for direct cell death or ADCC of target cells. Among the pathways of macrophage activity, inhibitory receptor-self ligand of the macrophages develops self-mediated recognition and self-mediated protection from phagocytosis, as the inhibitory signaling protects "self" from death. Macrophages also have inhibitory receptors upon binding to the ligands.

Many carbohydrate lectins such sialic acid-binding Siglecs carries such inhibitory signals on the surfaces. For the inhibitory signals, receptor for the signal inhibitory regulatory protein α (SIRPα) negatively regulates the macrophage phagocytosis activity via binding to CD47 known as its ligand. For example, the SIRPα expressed on monocytes and macrophages binds to the CD47 protein and this interaction prevents autologous itself phagocytosis. Like this signaling, the homeostatic responses of RBCs and hematopoietic cells are maintained by macrophages in spleen. The SIRPα-CD47 binding is interspecies-specific due to interspecies differences in glycosylation status and protein amino acids sequences of CD47 and SIRPα proteins [74]. Because pig CD47 and human SIRPα receptor are not recognized, pig cells are phagocytosed by human macrophage. To prevent the xenograft phagocytosis in human, transgenically expressed human CD47 on pig cells can eliminate phagocytic activity of human macrophages [75]. Significant prospect of SIRP-α-CD47 interaction in different species is documented in in vivo xenotransplantation models. For example, CD47 transgenic pig xenografts reduced phagocytic rejection of macrophages in rodent xenotransplantation and CD47 deletion disturbed prolonged survival [76]. CD47 expression can also reduce hematological disorders, because lowered expression level of CD47 protein of macrophages in bone marrow induces phagocytosis of autologous hematopoietic cells. In donor cells, expression of CD47 protein induces inhibitory signaling of the CD47-SIRPα DCs in the transplanted recipients. Genetically modified pigs with human CD47 expression in α1,3Gal-T KO pigs will greatly impact to the xenotransplantation.

17.4.1 Overcoming of Monocyte/Macrophage-Mediated Cellular Xenogeneic Rejection (CXR) via CD200/CD200R Signaling

Apart from CD47-SIRPα, some inhibitory receptors are known for CD200R, CD33-related receptors, Ig-like receptor B and Ig-like transcript 3 to increase the phagocytic activity even without their natural ligands. For example, CD200-Ig (OX2) fusion protein modified from the CD200R on macrophages and DCs prolonged survival in allo- and xenotransplantation of renal grafts [77]. CD200 and CD200 receptor family has been suggested to be a clue for overcoming of monocyte/macrophage-mediated xenogeneic rejection. Among innate immune cells, tissue-resident macrophages critically perform event of cellular xenogeneic rejection (CXR). Therefore, monocyte transition to macrophage type is a crucial step during monocyte-macrophagic activation-mediated xenogeneic rejection (MMXR) is another way to consider the event [78].

Tumor cells, which are CD200-positive, effectively inhibit in vitro and in vivo T-cell responses. Both of CD200 and CD200R (CD200 receptor) as type-1 membrane glycoproteins belongs to the Ig superfamily (IgSF) members. CD200 suppresses CD200R receptor signaling upon engagement. CD200-CD200R interaction mediates its distinct signaling during macrophage-mediated xenogeneic rejection or MMXR. CD200 protein is the typical ligand for its membrane receptor, CD200 receptor (CD200R). The CD200-CD200R complex has such inhibition. CD200R has an intracellular inhibitory motif of NPXY signaling and is expressed in myeloid/macrophage-bearing tissues. Allograft survival is tightly related with the CD200 expression in DCs because CD200 transduces suppressive signals to $CD200R^+$ macrophages [78]. In fact, macrophages, but not monocytes, are activated by α1,3Gal-independent response of immune reaction. Thus, the macrophage-mediated rejection should be overcome for xenotransplantation. CD200 is known as an independent type of prognostic factor for acute myeloid-lineage leukemia and multiple myeloma. Its expression indicates decreased in overall survival time of the myeloma and leukemia patients [79]. CD200R is known as an immune inhibitory receptor as an immunological checkpoint on BM-derived myeloid and lymphoid cells. Therefore, the CD200–CD200R signaling is known to prevent allograft rejection in animal models [80].

Blocking CD200 expressed on tumors reverses the suppression of effector cells, which are CD200R-positive, and usage of CD200-blocking agents is restricted in CD200-expressing tumors. Genetically engineered human CD200 proteins expressed on ECs of pigs effectively suppress cytotoxic killing mediated by xenogeneic macrophages via the interaction with surfaced CD200R. The CD200-elicited inhibition in macrophage-induced cytotoxicity is directly depended on the interaction between CD200 and CD200R. For example, THP-1 cells exhibit cytotoxic cell death of swine endothelial cells (SEC), while CD200R-transfected THP-1 cells reverse the cytotoxic effect on SEC.

Apart from the inhibitory signal, CD200/CD200R also regulates the allograft rejections, allergic reactions and Tregs as well as autoimmune, autoimmune-related and neuronal diseases. The CD200–CD200R signaling prevented allograft rejection in animal models. In xenogeneic rejection mediated by CD200/CD200R cellular downstream signaling of macrophage cells, CD200 of human cells elicits the macrophage MO to M2 polarization in pig ECs. Since polarized M2 macrophage cell types positively regulate inflammation and tissue repair [80], human CD200 is expected to be protective in the chronic xenogeneic rejection. Transgenically expressed CD200 protein in cells of pigs reduces xenogenic cell-killing cytotoxic activity mediated by $CD200R^+$ macrophage cells. In fact, M0 type of macrophage cells strongly express CD200R on their surfaces, but monocytes do not express at all. It is reminded that traditional macrophage cells can be classified to three types of pro-inflammatory macrophages (M1), anti-inflammatory macrophage (M2) or non-activated macrophage (M0) subtypes that involve in their distinctive and functional roles in the inflammation initiation and resolution. In cytotoxicity assay, transgene CD200 of human expressed on pig ECs inhibits cytotoxicity mediated by M0 macrophage cells. However, cytotoxicity-mediated monocytes are not observed. Thus, CD200 suppresses the M0 macrophage-specific cytotoxic cell death by the interaction with CD200R. Transgenically expressed human CD200 protein on pig ECs elicits M2 macrophagic polarization in M0 type of macrophage cells [80].

The human CD200 expressed on pig ECs inhibits the M0 macrophagic cytotoxicity because the human CD200 inhibits the M0 macrophage cell-mediated phagocytosis. M0 macrophage cells strongly upregulate the CD200R expression but monocytes do not. The artificial upregulation of human CD200 expression on pig ECs inhibits cytotoxicity and phagocytosis induced by M0 macrophages [78]. Human CD200 on pig ECs inhibits xenogeneic macrophages-induced cytotoxic cell death via the interaction with CD200R. The human CD200 protein expressed on pig ECs enhances macrophagic M2 type polarization in macrophage cells M0 types that regulates inflammation (Fig. 17.1). The phagocytotic extent can be expressed by calculation as ($CD14^+$ $CFSE^+$ cell numbers/$CD14^+$ cell numbers) × 100 (%), where carboxyfluorescein succinimidyl ester (CFSE)-labeled pig ECs and pig ECs/CD200 protein are added with macrophage M0 type and the

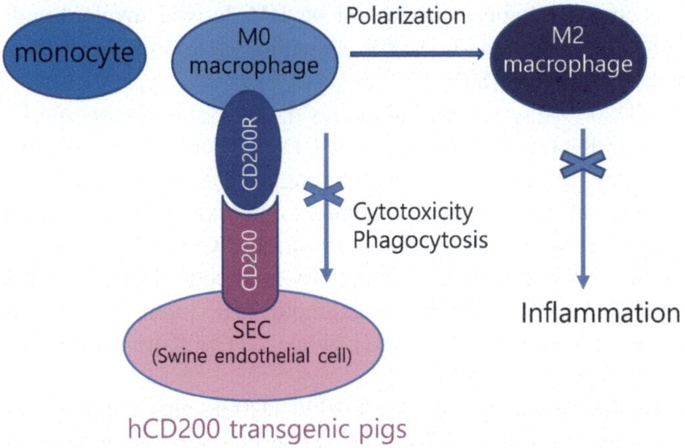

Fig. 17.1 hCD200 transgenic pigs prevent macrophage-mediated rejection

macrophage cells are then stained with anti-human CD14, which are APC-conjugated, as a macrophage protein. CFSE is a labeling dye color specific for fluorescent lymphocyte cells to be used and traces for their migrative mobility in experimental animals for several months. Human CD200 strongly enhances the Arginase-1 mRNA expression known as a major molecular marker and effector of myeloid-derived suppressor cells (MDSC) and M2a-type macrophage cells. The two MDSC and M2a macrophages are major mediators for T-cell suppression and also major IL-10 mediators in M0 type macrophage cells. Thus, these suggest that macrophages induce M2 polarization by human CD200 [80]. M2 type macrophage cells essentially resolve inflammatory status and involve in tissue repairs. Human CD200 exerts its protection activity of pig xenografts from macrophage-induced cytotoxic cell death by CD200R-binding. Moreover, the expressed CD200 of human in cells and tissues of pigs suppresses the inflammatory states by the M2 polarity induction in macrophage cells. The transgenic creation of human CD200-expressing pigs is another strategy to avoid macrophage-mediated rejection.

17.4.2 Transgenic hβ2m, HLA-E, and HLA-G-Expressing Pig Cells

Because human macrophage cells also express HLA-E-binding CD94/NKG2A inhibitory receptor with ITIMs, pig ECs which express hβ2m, HLA-E and HLA-G1 proteins were evaluated by macrophage-mediated pig EC cytolysis using the macrophages [81]. The transgenic HLA-E and HLA-G1 expressed on pig ECs suppress macrophage-mediated cytotoxicity. Transgenically expressed HLA-G1 protein elicits expression induction of the anti-inflammatory cytokines TGF-β and IL-10 and inhibits gene expression of iNOS gene at the mRNA level. Transgenic expression of HLA-G1 gene also shows inhibitory activities in xenotransplants [82]. Transgenic HLA-E expression also inhibits the cytotoxicity induced by macrophage cells. The expressed HLA-E protein inhibits the cytotoxicity induced by macrophage cells to the same extent of expressed CD47 protein. Moreover, HLA-E inhibits the pro-inflammatory cytokine expression by inflamed macrophage cells. HLA-E-expressing pig cells may show the protective effect on pig xenografts from NK-cell-induced cytotoxicity as well as macrophage-induced cytotoxicity. HLA-G1-expressing pigs also can be protected for pig xenografts from macrophage-induced cytotoxic cell death. The HLA-G1 protein inhibits monocyte and macrophage-induced cytotoxic cell death. When genes of HLA-E, HLA-G1, HLA-G3, and hβ2m are introduced into ECs of pigs, macrophage cells- and THP-1 cells-induced cytolytic cell death of pig ECs are suppressed [83]. The transgenic-expressed HLA-E and HLA-G1 ion ECs of pigs suppressed macrophage-mediated cytotoxicity to the same levels. While HLA-G1 gene is expressed in the cell surfaces of pig ECs, the HLA-G3 expression level is much weaker and intracellularly localized. Therefore, the HLA-G1 protein expressed in pig ECs significantly suppresses the THP-1 cells-induced cytotoxicity but relatively lowly suppresses it. Thus, HLA-G1 gene transgenic, but not HLA-G3 gene transgenic, pigs show efficient protection effect on pig xenografts from monocytes and macrophages-induced cytotoxic cell death [83].

17.5 Roles of Regulatory Macrophages (Mreg) and Immunomodulation of T-Cell Behavior

Macrophages as innate immune cells in self-defense and immunity bear diverse functions such as immune response, tissue homeostasis, pathogen recognition, adaptive immune cooperation, inflammation, and self-defense. The circulating macrophage precursors are heterogeneously naïve monocytes and their phenotypes and roles are changeable by the macrophage polarity state, which the functions, in turn, varies with two distinct macrophage subsets classified to two major types of classic-activated M1 macrophage cells and alternative-activated M2 macrophage cells, according to microenvironmental stimuli [84]. Macrophage M1 types are classically stimulated by two different cytokines TNF-α and IFN-γ. M1 type macrophages has a typical killing activity of pathogens and tumor cells via type-I pathological inflammatory response. To exert the type I inflammatory response, several pro-inflammatory IL-6, IL-12, IL-1β and TNF-α cytokines as well as inflammation-related molecules including IFN-γ-induced protein 10, Toll-like receptors (TLRs), iNOS (inducible nitric oxide synthase) and chemokine ligand-5 (CCL-5) are expressed by the M1 phenotype macrophages [85]. In contrast, M2 phenotype macrophages are activated by distinct IL-4 and IL-13 cytokines. These M2 type macrophage cells play key roles in adaptive immunity and Th-1 inflammatory responses as well as in Th-2 inflammatory responses, tissue remodeling and angiogenesis [85]. Macrophage M2 type cells highly express the mannose receptor (CD-206), MHC-II and the related genes such as arginase 1 (Arg 1) that deprives arginine via Arg 1, chitinase-like molecules, transglutaminase-2 (TG-2), and resistin-like molecule-α [86].

Apart from the M1 and M2 differentiation, interestingly, macrophages specifically polarize into distinct regulatory macrophages (Mreg) upon encouraging with immune complexes. Mreg are a macrophagic populations suggested in homeostasis and anti-inflammatory process. A distinct biochemical specificity of Mreg is its unusual synthesis of an anti-inflammatory IL-10 cytokine. The Mregs consequently suppress inflammation and experimental autoimmune encephalomyelitis (EAE) [87]. Cytokine IL-10-producing macrophage cells are therefore regarded as Mregs [87]. This cell type (Mψ(ind)) induction has recently been mentioned using by in vitro a specific mAb of anti-TNF Abs such as adalimumab and infliximab. However, the mechanism underlying this remains unclear. Infliximab and adalimumab are the major anti-TNF drugs; infliximab first approved, later followed by adalimumab. Infliximab is intravenously infused, whereas adalimumab is subcutaneously administered. Both agents are effective in ulcerative colitis (UC)-relieving activity, the mechanism underlying the activity remains some questions. The combinatory treatment effect of infliximab and azathioprine has been efficient on the Mreg differentiation and function. Upon infliximab or infliximab/azathioprine combination treatment, Mreg was induced in patients with mucosal healing and such elicited in vitro Mψ(ind) induced wound healing model and these macrophages are much more immunosuppressive than normal naïve macrophages [88]. The macrophages express a phenotype with immunosuppression because the macrophage cells inhibit activated T-cell proliferation and biosynthesize anti-inflammatory cytokines and the CD206 marker protein of Mregs [89]. Anti-TNF Abs stimulate the Mreg induction in an Fc region-depended mode and the anti-TNF functions can help to resolve inflammatory response that is involved in the unacceptability of xenografts. Hence, in contrast to the M1 and M2 macrophage phenotypes, Mregs are considered to regulate inflammation-related diseases and excessive-induced immune responses.

For xenotransplantation, cellular immunity-associated can be suppressed to downregulate the xenograft-specific immune response, minimizing immunosuppression and such filed in immunology is specific to immunoregulation. Representatively, Mregs are known as another type of immune cell regulators and exert their suppressive potentials to in vitro polyclonal T-cell proliferation, and resultantly giving the in vivo lasting allograft survivals [90], because Mregs indirectly inhibit growth of in vitro polyclonal T cells and prolong in vivo allograft survivals. Suppressive activity of Mregs on xenoimmune responses is therefore potentially applied to the human Mreg in immunomodulatory xenotransplantation. Mregs do not produce inflammatory mediators, but function as immunomodulator [91, 92]. Macrophages can also bind to the inhibitory receptors. A well-known case is CD200 and CD200R receptor, as described earlier, where both of CD200 and CD200R are classified to type 1 membrane glycoproteins, which belong to the Ig superfamily. CD200R1 expression is found in the surfaces of myeloid cells and its expression is upregulated in specific T-cell subsets [93]. CD200R1 specifically bind to CD200 also known as a transmembrane glycoprotein of Ig superfamily, in order to suppress functions of myeloid cells. CD200 is expressed in diverse surfaces of astrocytes, epithelial cells, ECs, fibroblasts, lymphoid cells, and neurons [93]. The CD200R1 signaling regulation is easily occurred by posttranslational modifications of protein. For example, Tyr residue phosphorylation present in the tail domain of CD200R1 protein cytoplasmic region and by the upregulation or downregulation of either CD200 protein or CD200R1 protein. Different from other immune inhibitory receptor proteins, CD200R1 protein lacks an ITIM sequence. Instead, CD200R1 protein of humans carries three distinct Tyr residues in cytoplasmic region of Y291, Y294, and Y302. One of them, Y302 or Y297 location is present within the known P-Tyr-binding domain recognition motif, termed NPXY sequence. In the mouse CD200R1 protein, the corresponding three distinct Tyr residues in cytoplasmic region to the human CD200R1 are Y286, Y289, and Y297. CD200 stimulation confers cellular Src kinase-mediated phosphorylation of the three Tyr residues and the process recruits the Tyr kinase-2 of adapter protein downstream signaling via its P-Tyr-binding domain [132].

Tyrosine Y302 and Y297 residues are essentially necessary for association between CD200R1 and Tyr kinase-2, and to a lesser extent tyr residues of Y291 and Y286 are also important for the association [94]. Tyr kinase-2 is essential factor of CD200R1 function because it is a signaling initiator via CD200R1 interaction, initially engaging with Ras-GTPase-activating protein recognition [94]. The signaling event is quietly contrasted to other inhibitory receptors, which contain ITIM sequences and recruit phosphatase enzymes such as SHPs and SHIP-1. The CD200 and CD200R1 interaction-exerted inhibitory signaling pathway limits inflammation events, as signaling of CD200R1 engagement inhibits the expression of pro-inflammatory molecules. Using CD200R1-based regulatory roles of multiple inflammatory responses, prolonged survival of xenografts can be exploited to apply this signaling to inhibit defense system of hosts. The CD200R1-regulated pathways are helpful to analyze clinical trial of transplantation by targeting the CD200-CD200R1 binding for transplantation benefits. How the present signaling pathway is utilized by a xenograft transplantation as representative and innovative strategy of xenograft targeting. CD200R has an inhibitory intracellular NPXY signaling motif and are expressed in myeloid/macrophage originated tissues. In allografts with portal vein immunization, the expressed CD200 on DCs and its receptor $CD200R^{+}$ on macrophages prolonged allograft survival [95]. Besides the inhibitory capacity on macrophages, the CD200/CD200R complex on macrophages regulate the allergic reactions, allograft rejections, autoimmune diseases, neurological diseases, Treg cells, and other pathogenic diseases. The human CD200 inhibits the xenoantigenic rejection responses by $CD200R^{+}$-positive macrophage cells, and hence, the creation of transgenic hCD200 pigs is suggested to attract for prevention of the macrophage cell-induced rejection during xenografts transplantation [91].

In the cardiac transplantation, Mregs cooperation with Tregs prolong the survival time of organ grafts, as Tregs act as Mreg's regulatory counterpart in the immune system. Mregs in living donor renal transplantation [96] suppress the T-cell-induced response of xenoimmunity. However, the exact mechanism(s) underlying the

Mreg-regulated Treg's suppression is/are not answered clearly yet. Guo et al. [90] have found that human Mreg suppresses the xenoantigenic response in vitro. From the experimental results, Mregs could be used for immunomodulation in xenotransplantation. In fact, human Mregs inhibit the in vitro xenoantigenic immune responses in a mode of modulation of the xenoantigen-reactive $CD4^+$ T-cell proliferation and $CD8^+$ T-cell proliferation. Considering that $CD4^+$ T cells predominately induce CR [68], Mreg is valuable to inhibit the T-cell responses of xenograft rejection in xenotransplantation. Future more scientific explanation and in vivo Mreg evaluation for clinical xenotransplantation are required for further generalization.

As a Mreg induction mechanism, indoleamine-2, 3-dioxygenase (IDO) inhibits the xenoantigenic response and prolong survival time of skin xenografts through suppression of xenoantigen-reactive proliferation of T cells and blocking of xenograft infiltration [97]. Heme-containing IDO enzyme catalyzes enzymatic conversion of amino acid Trp residue to *N*-formyl kynurenine as well as nicotinamide adenine dinucleotide as catabolites, totally referring to kynurenines [98]. Both the local Trp reduction and the accumulated Trp metabolites toxic, which are immunosuppressive, suppress T-cell growth by the T-cell arrest in the cell cycle G1 phase. Other known Trp-derived metabolites include L-kynurenine that suppresses antigen-specific T-cell growth and T-cell apoptosis induction. The IDO immunosuppression activity is caused by the sensitive T-cell growth to local Trp reduction, subsequently inhibiting T-cell growth. IDO expression is profound in hematopoietic cells of pDCs and immatured DCs, and also in non-hematopoietic BM-derived mesenchymal cells (BMDMC) [99]. IDO enzyme potently functions as an immune-regulatory molecule and IDO is primarily present in the plasmacytoid DCs (pDCs)-depleted lungs of mice. IDO expression is found in DCs and macrophage cell subsets, and the IDO enable to suppress priming of T cells and growth associated with enhanced differentiation of Tregs. IDO expression potentially prevents hyper-reactive responses of inflammation [100]. IDO also triggers to intracellular signaling to amplify and maintain the regulatory phenotype in pDCs that promotes immune tolerance. The IDO-Treg axis promotes immune tolerance]. IDO activity and IDO-produced metabolites, kynurenines, stimulate the Treg cell expansion. Considering the IDO roles in the pDC tolerogenic activity, the Treg cell number was decreased in the IDO-defected mice-derived pDCs [101]. The major IDO activity is reported to inhibit the immune response of susceptible and resistant mice. For example, the enzyme activity of IDO is dominantly enhanced by IFN-γ of B10.A mice, while IDO enzyme is upregulated by TGF-β signaling pathway of the resistant A/J mice [101]. But the IDO-based activity is likely not the sole mechanism provoked by Mregs to inhibit the xenoantigenic rejection. Mregs exhibit an immune-tolerance role via an IDO-induced action that enhances Tregs capacity. However, IDO-expressing Mreg can be tested for adjunctive therapy to reduce the immunosuppression in xenotransplantation.

17.6 Immunomodulation of Regulatory T Cells (Tregs)

To prolong the xenograft survivals in the transplantation, immune tolerance induction is another strategy. Natural Tregs are powerful therapeutic candidates to induce immune tolerogenicity in allogeneic cells and allotransplantation of organoid. T-cell-mediated xenoantigenic rejection is a key barrier for clinical application of xenotransplantation. Tregs-based tolerance is also significant in their capacity. In clinical transplantation, T cells directly regulate the adaptive immunity of the host cellular immunity for transplanted allogenic grafts [102] and xenogenic grafts [103]. The host immune tolerance to xenoantigenic grafts are being understood, because human Treg cells modulate the xenoimmune response. The immune tolerance is obtained by xenoantigen-responsive Tregs. Thus, induction of immune tolerance is the next medium of research in xenotransplantation. Tregs-mediated immunotherapy is suggested as an alternative tool to block and control graft-versus-host disease (GVHD) to modulate transplanted graft rejection of organs or to cure autoimmune diseases. The current clinical transplantation and trials are combined with the freshly isolated Tregs injection or polyclonal expansion-generated Tregs because the antigen-specific Tregs genesis is important in hosts. Tregs specific for recipient alloantigens efficiently function rather than polyclonal Tregs in controlling the GVHD process [104]. Recently, clinical-leveled allospecific Tregs are produced from either CD40 ligand-induced B cells or immortalized B cells of human as APCs. Resultantly, antigen-specific Tregs-treated trials are clinically tried in liver and kidney transplantations, by controlled approvals of NCT01624077 and NCT02145325, respectively. Such approaches are based on future development.

Tregs property has been specifically defined by enhanced forkhead box P3 (FOXP3) expression as a powerful transcription factor crucial for the Tregs activity [105], enhanced IL-2 receptor (IL-2R) α-chain (termed CD25) expression. In humans, the FOXP3 regulates to weak the IL-7 receptor (IL-7R) α-chain (termed CD127) expression or without IL-7R α-chain expression [106]. Tregs occupy a limited subset of $CD4^+CD25^+$ cells, which occupy 3–5% population of total $CD4^+$ T cells, and they downregulate autoreactive T-cell population in humans and mice. Treg-specific markers are known for FOXP3, Helios, CD39, CD152, CD278, GARP, and LAP are functional markers of Tregs. The most clinical background of the FOXP3 is to protect fetuses during pregnancy in mammals. In a murine GVHD model, donor Tregs co-transferred with effector T cells (Teffs) inhibit GVHD reaction and protect alloantigenic engraftments [104]. Tregs-mediated therapeutic approaches for GVHD blocking, therefore, utilize fresh Tregs, which are newly isolated, in patients [107] or Tregs isolated from umbilical cord blood using by CD25-specific Abs-based immune magnetic beads, and expanded using CD3-/CD28-specific Abs-conjugated beads. In mice, alloantigen-recognizing Tregs are shown to efficiently function rather than polyclonal Tregs to protect GVHD reaction. Alloantigen-recognizing Tregs significantly suppress in vitro and in vivo xenoantigenic GVHD mice of immunodeficiency model.

Successful xenotransplantation requires immunosuppression but side effect of immunosuppression contributes to increases in pathogenic infectious opportunity and cancer. In addition, systemic immunosuppression by life-long treatment reduces the xenograft rejection incidence, but pathogenic infection and cancer outcome are accompanyingly increased [108]. Thus, immunomodulation strategies are necessary for achieving immune tolerance to transplanted grafts without immunosuppression. For example, large numbers of xenoantigenic epitope-reactive Tregs of human is required for induction of immunotherapeutical success of organ xenotransplantation. Tregs are necessary for the autoimmune response modulation and the allograft tolerance maintenance [109]. For example, Treg suppress direct and indirect xenoantigenic reactions of APCs, B cells, T cells, and macrophages. Polyclonal Tregs induce and maintain immune tolerance. In addition, they have protective activity of immune rejection of islet xenografts in vivo. Nevertheless, polyclonal Tregs-induced immunotherapeutics globally suppress immune responses and in vitro xenoantigenic epitope-reactive Tregs expansion is necessary.

Among all Treg cell classes, self-antigen-reactive $CD4^+Foxp3^+$ Tregs cell population in the thymus peripherally suppress such xenoimmune response. Hence, for the therapeutic application, the $CD4^+Foxp3^+$ Tregs can overcome the risk such as GVHD. To date, Tregs population is diverse in the Tregs type subsets. For the suppressive cellular therapy of the allograft and xenograft rejection, the specifically affordable Treg cells should be defined and the potent subset population of xenoantigen-specific Treg cells should be used. The tolerant Treg cells are also classified to 2 subsets of the activated and memory Treg cells, depending on the xenoantigen-accounting history of the Treg cells. Thus, induction of a specific pig xenoantigen-tolerant Treg cells population is preferentially required for a clinical efficacy strategy. Consequently, such induced tolerant Tregs cells are desired in their immune-regulation function than naive Tregs cell population. Activated T cells, which are deficient for regulatory functions, induce the phenotypic $FOXP3^+$-positive population. Thus, $FOXP3^+$ cells reflect a protection of xenoantigenic pigs in the host and $CD4^+Foxp3^+$ Treg cells sustain their tolerance against xenoantigenic grafts. Memory Treg cells are initially found in pathogen-responsive immunity and autoimmunity [110]. However, it is difficult to phenotypically distinguish the memory Treg cells from activated Treg cells in the in vivo immune cells. As an acting $CD4^+Foxp3^+$ Tregs cell mediator, which suppress xenografts behavior, a cytokine IL-10 is known and IL-10 functions in the tolerant Tregs cells, where the tolerant $CD4^+Foxp3^+$ Tregs transmit IL-10. IL-10 potentiates function of $CD4^+Foxp3^+$ Tregs cells at both the in vitro/in vivo levels. Therefore, the xenoantigen-tolerant Treg cells express higher IL-10 than naive Treg cells [111]. Similarly, pig-islets xenotransplanted NOD-SCID mice, which were co-transferred with splenocytes and tolerant Tregs remarkably produced plasma IL-10 rather than mice co-transferred with naive Treg cells only. More precisely, the 2 cytokines of TGF-β and IL-10 mediate the graft rejection prevention and allografts tolerance [112]. Moreover, IFN-γ is known to generate allospecific $CD4^+Foxp3^+$ Treg cells to induce alloantigen tolerance in vivo. Xenoantigen-reactive $CD4^+Foxp3^+$ Tregs are stimulated by the B7-CD28 and CD40-CD40L signaling pathway blockade.

Hampered cellular rejection is known to be attenuated by Foxp3$^+$ Tregs in co-stimulative blockade-exerted dominant tolerance to pig xenografts in mice. Previously, co-stimulation blockade of CTLA-4Fc and CD154-specific monoclonal Abs induced tolerance of xenografts against the co-stimulatory pathways of CD40-CD40L (known as CD154) and B7-CD28 signalings [113].

The co-stimulation blockade-induced xenograft tolerance also clonally expands the potent CD4$^+$Foxp3$^+$ Treg cells population. Also, CTLA-4 regulates function of Foxp3$^+$ Tregs cells, as CTLA4-Ig inhibits tolerance of skin allografts mediated by CD28-dependent co-stimulative blockade and suppresses cell expansion of Tregs in MHC-mismatching mice [114]. CD4$^+$Foxp3$^+$ Treg cells stimulated by MR.1 and CTLA4-Ig inhibit alloantigenic immune reaction in vitro. Despite such progress, the CD4$^+$Foxp3$^+$ Tregs-induced tolerance of xenografts remains clearly unknown yet for the cellular mechanism. FOXP3$^+$ expression also reflects both tolerance and rejection. For example, human CD4$^+$ T cells display pig xenograft rejection, while human CD4$^+$CD25$^+$CD127^{low}FOXP3$^+$ Tregs inhibit these type responses of rejection. Also, CD4$^+$Foxp3$^+$ Tregs induce tolerances to xenografts and allografts [115] and in vitro adoptively transferred the expanded CD4$^+$CD25$^+$CD127^{low}Foxp3$^+$ Tregs block xenografts rejection of pigs in mice reconstituted with human PBMCs. Recently, Foxp3$^+$ Treg cell induce tolerance to pig-islets xenografts transplantation to mice treated with co-stimulative blockade of CTLA4-Fc and CD154-reactive monoclonal Abs [111]. This suppression was specific to pig-islets xenograft in the tolerant populations of Tregs cells. CD4$^+$Foxp3$^+$ Treg cells maintain and propagate tolerance to pig islets xenografts upon transplantation of the renal capsulated mice treated with a CTLA4-Fc. The CD4$^+$Foxp3$^+$ Tregs present in the tolerant islets xenografted spleens of mice are tolerant to pig-islets xenografts, as induced by the co-stimulation blockade in a mode of the CD4$^+$Foxp3$^+$ Tregs phenotype. The Foxp3$^+$ Tregs number is increased in the pig spleen, pig draining lymph nodes and pig xenografted islets, with potent functional regulating activity in vivo than that of naive Tregs. The pig islets xenografts from treated mice also exhibited the increased IFN-γ, TGF-β and IL-10 levels. Tolerant Tregs have diverse TCR Vβ repertoire with the TCR Vβ14 CDR3 expansion. Reversely, depletion of Foxp3$^+$ Treg was reported to reduce pig islets xenograft tolerance in the mice [111].

Antigen-specific Treg improves therapeutic efficacy level of Tregs and decreases the nonspecific immune suppression-caused risks enhanced by functional polyclonal Tregs transfer. Ex vivo expanded Treg cells, which are specific for alloantigen, upregulate immune suppression activity in in vitro allogeneic response rather than polyclonal Tregs in their protection potentials against allograft-immune injuries of human grafted skins in humanized model mice [116]. Excessive Tregs number with the increased xenoantigen specificity and suppression capacity are detected using by known Treg markers such as CD45RO and HLA-DR, as they are easily observed in xenoantigen expanded Treg. The Tregs' suppressive capacity seems to associate with Treg functional marker expression, named CTLA-4 and Foxp3. Recently, cobra venom factor, infusion of natural Tregs, T-cell-targeting drug therapies using ATG and sirolimus, CD154-specific Abs-based co-stimulation blockade, and

TNF-specific Abs are used, as such Tregs-based application develops tolerance in xenotransplantation. Although xenotransplantation requires systemic immunosuppression [117], the polyclonal Tregs infusion also induces such similar immunosuppressive effects against opportunistic infections and tumor-growth. Specific inhibition of the donor-alloresponsive effector T-cell expansion can eliminate any possible side effects after donor allospecific Tregs transplantation to prevent allograft, compared to polyclonal Treg usage [116]. For the xenotransplantation, ex vivo amplification of human Tregs, which are xenoantigen-specific, using natural-preformed Tregs is not easy. In the recent study [118], such amplification of xenoantigen specific human Tregs was developed for application of xenotransplantation. Those Tregs suppress the responder cells that are xenoantigens-reactive, compared to that of normally regular polyclonal Tregs. To obtain Treg population from human, human Treg from PBMCs can be proliferated with IL-2, CD3/CD28-specific Abs beads and rapamycin and consequently polyclonal Tregs expansion are populated. If Tregs are expanded and amplified in the xenoantigenic stimulative condition, suppressive Tregs phenotypes are retained without xenoantigenic specificity.

17.7 T-Cells-Mediated CR in Xenotransplantation

As the main hurdle to successfully sustainable clinical xenotransplantation, the T-cell-associated adaptive immune response has been marked because xenoantigenic T-cell responses control CR. Also, T cells involve largely in innate immunity. Vasculopathy observed in the CR of long surviving allografts is similar to that of the xenografts [119]. To date, there is a question on whether genesis of natural Abs and induced Abs is the key barrier to xenotransplantation because T cells involve in in vivo xenograft rejection [120]. In pig-to-primate xenotransplantation of solid organs, two different $CD4^+$ T cells or $CD8^+$ T cells were demonstrated to collaborate with innate immune cells of macrophages, monocytes, NK cells, and B cells. Th1-specific $CD4^+CD28^-$ cells infiltrate to pig cardiac xenografts baboons. In the xenotransplantation of pig islet non-vascularized xenografts, T cells displayed the xenografts rejection. Pig fetal islets xenotransplanted to mice are fastly rejected in wild-type mice. However, non-thymic mice or TCR-KO mice are not rejective for the pig fetal islets xenotransplanted to mice. The xenografts-infiltrating T-cell subpopulation is mainly $CD4^+$ T cells, not $CD8^+$ T cells, confirming a T-cell-dependent xenografts rejection, and this process is presumably operated by the helps of macrophages. In the pig neonatal islets [8] and pig adult islets xenotransplantation to mice pigs, by means of the intraportal and engrafted liver in non-immunosuppressed primates, the xenografted islets are infiltrated by $CD4^+$ and $CD8^+$ T cells as well as macrophages, consequently disrupting the xenografts within 3–5 days [121]. However, immunosuppression with basiliximab, FTY720, everolimus, and CD154-specific monoclonal Abs give the prolonged survival time of adult xenografted islets up to 187 days [122]. In addition, Abs of IgM and IgG isotypes as well as complements are not detected in the islet xenografts; however, tissue regional-infiltrated CD4$^+$ T cells and CD8$^+$ T cells as well as macrophages are observed in the xenografts that were rejected [122].

If CD4$^+$ T cells are depleted in mouse strains, the pig-specific Abs responses are also depleted. Therefore, the pig-reactive Abs responses T-cell-dependently elicited in wild-type mouse. The transplanted grafts are dramatically reduced in the mononuclear cell infiltrate. CD4$^+$ T cells cooperatively contribute to other xenoantigenic tissue rejection [123]. Xenograft rejection is predominantly caused by the Th2 immune responses, and T-cell-induced humoral responses reject the xenografts, because humoral antibodies directly reject xenoantigenic tissue. For example, xenografts activate innate immune cells such as macrophages when host inhibitory receptor-binding ligand is absent [124]. If CD47 known as an immune inhibitory receptor-SIRPα ligand is absent or incompatible in the donor cells, the cells are phagocytosed [125]. In a minor part, the typically known, Fas/FasL-mediated apoptotic cell death also rejects the xenografts in the human anti-pig killing mode, because anti-Fas-L antibody diminishes the anti-pig lytic signaling. Therefore, development of the xenoorgans having T-cell tolerance with donor specificity across xenoantigenic hurdles is a central issue in the successful xenotransplantation. Pig-primed human T cells exhibits aggressive cytotoxic killing activity of in vitro T cells. T cells also stimulate IL-2 biosynthesis, and CD4$^+$ T cells and NK cells mediated direct cellular killing activities. The T-cell cytotoxic killing activity is dependent on MHC C-I because the anti-CD8 Abs inhibit the cytotoxic killing activity. In the anti-pig cytotoxicity of human, which is induced by xenografts, approximately 60% of the anti-pig cytotoxic responses of human against xenografted xenoantigens are induced by CD4$^+$ T cells, and $< 20\%$ of the anti-pig cytotoxic responses of human are caused by CD8$^+$ T cells. Therefore, human T cells directly recognize pig antigens to bind them. The structural basis between MHC of alloantigenic donor pig and recipient host human will exhibits the long distance due to difference in species in evolutionary tree. From this immunological reason, it is consequently considered at the initial stage of immunological progress that the T-cell responses of human host to the pig-derived xenoantigen are likely less severe than that of allogeneic donors [126].

In the xenografts-specific T-cell immunity, when human or NHP lymphocytes were incubated with pig cells, like T-cell recognition during allotransplantation, T cells of human directly bind to pig SLA proteins and also react to pig lymphocytes or ECs. Further, they indirectly bind to xenoantigenic epitopes expressed in the self-MHC proteins. The CD4$^+$ T-cell and CD8$^+$ T-cell growth levels of human and baboon under direct stimulation with pig PBMCs or pig EC are higher than the stimulation with allogenic species cells. Allo or xenoantigen-reacting T-cell precursors are parallel to 1–10% of circulating lymphocytes. Direct binding of CD4$^+$ T cells to SLA C-II proteins is much stronger than the CD8$^+$ T-cell-SLA class I interaction. Although human memory and naïve CD4$^+$ T cells respond to pig endothelium, human naïve CD8$^+$ T cells only strongly respond to pig cells. Additionally, the CD4$^+$ T cells are required for potentiation of the CD8$^+$ T-cell-induced xenoreaction and the CD8$^+$ T-cell-induced xenoreaction is strongly eliminated in a

$CD4^+$ T-cell-depleted $CD8^+$ T-cell subpopulation. Thus, a basic actor for the T-cell anti-pig cytotoxic activity is caused by the $CD4^+$ T-cell subpopulation, being 65% of cell lysis activity of $CD4^+$ T-cell population and cell lysis 20% by $CD8^+$ T-cell population [127].

In antigen presentation of innate immune cells, two different pathways of direct and indirect pathways are associated with the anti-donor rejection. Xenoantigenic responses of T cells are a major type of rejections. Additionally, T-cell responses are not suppressed by immunosuppression. Allotransplantation is known to be recognized by the host recipient with the two different modes of pathway for alloantigenic presentation, which are the direct and indirect antigen presentations. The pig endothelium-specific $CD8^+$ and $CD4^+$ T-cell actions are derived by direct antigen binding to pig endothelial SLA C-I or C-II antigenic molecules in cooperation with co-stimulating signalings by CD80 and CD86 to activate the T cells [128]. In contrast to typical innate immunity, upon binding to pig endothelial α1, 3Gal epitope antigens, the direct responses of T cells to xenografts is increased with unknown explanation. For example, α1,3Gal-T KO endothelial and α1,3Gal antigen-deficient PBMCs reduces growth levels of $CD4^+$ T cells and $CD8^+$ T cells rather than a wild-type ECs. When $CD4^+$ and $CD8^+$ T cells are incubated with α1,3Gal-T KO pig cells, the IL-17A, INF-γ, and T cells are activated.

NF-α expression levels in $CD4^+$ T cells are decreased and the synthesis levels of IP-10, INF-γ and granzyme B are also decreased in $CD8^+$ T cells rather than exposed wild-type ECs. In the direct recognition pathway, T cells of host recipients recognize and are responded to donor APCs' MHC proteins, while in the indirect recognition pathway, the donor-derived peptides are presented to APCs via MHC C-II proteins of host recipients and the host recipient T cells are stimulated. Although a new concept of the "semi-direct" pathway was reported [129], the recipient host DCs recognize the intact MHC–antigen complex presented by the donor DCs and ECs itself for antigen presentation. Consequently, direct-presented antigens presented to the T cells of hosts are alloreactive. The same DCs can simultaneously present the antigens in an indirect fashion, and thus, these DCs can stimulate both of $CD4^+$ T and CD 8^+ T cells together [129]. It is known that interaction between human CD4 T cells and murine MHC C-II is defective and this easily eliminate T-cell activation. Thus, it seems that once the direct pathway-mediated humoral rejection is overcome, the next cell-mediated rejection is much lower giving a minor leveled obstacle that normal case. However, the direct activation pathway is present in the pig-to-human xenotransplantation. In chronic rejection against xenoantigen-specific $CD4^+$ T-cell activation, pig xenoantigenic epitopes are indirectly bound by human T cells. $CD4^+$ T cells of human display the indirect anti-pig T-cell reaction. The indirect anti-xenograft reaction is stronger than that of its allogenic grafts, presumably due to the human APC-presented xenogenic peptides. Human $CD4^+$ T cells bind to SLA C-I peptide and SLA C-I proteins present pig cell. Through the immunogenic SLA C-I epitope identification, peptide-specific immune tolerance can be potentiated [130]. The major xenogenic epitopes, which have indirectly been presented by APCs to $CD4^+$ T cells, include pig SLA C-I protein peptides or SLA C-I peptides, which have their HLA-DR affinities. When human HLA-DR4 transgenic mice were sensitized with the peptides, Th-1 immune response is strongly induced. Although $CD4^+$ T cells indirectly respond to xenoantigenic molecules, $CD8^+$ T cells indirectly recognize peptide derived from SLA C-I protein, which are corresponded to human HLA C-I molecules and this event is similar to that of allotransplantation.

From the above information of T cells, the strong anti-xenograft reaction of T cells should be overcome to obtain the better xenograft survival by following strategies. For construction of SLA C-I and C-II-defected pigs as an MHC-restricted phenomenon, inhibitory CIITA mutant (CIITA-DN) gene-transgenic and SLA II-KO pigs were created [131]. The pigs express CIITA-DN gene that induces the decreased $CD4^+$ T cells and $CD8^+$ T-cell behaviors. T-cell co-stimulatory blockade is also applicable. Activation blocking of T cells is possible by inhibition of the CD80/CD86-CD28 and CD40–CD154 signaling pathways [132]. Blocking of CD40–CD154 signaling seems to powerful, as it prevents the elicited antibody production, which is dependent on T cells. CD40-specific monoclonal Abs are also effective like anti-CD154 antibodies. In addition, some anti-CD154 agents is not causative for thromboembolic side-effects. The T-cell tolerance can be induced. For example, the pig thymus xenotransplantation to thymectomized baboon primates induces a tolerance in baboon when receives α1,3Gal-T KO pig kidneys, giving a renal function for 83 days more.

Therefore, it is important to overcome the acute rejection that T cells mediate in clinical xenotransplantation. For example, a limited T-cell number initiates cellular T-cell-mediated xenoantigenic rejection in islet xenotransplantation [133]. Pig xenoislet survival was increased by CTLA4 Ig or anti-CD40L treatment, even though $CD4^+$ T cells display the xenograft rejection of islet. T-cell depletion is one of the tolerance-inducing ways. The circulating T-cell "optimal number" as T-cell depletion level is crucial for prolonged survival period of α1,3Gal-T KO organs in baboon hosts [134]. Therefore, to avoid cellular rejection, tolerance induction and T-cell depletion are considered. T-cell-derived immune responses affect the next immune response to diverse immune cells of B/NK cells and innate immune cells. Thus, to control xenogeneic anti-pig xenogeneic cellular rejection, the immunosuppression prolongs xenograft survival. For tolerance induction, indeed donor chimerism is reported. T-cell tolerance induction is also activated by purinergic signaling by extracellular nucleotides and nucleoside derivatives are also crucial for tolerance induction of T cells in xenotransplantation [135]. CD39 molecule distinguishes the phenotype Tregs of activated and resting $FoxP3^+CD4^+CD25^+$ Tregs from other various T-cell subsets present in pig T cells. When hematopoietic stem cells isolated from miniature pig α1,3Gal-T KO strains or hDAF-transgenic pig strains are xenotransplanted to baboon hosts that are irradiated, the most xenogeneic cells are phagocytized within hours by macrophages [136]. However, human CD47-transgenic pig α1,3Gal-T KO BM cells exert immune tolerance, because these cells are protected from host macrophages, because the human-CD47-transgenic pig α1,3Gal-T KO BM cells synthesize CD47 molecule of hosts, escaping from phagocytosis by the host reticuloendothelial system. Therefore, tolerance induction with gene editing and engineering of donors will allow the future clinical success in transplantation of xenogeneic organs to host human patients.

17.8 Protection of Xenografts from T-Cell Responses

Despite xenotransplantation of pig xenografts as a new alternative source, meaningful xenografted cell and engraftment loss, rapid blood-mediated responses of immunity, and cellular rejective responses cause for the actual setting restriction. Consequently, it is accepted that suppression of immune responses is important for success in xenotransplantation. Apoptosis during engraftment induces xenograft failure and poor xenograft survival. T cells are demonstrated directly to mediate the xenograft rejection in the studies on cord blood-derived $CD34^+$ human stem cells or $CD34^+$ progenitor cells. Therefore, for the protection of immune rejection, co-stimulatory blockades of T cell prevent the reaction. The T-cell importance in xenoantigenic cellular rejection has well been evidenced in human $CD34(^+)$ progenitor cells of the hematopoietic system that maintains the systemic oxygen supply and immunological surveillance. In humans, the hematopoietic system initially occurs at the yolk sac of the fetal development to start hematopoiesis. After the fetal liver function till birth, hematopoietic stem cells (HSCs) settled down to in the niche BM region [137]. The immunophenotype is characterized as $CD34^+/CD38^-$/low/$CD59^+$/CD90(THY1) $^+$/CD117(KIT) $^+$/$CD133^+$/CD135(FLT3) $^+$/$CD164^+$/$CD338^+$/$GATA2^+$/TdT^+/Lin^- [138]. The HSCs are evaluated in the NOD/Shi-scid-IL2Rc null (NOG) mice, lacking functions of NK DCs, T-, or B-cell immunity. The mice have been generated by crossing NOD-scid mice with an IL2Rc-KO background, due to the efficient human HSC engraftment [139]. Cord blood (CB) contains young human HSCs that move to the bone marrow at birth and is the most progenitor HSCs sources. Therefore, HSCs-transplanted NOG mice are evaluated the development of human hematopoiesis.

In the recent report in islets xenografts, the human $CD34^+$ T cells and mouse alloantigenic $CD3^+$ T cells were subjected to co-transplantation to NOD-scid G-chain-null mice as a stem cell xenograft rejection model. Allograft $CD34^+$ progenitor cell transplantation is known to reduce the acute GVHD risk in the condition of donor T-cell depletion in stem cell preparation. Similarly, such allotransplantation can be engrafted to HLA matched transplants. However, the present such application is forwarded to patients having hematologic malignant or non-malignant disorders from HLA incompatible haploidentical donors [140]. For a demerit of the progenitor $CD34^+$ cells, $CD34^+$ cells in the transplantation-based immune tolerance induction are not such homogeneous cell population, rather the $CD34^+$ cell population is heterogeneous as in progenitors with APC activity and monocytic/DCs-differentiable cells when allogenic T cells are encountered. But currently, in vivo model is not available for alloreactivity tests of anti-$CD34^+$ cell response, and thus, NOD/SCID mouse is being used [141]. NOD/SCID mice are mutated in the common gamma chain gene of IL-2R. Thus, they are IL-2RG null (or NSG) and thus negative for production of B, T, and NK cells. Such phenotypic characters enabled to use human T-cell engraftment in xenogenic GVHD animal and also $CD34^+$ cells. Recently, a rejection in vivo model of human $CD34^+$ cells after HLA incompatible T-cell co-transplantation was designed. T cells reject the $CD34^+$ cells of human in

NSG Mice. After human $CD34^+$ transplantation, BM-engrafted human $CD34^+$ cells and T cells expanded in the spleen, as engraftments of human cells are assayed by the anti-hCD45 PerCP expression. Later, memory T cells moved and located on the sites, and rejected $CD34^+$ cells. NSG mouse BM and spleen are capable of engraftment with hematopoietic cells of human after transplantation. The human $CD34^+$ cell engraftment after transplantation in the presence or absence of HLA incompatible $CD3^+$ T cells are confirmed. Xenogenic GVHD is occurred only in high T-cell doses-received mice. In the BM and spleen of those mice, the $hCD45^+$ cells include $CD33^+$ myeloid cells, $CD14^+$ monocytes, $CD1c^+$ or $CD1a^+$ myeloid DCs, $CD19^+$ B cells, CD19-$BDCA2^+$ pDCs, and allogeneic T-cell-expanded $CD34^+$ cells. However, T-cell-negatively transplanted control mice show the engrafted $CD34^+$ cells and B cells.

Regarding the T-cell co-stimulation blockade with abatacept (CTLA4-IgG1), abatacept (CTLA4-IgG1) perfectly prevents immune rejection, even though abatacept (CTLA4-IgG1) induces lower engraftment of stem cells than negative controls. Mouse re-transplantation with human $CD34^+$ cells completely reconstituted human B-cell lineage and myeloid lineage. When in vivo co-stimulatory blockade abatacept was injected to NSG mice transplanted with 1:1 ratio of allogeneic T cells and $CD34^+$ T cells and the NSG mice were then $CD34^+$ cell-retransplanted, the $hCD45^+$ cell engraftment was analyzed. In this stem cell rejection, abatacept protected HLA mismatched human $CD34^+$ cell transplantation. Co-stimulatory blockade drug, abatacept of CTLA4-IgG1 molecule prevents human $CD34^+$ cell rejection, as abatacept prevented mouse MHC mismatched bone marrow transplant rejection and also human T cells in vitro [142]. After persisted T-cell co-stimulation blocking, stem cells boosting potentiated the stable incompatible human $CD34^+$ cell engraftment. Thus, a xenograft stem cell transplants reject $CD34^+$ hematopoietic stem cells of human by allogeneic T cells. Human $CD34^+$ cell-induced alloreactivity of T cells was also in vitro proved. Repopulation activity of human stem cells in the allogeneic T-cell presence showed the T-cell alloreactivity in vivo against human $CD34^+$ hematopoietic progenitor cells. Human $CD34^+$ cell and allogeneic T-cell co-transplantation leads to grafts failure of stem cells. Additionally, both $CD4^+$ T-cell and $CD8^+$ T-cell types cause immune rejection of $CD34^+$ cells. Thus, in xenotransplantation, even low allogeneic T-cell doses perfectly prevent, affecting $CD34^+$ cell engraftment.

During xenotransplantation, human T cells bind to pig SLA or MHC molecules. Human T-cell responses to pig xenoantigens grafts are much higher than that on allografts. Macrophages and $CD4^+$ T cells and $CD8^+$ T cells infiltrate into grafts. Among them, $CD4^+$ T cells are abundant cell population during rejection of xenografts]. Activated $CD4^+$ T cells infiltrate to the pig xenografts to induce rejection and consequently lead to IFN-γ-induced macrophages and NK-cell activation and infiltration [68]. Macrophage cells recruit effector T cells with antigen presentation and cytokine production such as IL-6 and TNF-α, tissue factor upregulation, and enhanced coagulation inflammation. The xenoantigenic response of T cells is strong. CD154-specific Abs treatment with conventional immunosuppression enables prolonged and long-term survivals of α1,3Gal-T KO/hCD46 tg islets of pigs in cynomolgus monkeys [143]. A CD154-specific Abs-based modified immunosuppression to α1,3Gal-T KO-hCD46 pig islets (α1,3Gal-T KO/CD46 pigs) with additional human TFPI, human CD39 and pig CTLA4-lg improved much the response. In humans, anti-CD154 alone is not utilized systemically because of its thrombotic side-effects. Moreover, a CD40 (the receptor for CD154)-specific antibody protects heart and kidney organ transplants of pigs transplanted to baboon hosts, indicating the meaningfully advanced progress tried to xenograft survival of vascularized kidney organs. In pig-to-baboon xenotransplantation of artery patch model, pig arteries and dominant negative MHC II transactivator-expressing ECs decrease MHC II expression. Targeted pig MHC I gene disruption was created via the CRISPR/Cas9 gene editing technology [144], although leading to an unwanted NK response. Also, the immunosuppression Abs including CD154-specific Abs or CD2-specific Abs prolong the survival of recipients.

From the importance of T-cell responses against pig xenoantigen protection to avoid xenoantigenic barriers, thymic transplantation is an alternatively promising challenge for T-cell tolerance induction. With regard to humanized mice, co-transplantation of $CD34^+$ cells and fetal thymus of human long-termly repopulates the cells with multiple lineages of human lymphohematopoietic cells and allows secondary lymphoid organ formation in immune-deficient mice. In addition, the constructed humanized mice show robust responses of antigen-specific immunity and xenograft rejection of pig islet and skin. Pig thymus can produce T cells of human tolerant to the thymic donor of pigs. Transplantation of pig thymus causes robust tolerance induction of xenografts in thymectomized mice or T-cell-depleted mice. The pig thymic grafts-received mice recipients exhibit tolerance specific for donor and become accept grafted skin of pig thymic donor. The pig thymus transplantation also prolongs pig α1,3Gal-defected kidney xenograft survival in NHPs. In addition, human T-cell development in thymic grafts of pig in immunodeficient mice show unresponsiveness against the pig thymic donor MHC, allowing human thymopoiesis in grafted thymus of pig and human T-cell reorganization after pig thymus transplantation. Therefore, there is a possibility of pig thymus acceptance for T-cell development and thymopoiesis of human. Recently, human T-cell tolerance was induced against pig xenoantigenic epitopes by thymus xenotransplantation in pre-established human immunity-based humanized mice. In mice, tolerance induction against pig xenografts and survival of α1,3Gal-T KO pig kidney xenografts were successfully demonstrated in NHPs. Pig thymus was xenotransplanted to a humanized mouse, which a human immunity is pre-established by human fetal thymus implantation via the kidney capsule and $CD34^+$ hematopoietic stem/progenitor intravenous (i.v.) injection.

Human T cells were depleted using by a CD2-specific mAb after human thymic graft removal. In these humanized mice, pig thymic grafts could allow development of human T cells and thymopoiesis, and tolerance induction of human T cells to pig xenoantigenic epitopes. These mice-derived human T cells do not respond to the pig stimulators matched with thymic donor SLA molecule. Pig-specific xenoreactive Abs are not produced in these humanized mice, while Abs levels increase in pig thymus graft-rejecting non-tolerant animals. Long-term pig thymic graft acceptance and donor-specific tolerance are observed in humanized mice with human thymic grafts removed by CD2-specific mAb injection. Pig thymus transplanted into humanized mice with a pre-established lymphohematopoietic system of human shows the thymopoiesis of human in pig thymic grafts in the humanized mice. Thus, lymphoid progenitors of human show the BM migration to the pig-grafted thymus, and this enables *de novo* human thymopoiesis. This indicates that pig thymus induces the human lymphoid progenitor recruitment and development of T cells. In addition, pig thymic transplantation can prevent the non α1,3Gal-specific Abs synthesis in the humanized mice.

Anti-non α1,3Gal Abs synthesis is stopped upon pig thymus transplantation in tolerant animals that had the Abs before pig thymic transplantation. Thus, the thymic xenotransplantation-induced T-cell tolerance contributes to the lost xenospecific Abs producibility by T-cell dependency. The human natural Abs-generating B cells are also tolerant by help of pig thymus transplantation. In α1,3Gal-defected mice, pig thymic xenotransplantation-mediated T-cell tolerance don't tolerize α1,3Gal-specific antibody-producing B cells, but inhibits anti-α1,3Gal Abs production after thymus xenotransplantation. A question is raised whether pig thymic transplantation induces tolerance of human natural Abs-generation of B cells? Differently to T cells present in the tolerized humanized mice with donor SLA-specific unresponsiveness, anti-donor pig Abs are not produced in the mice. Likely, anti-donor pig Abs are present in pig thymic grafts-rejecting humanized mice. Human anti-pig Abs predominantly bind to highly conserved non-MHC antigenic peptides and SLA protein-derived peptides, as shown in pig-to-baboon xenotransplantation [145]. As the humanized mouse model for applicable in vivo system, human anti-pig response of xenoimmunity and immunosuppression or tolerance induction is desired. Pig thymic transplantation is thus an efficient technology for the clinical xenograft tolerance induction. Although the issue that circulating progenitors of human lymphoid lineage cells are capable of migration to a pig thymic organ likely in human thymus is not sure. Pig xenoantigen-specific tolerance of human T cells is obtained by thymus xenotransplantation in humanized mice as a potentially inducing tolerance in clinical xenotransplantation. Pig thymic xenotransplantation provokes donor tissue-specific tolerance in immunocompetently-humanized mice. Thus, it has been supported that tolerance is induced in clinical xenotransplantation. Therefore, the humanized mice model shows the pig thymal xenotransplantation possibility to raise tolerance of human T cells in humanized mice with a pre-established human immunity.

Finally, the broad spectrum to avoid T-cell-mediated immune responses is another strategy. HO-1 is known as a ubiquitous stress protein upon endotoxin, heat shock, hemolysis, inflammation and oxidative stress, and it is also an antioxidant, anti-apoptotic, and anti-inflammatory protein in many cells [146]. To avoid T-cell-mediated immune responses, the HO-1 is often used. In order to overcome such xenograft survival failure mediated T cells, several trials are made. For example, human HO-1 overexpression or soluble TNF-α receptor type I with human IgG1 Fc (sTNF-αR-hFc) overexpression in pig islets are reported to enhance survival of islet xenografts. Adult pig islets are humanized with sTNF-αR-hFc, sTNF-αR-hFc/

HO-1 or human HO-1 transduction by cord blood-derived CD34($^{+}$) stem cell of human injection into recipient NOD-scid-IL-2Rγ (null) mouse [147]. sTNF-αR-Fc and HO-1 genes decrease islet apoptotic cell death and raise the prolonged survival of islet xenografts in pig-to-humanized mouse transplantation, being the sTNF-αR-hFc/HO-1 set as the desired choice. Suppressed macrophages and T-cell infiltration is coupled with decreased IL-6, RANTES, and TNF-α expression.

17.9 NK-Cell-Mediated CR

Human NK cells involve in host defense system and also xenoantigenic cytotoxic responses. In 1990, Ljunggren and Kärre suggested that NK cells bind to the MHC C-I protein lost cells or MHC C-I altered cells as target-binding cells [148]. The NK cells are still treated as one of major immune cells that directly contribute to xenotransplantation. The HLA C-I present on the target cells triggers to induce inhibition signaling pathways to NK cells via the killer cell inhibitory receptors (KIR) known as human inhibitory receptors [149]. The two inhibitory receptors include Ig superfamily and the heterodimeric CD94/NKG2 complex as the type II C-type lectin superfamily. Binding of the HLA C-I to the NK inhibitory receptor proteins contributes to an inhibition signal transduction to the downstream of the ITIM phosphorylation. Two distinct HLA C-I molecules in humans inhibit the human NK-cell function: (1) the classical HLA C-I such as HLA-A, -HLAB, and HLA-C molecules bind the Ig superfamily, KIR, as the direct human NK-cell-inhibiting activity and (2) the non-classical HLA-C-I of HLA-G or classical HLA C-I peptides increase expression level of HLA-E known as the second non-classical HLA C-I [149].

Although NK cells are not associated with allograft rejection responses, NK cells of human are involved in pig-to-human xenotransplantation. Pig ECs targeting by human NK-cell-induced xenogeneic cytotoxic activity are caused due to the lacked HLA C-I molecule and also the lacked SLA C-I molecule to transduce inhibitory signalings to human NK-cell-inhibitory receptors [150]. In xenotransplantation, null strains, which lack for conventional killer T-cell markers, or NK cells involve in organ xenograft rejection. Human NK cells show adhesion capacity to target cells and lysis capacity of pig ECs. Therefore, human NK cells are associated with the cellular immune rejection in pig-to-human vascularized xenograft transplantation. The T-cell-derived xenoimmune disease response is also a powerful barrier of host immunity [120]. Because MHC C-I molecules suppress the NK-cell cytotoxic activity, the HLA gene expression in pig ECs is believed to be a potential solution way to overcome NK-cell-dependent cytotoxicity against xenoantigens. The NK-cell-dependent cytotoxicity against xenoantigens is a direct hurdle of the xenotransplantation. Thus, pig xenograft should be protected from polyclonal human NK cells. Pig SLA C-I molecular structures are highly similar to human HLA C-I molecules. The conserved amino acid motif necessary for human KIR interaction lack for the SLA C-I protein amino acid sequences [150]. In humans, the position

77 and 80 of HLA C-I amino acid residues are the KIR recognition region. But such motif is not located in the SLA C-I molecule which is corresponded to human HLA C-I. Hence, although the MHC-I molecules between pig and human ECs are similar, the pig EC susceptibility induces cytotoxic cell death by human NK cells. The human pig-specific xenogeneic cytotoxic activity is derived by NK cells, and the pig-specific xenogeneic cytotoxicity is not caused by T cells. In fact, $CD56^{+}CD3^{-}$ NK cells only show the cell lysis on pig target cells, but the $CD3^{+}$ T cells don't lyse pig target cells [151].

In xenotransplantation, host human TCRs directly recognizes SLA C-I and C-II present in pig APCs and host human T cells indirectly bind to pig donor peptides presented on host human MHC C-II. From the above two different events, the host human T cells are stimulated and consequently, contributing to T-cell-induced cytotoxicity to the pig xenografted vascular endothelium and activation of B cells to synthesize *de novo* antibodies for humoral xenograft rejection [152]. Moreover, the activated T-cell-produced cytokine proteins prime the NK cells and macrophages known as the innate immune cells, and they consequently reject the xenografts. Direct xenogeneic and allogeneic responses are similar, but the indirect response to xenoantigens is much stronger than that of allogeneic responses, needing immunosuppression. Therefore, for successfully long-term-survived xenografts, T-cell-mediated xenoantigenic immune responses must be controlled. From transplantation models, regulatory cells were emphasized for immunological unresponsiveness [153]. This is the transplantation tolerance induction.

The NK cells as lymphocyte subset directly or ADCC typically lyse cells and pathogens if they are recognized as non-self. For lysis, perforin and granzymes are factors. NK-cell adhesion ability to vascular endothelium of donors and lysis ability of target cells via Abs-dependent and Abs-independent action manners are known [154]. This process activates ECs or damaged injury, leading to impaired vascular permeability, thrombosis, and further inflammation. Activated NK cells synthesize chemokines and cytokines to stimulate immune cell infiltration. Subsequently, NK-cell pathway in pig-to-human xenotransplantation is crucial, as human NK-cell-mediated cytotoxicity kills pig ECs and injures human blood-perfused pig cells [155]. The direct cell-killing capacity of NK cells is mediated by specific receptors known as activator and inhibitor receptors that include LIRs, NCRs, KIRs, and KLRs. NK-cell-activating receptors bind to their self-MHC C-I and inhibitory signaling-deficient target cell ligands. Human NK cells fastly infiltrate pig-xenografted organs when they adhere and transmigrate the interstitial space of pig endothelium by CD49, CD11a/CD18, and CD11b/CD18 integrins, and CD99 bindings to pig ligands [156]. Human NK cells activated by pig endothelium produce INF-γ and this cytokine consequently activate T cells. Thus, in order to avoid pig EC lysis by NK cells of human, human classical and non-classical HLA C-I protein transgenic pigs can be created to enable to recognize human NK-cell-inhibitory receptors.

Pig α1,3Gal xenoantigens are not directly related to adhesion and cytotoxicity of NK cells of human; however, NK cells of human respond to the graft glycosylation phenotypes. This is based on the result that α1,3Fuc-T-expressing skin grafts was transplanted in the T- and B-cell-deficient rodent MHC-matched recipients, the skin grafts was acutely rejected by a host rodent NK cells [157]. But CMAH-KO and α1,3Gal-T KO pigs are not problematic matter in xenotransplantation. Apart from the direct xenoantigen recognition, human NK cells reject xenograft by complement-independent ADCC. This property is also general action in the antibody-mediated rejection of allogenic grafts. IgG receptors are predominant in the ADCC, while IgE and IgA-binding NK-cell Fc receptors are also engaged in the ADCC. In the xenotransplantation, Fc receptor for IgG1 subclass known as weak complement activator activates human NK cells, whereas IgG3 known as a strong complement activator do not activate NK cells [158]. Apart from ADCC, NK cells in the spleen reject xenografts in an antibody-driven manner by B-cell binding in marginal zones via the CD40 interaction with CD154 and by stimulation of non-α1,3Gal antibody production that is T-cell-independent. Therefore, blocking of the CD40 interaction with CD154 or splenectomy can eliminate the T-cell-independent Abs synthesis.

The human NK cells attack transplanted pig ECs because human NK-cell-inhibitory receptors don't recognize the pig ECs. Human NK cells have initially recognized for the cancer immunotherapy for more than 5 decades since 1980. Human NK cells specifically bind to xenografts and lyse them. Main receptors of cell surfaces suppress and stimulate human NK cells for target cell lysis, as have long been investigated. They are C-type lectins, killer cell Ig-like receptors and natural cytotoxic receptors. Then, inhibitory receptors of human NK cells can not bind MHC C-I molecules, due to difference between pig SLA and human HLA. A recent pig ECs-killing factor suggested the pig UL-16-binding protein 1 (ULBP1) interaction with NKG2D receptors, where ULBP1 is known as the pig cell primary ligand for the NKG2D receptor of human. Macrophage cells can also be stimulated by xenoantigen-specific T cells or donor endothelial antigens. Recipient macrophages reject α1,3-Gal in pig pancreatic islets and phagocytose pig red blood cells. Integrin-associated protein, CD47 surface antigen as an Ig superfamily can suppress phagocytosis. The SIRPα expressed on the cell surface is a CD47-inhibitory receptor. Binding of CD47 to SIRPα transduces a phagocytotic-inhibitory signals to the macrophage cells. CD47 present on pig EC surfaces is different from human CD47, and thus, pig CD47–human SIRPα interaction has no such inhibiting activity on human macrophages [159]. α1,3Gal-T KO/hCD46 pigs showed retarded macrophage function and phagocytosis [160]. In addition, they also proved that hematopoietic cell hCD47 expression α1,3Gal-T KO pigs elevated the transiently xenogeneic chimerism event in baboon hosts, prolonging grafts survival of pig skin [161] in the hCD47–SIRPα interaction. Enforced expression of human natural extrinsic coagulation pathway inhibitor in pig cells elevated the hCD47-SIRPα function and efficiency of signaling of hCD47-SIRPα complex and consequently leading to macrophage cells-derived immune rejective response [162].

17.10 Inhibition of Human NK-Cell Function by HLA-G and HLA-E

In contrast to classical HLA C-I, HLA-G exhibits the lowered polymorphism in its sequence and HLA-G expression shows seven distinct isoforms of 4 plasma membrane-anchored proteins ranged from HLA-G1, HLA-G2, HLA-G3, and HLA-G4 form as well as 3 soluble proteins ranged from HLA-G5, HLA-G6, and HLA-G7 forms. The HLA-G is originally described in relation with tolerance between mother's maternal immunity and fetus and also in acceptance of allografts as well as tumor escape from immunity. Thus, expression levels of HLA-G proteins are closely correlated with better survivals of allografts, reduction in autoimmune reactivity, and tumor immune escape from host immune surveillance [163]. The two forms of non-classical HLA-G form and HLA-E form as ligands for NK-cell-inhibitory receptors exhibit very limited polymorphism and HLA-G/E-expressing pig AECs activate inhibitory signals for protection of pig-grafted organs and cells from NK-cell-mediated cytotoxic cell death of human. HLA-E form raises inhibitory signalings to human NK cells to give indirect-inhibitory signalings received from HLA-A, HLA-B, HLA-C, and HLA-G forms to the human NK cells. The indirect-inhibitory signaling pathway is derived from the results that the HLA-G form induces the indirect presentation by HLA-E form from HLA-G peptides [164].

The classical HLA-I-expressing pig ECs are protected from a human NK-cell subset based on polyclonal human NK cells. For example, a Cw3 allele of the HLA class I-expressing pig cells partially protect from the human NK cells because the HLA-G-originated peptides increase the HLA-E expression and consequently negatively transduce to human NK cells. Both of two HLA-E form and HLA-G form induces the direct inhibition of human NK cells without other endogenous HLA-I molecules, preventing xenograft rejection. HLA-G sequence is minimally polymorphic among HLA C-I molecules and HLA-G has a wide-ranged NK-cell-inhibitory property. The HLA-G also protects the fetus from aggressive attacks of the mother's maternal immunity [165]. HLA-G also protects the transplanted and grafted tissues via all immune effector inhibition and their mediated rejection of grafts. The HLA-E expression or HLA-G expression without other HLA-I molecules protect the pig ECs from the immune attacks mediated by polyclonal NK cells of human. Although the inhibition mechanism underlying the two non-classical HLA C-I molecule actions is specific, such direct inhibitory activity of HLA-G form against human NK cells still remains unclear. Therefore, the direct HLA-G concept to suppress human NK cells is controversial.

To overcome the xenotransplantation hurdle, artificial abrogation of human NK-cell activation by pig cells activating ligand elimination or ligand molecule expression used for human NK-cell-inhibitory receptors are needed. Pig cells produce SLA molecule, but the swine HLA-E molecular analogs are deficient for the interacting capacity with human NKG2A/CD94. Thus, the NKG2A/CD94 inhibitory signaling on the circulated human NK cells is not induced. Activation of human NK cells is suppressed by interaction between HLA-E and the NKG2A/B-CD94 complex [166], and the HLA-E interaction with NKG2A protects the ECs and other-related cells from human NK-cell-mediated cytotoxicity [149]. But human NK cell's NKp44 or NKG2D receptor-pig ligand ULPB interaction leads to pig cell lysis. But swine MHC-I or leukocyte antigen-I (SLA-I) molecules are not bound with human NK inhibitory receptors. Swine SLA-E form present on pig cells has no capacity to bind to human NKG2A/CD94, consequently, the inhibitory receptor activity of NKG2A/CD94 present on the human NK cells is not functional.

Pig ECs expression of HLA-E form suppresses human NK cell- and macrophages-mediated cytotoxic cell death. HLA-E participates directly in leukocyte adhesion. HLA-E form functions basically through its binding to the receptor complex of CD94/NKG2 of human NK cells. HLA-E expression on human aortic ECs does not alter the NK-cell adhesion level of human to the endothelium. Prevalent-circulating NK cells of human are sequestrated by lungs that express HLA-E form in the organ perfusion *ex vivo* models. HLA-E negatively regulates the human NK-cell activation through binding to the NKG2A/B-CD946 complex. Also, binding of HLA-E to NKG2A protects ECs from cell lysis mediated by human NK cells [154, 167]. HLA-E decreases endothelial injuries and damages as well as EC death via the dominant ligation of NKG2A and CD94 inhibitory receptors and stimulates an anti-inflammation and endothelial sparing binding at the interfaces between blood and endothelium. In addition, HLA-E in part mitigates the inflammatory and coagulatory dysregulation that contribute to dysfunctional xenografts and eventually transplantation failure. HLA-E expressed on pig ECs suppresses human NK-cell- and macrophages-mediated cytotoxic cell death and inhibits the IFN-γ secretion by co-cultured human NK cells [168].

When primary aortic ECs and genetically engineered pig lungs for HLA-E expression were exposed to human bloods, the HLA-E effects on human NK-cell-associated ECs injuries in vitro and xenoantigenic-perfused *ex vivo* pig lungs were studied. NK cells of human in xenogeneic rejection can be restricted by pig transgenic expressed with HLA-G and HLA-E. The HLA-G and HLA-E facilitate the inhibitory receptor functions of human NK cells or macrophage cells in vitro [169, 170]. Maeda group [171] evidenced that HLA-E-expressing pig ECs attenuate the pro-inflammatory cytokine synthesis from the pro-inflammatory macrophage cells. However, total NK-cell reactivity was not eliminated because NK cells express a lot of non-MHC molecules-binding receptors. Similarly, transgenic pigs of human HLA-E/β2 microglobulin genes are also subjected in further application [169].

HLA-G or HLA in human NK cells as an NK-cell inhibitory ligand directly protect sensitive pig cells from human NK-cell-mediated xenogeneic cytotoxicity [151]. HLA-G on pig aortic ECs protect xenogeneic pig aortic ECs from human polyclonal NK-cell-mediated cytotoxic cell lysis when peripheral blood lymphocytes of pigs, having homozygous type of HLA-Cw3 or -Cw4 NK effector cells, are present. However, Fab-2 part of HLA C-I-reactive antibody inhibits such protection, suggesting that HLA-G expression directly protects the pig aortic ECs. From Abs (Fab-2 part of HLA-I-reactive mAb PA2.6)-blocking assay, the HLA-G and HLA-E on pig cells were demonstrated to directly protect it. The protection by HLA-E expression is suppressed by human CD94-specific Abs. HLA-E phosphorylates the

CD94/NKG2 complex forms and consequently, recruit adaptor protein of a specific phosphatase SHP-1 (SH2 domain-containing protein phosphatase-1) to the complexed form. Thus, HLA-E-based protection is mediated in a CD94/NKG2-depended signaling pathway. However, contrastly, HLA-G attenuates NK cells of human in the absent condition of phosphoryl-CD94/NKG2 form or in SHP-1 recruitment condition, because the anti-CD94 Abs inhibit the blocking. Therefore, HLA-G protection is performed by a CD94/NKG2-independent pathway. HLA-E on pig ECs phosphorylates CD94/NKG2 complexed form and activates downstream signaling pathway of human NK cells. Contrastly, however, HLA-G protects pig ECs via a CD94/NKG2-independent signalings. In the absent condition of other non-classical HLA C-I proteins, HLA-G form alone attenuates cytotoxic lysis capacity of human NK cells. HLA-G is therefore considered to protect and overcome grafts from human NK-cell cytotoxicity during cellular rejection of xenografts [172].

The HLA-E needs the leader peptides-derived nonameric peptide binding of the HLA-I alleles [149]. The HLA-E-binding peptide amino acids have the anchor amino acids conserved at the positions 2 and 9, as exampled by -VMAPRTLLL sequence [173]. Change in the conserved Met or Leu causes non-binding and therefore reduces surface HLA-E expression. Notably, the leader peptide of SLA class I (SLA C-I) is non-homologues with the conserved HLA-E peptide sequence. Because HLA-E inhibits xenoreactive human polyclonal NK cells, modified peptides might locate on pig cells. Although classical HLA-I allele-specific receptor-carrying NK cells are limited in quantity, a lot of quantity of human NK cells contain non-classical HLA-I-specific receptors. For instance, the HLA-E-derived tetrameric HLA-I molecule binds many human NK cells. The HLA-I requires the α-chain, β2m, and trimeric peptide complexes in the small organelle ER [174]. α-Chain class Is alone is rapidly degraded. In the condition without other HLA-I expression, HLA-G directly protect from human NK-cell-mediated cytotoxic cell lysis. HLA-G receptor-expressing human NK cells also exhibit the similar results [174]. However, currently, pig HLA-G expression inhibits human NK-cell-mediated lysis without reasonable mechanism. Thus, HLA-G form alone shows a direct induction of inhibitory signalings in human NK cells and HLA-G leader sequence is needed for HLA-E expression, providing a protection strategy of pig xenograft from human NK-cell-mediated cytotoxic cell death. Two non-classical HLA-E and HLA-G forms have a potentially efficient strategy for pig cell protection from human NK-cell-mediated cytotoxic cell lysis [151].

Expression of HLA-E advanced the success in the xenotransplantation research field. For example, α1,3Gal-T KO pigs co-coupled with complement-regulatory hCD46 and hCD55 expression as well as coagulation pathway regulatory hTBM expression prolonged xenografts survival in heart and kidney preclinical xenotransplantation, giving 2 years longer and 9 months longer survivals, respectively. However, pig lung xenografts were injured with damages within several minutes to several hours after human blood exposure despite genetic modification and engineering, presumably due to the incompatible differences between species in complement-regulating and coagulation-regulating pathways. Alternatively, the expressed HLA-E also initiates activation of human NK cells via the NKG2C/

CD94 complexed form. Pig cells transgenic with the inhibitory ligand HLA-E and β2m of human NK cells inhibit injuries of α1,3Gal-T KO.hCD46 pig cells or enhances the prolonged lung functions in two perfusion preclinical model animals. When pig aortic ECs were served in cytotoxic cell death assays of human NK cells, the expressed HLA-E on aortic ECs of α1,3Gal-T KO.hCD46 pigs strongly reduce antibody-dependent human NK-cell-mediated cytotoxicity and antibody-independent human NK-cell-mediated cytotoxicity. As evidenced in the α1,3Gal-T KO.hCD46 lungs of pigs, which were perfused with human bloods *ex vivo*, additive HLA-E expression increases median survival of xenografted lungs longer than 4 h, compared to 162 min, and suppress vascular resistant property, and decrease activation of platelets and elaboration of histamine [154]. The expressed HLA-E is not effective to in vitro the human NK-cell adhesion to pig ECs, sequestration of human NK cells and neutrophils during perfusion of organs. Thus, human NK-cell activation seems to contribute to endothelial injuries and lung inflammations of α1,3Gal-T KO.hCD46 pig organs. HLA-E expressed in α1,3Gal-T KO.hCD46.HLA-E pigs attenuates in vitro human NK-cell cytotoxicity and shows the prolonged survival of ex vivo α1,3Gal-T KO.hCD46 lungs. Adhesion of human NK cells to monolayered ECs and sequestration in lung perfusion are not altered [154].

HLA-E-expressing pig aortic ECs and lungs are protected from cytotoxic cell death and barrier function loss. Monolayered cultures of α1,3Gal-T KO.hCD46 aortic ECs coupled with HLA-E-expression exhibit protection from both cytotoxic injuries caused by antibody-dependent human NK cells and antibody-independent human NK cells. Human NK cells induce xenografted cellular injuries, and the HLA-E expression suppresses human NK-cell-involved cytotoxic cell death, presumably due to HLA-E-inhibitory NKG2A/CD94 signaling. On ex vivo-perfused platformed system, HLA-E expression physiologically protects the lung and increases the prolonged survival time of lings with delayed vascular barrier functional loss. HLA-E expression mitigates early damages of endothelium from human NK cells ex vivo. The resulted reduction in endothelial dysfunctionality decreases inflammation responses in such parameters of histamine release and coagulatory activation triggered by elaborated βTG decrease as well as platelet retention of the lung that expresses HLA-E. Non-NK immune cells of human including mast cells or monocytes as well as platelets are deficient for NKG2A. Therefore, this NKG2A lacking reduces inflammatory responses and activation of platelets is indirectly and secondary effect. In fact, reduction in endothelial damages or injuries allows the surfaces of ECs less adhesive to surrounded platelets. Human NK cells function importantly as an initial starter of activation of platelets and histamine release. Nevertheless, completely elaborated thromboxane and thrombin production despite expression of HLA-E indicates that unknown pathway of pro-inflammatory response is not suppressed even by expression of HLA-E in α1,3Gal-T KO.hCD46 pig lung in a human NK-independent manner.

Human NK cells affect importantly to injuries of pig xenografts and HLA-E has protective effects. HLA-E expressed in xenografts of pig cardiac organs improves survivals of pig xenografts and functional activities in an ex vivo heart perfusion animals [175]. In the amputated pig forelimb xenoperfusion system, which human blood was cannulated and perfused, decreased levels of complement activation, pro-inflammatory cytokine synthesis, and infiltration of human NK cells are detected in HLA-E gene-expressing transgenic pigs [176]. The first direct comparison between α1,3Gal-T KO.hCD46.HLA-E pig lung cells and control α1,3Gal-T KO. hCD46 pig lung cells proves that HLA-E increases survival time and function of pig xenografts in pig-to-human xenomodels. The HLA-E transgenic expression in pig ECs protects the pig ECs in vitro against cytotoxicity of human NK cells and also against cytotoxicity of macrophages [168]. In the transgenic HLA-E expression in α1,3Gal-T KO/hCD46/HLA-E pig lungs, which was triply constructed with α1,3-Gal-T KO/hCD46 pig lungs, the triple HLA-E expression in α1,3Gal-T KO/hCD46/HLA-E pig lungs raised increased survival in terms of pulmonary vascular resistance decrease and platelet activation decrease [177].

In organ xenotransplantation, the HLA-E-expressing cardiac and lung xenografts are also prevented from endothelial damage through inhibition of NK-cell activation and cytotoxicity [178]. The inhibitory ligands such as HLA-E and β2m of human NK cells additionally expressed in pigs suppress cell injury of α1,3Gal-T KO. hCD46 pigs or enhance pig lung function in two different models of preclinical perfusion [179]. HLA-E expressed on α1,3Gal-T KO.hCD46 pig aortic ECs importantly decreases cytotoxicity mediated by antibody-dependent human NK cells and antibody-independent human NK cells in the physiologic shear stress-inducing condition in vitro. Relatively to ex vivo-perfused human blood to α1,3Gal-T KO. hCD46 pig lungs, additively expressed HLA-E enhances median survival time of lungs longer than 4 h, compared to control 162 min, and suppresses pulmonary vascular resistant potentials, and decreases activation of platelets and secretion of histamine. The forced expression of HLA-E is not associated with in vitro adhesion of human NK cells to pig ECs, or neutrophils and NK-cell sequestration. Human NK-cell activation leads to ECs injury and lung inflammation of α1,3Gal-T KO. hCD46 pigs. Expression of HLA-E is involved in physiological and biochemical protection of human blood-exposed α1,3Gal-T KO.hCD46 cells and organs [179]. Recently, HLA-E substituted with S147C has been developed for gene tool of xenotransplantation and synthesized HLA-Ev gene was optimized with a mammalian codon usage for transgenic animal generation [180].

17.11 Neutrophils

The success of xenotransplantation is disturbed by immune reactive system driven to protect the host from xenoantigenic foreign cells. Activation of innate immune system, which includes complement and coagulation cascade, occur at xenogeneic cell sites shortly after xenotransplantation. Human leukocyte adhesion to pig endothelium is also early step triggering the immunological attacks. Successful xenotransplantation is possible by multi-gene editing of the pig organ with the thrombosis-resistant vasculature formation. The anti-xenoimmune problems are comprehensive and representatively due to the innate immunity including Abs, complementation, donor and recipient macrophage cells, neutrophils, monocytes, and NK cells, T cells/B cells as adaptive cells, blood coagulation issue, and inflammation issue raised by IL-6, TNF-α, and CRP, etc. Xenotransplantation is basically failed by rapid innate immune reactions as pig xenografts are exposed to human blood. The rapid reactions are known for thrombotic and inflammatory responses, representatively by the IBMIR, where complement activation and coagulative cascade with activation of platelets and graft surface interaction as well as leukocyte adhesion to grafts causing for grafts disruption [181]. The vascular endothelium of xenografts comprises of the donor and recipient interfaces, and the endothelium is the basic target site for primary immune response. Endothelium activation increases surfaced cellular adhesion molecules (CAMs) expression to trigger the recipients' leukocyte adhesion to the endothelium of donor tissues. Recipient adhesion of leukocytes to the pig endothelium initiates the graft immune attack of host.

In the circulatory leukocytes, the most abundant population is neutrophils and the neutrophils infiltrate at the initial stage of inflammation and transplantation of solid organ xenografts to the host human [182]. Neutrophils initiate the acute inflammatory responses and release pro-inflammatory mediators and producing ROS. Then, additional inflammatory cells are invaded and amplified toward the immune responses. The deleterious neutrophils function is also known in IBMIR upon xenotransplantation [183]. Moreover, exposure of human blood to pig aortic ECs causes the increased CAMs expression for ICAM-1, ICAM-2, VCAM-1, and E-selectin, depending on the complement activation. Vasculature recruitment of leukocytes to xenografts influences the transplantation direction. Rolling and adhesion of the recipients' leukocytes to pig endothelium are induced by interaction of cross-species. Indeed, E-selectin of pig binds to the PSGL-1 expressed on neutrophils. Likely, pig ICAM-1 and ICAM-2 recognize β2 integrins. Pig VCAM-1 recognizes the β1-integrin VLA-4 expressed on monocytes [184]. The pig VCAM-1 and E-selectin are complement-dependently expressed in different xenotransplantation models [185].

Complement activation upon human blood exposure to pig endothelium modulates the adhesive leukocyte behaviors to the pig endothelium. Also, the specific β2-integrin Mac-1 suppression attenuates leukocyte adhesiveness. C3 inhibition of complement activation decreases the Mac-1 expressed on the neutrophil surfaces. Thus, inhibition of complement by complement C3 inhibitors such as compstatin analog Cp40 abrogates the leukocyte adhesiveness and the neutrophil attachment to pig endothelium. In addition, C3 inhibitor Cp40 inhibit the pig aortic EC activation and leukocyte activation, since expression levels of ICAM-1, ICAM-2, VCAM-1, and E-selectin known as the adhesion molecules on pig aortic ECs and CD11b as the surfaced integrin expressed on neutrophils are downregulated. Inhibition of CD11b also decreases the leukocyte adhesion. Depending on the symptom progression and transplantation time, the neutrophils are decreasingly located the inflammation sites and grafted organs, instead, macrophages and T cells are increasingly accumulated in the inflammation organ sites. Therefore, the neutrophils are replaced by the macrophages and T cells during the responses. The binding levels of human neutrophils to naïve xenoantigenic ECs are higher than those of the allogenic ECs.

Inflammation status with pro-inflammatory cytokines increases the binding capacity of neutrophils to pig ECs. Increased chemotaxis and homing of human neutrophils are observed when they are encountered with pig ECs. Therefore, human phagocytic immune cells can easily distinguish the surfaces of xenoantigenic cells and allogenic cells. Recognition and homing migration upon interaction with xenoantigenic cellular surfaces are α1,3Gal-independently carried out.

Because the masked endothelial α1,3Gal epitope by α1,3Gal-specific Abs and α1,3Gal-T KO islets cannot prevent the infiltration of neutrophils to the xenoantigenic grafts [186] and neutrophil binding to xenoantigenic cells is increased when LFA-1 (or CD11a), Mac-1 (or CD11b/CD18, CR-3 or iC3b receptor) known as the β2 integrins [187] are bound with their ligands. The iC3b receptor on neutrophils regulates complementation to potentiate the adhesion to xenoantigenic cells. Upon solid organ and non-vascularized xenotransplantation, inflammation is associated with the iC3b receptor expression in neutrophils. During leukocyte adhesion and activation in islet xenotransplantation, complementation has been known to activate in IBMIR in islet xenografting [188]. Strong iC3b accumulation is found in decellularized porcine heart valves when they were incubated with human sera, contributing to adhesion capacity of polymorphonuclear leukocytes [189]. The above two integrins-specific antibodies suppress neutrophil adhesion in vitro. As α4 integrin subunits-membrane organizing and immune synapse forming protein, CD82 on naïve neutrophils regulates neutrophil to both α1,3Gal epitope-positive and α1,3Gal-T KO ECs adhesion [190]. Therefore, anti-complement approach can be used for blocking of the initial inflammatory injury and damage. Accordingly, C3 complement inhibitors have been developed, as compstatin derivative Cp40 downregulate the iCb3 receptor α-integrin subunit expression on neutrophils and neutrophil to endothelium adhesion [191]. The neutrophil interaction with ECs increases in the endothelial P-selectin and VCAM-1 expression levels. Among them, VCAM-1 potentiates human NK cells' cytotoxicity, with secreted reactive oxygen metabolites and pro-inflammatory IL-1α, IL-β, IL-6, and IL-8 cytokines in neutrophils [192].

Moreover, human neutrophils bind to pig cells to induce injuries and damages of xenografts. Knowledges and understanding of the natural immunity involved in xenoantigenic reactivity, concentrating on clinical xenotransplantation in forms of pig-to-human or to NHP systems, will propose strategies to overcome the barriers.

17.12 T-Cell-Co-Stimulation Blockade in Overcoming Xenotransplantation

Currently immunosuppressive agents have been used to reduce graft rejection. However, such immunosuppression consequently leads to increased infection susceptibilities and cancers due to defected immune surveillance. In addition, the immunosuppression agents show detrimental off-target activities. In fact, tacrolimus and cyclosporine as the immunosuppressive drugs have renal and islets toxicities [193]. Therefore, the only immune system-specifically targeting agents without immunosuppression are desired. Genetic mice editing to synthesize its own immunosuppression agent is known to help immunosuppression of human T cells against allograft or pig xenograft rejection [194]. Such immunosuppressive "drugs" are recently known as co-stimulation blockades or T-cell depletors, which are being considered importantly from the studies on co-stimulation to alloimmune response. Currently, blockades of co-stimulatory molecules are believed to be promising immunosuppressive drugs to prolong graft survival after allo- and xenografts transplantation [195–197].

The co-stimulation concept is originated from the comparative xeno- and alloimmune response study in the late 1960s, giving the distinct genetic relationship between donor and recipient. In the species compatibility, the allograft response is stimulated and the allograft-stimulating signal is expanded to the co-stimulation signal in the T-cell response. Co-stimulation of T cells occurs by a complex interplay of multiple pathways. Among the co-stimulatory pathways, CD28-CD80/CD86 pathway is the most precisely known. Constitutive CD28 expression is present on naïve T cells, but another CD40-CD40L pathway is rather inducible. Even though CD28, CD40L or LFA-3-blocking molecules are explored, they are not commercialized yet because of uncertain issues in human trials [198]. In animal and non-NHP models, successful engraftment was reported using CTLA4Ig with minimal conditions [199]. As the first clinical administration, a co-stimulation blockade named CTLA4Ig was used to BM co-cultured with leucocytes of recipients to form alloantigen-reactive anergy state with combined conventional prophylaxis of GVHD [200].

The two representative pathways of CD28-B7 (CD80-CD86) interaction and CD40-CD154 interaction are well known to be associated with both allo transplantation and xenotransplantation. In the co-stimulatory molecular blockades, CD28/B7 blockades are representatively known for human cytotoxic T-cell-associated protein 4-Ig (CTLA4-Ig) or belatacept. CD28 and CTLA-4 as co-stimulatory cell surface signaling molecule and co-inhibitory signaling molecule of cell surfaces recognize CD80-CD86, critically functioning for initiation and regulation of immune responses, respectively [201]. Therefore, cell surfaced CTLA-4 is an inhibitory molecule and inhibits CD28 behavior on APCs. Their ligand is CD80/86. This is important in the transplantation and CTLA4-Ig is developed. The fusion protein CTLA4-Ig recognizes CD80 and CD86 and inhibits CD28 ligation. Belatacept is a CTLA4-Ig molecule with high-affinity. CTLA4-Ig is a CD80/86 blockade agent to prevent T-cell co-stimulation. As a co-stimulation blockade, a belatacept allowed for survival to improve in both cardiac and renal kidney xenotransplantation model animals [132]. Belatacept and CD40/CD154 pathway blockade are classified to the so-called T-cell-suppressing agents. CD154 (or known as CD40L) is a known CD40 ligand. Indeed, in pig-to-baboon xenotransplantation model of heart/artery patch, such co-stimulation blockades prevent a T-cell immune response. Anti-CD4 or CTLA4Ig expression at the graft sites prolongs graft survival in an allograft model. The local administration of immunomodulatory agents can also protect grafts

[202]. However, although it is a rare case in the kidney transplantation with belatacept, high acute immune and cell-mediated immune rejection incidences were still observed due to heterologous immunity [203].

On the other hand, CD40, known as CD154-binding ligand, is a TNF receptor (TNFR) superfamily member. CD40 is associated with inflammation and broad immune responses, including B cells and DCs activation, B cell and DC differentiation and proliferation, class-switching of Ig isotype, development of memory B cells, and formation of germinal centers [204]. The CD40 and CD154 signaling regulates activation of B cells by T-cell-dependent or T-cell-independent way, leading to the in increased responses of humoral immunity including the donor-specific Abs generation [204]. Particularly, soluble CD154 form (sCD154) contributes to T-cell-independent allograft rejection. sCD154 form is derived from platelets known as a major source. The sCD154 protein binds to CD40 present on ECs to upregulate cellular adhesion molecules (CAMs) [204, 205]. The CD40 expression is constitutive in APCs of B cells, DCs, eosinophils, macrophage cells, monocytes and platelets, an also in non-hematopoietic cells including ECs and fibroblasts [206]. Activated $CD8^+$ T cells highly express CD40 molecule.

In addition, although CD154 (CD40L) protein expression is present on platelets, two cell subsets of activated B cells and activated $CD4^+/CD8^+/\gamma\delta$ T cells also express the CD154 (CD40L) protein. During inflammation, CD154 protein expression is upregulated in basophils, mast cells, monocytes, and NK cells and also in non-hematopoietic ECs. Therefore, CD40-specific Ab blockade is no effective on the activated $CD4^+$ T cells unlikely to the CD154-specific Ab blockade. Moreover, interaction between sCD154 and CD40 present on B cells activates formation of germinal centers, activation of B cells, and production of anti-donor IgG Abs. Thus, CD40/CD154 signaling is crucial in transplantation immunity. CD40/CD154 pathway-targeting CD154-specific mAbs causes thrombosis complications in a CD40-independent way. Thus, CD40-specific blocker, compared to CD154-specific blocker, is suggested as an alternatively effective target. Several anti-CD40 mAbs including 3A8, 2C10, Chi220, 5D12, and 4D12 are such cases without thromboembolic complication [196–199]. However, some anti-CD40 mAbs such as Chi220, 3A8 develop adverse effects as B-cell agonists or B-cell depletors [199]. Among them, clone 4D11 (IgG4 isotype) is reported to give the prolonged survivals of hepatic, islet, and renal allografts in cynomolgus monkeys, currently giving a Phase I trials in clinical renal transplantation. Recently, a CD40-specific mAb (2C10R4) was dosed to cardiac xenotransplantation in transgenic α1,3Gal-T KO pig with ATG and CD20 mAb-derived induction and anti-CD40 prolonged survival [207]. The anti-CD40 blockade suppressed both cellular and humoral alloimmune responses in a hepatic NHP allotransplantation model. In the liver model, the anti-CD40 mAb inhibits synthesis of non-α1,3Gal anti-pig cytotoxic Abs. The immunomodulatory aCD40 antibody suppressed anti-pig antibody elaboration, thrombotic microangiopathy in the grafted organs and systemic *CC* incidence [208]. Consequently, the treatment with αCD40 (2C10R4)-specific Abs prevents immune rejection of xenografts and allowed the prolonged survivals of grafts beyond 2 years.

During anti-CD40 Abs treatment, immunologic tolerance induction is not well succeeded, donor-specific hyporesponsiveness is detected, and sustainable immune-regulation is also observed in the T- and B-cell subpopulations without consumptive coagulopathy and thrombotic microangiopathy. B cells don't synthesize Abs after anti-CD20 antibody treatment. The anti-CD40 antibody sustains suppression of B-cell function because IgM-promoted Abs synthesis of B cells in vitro and pig-specific Abs elaboration is in vivo undetectable in the serum. For the immunomodulation, anti-thymocyte globulin (ATP) and anti-αCD20 antibody are used [209] because the co-stimulation signaling of CD40-CD154 complex is important in the immunopathogenic rejection of xenografts. Like hCD46, similarly, the immunosuppressive anti-CD154/CD40 antibody and anti-CD20 antibody prolong xenograft survival in baboons [210]. The pig α1,3Gal-T KO.hCD46.hTBM strain and a CD40-specific antibody (2C10R4)-treated immunomodulation gave prolonged graft survival, giving tolerance in baboons. A non-depleting primatized anti-CD40 mAb (2C10R4) is a co-stimulation pathway-blockade [211]. Some immunosuppressive CD20-specific Abs cause for depletion of B cells and ATG reduces the T-cell proliferation level. Anti-CD40 antibody as a co-stimulation blockade acts as suppressive agent for both B cells and T cells and also a complement activity inhibitor. CD20-specific Abs deplete B cells and improve xenograft stability and recipient survivals when co-stimulation is blocked by blockade. Blockade treatment, targeting for the CD40/154 pathway combined with ATG and CD20-specific Abs [212], α1,3Gal-T KO.hCD46.hTBM xenografts of heart with immunomodulatory αCD40 gave long xenograft survival. aCD40-specific antibody mediates pig-specific maturation of B cells and the transgenic hTBM reduces the interspecies coagulation incompatibilities to prolong xenograft survivals. Removal of CD40-specific Abs resulted in graft loss. As an important role for hTBM, the CD154-CD40-specific co-stimulation blockade Abs inconsistently protect α1,3Gal-T KO.hCD46 xenografts from thrombotic microangiopathy with survival of grafts. Interplay between coagulation, complement, pig-specific Abs production, and T cells is suggested to be a rejection reason of xenografts. At present, non-depleted CD40-specific Abs are found at the late stages of preclinical development [209].

Currently, belatacept or anti-CD154 is known as the most effective immunosuppressive agents that block T-cell co-stimulation. Belatacept is currently the sole approved co-stimulation blockade agent for transplantation, as reported in NHP allotransplantation models [213]. Despite pig-specific low titer of Abs, belatacept-received xenograft recipients undergo rejection outcome at early stage of post-transplantation. Despite the belatacept protection effects, a high rejection rate is reported in belatacept-administered patients [214]. Belatacept recognizes both CD80 and CD86 receptors, blocking their bindings to the T-cell co-stimulatory CD28 molecule. The caused recognition limits CTLA-4 binding and consequently leads to co-inhibitory signalings. The caused imbalanced co-stimulatory and co-inhibitory signalings leads to induction of differentiation of T cells and T-cell function, especially to distinct T-cell subpopulations such as Treg sub-populations that depends on the co-inhibitory signalings. Also, the belatacept depletes $CD4^+$ T-cell population, although the deplete extent is lower than that with anti-CD154. Another promising candidate is a type I surface glycoprotein CD2 is T-cell adhesion molecule, also known as a co-stimulation molecule. All human T cells including both CD4 and CD8 abundantly express CD2 and the CD2 is expressed during T-cell activation [215]. CD2 functions as a co-stimulation molecule of T cells as an adhesion molecule. Human CD2 interacts with CD3 in a cis way [216] and LFA-3 (known as CD58) in trans way. Human CD2 also binds to sheep LFA-3. LFA-3Ig (Alefacept) is an immunosuppressive drug for transplantation [215], because LFA-3Ig is known to protect islet xenograft from T cells. By binding to the memory T-cell CD2, LFA-3Ig alefacept depletes memory T cells.

Although only human NK cells and T cells express CD2, B cells in mouse express CD2 molecule. Anti-CD2 targets and suppresses T cells in immune rejection. Anti-CD-2 monoclonal Ab recognizes CD2 in primates including human. Therefore, CD2-specific Abs are expected to deplete T cells or block T-cell co-stimulation. The newly developed mAb, CD2hb11, depletes T cells and suppresses activation of T cells without activation activity [217]. Prior to anti-CD2 Ab CD2hb11-transgenic pigs, CD2hb11-expressing adenoviral vector has been established for pig tissue transduction [218]. The CD2-specific Ab, CD2hb11-coding gene has been cloned and fused for a human IgG3 Fc-fused chimeric Abs and the humanized Abs have mouse complementarity-determining regions (CDR) only. Human-produced IgG3 interacts with complement and FcR. The anti-human CD2-expressing pig neonatal islet cell clusters, which are gene edited for modification by adenovirus, synthesize and secrete human CD2-specific Abs and protect them from T-cell-mediated cytotoxicity. Anti-human CD2 antibody-based systemic immunosuppression is likely to mimic anti-thymocyte globulin induction and be beneficial in clinical transplantation. This does not cause immune suppression, too. Diliximab is an anti-human CD2 antibody and diliximab-generating xenografts decrease infiltration of T cells, and this represents a key step to avoid and overcome the immune hurdle and barrier during xenotransplantation.

Development Tregs critically acquire immune tolerance after transplantation because Tregs are associated with reduced alloreactivity, although CTLA4Ig effect on Tregs is controversial [219]. Two different results of the prolonged survival of grafts in MHC-different mismatching mice as well as the survival time reduced in MHC C-II-mismatched mice are attributed to Treg-dependency [220]. In addition, CD28 and CTLA4Ig ligation is crucial on the T cells, because $CD4^{+}CD25^{-}$ T cells are converted to the differentiated $CD4^{+}CD25^{+}FoxP3^{+}$ subpopulation during CTLA4Ig exposure [221]. CD28 ligation and CTLA4 expression on T cells are necessary for Tregs [219, 222]. Therefore, in full MHC-mismatched model mice, sirolimus combined with co-stimulation blockade can contribute to successful engraftment of donor organs.

17.13 Surface Modification of Xenograft Endothelial Cells to Improve the Compatibility with Human Blood

In the pig liver, which human blood has been subjected to extracorporeal perfusion, pig Kupffer cells binding to human erythrocyte NeuAc residues destruct most of them. This is a critical outcome of the xenotransplantation. The pig sialoadhesin-specific mAb addition decreases the human erythrocyte loss and prolongs pig liver function [223]. In endothelium, the cells are protected by membrane regulatory proteins and the multicomponent layered proteoglycans and glycoproteins named glycocalyx. Such endothelial- and serum molecules integrate into the cell layers. For example, chondroitin sulfate (10–20%) and soluble heparan sulfate (50–90%) are abundantly integrated poly- and oligosaccharides in the glycocalyx. During the transplantation, the grafted vascular endotheliums are activated, damaged, and injured, and consequently, the glycocalyx are shed to cause a pro-inflammatory and pro-coagulating condition [224]. For example, the pig vascular endothelium of xenografts is sensitively responded to human blood, compared to that of human allografts due to pig-specific Abs and molecular incompatibilities for inflammation and coagulation. Although α1,3Gal-T KO technology gave the improvement of the pig xenograft survival in hosts, the endothelial activation issue and the glycocalyx shedding are not explained yet. Heparin, a plasma natural sulfated polysaccharide, is potentially anti-coagulative and anti-inflammatory, as frequently utilized for prevention of blood clotting and coagulation during operation. Its anti-coagulative capacity is in thrombin inactivation and antithrombin-caused activated factor X [225]. The heparin limitation over the antithrombin binding is that it binds to plasma proteins positively charged and platelets-derived proteins and ECs. This leads to anticoagulation named heparin resistance phenomenon.

Therefore, the endothelium function-modifying compounds by direct binding to the living human EC surfaces are desired toward the improved blood compatibility with the inhibited coagulation and inflammation. For example, a macromolecular heparin chain has been modified to Coraline-heparin conjugate (CHC) as the case. CHCs immobilized to surfaces strongly improve the biomaterial surface blood compatibility. CHCs are also coated on layered aortic ECs of pigs that protect themselves from the coagulant activation and inflammatory responses caused environmentally by xenotransplantation. CHCs-coated glycocalyx-like cell layers attenuate xenotransplantation-induced complementation activation. In addition, CHCs-coated wild-type aortic ECs and α1,3Gal-T KO.hCD46.hTBM aortic ECs of pigs xenografts suppress coagulation activation and inflammation responses.

17.14 Human A20 Prevention of Hypoxia Raised by Innate Immune Response

Hypoxia occurs due to the loss of vascularization and once engrafted organs are immune-mediated xenograft rejected requiring relevant immunosuppression. For example, vasculature is completed lost when islets are isolated and cultured. Such successful preparation and culture of isolated islet cells rely on environments-diffused oxygen and nutrients from blood. The lasted hypoxia status leads to early graft loss and apoptosis via hypoxia inducible factor (HIF)-1α-target genes. HO-1 or A20 transgenic expression mitigates from inflammation. The transgenic human HO-1 pigs importantly protect aortic ECs of human HO-1 transgenic pigs from TNF-α-induced apoptotic cell death and increase the prolonged human HO-1 transgenic kidney survival in human blood xenoperfusion system *ex vivo* [226]. The human HO-1 functions for anti-apoptosis and anti-inflammation are exerted by catabolic conversion of cytotoxic heme molecule to its catabolites such as biliverdin, CO (carbon monoxide), and iron ion. The CO molecule is served as a signaling inducer that activated the p38 MAPK signaling to upregulate anti-apoptosis and anti-inflammation activities. Free iron ions upregulate ferritin, an iron storage, that has a protection activity of cells from oxidative damages of cell membranes.

The human HO-1 attenuates the detrimentally converted microenvironments caused by immune rejection responses of xenografts. Human A20 (hA20) is also initially known as a TNF-α-induced protein 3 and a Zn-finger protein in cytoplasm, which prevents inflammatory responses via functional roles of activation blockade of NF-κB and suppression of TNF-α-induced apoptotic death. Transgenic hA20 pigs exhibit immunomodulatory activities in aortic ECs by reduced susceptibility to CD95L (or known as FasL)-induced apoptotic cell death [227]. Therefore, two independent xenoprotective genes of hHO-1 and hA20 exhibit the protection activities of grafted tissues and organs from ischemic damages and injuries. In addition, the hHO-1-catabolized CO molecule attenuates EC apoptotic cell death via the p38 MAPK signaling activation as well as CO stimulates vasodilation and suppresses activated guanylate cyclase-mediated platelet aggregation. Furthermore, both hA20 and hHO-1 molecules suppress NF-κB-activating signaling that inhibits apoptotic cell death and pro-inflammatory molecule synthesis. The triple gene-edited GGTA1-KO/hHO-1/hA20 in pig strains are assayed using ex vivo human blood-perfused kidney system. The triple gene edited GGTA1-KO/hHO-1/hA20 in pig organs alleviated ischemia-reperfusion injuries and damages as well as grafts rejection in pig xenografted organs [228]. The triple gene edited GGTA1-KO/hHO-1/hA20 organs of pig strains can prove potential clinical relevance in pig-to-NHP xenotransplantation.

The donor pig genetically modified to improve hypoxia by human A20 transgene protected it from apoptosis and inflammation, although the human A20 gene is expressed in only heart, skeletal muscle and pig aortic ECs with unknown reason. The hA20-transgenic pig aortic ECs protect the cells from TNF-α-derived apoptotic death and CD95 (Fas)L-induced apoptosis. hA20-transgenic pig cardiomyocytes partially protected ischemia/reperfusion. Thus, hA20 is regarded as hypoxia- and innate immune-protection molecule in xenotransplantation. Another human HO-1 is a cytoprotecting enzyme by scavenging ROS and an anti-apoptotic molecule from inflammatory stress in pig islet-like cluster cells [229]. CRISPR/Cas9 system generated human cells defecting for manganese superoxide dismutase (MnSOD) [230] that remove the hypoxia-produced oxygen radicals upon xenotransplantation. The investigation suggests the importance of gene edited pig strains genetically modified by recent CRISPR/Cas9 editing strategy. To protect against hypoxia is important in pig to human xenotransplantation.

References

1. Ierino FL, Sandrin MS. Spectrum of the early xenograft response: from hyperacute rejection to delayed xenograft injury. Crit Rev Immunol. 2007;27:153–66. https://doi.org/10.1615/CritRevImmunol.v27.i2.30.
2. Lin CC, Ezzelarab M, Shapiro R, Ekser B, Long C, Hara H, Echeverri G, Torres C, Watanabe H, Ayares D, Dorling A, Cooper DK. Recipient tissue factor expression is associated with consumptive coagulopathy in pig-to-primate kidney xenotransplantation. Am J Transplant. 2010;10:1556–68. https://doi.org/10.1111/j.1600-6143.2010.03147.x.
3. Cowan PJ, Cooper DK, d'Apice AJ. Kidney xenotransplantation. Kidney Int. 2014;85:265–75. https://doi.org/10.1038/ki.2013.381.
4. Matsuda D, et al. Thrombotic microangiopathy caused by severe graft dysfunction after living donor liver transplantation: report of a case. Clin J Gastroenterol. 2014;7:159–63. https://doi.org/10.1007/s12328-013-0446-2.
5. Jiang X, et al. Graft microvascular disease in solid organ transplantation. J Mol Med. 2014;92:797–810. https://doi.org/10.1007/s00109-014-1173-y.
6. Sakamoto K, et al. Hematopoietic stem cell transplantation-associated thrombotic microangiopathy accompanied by renal arteriolar C4d deposition. Pediatr Transplant. 2015;19:E97–100. https://doi.org/10.1111/petr.12450.
7. Schuurman HJ. Pig-to-nonhuman primate solid organ xenografting: recent achievements on the road to first-in-man explorations. Xenotransplantation. 2016;23:175–8. https://doi.org/10.1111/xen.12244.
8. Janowski M, et al. Neonatal desensitization does not universally prevent xenograft rejection. Nat Methods. 2012;9:856–8. https://doi.org/10.1038/nmeth.2146. author reply 858.
9. Jiang X, Sung YK, Tian W, Qian J, Semenza GL, Nicolls MR. Graft microvascular disease in solid organ transplantation. J Mol Med (Berl). 2014;92(8):797–810.
10. Chen RH, Kadner A, Tracy J, Santerre D, Adams DH. Differential detection of alpha-gal and human CD59 molecules on pig-to-primate cardiac xenotransplantation: a marker of delayed xenograft rejection. Transplant Proc. 2001;33(1–2):732–5.
11. Ahrens HE, Petersen B, Herrmann D, Lucas-Hahn A, Hassel P, Ziegler M, Kues WA, Baulain U, Baars W, Schwinzer R, Denner J, Rataj D, Werwitzke S, Tiede A, Bongoni AK, Garimella PS, Despont A, Rieben R, Niemann H. siRNA mediated knockdown of tissue factor expression in pigs for xenotransplantation. Am J Transplant. 2015;15(5):1407–14.
12. Zhang J, Xie C, Lu Y, Zhou M, Qu Z, Yao D, Qiu C, Xu J, Pan D, Dai Y, Hara H, Cooper DKC, Ma S, Li M, Cai Z, Mou L. Potential antigens involved in delayed xenograft rejection in a Ggta1/Cmah Dko pig-to-monkey model. Sci Rep. 2017;7(1):10024.
13. Knosalla C, Yazawa K, Behdad A, Bodyak N, Shang H, Bühler L, Houser S, Gollackner B, Griesemer A, Schmitt-Knosalla I, Schuurman HJ, Awwad M, Sachs DH, Cooper DK, Yamada K, Usheva A, Robson SC. Renal and cardiac endothelial heterogeneity impact

acute vascular rejection in pig-to-baboon xenotransplantation. Am J Transplant. 2009;9:1006–16. https://doi.org/10.1111/j.1600-6143.2009.02602.x.
14. Ekser B, Ezzelarab M, Hara H, et al. Clinical xenotransplantation: the next medical revolution? Lancet. 2012;379:672–83.
15. Ezzelarab M, Ekser B, Azimzadeh A, et al. Systemic inflammation in xenograft recipients precedes activation of coagulation. Xenotransplantation. 2015;22:32–47.
16. Barrett TD, Hennan JK, Marks RM, Lucchesi BR. C-reactive-protein-associated increase in myocardial infarct size after ischemia/reperfusion. J Pharmacol Exp Ther. 2002;303:1007–13.
17. Ezzelarab MB, Cooper DKC. Systemic inflammation in xenograft recipients (SIXR): a new paradigm in pig-to-primate xenotransplantation? Int J Surg. 2015;23(0 0):301–5. https://doi.org/10.1016/j.ijsu.2015.07.643.
18. Esmon CT, Xu J, Lupu F. Innate immunity and coagulation. J Thromb Haemost. 2011;9(Suppl 1):182–8.
19. Cowan PJ, Robson SC, d'Apice AJ. Controlling coagulation dysregulation in xenotransplantation. Curr Opin Organ Transplant. 2011;16:214–21.
20. Chong AS, Alegre ML. The impact of infection and tissue damage in solid-organ transplantation. Nat Rev Immunol. 2012;12:459–71.
21. Ezzelarab MB, Ekser B, Azimzadeh A, Lin CC, Zhao Y, Rodriguez R, et al. Systemic inflammation in xenograft recipients precedes activation of coagulation. Xenotransplantation. 2015;22:32–47.
22. Ezzelarab M, Garcia B, Azimzadeh A, Sun H, Lin CC, Hara H, et al. The innate immune response and activation of coagulation in alpha1,3-galactosyltransferase gene-knockout xenograft recipients. Transplantation. 2009;87:805–12.
23. Marcucci R, Gori AM, Giannotti F, Baldi M, Verdiani V, Del Pace S, et al. Markers of hypercoagulability and inflammation predict mortality in patients with heart failure. J Thromb Haemost. 2006;4:1017–22.
24. Chandra AP, Ouyang L, Yi S, Wong JK, Ha H, Walters SN, et al. Chemokine and toll-like receptor signaling in macrophage mediated islet xenograft rejection. Xenotransplantation. 2007;14:48–59.
25. Li X, Syrovets T, Paskas S, Laumonnier Y, Simmet T. Mature dendritic cells express functional thrombin receptors triggering chemotaxis and CCL18/pulmonary and activation-regulated chemokine induction. J Immunol. 2008;181:1215–23.
26. Lin CC, Chen D, McVey JH, Cooper DK, Dorling A. Expression of tissue factor and initiation of clotting by human platelets and monocytes after incubation with porcine endothelial cells. Transplantation. 2008;86:702–9.
27. Geissmann F, Jung S, Littman DR. Blood monocytes consist of two principal subsets with distinct migratory properties. Immunity. 2003;19:71–82.
28. Lin CC, Ezzelarab M, Shapiro R, Ekser B, Long C, Hara H, et al. Recipient tissue factor expression is associated with consumptive coagulopathy in pig-to-primate kidney xenotransplantation. Am J Transplant. 2010;10:1556–68.
29. Wullstein C, Drognitz O, Woeste G, Schareck WD, Bechstein WO, Hopt UT, et al. High levels of C-reactive protein after simultaneous pancreas-kidney transplantation predict pancreas graft-related complications and graft survival. Transplantation. 2004;77:60–4.
30. Wu J, Stevenson MJ, Brown JM, Grunz EA, Strawn TL, Fay WP. C-reactive protein enhances tissue factor expression by vascular smooth muscle cells: mechanisms and in vivo significance. Arterioscler Thromb Vasc Biol. 2008;28:698–704.
31. Knosalla C, Yazawa K, Behdad A, Bodyak N, Shang H, Buhler L, et al. Renal and cardiac endothelial heterogeneity impact acute vascular rejection in pig-to-baboon xenotransplantation. Am J Transplant. 2009;9:1006–16.
32. Byrne GW, Davies WR, Oi K, Rao VP, Teotia SS, Ricci D, et al. Increased immunosuppression, not anticoagulation, extends cardiac xenograft survival. Transplantation. 2006;82:1787–91.

33. Matsuda N, Hattori Y. Systemic inflammatory response syndrome (SIRS): molecular pathophysiology and gene therapy. J Pharmacol Sci. 2006;101:189–98.
34. Iwase H, Ekser B, Zhou H, Liu H, Satyananda V, Humar R, Humar P, Hara H, Long C, Bhama JK, Bajona P, Wang Y, Wijkstrom M, Ayares D, Ezzelarab MB, Cooper DK. Further evidence for sustained systemic inflammation in xenograft recipients (SIXR). Xenotransplantation. 2015;22(5):399–405. https://doi.org/10.1111/xen.12182.
35. Robson SC, Shephard EG, Kirsch RE. Fibrin degradation product D-dimer induces the synthesis and release of biologically active IL-1 beta, IL-6 and plasminogen activator inhibitors from monocytes in vitro. Br J Haematol. 1994;86:322–6.
36. Shimizu A, Hisashi Y, Kuwaki K, et al. Thrombotic microangiopathy associated with humoral rejection of cardiac xenografts from alpha1,3-galactosyltransferase gene-knockout pigs in baboons. Am J Pathol. 2008;172:1471–81.
37. Wu G, Pfeiffer S, Schroder C, et al. Coagulation cascade activation triggers early failure of pig hearts expressing human complement regulatory genes. Xenotransplantation. 2007;14:34–47.
38. Bari SF, Khan A, Lawson T. C reactive protein may not be reliable as a marker of severe bacterial infection in patients receiving tocilizumab. BMJ Case Rep. 2013;2013: bcr2013010423.
39. Kanterman J, Sade-Feldman M, Baniyash M. New insights into chronic inflammation-induced immunosuppression. Semin Cancer Biol. 2012;22:307–18.
40. Conlon TM, Saeb-Parsy K, Cole JL, Motallebzadeh R, Qureshi MS, Rehakova S, Negus MC, Callaghan CJ, Bolton EM, Bradley JA, et al. Germinal center alloantibody responses are mediated exclusively by indirect-pathway CD4 T follicular 584 helper cells. J Immunol. 2012;188:2643–52.
41. Morelli AE, Bracamonte-Baran W, Burlingham WJ. Donor-derived exosomes: the trick behind the semidirect pathway of allorecognition. Curr Opin Organ Transplant. 2017;22:46–54.
42. Zuber J, Shonts B, Lau S, Obradovic A, Fu J, Yang S, Lambert M, Coley S, Weiner J, Thome J, et al. Bidirectional intragraft alloreactivity drives the repopulation of human intestinal allografts and correlates with clinical outcome. Sci Immunol. 2016;1:1–11. -623.
43. Senev A, Coemans M, Lerut E, van Sandt V, Kerkhofs J, Daniëls L, Vanden Driessche M, Compernolle V, Sprangers B, van Loon E, et al. Eplet mismatch load and De novo occurrence of donor-specific anti-HLA antibodies, rejection, and graft failure after kidney transplantation: an observational cohort study. J Am Soc Nephrol. 2020;31:2193–204.
44. Fu J, Sykes M. Emerging concepts of tissue-resident memory T cells in transplantation. Transplantation. 2022;106:1132–636.
45. Matsumoto CS, Subramanian S, Fishbein TM. Adult intestinal transplantation. Gastroenterol Clin N Am. 2018;47:341–54. 552, 553.
46. Meier D, Docena GH, Ramisch D, Toscanini U, Berardi G, Gondolesi GE, Rumbo M. Immunological status of isolated lymphoid follicles after intestinal transplantation. Am J Transplant. 2014;14:607.
47. Meier D, Cagnola H, Ramisch D, Rumbo C, Chirdo F, Docena G, Gondolesi GE, Rumbo M. Analysis of immune cells 603 draining from the abdominal cavity as a novel tool to study intestinal transplant immunobiology. Clin Exp Immunol. 2010;604(162):605.
48. Sykes M. Immune monitoring of transplant patients in transient mixed chimerism tolerance trials. Hum Immunol. 2018;79:334–42. 625, 626.
49. Fu J, Zuber J, Shonts B, Obradovic A, Wang Z, Frangaj K, Meng W, Rosenfeld AM, Waffarn EE, Liou P, et al. 633 Lymphohematopoietic graft-versus-host responses promote mixed chimerism in patients receiving intestinal transplantation. J Clin Invest. 2021;131:635.
50. Stephens WZ, Kubinak JL, Ghazaryan A, Bauer KM, Bell R, Buhrke K, Chiaro TR, Weis AM, Tang WW, Monts JK, et al. Epithelial-myeloid exchange of MHC class II constrains immunity and microbiota composition. Cell Rep. 2021;37:109916.

51. Lyu M, Suzuki H, Kang L, Gaspal F, Zhou W, Goc J, Zhou L, Zhou J, Zhang W, Artis D, et al. ILC3s select microbiota-specific regulatory T cells to establish tolerance in the gut. Nature. 2022;610:744–51. 664, 665
52. Pucci Molineris M, González Polo V, Rumbo C, Fuxman C, Lowestein C, Nachman F, Rumbo M, Gondolesi G, Meier D. 666 D. Acute cellular rejection in small-bowel transplantation impairs NCR+ innate lymphoid cell subpopulation 3/interleukin 22 667 axis. Transpl Immunol. 2020;60:668.
53. Pappas PA, Tzakis AG, Gaynor JJ, Carreno MR, Ruiz P, Huijing F, Kleiner G, Rabier D, Kato T, Levi DM, et al. An analysis of the association between serum citrulline and acute rejection among 26 recipients of intestinal transplant. Am J Transplant. 2004;4:1124–32. 677, 678, 679.
54. Morrissey PE, Gollin G, Marks WH. Small bowel allograft rejection detected by serum intestinal fatty acid-binding protein 687 is reversible. Transplantation. 1996;61:1451–5.
55. Klaus A, Weiss H, Nguyen JH, Margreiter R, Obrist P, Schwelberger HG. Histamine-degrading enzymes as cellular 694 markers of acute small bowel allograft rejection. Transpl Int. 2003;16:572–7.
56. Pastorelli L, Garg RR, Hoang SB, Spina L, Mattioli B, Scarpa M, Fiocchi C, Vecchi M, Pizarro TT. Epithelial-derived 711 IL-33 and its receptor ST2 are dysregulated in ulcerative colitis and in experimental Th1/Th2 driven enteritis. Proc Natl Acad Sci U S A. 2010;107: 8017–22. 712, 713.
57. Colonna M, Navarro F, Bellón T, Llano M, García P, Samaridis J, Angman L, Cella M, López-Botet M. A common inhibitory receptor for major histocompatibility complex class I molecules on human lymphoid and myelomonocytic cells. J Exp Med. 1997;186:1809–18.
58. Liang S, Ristich V, Arase H, Dausset J, Carosella ED, Horuzsko A. Modulation of dendritic cell differentiation by HLA-G and ILT4 requires the IL-6—STAT3 signaling pathway. Proc Natl Acad Sci U S A. 2008;105:8357–62.
59. Kawahara T, Ohdan H, Zhao G, Yong-Guang Y, Sykes M. Peritoneal cavity B cells are precursors of splenic IgM natural antibody-producing cells. J Immunol. 2003;171:5406–14.
60. Ohdan H, Swenson KG, Kruger Gray HS, Yong-Guang Y, Xu Y, Thall A, Sykes M. Mac-1 negative B-1b phenotype of natural antibody-producing cells, including those responding to gal α1,3 galactosyltransferase-deficient mice. J Immunol. 2000;165:5518–29.
61. Xu Y, Yong-Guang Y, Ohdan H, et al. Characterization of anti-gal antibody-producing cells of baboons and humans. Transplantation. 2006;81:940–8.
62. Fischer-Lougheed J, Gregory C, White Z, Shulkin I, Gunthart M, Kearns-Jonker M. Identification of an anti-idiotypic antibody that defines a B-cell subset(s) producing xenoantibodies in primates. Immunology. 2008;123(3):390–7.
63. McElveen JE, Furtado PB, Smith SJ, Clark MR, Spendlove I, Sewell HF, Shakib F. Characterisation of a mouse monoclonal anti-idiotype reactive with a V region sequence commonly used by human immunoglobulins. Mol Pathol. 2000;53:77–82.
64. Ramirez P, Montoya MJ, Rios A, et al. Prevention of hyperacute rejection in a model of orthotopic liver xenotransplantation from pig to baboon using polytransgenic pig livers (CD55, CD59, and H-transferase). Transplant Proc. 2005;37:4103–6.
65. Lang J, Zhan J, Xu L, Yan Z. Identification of peptide mimetics of xenoreactive alpha-gal antigenic epitope by phage display. Biochem Biophys Res Commun. 2006;344:214–20.
66. Chen G, Qian H, Starzl T, et al. Acute rejection is associated with antibodies to non-gal antigens in baboons using gal-knockout pig kidneys. Nat Med. 2005;11:1295–8.
67. Ezzelarab M, Garcia B, Azimzadeh A, et al. The innate immune response and activation of coagulation in alpha1,3-galactosyltransferase gene-knockout xenograft recipients. Transplantation. 2009;87:805–12.
68. Yi S, Hawthorne WJ, Lehnert AM, et al. T cell-activated macrophages are capable of both recognition and rejection of pancreatic islet xenografts. J Immunol. 2003;170:2750–8.

69. Vaithilingam V, Fung C, Ratnapala S, et al. Characterisation of the xenogeneic immune response to microencapsulated fetal pig islet-like cell clusters transplanted into immunocompetent C57BL/6 mice. PLoS One. 2013;8:e59120.
70. Thompson P, Badell IR, Lowe M, et al. Islet xenotransplantation using gal-deficient neonatal donors improves engraftment and function. Am J Transplant. 2011;11:2593–602.
71. Greenwald AG, Jin R, Waddell TK. Galectin-3-mediated xenoactivation of human monocytes. Transplantation. 2009;87:44–51.
72. Wang H, Yang YG. Innate cellular immunity and xenotransplantation. Curr Opin Organ Transplant. 2012;17:162–7.
73. Hauzenberger E, Hauzenberger D, Hultenby K, Holgersson J. Porcine endothelium supports transendothelial migration of human leukocyte subpopulations: anti-porcine vascular cell adhesion molecule antibodies as species-specific blockers of transendothelial monocyte and natural killer cell migration. Transplantation. 2000;69:1837–49.
74. Subramanian S, Parthasarathy R, Sen S, Boder ET, Discher DE. Species- and cell type-specific interactions between CD47 and human SIRPalpha. Blood. 2006;107:2548–56.
75. Ide K, Wang H, Tahara H, et al. Role for CD47-SIRPalpha signaling in xenograft rejection by macrophages. Proc Natl Acad Sci U S A. 2007;104:5062–6.
76. Navarro-Alvarez N, Yang YG. Lack of CD47 on donor hepatocytes promotes innate immune cell activation and graft loss: a potential barrier to hepatocyte xenotransplantation. Cell Transplant. 2014;23:345–54.
77. Gorczynski RM, Cattral MS, Chen Z, et al. An immunoadhesin incorporating the molecule OX-2 is a potent immunosuppressant that prolongs Allo- and xenograft survival. J Immunol. 1999;163(3):1654–60.
78. Wang HT, Matsuura R, Kodama T, Eguchi H, Okuyama H, Miyagawa S. Human CD200 suppresses macrophage-mediated xenogeneic cytotoxicity and phagocytosis. Surg Today. 2018;48(1):119–26.
79. Gorczynski RM, Hadidi S, Yu G, Clark DA. The same immunoregulatory molecules contribute to successful pregnancy and transplantation. Am J Reprod Immunol. 2002;48:18–26.
80. Scheurlen KM, Snook DL, Gardner SA, Eichenberger MR, Galandiuk S. Macrophage differentiation and polarization into an M2-like phenotype using a human monocyte-like THP-1 Leukemia cell line. J Vis Exp. 2021;174:3850.
81. Esquivel EL, Maeda A, Eguchi H, Asada M, Sugiyama M, Manabe C, Sakai R, Matsuura R, Nakahata K, Okuyama H, Miyagawa S. Suppression of human macrophage-mediated cytotoxicity by transgenic swine endothelial cell expression of HLA-G. Transpl Immunol. 2015;32(2):109–15.
82. Maeda A, Kawamura T, Ueno T, Usui N, Eguchi H, Miyagawa S. The suppression of inflammatory macrophage-mediated cytotoxicity and proinflammatory cytokine production by transgenic expression of HLA-E. Transpl Immunol. 2013;29(1–4):76–81. https://doi.org/10.1016/j.trim.2013.08.001.
83. Eguchi H, Maeda A, Lo PC, Matsuura R, Esquivel EL, Asada M, Sakai R, Nakahata K, Yamamichi T, Umeda S, Deguchi K, Ueno T, Okuyama H, Miyagawa S. HLA-G1, but not HLA-G3, suppresses human monocyte/macrophage-mediated swine endothelial cell Lysis. Transplant Proc. 2016;48(4):1285–7.
84. Chung TW, Choi HJ, Park MJ, Choi HJ, Lee SO, Kim KJ, Kim CH, Hong C, Kim KH, Joo M, Ha KT. The function of cancer-shed gangliosides in macrophage phenotype: involvement with angiogenesis. Oncotarget. 2017;8(3):4436–48. https://doi.org/10.18632/oncotarget.13878.
85. Gorden S, Martinez FO. Alternative activation of macrophages: mechanism and functions. Immunity. 2010;32:593–604. https://doi.org/10.1016/j.immuni.2010.05.007.
86. Sandler NG, Mentink-Kane MM, Cheever AW, Wynn TA. Global gene expression profiles during acute pathogen-induced pulmonary inflammation reveal divergent roles for Th1 and Th2 responses in tissue repair. J Immunol. 2003;171:3655–67. https://doi.org/10.4049/jimmunol.171.7.3655.

87. Fleming BD, Mosser DM. Regulatory macrophages: setting the threshold for therapy. Eur J Immunol. 2011;41:2498–502. https://doi.org/10.1002/eji.201141717.
88. Vos AC, Wildenberg ME, Arijs I, Duijvestein M, Verhaar AP, de Hertogh G, Vermeire S, Rutgeerts P, van den Brink GR, Hommes DW. Regulatory macrophages induced by infliximab are involved in healing in vivo and in vitro. Inflamm Bowel Dis. 2012;18(3):401–8. https://doi.org/10.1002/ibd.21818. Epub 2011 Sep 20.
89. Vos AC, Wildenberg ME, Duijvestein M, Verhaar AP, van den Brink GR, Hommes DW. Anti-tumor necrosis factor-α antibodies induce regulatory macrophages in an fc region-dependent manner. Gastroenterology. 2011;140(1):221–30. https://doi.org/10.1053/j.gastro.2010.10.008. Epub 2010 Oct 16.
90. Guo F, Hu M, Huang D, Zhao Y, Heng B, Guillemin G, Lim CK, Hawthorne WJ, Yi S. Human regulatory macrophages are potent in suppression of the xenoimmune response via indoleamine-2, 3-dioxygenase-involved mechanism(s). Xenotransplantation. 2017;24(5):e12326. https://doi.org/10.1111/xen.12326.
91. Sakai R, Maeda A, Choi TV, Lo PC, Jiaravuthiasan P, Shabri AM, Wang HT, Matsuura R, Kodama T, Eguchi H, Okuyama H, Miyagawa S. CD200 suppresses macrophage-mediated xenogeneic cytotoxicity and phagocytosis. Surg Today. 2017;48(1):119–26. https://doi.org/10.1007/s00595-017-1546-2.
92. Riquelme P, Tomiuk S, Kammler A, et al. IFN- gamma- induced iNOS expression in mouse regulatory macrophages prolongs allograft survival in fully immunocompetent recipients. Mol Ther. 2013;21:409–22.
93. Caserta S, Nausch N, Sawtell A, Drummond R, Barr T, Macdonald AS, Mutapi F, Zamoyska R. Chronic infection drives expression of the inhibitory receptor CD200R, and its ligand CD200, by mouse and human CD4 T cells. PLoS One. 2012;7(4):e35466.
94. Mihrshahi R, Barclay AN, Brown MH. Essential roles for Dok2 and RasGAP in CD200 receptor-mediated regulation of human myeloid cells. J Immunol. 2009;183(8):4879–86.
95. Gorczynski RM. CD200 and its receptors as targets for immunoregulation. Curr Opin Investig Drugs. 2005;6:483–8.
96. Sahu D, Dey S, Pathak T, Ganguly B. Regioselectivity of vinyl sulfone based 1,3-dipolar cycloaddition reactions with sugar azides by computational and experimental studies. Org Lett. 2014;16:2100–3.
97. Li Y, Tredget EE, Ghaffari A, Lin X, Kilani RT, Ghahary A. Local expression of indoleamine 2,3-dioxygenase protects engraftment of xenogeneic skin substitute. J Invest Dermatol. 2006;126:128–36.
98. Mellor AL, Munn DH. Tryptophan catabolism and T-cell tolerance: immunosuppression by starvation? Immunol Today. 1999;20:469–73.
99. Arandi N, Ramzi M, Safaei F, Monabati A. Overexpression of indoleamine 2,3-dioxygenase correlates with regulatory T cell phenotype in acute myeloid leukemia patients with normal karyotype. Blood Res. 2018;53(4):294–8. https://doi.org/10.5045/br.2018.53.4.294.
100. Munn DH, Shafizadeh E, Attwood JT, Bondarev I, Pashine A, Mellor AL. Inhibition of T cell proliferation by macrophage tryptophan catabolism. J Exp Med. 1999;189:1363–72.
101. Araújo EF, Medeiros DH, Galdino NA, Condino-Neto A, Calich VL, Loures FV. Tolerogenic plasmacytoid dendritic cells control Paracoccidioides brasiliensis infection by inducting regulatory T cells in an IDO-dependent manner. PLoS Pathog. 2016;12(12):e1006115. https://doi.org/10.1371/journal.ppat.1006115.
102. Bolton EM, Gracie JA, Briggs JD, et al. Cellular requirements for renal allograft rejection in the athymic nude rat. J Exp Med. 1989;169:1931–46.
103. Sebille F, Dorling A, Lechler RI. The cellular rejection of xenografts—recent insights. Xenotransplantation. 2003;10:4–6.
104. Trenado A, Charlotte F, Fisson S, Yagello M, Klatzmann D, Salomon BL, Cohen JL. Recipient-type specific CD4+CD25+ regulatory T cells favor immune reconstitution and control graft-versus-host disease while maintaining graft-versus-leukemia. J Clin Invest. 2003;112(11):1688–96.

105. Hori S, Nomura T, Sakaguchi S. Control of regulatory T cell development by the transcription factor Foxp3. Science. 2003;299(5609):1057–61.
106. Liu W, Putnam AL, Xu-Yu Z, Szot GL, Lee MR, Zhu S, Gottlieb PA, Kapranov P, Gingeras TR, de St F, Groth B, Clayberger C, Soper DM, Ziegler SF, Bluestone JA. CD127 expression inversely correlates with FoxP3 and suppressive function of human CD4+ T reg cells. J Exp Med. 2006;203(7):1701–11.
107. Di Ianni M, Falzetti F, Carotti A, Terenzi A, Castellino F, Bonifacio E, Del Papa B, Zei T, Ostini RI, Cecchini D, Aloisi T, Perruccio K, Ruggeri L, Balucani C, Pierini A, Sportoletti P, Aristei C, Falini B, Reisner Y, Velardi A, Aversa F, Martelli MF. Tregs prevent GVHD and promote immune reconstitution in HLA-haploidentical transplantation. Blood. 2011;117(14): 3921–8.
108. Lechler RI, Sykes M, Thomson AW, Turka Lechler RI, Sykes M, Thomson AW, Turka LA. Organ transplantation—how much of the promise has been realized? Nat Med. 2005;11: 605–13.
109. Brusko TM, Putnam AL, Bluestone JA. Human regulatory T cells: role in autoimmune disease and therapeutic opportunities. Immunol Rev. 2008;223:371–90.
110. Gratz IK, Rosenblum MD, Maurano MM, et al. Cutting edge: self-antigen controls the balance between effector and regulatory T cells in peripheral tissues. J Immunol. 2014;192:1351–5.
111. Wu J, Hu M, Qian YW, Hawthorne WJ, Burns H, Liuwantara D, Alexander SI, Yi S, O'Connell PJ. In vivo costimulation blockade-induced regulatory T cells demonstrate dominant and specific tolerance to porcine islet xenografts. Transplantation. 2017;101(7):1587–99.
112. Walsh PT, Taylor DK, Turka LA. Tregs and transplantation tolerance. J Clin Invest. 2004;114: 1398–403.
113. Lehnert AM, Mottram PL, Han W, et al. Blockade of the CD28 and CD40 pathways result in the acceptance of pig and rat islet xenografts but not rat cardiac grafts in mice. Transpl Immunol. 2001;9:51–6.
114. Charbonnier LM, Vokaer B, Lemaitre PH, et al. CTLA4-Ig restores rejection of MHC class-II mismatched allografts by disabling IL-2-expanded regulatory T cells. Am J Transplant. 2012;12:2313–21.
115. Edozie FC, Nova-Lamperti EA, Povoleri GA, et al. Regulatory Tcell therapy in the induction of transplant tolerance: the issue of subpopulations. Transplantation. 2014;98:370–9.
116. Sagoo P, Ali N, Garg G, Nestle FO, Lechler RI, Lombardi G. Human regulatory T cells with alloantigen specificity are more potent inhibitors of alloimmune skin graft damage than polyclonal regulatory T cells. Sci Transl Med. 2011;3:83ra42.
117. Muller YD, Ehirchiou D, Golshayan D, Buhler LH, Seebach JD. Potential of T-regulatory cells to protect xenografts. Curr Opin Organ Transplant. 2012;17:155–61.
118. Jin X, Lu Y, Zhao Y, Yi S. Large-scale in vitro expansion of human regulatory T cells with potent xenoantigen-specific suppression. Cytotechnology. 2016;68:935–45.
119. Ekser B, Cooper DK. Overcoming the barriers to xenotransplantation: prospects for the future. Expert Rev Clin Immunol. 2010;6(2):219.
120. Vadori M, Cozzi E. The immunological barriers to xenotransplantation. Tissue Antigens. 2015;86(4):239–53. https://doi.org/10.1111/tan.12669.
121. Cardona K, Korbutt GS, Milas Z, et al. Long-term survival of neonatal porcine islets in nonhuman primates by targeting costimulation pathways. Nat Med. 2006;12:304–6.
122. Hering BJ, Wijkstrom M, Graham ML, et al. Prolonged diabetes reversal after intraportal xenotransplantation of wild-type porcine islets in immunosuppressed nonhuman primates. Nat Med. 2006;12:301–3.
123. Uchida T, Tomita Y, Anzai K, Zhang QW, Yoshikawa M, Kishihara K, Nomoto K, Yasui H. Roles of CD4+ and CD8+ T cells in discordant skin xenograft rejection. Transplantation. 1999;68:1721–7.
124. Fu Y, Lu X, Yi S, et al. Selective rejection of porcine islet xenografts by macrophages. Xenotransplantation. 2008;15:307–12.

125. Ide K, Wang H, Tahara H, et al. Role for CD47-SIRPα signaling in xenograft rejection by macrophages. Proc Natl Acad Sci U S A. 2007;104:5062–6.
126. Deschamps JY, Roux FA, Sai P, Gouin E. History of xenotransplantation. Xenotransplantation. 2005;12:91–109.
127. Yi S, Feng X, Wang Y, Kay TW, O'Connell PJ. CD4+ cells play a major role in xenogeneic human anti-pig cytotoxicity through the Fas/Fas ligand lytic pathway. Transplantation. 1999;67:435–43.
128. Rogers NJ, Jackson IM, Jordan WJ, Hawadle MA, Dorling A, Lechler RI. Cross-species costimulation: relative contributions of CD80, CD86, and CD40. Transplantation. 2003;75: 2068–76.
129. Afzali B, Lombardi G, Lechler RI. Pathways of major histocompatibility complex allorecognition. Curr Opin Organ Transplant. 2008;13:438–44.
130. Park JY, Park MR, Bui HT, et al. α1,3-galactosyltransferase deficiency in germ-free miniature pigs increases Nglycolylneuraminic acids as the xenoantigenic determinant in pighumanxenotransplantation. Cell Reprogram. 2012;14(4):353–63.
131. Iwase H, Ekser B, Satyananda V, et al. Pig-to-baboon heterotopic heart transplantation - exploratory preliminary experience with pigs transgenic for human thrombomodulin and comparison of three costimulation blockade-based regimens. Xenotransplantation. 2015;22: 211–20.
132. Higginbotham L, Mathews D, Breeden CA, et al. Pre-transplant antibody screening and anti-CD154 costimulation blockade promote long-term xenograft survival in a pig-to-primate kidney transplant model. Xenotransplantation. 2015;22:221–30.
133. Wennberg L, Song Z, Bennet W, et al. Importance of the Gal alpha1-3 Gal antigen in discordant islet xenotransplantation: immunosuppression, which inhibits porcine islet xenograft rejection in ordinary mice, is equally effective in Gal-knockout mice. Transplantation. 2004;77:1275–80.
134. Nishimura H, Scalea J, Wang Z, et al. First experience with the use of a recombinant CD3 immunotoxin as induction therapy in pig-to-primate xenotransplantation: the effect of T-cell depletion on outcome. Transplantation. 2011;92:641–7.
135. Robson SC, Wu Y, Sun X, Knosalla C, Dwyer K, Enjyoji K. Ectonucleotidases of CD39 family modulate vascular inflammation and thrombosis in transplantation. Semin Thromb Hemost. 2005;31:217–33.
136. Tseng YL, Dor FJ, Kuwaki K, et al. Bone marrow transplantation from α-1,3-galactosyltransferase gene-knockout pigs in baboons. Xenotransplantation. 2004;11: 361–70.
137. Medvinsky A, Rybtsov S, Taoudi S. Embryonic origin of the adult hematopoietic system: advances and questions. Development. 2011;138(6):1017–31. https://doi.org/10.1242/dev.040998.
138. Kent D, Copley M, Benz C, Dykstra B, Bowie M, Eaves C. Regulation of hematopoietic stem cells by the steel factor/KIT signaling pathway. Clin Cancer Res. 2008;14(7):1926–30. https://doi.org/10.1158/1078-0432.ccr-075134. Epub 2008/04/03.
139. Ito M, Kobayashi K, Nakahata T. NOD/Shi-scid IL2rgamma(null) (NOG) mice more appropriate for humanized mouse models. Curr Top Microbiol Immunol. 2008;324:53–76.
140. Zecca M, Strocchio L, Pagliara D, et al. HLA-haploidentical Tcell-depleted allogeneic hematopoietic stem cell transplantation in children with Fanconi anemia. Biol Blood Marrow Transplant. 2014;20:571–6.
141. Shultz LD, Schweitzer PA, Christianson SW, et al. Multiple defects in innate and adaptive immunologic function in NOD/LtSz-scid mice. J Immunol. 1995;154:180–91.
142. Arpinati M, Terragna C, Chirumbolo G, et al. Human CD34(+)bloodcells induce T-cell unresponsiveness to specific alloantigens only under costimulatory blockade. Exp Hematol. 2003;31:31–8.

143. van der Windt DJ, Bottino R, Casu A, et al. Long-term controlled normoglycemia in diabetic nonhuman primates after transplantation with hCD46 transgenic porcine islets. Am J Transplant. 2009;9:2716–26.
144. Reyes LM, Estrada JL, Wang ZY, et al. Creating class I MHC-null pigs using guide RNA and the Cas9 endonuclease. J Immunol. 2014;193:5751–7. (Baltimore, Md: 1950).
145. Ito M, Hiramatsu H, Kobayashi K, Suzue K, Kawahata M, Hioki K, et al. NOD/SCID/gamma (c)(null) mouse: an excellent recipient mouse model for engraftment of human cells. Blood. 2002;100(9):3175–82. https://doi.org/10.1182/blood-2001-12-0207.
146. Ollinger R, Pratschke J. Role of heme oxygenase-1 in transplantation. Transpl Int. 2010;23: 1071–81.
147. Lee HS, Lee JG, Yeom HJ, Chung YS, Kang B, Hurh S, Cho B, Park H, Hwang JI, Park JB, Ahn C, Kim SJ, Yang J. The introduction of human heme oxygenase-1 and soluble tumor necrosis factor-α receptor type I with human IgG1 fc in porcine islets prolongs islet xenograft survival in humanized mice. Am J Transplant. 2016;16(1):44–57.
148. Ljunggren HG, K. Ka¨rre. In search of the 'missing self': MHC molecules and NK cell recognition. Immunol Today. 1990;11:237.
149. Borrego F, Ulbrecht M, Weiss EH, Coligan JE, Brooks AG. Recognition of human histocompatibility leukocyte antigen (HLA)-E complexed with HLA class I signal sequence-derived peptides by CD94/NKG2 confers protection from natural killer cell-mediated lysis. J Exp Med. 1998;187:813.
150. Sullivan JA, Oettinger HF, Sachs DH, Edge ASB. Analysis of polymorphism in porcine MHC class I genes. J Immunol. 1997;159:2318.
151. Sasaki H, Xu XC, Mohanakumar T. HLA-E and HLA-G expression on porcine endothelial cells inhibit xenoreactive human NK cells through CD94/NKG2-dependent and -independent pathways. J Immunol. 1999;163(11):6301–5.
152. Griesemer A, Yamada K, Sykes M. Xenotransplantation: immunological hurdles and progress toward tolerance. Immunol Rev. 2014;258:241–58.
153. Raker VK, Domogalla MP, Steinbrink K. Tolerogenic dendritic cells for regulatory T cell induction in man. Front Immunol. 2015;6:569.
154. Laird CT, Burdorf L, French BM, Kubicki N, Cheng X, Braileanu G, Sun W, O'Neill NA, Cimeno A, Parsell D, So E, Bähr A, Klymiuk N, Phelps CJ, Ayares D, Azimzadeh AM, Pierson RN. Transgenic expression of human leukocyte antigen-E attenuates GalKO.hCD46 porcine lung xenograft injury. Xenotransplantation. 2017;24(2):e12294.
155. Rieben R, Seebach J. Xenograft rejection: IgG1, complement and NK cells team up to activate and destroy the endothelium. Trends Immunol. 2005;26:2–5.
156. Schneider MK, Ghielmetti M, Rhyner DM, Antsiferova MA, Seebach JD. Human leukocyte transmigration across Galalpha(1,3)Gal-negative porcine endothelium is regulated by human CD18 and CD99. Transplantation. 2009;87:491–9.
157. Gock H, Murray-Segal LJ, Winterhalter AC, et al. Altered glycosylation in donor mice causes rejection of strain-matched skin and heart grafts. Am J Transplant. 2014;14:797–805.
158. Yin D, Zeng H, Ma L, et al. Cutting Edge: NK cells mediate IgG1-dependent hyperacute rejection of xenografts. J Immunol. 2004;172:7235–8.
159. Ide K, Wang H, Tahara H, Liu J, Wang X, Asahara T, et al. Role for CD47-SIRPalpha signaling in xenograft rejection by macrophages. Proc Natl Acad Sci U S A. 2007;104:5062–6.
160. Tena A, Kurtz J, Leonard DA, Dobrinsky JR, Terlouw SL, Mtango N, et al. Transgenic expression of human CD47 markedly increases engraftment in a murine model of pig to-human hematopoietic cell transplantation. Am J Transplant. 2014;14:2713–22.
161. Tena AA, Sachs DH, Mallard C, Yang YG, Tasaki M, Farkash E, et al. Prolonged survival of pig skin on baboons after administration of pig cells expressing human CD47. Transplantation. 2016;101(2):316–21.
162. Jung SH, Hwang JH, Kim SE, Young KK, Park HC, Lee HT. The potentiating effect of hTFPI in the presence of hCD47 reduces the cytotoxicity of human macrophages. Xenotransplantation. 2017;24:e12301.

163. Carosella ED, Moreau P, Lemaoult J, Rouas-Freiss N. HLA-G: from biology to clinical benefits. Trends Immunol. 2008;29:125–32.
164. Sodersstrom K, Corliss B, Lanier LL, Phillips JH. CD94/NKG2 is the predominant inhibitory receptor involved in recognition of HLA-G by decidual and peripheral blood NK cells. J Immunol. 1997;159:1072.
165. Le Maoult J, Rouas-Freiss N, Le Discorde M, Moreau P, Carosella ED. HLA-G in organ transplantation. Pathol Biol (Paris). 2004;52(2):97–103.
166. Lee N, Llano M, Carretero M, et al. HLA-E is a major ligand for the natural killer inhibitory receptor CD94/NKG2A. Proc Natl Acad Sci U S A. 1998;95:5199–204.
167. Sommaggio R, Cohnen A, Watzl C, Costa C. Multiple receptors trigger human NK cell-mediated cytotoxicity against porcine chondrocytes. J Immunol. 2012;188:2075–83.
168. Maeda A, Kawamura T, Ueno T, Usui N, Eguchi H, Miyagawa S. The suppression of inflammatory macrophage-mediated cytotoxicity and proinflammatory cytokine production by transgenic expression of HLA-E. Transpl Immunol. 2013;29:76–81.
169. Weiss EH, Lilienfeld BG, Muller S, et al. HLA-E/human beta2-microglobulin transgenic pigs: protection against xenogeneic human antipig natural killer cell cytotoxicity. Transplantation. 2009;87:35–43.
170. Esquivel EL, Maeda A, Eguchi H, et al. Suppression of human macrophagemediated cytotoxicity by transgenic swine endothelial cell expression of HLA-G. Transpl Immunol. 2015;32:109–15.
171. Maeda A, Kawamura T, Ueno T, et al. The suppression of inflammatory macrophage-mediated cytotoxicity and proinflammatory cytokine production by transgenic expression of HLA-E. Transpl Immunol. 2013;29:76–81.
172. Sasaki H, Xu XC, Smith DM, Howard T, Mohanakumar T. HLA-G expression protects porcine endothelial cells against natural killer cell-mediated xenogeneic cytotoxicity. Transplantation. 1999;67(1):31–7.
173. Llano M, Lee N, Navarro F, Garcia P, Albar JP, Geraghty DE, Lopez-Botet M. HLA-E-bound peptides influence recognition by inhibitory and triggering CD94/NKG2 receptors: preferential response to an HLA-G-derived nonamer. Eur J Immunol. 1998;28:2854.
174. Navarro F, Llano M, Bellon T, Colonna M, Geraghty DE, Lopez-Botet M. The IL2 (LIR1) and CD94/NKG2A NK cell receptors respectively recognize HLA-G1 and HLA-E molecules co-expressed on target cells. Eur J Immunol. 1999;29:277.
175. Abicht J, Bongoni A, Mayr T, et al. Ex- vivo perfusion of alpha- gal knockout, hCD46/HLA-E double transgenic pig hearts [abstract]. Xenotransplantation. 2015;22:933.
176. Bongoni AK, Kiermeir D, Constantinescu M, et al. Functional evaluation of hTBM, CD46 and HLA-E transgenes in vitro and in pig leg xenoperfusion experiments. Xenotransplantation. 2014;21:189.
177. Laird CT, Burdorf L, French BM, Kubicki N, Cheng X, Braileanu G, et al. Transgenic expression of human leukocyte antigen-E attenuates GalKO.hCD46 porcine lung xenograft injury. Xenotransplantation. 2017;24:e12294.
178. Kubicki N, Laird C, Burdorf L, et al. The effect of human leukocyte antigen-E expression on GalTKO.hCD46 lung xenograft survival and injury in an ex-vivo xenoperfusion model [abstract]. Xenotransplantation. 2015;22(S193):1051.
179. Klymiuk N, Phelps CJ, Ayares D, Azimzadeh AM, Pierson RN. Transgenic expression of human leukocyte antigen-E attenuates GalKO.hCD46 porcine lung xenograft injury. Xenotransplantation. 2017;24(2):e12294. https://doi.org/10.1111/xen.12294.
180. Matsuura R, Maeda A, Sakai R, Eguchi H, Lo PC, Hasuwa H, Ikawa M, Nakahata K, Zenitani M, Yamamichi T, Umeda S, Deguchi K, Okuyama H, Miyagawa S. Human HLA-Ev (147) expression in transgenic animals. Transplant Proc. 2016;48(4):1323–5.
181. Nilsson B, Ekdahl KN, Korsgren O. Control of instant blood-mediated inflammatory reaction to improve islets of Langerhans engraftment. Curr Opin Organ Transplant. 2011;16:620–6.
182. Shimizu A, Yamada K, Robson SC, Sachs DH, Colvin RB. Pathologic characteristics of transplanted kidney xenografts. J Am Soc Nephrol. 2012;23:225–35.

183. Moberg L, Korsgren O, Nilsson B. Neutrophilic granulocytes are the predominant cell type infiltrating pancreatic islets in contact with ABO-compatible blood. Clin Exp Immunol. 2005;142:125–31.
184. Robinson LA, Tu L, Steeber DA, Preis O, Platt JL, Tedder TF. The role of adhesion molecules in human leukocyte attachment to porcine vascular endothelium: implications for xenotransplantation. J Immunol. 1998;161:6931–8.
185. Bongoni AK, Kiermeir D, Jenni H, Bähr A, Ayares D, Klymiuk N, Wolf E, Voegelin E, Constantinescu MA, Seebach JD, Rieben R. Complement dependent early immunological responses during ex vivo xenoperfusion of hCD46/HLA-E double transgenic pig forelimbs with human blood. Xenotransplantation. 2014;21:230–43.
186. Martin BM, Samy KP, Lowe MC, et al. Dual islet transplantation modeling of the instant blood-mediated inflammatory reaction. Am J Transplant. 2015;15:1241–52.
187. Al-Mohanna F, Collison K, Parhar R, et al. Activation of naïve xenogeneic but not allogeneic endothelial cells by human naïve neutrophils: a potential occult barrier to xenotransplantation. Am J Pathol. 1997;151:111–20.
188. Goto M, Tjernberg J, Dufrane D, et al. Dissecting the instant blood-mediated inflammatory reaction in islet xenotransplantation. Xenotransplantation. 2008;15:225–34.
189. Bastian F, Stelzmuller ME, Kratochwill K, Kasimir MT, Simon P, Weigel G. IgG deposition and activation of the classical complement pathway involvement in the activation of human granulocytes by decellularized porcine heart valve tissue. Biomaterials. 2008;29:1824–32.
190. Saleh SM, Parhar RS, Al-Hejailan RS, et al. Identification of the tetraspanin CD82 as a new barrier to xenotransplantation. J Immunol. 2013;191:2796–805.
191. Kourtzelis I, Ferreira A, Mitroulis I, et al. Complement inhibition in a xenogeneic model of interactions between human whole blood and porcine endothelium. Horm Metab Res. 2015;47:36–42.
192. Li S, Waer M, Billiau AD. Xenotransplantation: role of natural immunity. Transpl Immunol. 2009;21(2):70–4. https://doi.org/10.1016/j.trim.2008.10.004.
193. Knoll G. Trends in kidney transplantation over the past decade. Drugs. 2008;68(Suppl 1): 3–10.
194. Londrigan SL, Sutherland RM, Brady JL, et al. In situ protection against islet allograft rejection by CTLA4Ig transduction. Transplantation. 2010;90:951–7.
195. Buermann A, Römermann D, Baars W, Hundrieser J, Klempnauer J, Schwinzer R. Inhibition of B-cell activation and antibody production by triggering inhibitory signals via the PD-1/PD-ligand pathway. Xenotransplantation. 2016;23(5):347–56.
196. Mohiuddin MM, Singh AK, Corcoran PC, Hoyt RF, Thomas ML 3rd, Lewis BG, et al. Role of anti-CD40 antibody-mediated costimulation blockade on non-Gal antibody production and heterotopic cardiac xenograft survival in a GTKO.hCD46Tg pig-to-baboon model. Xenotransplantation. 2014;21:35–45.
197. Choi HJ, Lee JJ, Kim DH, Kim MK, Lee HJ, Ko AY, et al. Blockade of CD40-CD154 costimulatory pathway promotes long-term survival of full-thickness porcine corneal grafts in nonhuman primates: clinically applicable xenocorneal transplantation. Am J Transplant. 2015;15:628–41.
198. Yamada AA, Sayegh MH. The CD154-CD40 costimulatory pathway in transplantation. Transplantation. 2002;73:S36–9.
199. Jonker M, Ossevoort And MA, Vierboom M. Blocking the CD80 and CD86 costimulation molecules: lessons to be learned from animal models. Transplantation. 2002;73:S23–6.
200. Guinan EC, Boussiotis VA, Neuberg D, Brennan LL, Hirano N, Nadler LM, Gribben JG. Transplantation of anergic histoincompatible bone marrow allografts. N Engl J Med. 1999;340:1704–14.
201. Ville S, Poirier N, Blancho G, Vanhove B. Co-stimulatory blockade of the CD28/CD80-86/CTLA-4 balance in transplantation: impact on memory T cells? Front Immunol. 2015;6:411.
202. Miao G, Ito T, Uchikoshi F, et al. Development of donor-specific immunoregulatory T-cells after local CTLA4Ig gene transfer to pancreatic allograft. Transplantation. 2004;78:59–64.

203. Selin LK, Brehm MA. Frontiers in nephrology: heterologous immunity, T cell cross-reactivity, and alloreactivity. J Am Soc Nephrol. 2007;18:2268–77. https://doi.org/10.1681/ASN.2007030295.
204. van Kooten C, Banchereau J. CD40-CD40 ligand. J Leukoc Biol. 2000;67:2–17.
205. Xu H, Zhang X, Mannon RB, Kirk AD. Platelet-derived or soluble CD154 induces vascularized allograft rejection independent of cell-bound CD154. J Clin Invest. 2006;116: 769–74.
206. Pinelli DF, Ford ML. Novel insights into anti-CD40/CD154 immunotherapy in transplant tolerance. Immunotherapy. 2015;7:399–410.
207. Mohiuddin MM, Singh AK, Corcoran PC, et al. Chimeric 2C10R4 anti-CD40 antibody therapy is critical for long-term survival of GTKO.hCD46.hTBM pig-to-primate cardiac xenograft. Nat Commun. 2016;7:11138.
208. Mohiuddin MM, et al. Genetically engineered pigs and target-specific immunomodulation provide significant graft survival and hope for clinical cardiac xenotransplantation. J Thorac Cardiovasc Surg. 2014;148:1106–13. discussion 1113–1104.
209. Mohiuddin MM, Singh AK, Corcoran PC, Thomas ML 3rd, Clark T, Lewis BG, Hoyt RF, Eckhaus M, Pierson RN 3rd, Belli AJ, Wolf E, Klymiuk N, Phelps C, Reimann KA, Ayares D, Horvath KA. Chimeric 2C10R4 anti-CD40 antibody therapy is critical for long-term survival of GTKO.hCD46.hTBM pig-to-primate cardiac xenograft. Nat Commun. 2016;7:11138. https://doi.org/10.1038/ncomms11138.
210. Mohiuddin MM, et al. Role of anti-CD40 antibody-mediated costimulation blockade on non-Gal antibody production and heterotopic cardiac xenograft survival in a GTKO. hCD46Tg pig-to-baboon model. Xenotransplantation. 2014;21:35–45.
211. Lowe M, et al. In vivo characterization of novel non-depleting anti-CD40 monoclonal antibodies in nonhuman primates. Am J Transplant. 2011;11:474.
212. Iwase H, et al. Pig-to-baboon heterotopic heart transplantation—exploratory preliminary experience with pigs transgenic for human thrombomodulin and comparison of three costimulation blockade-based regimens. Xenotransplantation. 2015;22:211–20.
213. Larsen CP, Pearson TC, Adams AB, et al. Rational development of LEA29Y (belatacept), a high-affinity variant of CTLA4-Ig with potent immunosuppressive properties. Am J Transplant. 2005;5:443–53.
214. Vincenti F, Charpentier B, Vanrenterghem Y, et al. A phase III study of belatacept-based immunosuppression regimens versus cyclosporine in renal transplant recipients (BENEFIT study). Am J Transplant. 2010;10:535–46.
215. Lo DJ, Weaver TA, Stempora L, et al. Selective targeting of human alloresponsive CD8+ effector memory T cells based on CD2 expression. Am J Transplant. 2011;11:22–33.
216. Brown MH, Cantrell DA, Brattsand G, Crumpton MJ, Gullberg M. The CD2 antigen associates with the T-cell antigen receptor CD3 antigen complex on the surface of human T lymphocytes. Nature. 1989;339:551–3.
217. Brady JL, Mannering SI, Kireta S, et al. Monoclonal antibodies generated by DNA immunization recognize CD2 from a broad range of primates. Immunol Cell Biol. 2009;87:413–8.
218. Brady JL, Sutherland RM, Hancock M, Kitsoulis S, Lahoud MH, Phillips PM, Hawthorne WJ, d'Apice AJ, Cowan PJ, Harrison LC, O'Connell PJ, Lew AM. Anti-CD2 producing pig xenografts effect localized depletion of human T cells in a huSCID model. Xenotransplantation. 2013;20(2):100–9. https://doi.org/10.1111/xen.12025.
219. Povoleri GA, Scotta C, Nova-Lamperti EA, John S, Lombardi G, Afzali B. Thymic versus induced regulatory T cells – who regulates the regulators? Front Immunol. 2013;4:169.
220. Riella LV, Liu T, Yang J, Chock S, Shimizu T, Mfarrej B, Batal I, Xiao X, Sayegh MH, Chandraker A. Deleterious effect of CTLA4-Ig on a Treg-dependent transplant model. Am J Transplant. 2012;12:846–55.
221. Razmara M, Hilliard B, Ziarani AK, Chen YH, Tykocinski ML. CTLA-4 x Ig converts naive CD4+. Int Immunol. 2008;20:471–83.

222. Pilat N, Klaus C, Schwarz C, Hock K, Oberhuber R, Schwaiger E, Gattringer M, Ramsey H, Baranyi U, Zelger B, Brandacher G, Wrba F, Wekerle T. Rapamycin and CTLA4Ig synergize to induce stable mixed chimerism without the need for CD40 blockade. Am J Transplant. 2015;15:1568–79.
223. Waldman JP, Vogel T, Burlak C, et al. Blocking porcine sialoadhesin improves extracorporeal porcine liver xenoperfusion with human blood. Xenotransplantation. 2013;20:239–51.
224. Ihrcke NS, Platt JL. Shedding of heparan sulfate proteoglycan by stimulated endothelial cells: evidence for proteolysis of cellsurface molecules. J Cell Physiol. 1996;168:625–37.
225. Rezaie AR, Olson ST. Calcium enhances heparin catalysis of the antithrombin-factor Xa reaction by promoting the assembly of an intermediate heparin-antithrombin-factor Xa bridging complex. Demonstration by rapid kinetics studies. Biochemistry. 2000;39:12083–90.
226. Petersen B, Ramackers W, Lucas-Hahn A, et al. Transgenic expression of human heme oxygenase-1 in pigs confers resistance against xenograft rejection during ex vivo perfusion of porcine kidneys. Xenotransplantation. 2011;18:355–68.
227. Oropeza M, Petersen B, Carnwath JW, et al. Transgenic expression of the human A20 gene in cloned pigs provides protection against apoptotic and inflammatory stimuli. Xenotransplantation. 2009;16:522–34.
228. Ahrens HE, Petersen B, Ramackers W, Petkov S, Herrmann D, Hauschild-Quintern J, Lucas-Hahn A, Hassel P, Ziegler M, Baars W, Bergmann S, Schwinzer R, Winkler M, Niemann H. Kidneys from α1,3-galactosyltransferase knockout/human heme oxygenase-1/human A20 transgenic pigs are protected from rejection during ex vivo perfusion with human blood. Transplant Direct. 2015;1(6):e23. https://doi.org/10.1097/TXD.0000000000000533.
229. Yeom HJ, Koo OJ, Yang J, et al. Generation and characterization of human heme oxygenase-1 transgenic pigs. PLoS One. 2012;7:e46646.
230. Cramer-Morales K, Heer CD, Mapuskar KA, et al. SOD2 targeted gene editing by CRISPR/Cas9 yields human cells devoid of MnSOD. Free Radic Biol Med. 2015;89:379–86.

Chapter 18
Induction of Xenograft Tolerance and Chimerism as an Alternative Prevention of Xenograft Rejection

18.1 Introduction

As an alternative prevention of xenograft rejection, induction of immunological tolerance status is another strategy in which the host immune system of human recipients is specifically not responsive to the xenografted organs of pigs [1–3], but the human immune system is normally with the responding ability to infectious pathogens. There are huge progresses in strategies to induce tolerance across xenoantigenic hurdles and rejection barriers. Also, immune tolerance induction is potentially beneficial in xenotransplantation. By advanced progress in recent technologies of gene editing, survival time of xenotransplanted grafts of multiple-transgenic α1,3Gal-T KO pigs has greatly been prolonged, which is ranged from several days to months for transplanted organ grafts, allowing life-supporting healthy. For example, 2 years more survivals have been observed in non-life-supporting heterotopic model of xenotransplantation of cardiac xenografts. To achieve such success, continuous treatment with immunosuppressive drugs is prerequisite. In fact, to obtain much improved outcome, immunosuppression is required in host recipients and hence, anti-CD40 or anti-CD154 monoclonal Abs as immunosuppression agents are continuously administered. The status eventually causes death of recipient host from infectious diseases involved in chronically administered immunosuppression or from immune rejection of xenografts [4, 5], because immunosuppression is unsuccessful [6]. Human T-cell responses against xenoantigenic barriers of pigs are much strong rather than those of allogeneic antigens [7]. This invites a new strategic option for the tolerance induction in human host. Because two different immune responses including low T-cell-dependent Abs levels and innate immune response activation are directly associated with responsible damages of xenografts, immune tolerance against xenografts has been achieved using by approaches of mixed chimerism and thymus transplantation in pig-to-mouse xenotransplantation model. Such successful application has also been applied to pig-to-baboon xenotransplantation model as an experimental extension. To achieve such

C.-H. Kim, *Glycoimmunology in Xenotransplantation*,
https://doi.org/10.1007/978-981-99-7691-1_18

immune tolerance, alternative studies are cooperatively required for macro-chimerism persistence, prolonged survival of pig xenografts, and donor unresponsiveness [8]. Successful tolerance induction reflects success in xenotransplantation.

18.2 Tolerance Induction of Thymic Transplantation

Thymic transplantation can achieve tolerance induction in xenogeneic T cells. Before HR, xenogeneic responses of xenogeneic T cells are thought to be non-severe, compared to allogeneic responses of T cells [9]. The T cells of human initially respond to xenogeneic MHC antigens as main targets, which similarly responds to allogeneic MHC antigens. Moreover, human T-cell responses against pig xenoantigens are similar to those appeared in APC requirements necessary for stimulating factor (termed direct pathway) or responding factors (termed indirect pathway). The primary and major xenoantigen-specific responses of human are directed to target MHC C-II antigens of pigs. These responses are begun by binding to human $CD4^+$ T cells [10]. Activation of CD4 T cells leads to stimulation of B cells and NK cells. In addition, activation of CD4 T cells helps progenitors of cytotoxic $CD8^+$ T cells [11]. Now, thymic transplantation is a reliable strategic approach to potentiate tolerance induction of T cells across both xenogeneic and allogeneic barriers in transplantations including pigs-to-humanized mice, pigs-to-baboons, pigs-to-pigs, and pigs-to-mice organ transplantation models [12–14]. This is because education of T cells occurs in the thymus that can induce self-tolerance. Hence, thymic transplantation contributes to tolerance induction of transplantation through the known central-acting mechanisms.

Usage of mixed chimerism is also another promising strategy for tolerance induction against allografts and xenografts. The mixed chimerism state refers to a state where host and donor hematopoietic cells coexist together [15]. Mixed chimerism with transient chimerism is achieved by bone marrow (BM) infusion of donors. This status of mixed chimerism efficiently enables the tolerance induction in renal transplantation of clinical trials [16, 17]. Nevertheless, the mixed chimerism status eventually activate T cells, and this again continuously activates NK cells and B cells [18, 19]. Thus, so-called durable chimerism concept has been generated as durable chimerism is necessary for tolerance induction of stable T cells/B cells/NK cells upon encountering xenografts as immune rejection barriers. Mixed chimerism also can tolerize NK cells [20]. In the durable chimerism of rat to mouse model of xenogeneic transplantation, mice NK cells are not responsive to the rat chimerism [21].

For the mixed chimerism model of pig and human in humanized mouse, NK cells of human are specifically tolerant to mixed chimerism of pigs [20]. The expressed inhibitory ligands of human NK cells inhibit NK-cell-mediated rejection in pig hematopoietic cells. Unfortunately, such macro-chimerism in NHPs was not observed [23]. Most pig cells are cleared at early time within 24–48 h of infusion of BM [22]. This indicates that additional strategies for donor chimerism are needed. To block clearance of BM cells, the known SIRPα protein, which is a transmembrane signal regulatory protein with an intracellular Tyr kinase activity is upregulated in DCs, macrophage cells and neutrophils. Human CD47 (hCD47) protein has been known as the SIRPα receptor ligand, and SIRPα–CD47 interaction inhibits phagocytosis by macrophage cells [24]. In human hematopoietic stem cells-transplanted animal model, which was introduced to severely immunocompromised mice, to provoke human hematopoietic genesis, the hematopoietic genesis and engraftments were much improved in human SIRPα gene-transgenic mouse [25]. In addition, hCD47 also improves engraftment level of donor cells of pig in a similar manner in recipient baboons, when α1,3Gal-T KO-hCD47 tg pigs were used [26]. When the transgenic pig peripheral blood stem cells (PBSCs) were transplanted into conditioned baboons, the survival time of pig hematopoietic cells was increased by the expressed hCD47 during pig-to-primates transplantation. Therefore, the CD47 expression is correlated with prolongations of peripheral pig chimerism and pig grafts [26].

18.3 Tolerance Induction of Hematopoietic Cell Transplantation-Based Mixed Chimerism

Hematopoietic cell transplantation-based mixed chimerism of xenogeneic type effectively prevents xenograft rejection in mouse models [27]. Mouse model of the mixed chimerism of xenogeneic and allogeneic types allows hyporesponsiveness and specific tolerance induction in NK cells of mouse host, respectively. For the pig-to-mouse or rat-to-mouse transplantations, the mixed xenoantigenic chimerism tolerizes T cells. For rat-to-mouse transplantation, chimerism also induces B-cell tolerance [28, 29]. NK cells obtained from chimeric mouse of pig and human exhibit the reduced cytotoxic activity against cells of pigs or systemic hyporesponsiveness activity. Mixed xenogeneic chimerism has no effect on the NK-cell maturation of human, while influencing distribution of NK-cell subsets and BM IFN-γ synthesis. NK cells are associated to rodent and primate rejections of xenografts [30, 31]. In xenografts rejection in a mixed chimerism of allogeneic model, host NK-cell tolerance is observed [32]. In rat-to-mouse xenoantigenic model of transplantation, mixed chimerism of xenogeneic cells allows the host systemic unresponsiveness induction of NK cells without rat rejection of donors or MHC C-I or known β2 microglobulin (β2m)-defected mice BM cells [33]. Human peripheral blood $CD56^{dim}CD16^{bright}$ NK cells are highly cytotoxic with low cytokine production, while $CD56^{bright}CD16^{dim}$ NK cells are low cytotoxic with high cytokine production [34]. Chimeric mice BM NK cells reduced IFN-γ production, even though $CD56^{bright}CD16^{dim}$ NK cells occupy high levels rather than non-chimeric mice [35]. As described in target cells-encountered NK cells of human, human BM NK cells $CD56^{bright}CD16^{dim}$ of chimeric mouse refer to the reflection of shedding CD16 from NK cells when bind to pig cells [36].

Reduction in responses between pig cells and human NK cells is related with systemic hyporesponsiveness event in chimeric mice mixed with pig and human cells. Induction of allograft and xenograft tolerance indicates systemic hyporesponsiveness event of NK cells of host. Technically, mixed chimerism with allogeneic and xenogeneic cells has been tried in mice models of mixed chimerism with pig and human to induce tolerance of NK cells of human. When NK cells have been isolated from most mixed chimeric mice with pig and human, the NK cells are weak in systemic hyporesponsiveness event and cytotoxic activity against pig cells. Mixed xenogeneic chimerism has been suggested to elevate hyporesponsiveness event of human NK cells against pig cells [36]. As described earlier, xenogeneic organs are the most reliable sources to overcome the organ shortage for transplantation solution and the animal pigs are absolute to use as a promising supplier [2]. But strong immune rejection as a key hurdle is obstacle to overcome xenotransplantation [38]. The most global approach of xenograft rejection prevention is to remove the xenoantigens. As a second strategy to overcome rejection of xenografts, induction of immune tolerance is considered. The xenograft rejection tolerance is the status that the immunity of host is hyporesponsive to the organ xenografts of pigs [3]. The merit of xenograft rejection tolerance is that this does not need the long-term immunosuppression. Because rejection of xenografts is largely progressed by NK cells [30, 31], mixed chimerism model of allogeneic organs triggers the specific tolerance induction of tolerance of NK cells in hosts [31], as human NK-cell responses to pig cells are downregulated [37].

Although mixed chimerism with xenogeneic organs exerts hyporesponsiveness induction of NK cells of human against pig cells and thus such mixed xenogeneic chimerism may induce xenogeneic tolerance, however, the NK-cell efficacy of tolerance induction is not evident yet [35]. However, at present, it is not studied yet whether mixed chimerism exerts tolerance induction of NK cells of human to xenografted organs of pig. In a humanized model of mouse, mixed hematopoietic chimerism with pig and human is potentiated [39]. Induction of the human NK-cell development in mixed chimeras with pig and human is not influential to chimerism of pigs, although mixed chimerism with pig and human exerts tolerance induction of human NK cells in a humanized mice model [35]. NK cells of human obtained from the chimeric mice mixed with pig and human are not cytotoxic to pig cells and do not display systemic hyporesponsiveness. Thus, mixed chimerism with xenogeneic hematopoietic cells easily suppresses responses of human NK cells to pig-derived cells [35]. SLA and HLA show high sequence identity [40]. Mixed chimerism with xenogeneic hematopoietic cells contributes to eithers pig cells-specific hyporesponsiveness or systemic hyporesponsiveness of NK cells of human. Despite the possibility of xenograft tolerance, human NK-cell tolerance with normal function is defined. For example, mixed xenogeneic chimerism is correlated with human NK-cell tolerization induction in immune system. Currently, relation between pre-existing NK cells and engraftment of pig BM cells is required to be clear in future. From hyporesponsiveness of the human NK cells to pig cells, the HLA C-I-expressing donor pigs with NK cells-inhibitory peptides like human HLA-E, β2m, and leader peptide trimers can be searched for their activity [41].

When human NK cells recognize target cells, the cells are activated to kill the target cells. The human NK cells have various surface receptor proteins like ICAM-1 and LFA-1 [42]. Human NK-cell LFA-1 binds mouse ICAM-1 [43] and induces cytolytic granulation, but not degranulation [44]. Murine NK-cell inhibitory receptors are associated with the class-I allorecognition [45], and wholly allogeneic C-I molecules strongly decrease the BM destruction level, in which NK cells are mediated, rather than MHC-I-deficient cells [46]. Pig inhibitory signals are transduced to human NK cells to accept pig's inhibitory signals. For example, pig aortic ECs can express unknown ligands specifically bound to human inhibitory killer cell Ig-like receptor (KIR), to protect the pig aortic ECs from cytotoxicity of human NK cells [47]. Genetic polymorphism of human KIR gene can suggest the differential importance on human NK cells for mixed chimerism with pig and human. Family of the human KIR genes, encoding for both activating and inhibitory receptor proteins, are comprised of 14 different genes, which are significantly polymorphic. Inhibitory KIRs bind to specific HLA C-I proteins. Xenogeneic molecules of MHC are bound to inhibitory receptors of NK cells, while foreign xenogeneic receptors are activated via binding to activating ligands. Hence, in mixed xenogeneic chimerism with rat and mouse, xenogeneic hematopoietic cells of rat stimulate murine NK cells to result in an "anergy state" in NK cells and systemic hyporesponsiveness state. These states are called "disarming" NK-cell tolerance model and dominantly tolerant induction of MHC-I-defected cells in mixed chimerism. Matured human NK-cell inhibitory receptors are not recognizable for MHC molecules of pigs, whereas activating receptors of human NK cells specifically bind to their target molecules as their ligands expressed on pig cells.

18.4 Receptor and Ligand Incompatibilities Between Xenogenic Organisms

Receptor and ligand incompatibilities between mouse cells and human cells reflect the weak human NK-cell cytotoxic activity against mouse-derived cells. Because systemic unresponsiveness state is mediated by tolerance of NK cells in mouse and rat chimerism, interaction failure between mouse cells and human cells reflects the failed hyporesponsiveness state induction of NK cells of human in the mouse. In the mixed xenogeneic chimerism, NK cells of human lowery respond to target cells of pig with the reduced cytotoxic cell death level against target cells of pig. Thus, mixed xenogeneic chimerism induction reflects the induced human NK-cell tolerance with B-cell tolerance induction and T cells against xenografted organs in human recipients. In a rat-to-mouse BM transplantation, NK cells of host mouse are tolerantly induced against targeted rat cells via systemic hyporesponsiveness state [33]. However, in mixed allogeneic chimerism, NK-cell tolerance of host mouse is induced to tolerize donor cells [32]. In case of pig-to-baboon chimerism, to improve pig/baboon chimerism, another strategy termed "intrabone BM transplantation (abbreviated to IBBMTx)" has been developed. Recently, IBBM-Tx administration of BM cells of donors to BM spaces of recipients by direct injection rapidly reconstitutes with high survival outcomes in allogeneic animal. The modified IBBM-Tx combined with 3D collagen gel matrix was also applied for a pre-clinical approaches of pig-to-baboon xenotransplantation.

References

1. Yang Y-G, Sykes M. Xenotransplantation: current status and a perspective on the future. Nat Rev Immunol. 2007;07(07):519–31.
2. Griesemer A, Yamada K, Sykes M. Xenotransplantation: immunological hurdles and progress toward tolerance. Immunol Rev. 2014;258(1):241–58.
3. Sachs DH, Sykes M, Yamada K. Achieving tolerance in pig-to-primate xenotransplantation: reality or fantasy. Transpl Immunol. 2009;21(2):101–5.
4. Iwase H, Hara H, Ezzelarab M, et al. Immunological and physiological observations in baboons with life-supporting genetically engineered pig kidney grafts. Xenotransplantation. 2017;24: e12293. https://doi.org/10.1111/xen.12293.
5. Mohiuddin MM, Singh AK, Corcoran PC, et al. Chimeric 2C10R4 anti-CD40 antibody therapy is critical for long-term survival of GTKO.hCD46.hTBM pigto-primate cardiac xenograft. Nat Commun. 2016;7:11138.
6. Shin JS, Min BH, Kim JM, et al. Failure of transplantation tolerance induction by autologous regulatory T cells in the pig-to-nonhuman primate islet xenotransplantation model. Xenotransplantation. 2016;23:300–9.
7. Yamada K, Sachs DH, DerSimonian H. Human antiporcine xenogeneic T cell response. Evidence for allelic specificity of mixed leukocyte reaction and for both direct and indirect pathways of recognition. J Immunol. 1995;155:5249–56.
8. Yamada K, Sykes M, Sachs DH. Tolerance in xenotransplantation. Curr Opin Organ Transplant. 2017;22(6):522–8. https://doi.org/10.1097/MOT.0000000000000466.
9. Moses RD, Pierson RN 3rd, Winn HJ, Auchincloss H Jr. Xenogeneic proliferation and lymphokine production are dependent on CD4þ helper T cells and self antigen-presenting cells in the mouse. J Exp Med. 1990;172:567–75.
10. Lee RS, Yamada K, Womer KL, Pillsbury EP, Allison KS, Marolewski AE, Geng D, Thall AD, Arn JS, Sachs DH, Sayegh MH, Madsen JC. Blockade of CD28-B7, but not CD40-CD154, prevents costimulation of allogeneic porcine and xenogeneic human anti-porcine T cell responses. J Immunol. 2000;164(6):3434–44. https://doi.org/10.4049/jimmunol.164.6.3434.
11. Seebach JD, Yamada K, McMorrow IM, et al. Xenogeneic human antipig cytotoxicity mediated by activated natural killer cells. Xenotransplantation. 1996;3:188–97.
12. Tanabe T, Watanabe H, Shah JA, et al. Role of intrinsic (graft) versus extrinsic (host) factors in the growth of transplanted organs following allogeneic and xenogeneic transplantation. Am J Transplant. 2017;17:1778–90.
13. Griesemer AD, Hirakata A, Shimizu A, et al. Results of gal-knockout porcine thymokidney xenografts. Am J Transplant. 2009;9:2669–78.
14. Fudaba Y, Onoe T, Chittenden M, et al. Abnormal regulatory and effector T cell function predispose to autoimmunity following xenogeneic thymic transplantation. J Immunol. 2008;181:7649–59.
15. Sykes M. Mixed chimerism and transplant tolerance. Immunity. 2001;14(4):417–24.

16. Kawai T, Cosimi AB, Spitzer TR, et al. HLA-mismatched renal transplantation without maintenance immunosuppression. N Engl J Med. 2008;358:353–61.
17. Scandling JD, Busque S, Dejbakhsh-Jones S, et al. Tolerance and chimerism after renal and hematopoietic-cell transplantation. N Engl J Med. 2008;358:362–8.
18. Tonomura N, Shimizu A, Wang S, et al. Pig islet xenograft rejection in a mouse model with an established human immune system. Xenotransplantation. 2008;15:129–35.
19. Lan P, Tonomura N, Shimizu A, et al. Reconstitution of a functional human immune system in immunodeficient mice through combined human fetal thymus/liver and CD34þ cell transplantation. Blood. 2006;108:487–92.
20. Li HW, Vishwasrao P, Holzl MA, et al. Impact of mixed xenogeneic porcine hematopoietic chimerism on human NK cell recognition in a humanized mouse model. Am J Transplant. 2017;17:353–64.
21. Kawahara T, Rodriguez-Barbosa JI, Zhao Y, et al. Global unresponsiveness as a mechanism of natural killer cell tolerance in mixed xenogeneic chimeras. Am J Transplant. 2007;7:2090–7.
22. Griesemer A, Liang F, Hirakata A, et al. Occurrence of specific humoral nonresponsiveness to swine antigens following administration of GalT-KO bone marrow to baboons. Xenotransplantation. 2010;17:300–12.
23. Liang F, Wamala I, Scalea J, et al. Increased levels of antinon-Gal IgG following pig-to-baboon bone marrow transplantation correlate with failure of engraftment. Xenotransplantation. 2013;20:458–68.
24. Ide K, Wang H, Tahara H, et al. Role for CD47-SIRPalpha signaling in xenograft rejection by macrophages. Proc Natl Acad Sci U S A. 2007;104:5062–6.
25. Wang C, Wang H, Ide K, et al. Human CD47 expression permits survival of porcine cells in immunodeficient mice that express SIRPalpha capable of binding to human CD47. Cell Transplant. 2011;20:1915–20.
26. Tena AA, Sachs DH, Mallard C, et al. Prolonged survival of pig skin on baboons after administration of pig cells expressing human CD47. Transplantation. 2017;101:316–21.
27. Ildstad ST, Wren SM, Bluestone JA, Barbieri SA, Stephany D, Sachs DH. Effect of selective T cell depletion of host and/or donor bone marrow on lymphopoietic repopulation, tolerance, and graft-vs-host disease in mixed allogeneic chimeras (B10 + B10.D2----B10). J Immunol. 1986;136(1):28–33.
28. Abe M, Qi J, Sykes M, Yang Y-G. Mixed chimerism induces donor-specific T-cell tolerance across a highly disparate xenogeneic barrier. Blood. 2002;99(10):3823–9.
29. Ohdan H, Yang Y-G, Shimizu A, Swenson KG, Sykes M. Mixed chimerism induced without lethal conditioning prevents T cell- and anti-Galα1,3Gal-mediated graft rejection. J Clin Invest. 1999;104(3):281–90.
30. Forte P, Lilienfeld BG, Baumann BC, Seebach JD. Human NK cytotoxicity against porcine cells is triggered by NKp44 and NKG2D. J Immunol. 2005;175(8):5463–70.
31. Lilienfeld BG, Garcia-Borges C, Crew MD, Seebach JD. Porcine UL16-binding protein 1 expressed on the surface of endothelial cells triggers human NK cytotoxicity through NKG2D. J Immunol. 2006;177(4):2146–52.
32. Zhao Y, Ohdan H, Manilay JO, Sykes M. NK cell tolerance in mixed allogeneic chimeras. J Immunol. 2003;170(11):5398–405.
33. Kawahara T, Rodriguez-Barbosa JI, Zhao Y, Zhao G, Sykes M. Global unresponsiveness as a mechanism of natural killer cell tolerance in mixed xenogeneic chimeras. Am J Transplant. 2007;7(9):2090–7.
34. Cooper MA, Fehniger TA, Caligiuri MA. The biology of human natural killer-cell subsets. Trends Immunol. 2001;22(11):633–40.
35. Li HW, Vishwasrao P, Hölzl MA, Chen S, Choi G, Zhao G, Sykes M. Impact of mixed xenogeneic porcine hematopoietic chimerism on human NK cell recognition in a humanized mouse model. Am J Transplant. 2017;17(2):353–64. https://doi.org/10.1111/ajt.13957.
36. Grzywacz B, Kataria N, Verneris MR. CD56dimCD16+ NK cells downregulate CD16 following target cell induced activation of matrix metalloproteinases. Leukemia. 2007;21(2):356–9.

37. Li HW, Vishwasrao P, Hölzl MA, Chen S, Choi G, Zhao G, Sykes M. Impact of mixed xenogeneic porcine hematopoietic Chimerism on human NK cell recognition in a humanized mouse model. Am J Transplant. 2017;17(2):353–64.
38. Cooper DKC, Ekser B, Tector AJ. Immunobiological barriers to xenotransplantation. Int J Surg. 2015;23(Part B):211–6.
39. Lan P, Wang L, Diouf B, Eguchi H, Su H, Bronson R, et al. Induction of human T-cell tolerance to porcine xenoantigens through mixed hematopoietic chimerism. Blood. 2004;103(10): 3964–9.
40. Chardon P, Rogel-Gaillard C, Cattolico L, Duprat S, Vaiman M, Renard C. Sequence of the swine major histocompatibility complex region containing all non-classical class I genes. Tissue Antigens. 2001;57(1):55–65.
41. Weiss EH, Lilienfeld BG, Müller S, Müller E, Herbach N, Keler B, et al. HLA-E/human beta2-microglobulin transgenic pigs: protection against xenogeneic human anti-pig natural killer cell cytotoxicity. Transplantation. 2009;87(1):35–43.
42. Bryceson YT, March ME, Ljunggren H-G, Long EO. Activation, coactivation, and costimulation of resting human natural killer cells. Immunol Rev. 2006;214(1):73–91.
43. March ME, Long EO. β2 integrin induces TCRζ-Syk-phospholipase C-γ phosphorylation and paxillin-dependent granule polarization in human NK cells. J Immunol. 2011;186(5): 2998–3005.
44. Bryceson YT, March ME, Barber DF, Ljunggren H-G, Long EO. Cytolytic granule polarization and degranulation controlled by different receptors in resting NK cells. J Exp Med. 2005;202 (7):1001–12.
45. Kumar S. Natural killer cell cytotoxicity and its regulation by inhibitory receptors. Immunology. 2018;154(3):383–93. https://doi.org/10.1111/imm.12921.
46. Höglund P, Sundbäck J, Olsson-Alheim MY, Johansson M, Salcedo M, Öhién C, et al. Host MHC class I gene control of NK-cell specificity in the mouse. Immunol Rev. 1997;155(1): 11–28.
47. Kwiatkowski P, Artrip JH, John R, Edwards NM, Wang S-F, Michler RE, et al. Induction of swine major histocompatibility complex class I molecules on porcine endothelium by tumor necrosis factor-alpha reduces lysis by human natural killer cells. Transplantation. 1999;67(2): 211–8.

Chapter 19
Genome Editing and Transgenes in Pigs

19.1 Introduction

Disruption of specific target genes is a first step to uncover the gene function. Large-sized animals are difficult to be targeted in embryo level, causing damages in early embryo developments. The recent development of genome editing techniques is being evolved to the easier ZFN, TALEN, and CRISPR/Cas9 technologies than the conventional laborious one in various animals, giving generation of gene knock-out pigs. The pig genome is about 2.7 Gb size in length, which is packaged in 18 autosomes with sex chromosomes X and Y [1, 2]. Genetic manipulation in pigs was conventionally performed by breeding selection to keep specific segments of the genome. The first genetic modification in the pig was the transgenic case [3]. Then, homologous recombination-based gene knocking out [4, 5] and knocking in were made in pigs and knock-ins [6], respectively. Gene KO disrupts specific genetic sequence to inactivate the genes that encode proteins, while knock-in gene is reversed with the gene's encoded protein.

A recent CRISPR/Cas9 technology much improved the current targeting mutation level during embryogenesis in large animals such as pigs, without the time-consuming breeding to produce gene-targeted embryos [7, 8]. The CRISPR (clustered, regularly interspaced short palindromic repeat)/CRISPR-associated 9 (Cas9) (referred to CRISPR/CAS9) resulted in the significantly improved gene manipulations in pig xenograft animals [9]. In addition, there are also several developed technologies with gene editing, too. The genome editing CRISPR/Cas9 system is now frequently utilized for cell transplant applications using corrected autologous stem cells of patients and gene-inherited genetic diseases as well as the tailored xenogeneic donor pigs in transplantation. Utilizing such technologies, xenotransplantation potentiated prolonged survival time of organ xenografts within several days to several months in transplantation [10, 11], over 6 months in NHPs [12–16] and 2 years and more in a non-life-supporting heterotopic model of cardiac xenografts [17]. The long survival time more than 4 months was observed in the

C.-H. Kim, *Glycoimmunology in Xenotransplantation*,
https://doi.org/10.1007/978-981-99-7691-1_19

rhesus macaques model that CD154-specific Abs were treated and α1,3Gal-T KO/CD55 transgenic pig kidneys were transplanted, where anti-CD154-drug decline the production of natural antibodies [10]. Thus, xenotransplantation is lightening for resolution of the organ shortage issue. The CRISPR/CAS9 editing tools are now a promising strategy as an emerging, powerful and efficient system [1–14, 18–21].

The CRISPR/CAS9 technology has now been successfully utilized in many organisms of mice, monkeys, pigs and rats [22–27]. B-cell-deficient pigs have also been created by the CRISPR/Cas9 gene editing technology [28], by combination of CRISPR/Cas9 system with SCNT to generate human antigen-reactive Abs. The CRISPR/Cas9 editing system can create genetically targeted pig strains that are specifically and genetically modified either by modification of somatic cells and SCNT or direct modification of in vitro generated zygotes. Targeted and specifically mutated cells utilized as donor nuclei can create targeted gene-modified pig strains with stably identic mutation of the cloned piglet genotype without embryo toxicity during the mRNA injection strategy. Plasmids of CRISPR/Cas9 gene editing tool can be transferred to the cellular nucleic region and the CRISPR/Cas9 component expression is enhanced with high targeting efficiency [28].

19.2 Scientific History of Genome Editing

The recent xenotransplantation requires the pig genome engineered donor organ prior to transplantation. Genetic modifying CRISPR/Cas9 strategy provide the humanized multiple transgenic pigs for xenotransplantation. Pig xenotransplantation using genetically modified pigs opened new way and vista because the barriers to successful xenotransplantation are overcome by genetic modifications. Currently, livestock are being subjected to complex combined genetic modifications with efficient engineering outcomes. Indeed, xenoprotective pigs are modified and generated to suppress short- to mid-term rejection of pig xenografts. Genetically targeted editing of house miniature pigs are now available to apply because somatic cell nuclear transfer is now possible to allow cell-mediated transgenic insertion and targeted gene, alternatively replacing the lacked pluripotent stem cells as functional cells [29, 30]. In fact, a lot of pig strains have been produced by stepwise nuclear transfer. The genetically modified strains were examined at each step to check and overcome HR and AVR by the recipient. The gene targeting efficiency has been gradually improved by developed gene editing sources of CRISPR/Cas9 system, Tal effector nuclease, zinc finger nuclease (ZFN) and transcription activator-like effector nuclease (TALEN). Those gene-editing systems are efficient in creating the targeted genetic change.

The genetic engineering technologies are serially progressed with the ZFN [31, 32] and rapid targeted gene KO using TALEN [33–35]. CRISPR/Cas9- or ZFN-based homozygous CMAH and GGTA1 genes KO-eliminated α-1,3Gal antigens and Neu5Gc epitope antigens [36]. ZFNs and TALENs are established using specific nucleotides- or nucleotide strings-recognizing protein domains. ZFNs and

TALENs are also nuclease-dependent, which catalyzes dimerization for cleavage of the DNA sequences. Monomeric form of each nuclease enzyme, known as Fok1, from zinc finger pairs or Tal effector sis aligned on the targeted cutting site. When the stranded DNAs once are cleaved, non-homologous end joining repairs the sequence. If the DNA sequences are repaired correctly, the cleavage DNA site is subjected to be repeated cutting until the cutting error is introduced. The error preventively blocks one or both of the two TALEN and ZFN nuclease pair binding to DNA, allowing obstructed dimerization of these nucleases of ZFN and TALEN. Single or base pair insertion or deletion leads to a frame-shift and this status is a gene knockout [36]. ZFNs were used in pigs to knock out [37] and endogenously expressed glycoprotein or α1,3Gal-T-1 genes in somatic cells are subjected to somatic cell nuclear transfer and also injected to zygotes [38]. Seemingly, TALEN nuclease is utilized in both of somatic [39] and zygotic cells [40].

Apart from production of the homozygous single-gene KO pigs by homologous recombination technology, a genome editing technology has been developed. Recently, multiple genetic modifications are applied for normal animal breeding and for transplants. The CRISPR/Cas9 system is utilized to targeted-delete specific genes and to generate genetically modified animals. Differently to TALEN and ZFN, the gene editing technology, CRISPR/Cas9 tool, is depended on a Cas9 single nuclease and a DNA sequence homologous to the DNA sequence of target genome. The CRISPR/Cas9 system is known to have a great merit compared to the above gene editors because this system is very easy to construct. Targeting genomic DNA location is created in a short time of several days. The CRISPR/Cas9 was used in pigs for in vivo-produced zygotes [41], and somatic cells and zygotes were further used to produce the target gene KO by utilization of eggs and zygotes in vitro [42]. Since the pig report, many pig genes were inactivated via the CRISPR/Cas9 agent injection to zygotes and all the targeted pigs were edited at the target site. The techniques use the ZFNs and TALENs at the initial technology development stage; however, the CRISPR/Cas9 system was developed to generate simultaneously multigene inactivated genes [43]. Pig organs and tissues have genetically modified to the lowered antigenicity and increased compatibility between donor and recipient humans. For example, homozygous triple knockouts for SLA class I genes [9] and xenoantigenic glycan genes [44] have been generated. CRISPR/Cas9 technology does not require somatic cell nuclear transfer stage because of component injection of the CRISPR/Cas9 into pig zygotes and finally transferring to mothers [45]. The most recently developed technology, a type II CRISPR)/Cas9 system, was reported to efficiently generate such modified pigs [46–49]. Such rapid progress enabled genome editing of mammalian genomes including multiple transgenes for xenotransplantation [16, 31, 32, 34, 35, 46–58].

At present, 40 more different gene-edited pigs are estimated worldwide up to seven modified genes by combination with a single pig. The pig organ and cell transplantations to human recipients are anticipated in successful achievement in the near future together with the glycobiology progress, as did during the past 30 years. For xenograft survival, multiple genetic modifications are effective [59] and such pigs were created [60] with more modifications in future, as most likely needed.

Removal of the single α1,3Gal-T in pigs has not exhibited any negative phenotype [60, 61]; however, more multiple genetic modifications in pigs may display any unexpected phenotype, as cataract was developed in pigs [62]. CRISPR/Cas9 technology gives a big promising for the clinically successful transplantation. The nucleases-mediated homologous DNA recombination can effectively create mutation defects by targeting in pig cells [63]. The CRISPR/Cas9 technology creates pig gene-targeted lineages [46, 64, 65]. CRISPR/Cas9 genome editing technology as a revolutionarily creation way modifies precise pig sources as donors required for future xenotransplantation [9]. Compared to ZFNs and TALENs, CRISPR/Cas9 is efficient and multiplies target genes [66]. The merit points of the efficient targeting and easy CRISPR/Cas9 gene-editing tool facilitate the superior pig gene KO. Non-effected targets are a negative point of the CRISPR/Cas9 gene targeting system [67–69]. CRISPR/Cas9 editing technology confers potential possibilities to better study the target genotype functions.

Genetically modified pigs are a future promising technology in xenotransplantation, giving donor cells, tissues and organs to cover organ donor shortage. The combinatory genome editing via gene modification of donor animals can solve the barriers to xenogeneic transplantation with the developed immunosuppressive drugs to use in xenograft rejection inhibition. When three glycan epitopes are removed, and additionally human coagulation-regulatory proteins and CRPs are expressed in pigs, a clinical organ xenotransplantation will be potentiated. However, there are numerous genetic manipulations for expression of anti-inflammatory genes that are beneficial for the xenotransplant outcome and that decrease or inhibit the adaptive response of immunity. To genetically modify and prevent HR, the CRISPR/Cas9 gene editing is applied to efficiently crease a biallelic α1,3Gal-T KO gene in pigs [56]. Because HR is caused by binding capacities of premade natural antibodies and artificial depletion of anti-pig Abs in recipients reflects the HR prevention.

GGTA1 and CMAH genes were simultaneously inactivated by the targeting technology [46, 47]. As reported, the CRISPR/Cas9 can afford human transgene delivery and mutation at the same time, as CRISPR/Cas9 editing is utilized to remove the GGTA1, CMAH, and β4GalNT2 genes in pigs [46]. To date, GGTA1/CMAH/β4GALNT2 triple KO pigs, phenotypically deleting the Gal-α1,3-Gal, *N*-NeuGc and an SDa antigen-like glycan, have been produced by the CRISPR/Cas9. The resultant bioprosthetic heart valves obtained from the GTKO/CMAHKO/β4GalNT2KO pigs were functionally active for longer periods than the control pigs. Such pig carbohydrate profile alteration prevents human antibody-dependent cytotoxicity and CDC in the rigorous NHP model of xenotransplantation [49]. However, it is considered that pigs are necessary to modify numerous genes to escape xenografted organs from the serial rejection immune surveillance operated by the host recipients. For example, xenoprotection is further improved by multiple combination, depending on the xenoantigenicity [70, 71]. Many complement regulators-expressing transgenic pigs are therefore created but most of them consist of complement regulatory factors like cDNAs or mini-genes. Multiple genetic modified pigs only, which are generated via single modification-based inter-breeding of animals or via several gene construct co-transfections can be an actual supplier of beneficial xenoorgans for human uses. Genetically modified pig crossbreeding gives us future hope for xenotransplantation. Thus, glycoimmune-xenobiology of xenotransplantation includes all the transplantation. Genetic modifying pigs are technically affordable to creation for any type of genetic modification in a short time and cheap levels. The knocking out and modifying select nucleotides are possible at a time [72]. Genetically modified pigs will be on hands to further study for humans. NSRRC (National Swine Resource and Research Center) of the University of Missouri is an established organization toward biomedical utilization with currently over 60 strains available (http://nsrrc.missouri.edu).

19.3 Recent Progress in Gene Editing: "Gene Stacking" and "Combineering" as Well as FokI-dCas9-Targeted Insertion

Sequential targeted gene editing technology named "gene stacking" has been described as a tool of transgene co-integration [73]. The "Gene stacking" was initially performed using site-specific recombinases [37]. Recently, co-location strategy at one genomic locus of multiple genes, including xenoantigen-protecting transgenes and sequentially targeting transgene named "gene stacking," and multiple co-integrated genes named "combineering," were designed to create genetically modified pigs. Such produced pigs can suppress rejection of pig xenografted organs within short to mid-term periods. Technically, combination with two alternative strategies was used for multiple transgenes at one genomic locus for xenoprotection. The co-located transgenes with sequentially-targeted placement of transgenes, named "gene stacking" at one genomic-targeted locus is combined with co-integrated bacterial artificial chromosome (BAC) vector and phage AC (PAC) vector. Thus, this combined recombination is called "combineering." Such produced cells were then used to inactivate xenoreactive gene [73].

Co-integration of multiply-engineered vectors with high capacity of the term "combineering," which is combined with gene editing modification and nuclear-transferring technology are serially utilized [74, 75]. The *E. coli* BAC and PAC constructs were used to incorporate multiple xenoprotective transgenes in pigs. Alternative combination of 2A peptide cDNAs or polycistronic system designed by IRES to create xenoreaction-protective donor pigs with multiple transgenes [76, 77] and serial nuclear transfer was employed to generate pigs. Then, genomic multiple-transgenic pigs which carry the expressed complement regulatory proteins of human including CD46, CD55, and CD59, and also human A20 and HO-1 cDNA cassettes have been created to protect endothelium at a single chromosomal locus [73]. Highly expressing-2 transgenic CD55 and HO-1 genes is inserted by homologous-recombination technology as each single copy at the pig *ROSA26* genomic locus and additional transgenes has been inserted to the array. Biallelic KO animals that targeted with two different genes of *GGTA1* and *CMAH* were therefore examined in the multiple tg-based animals to block HR [78]. Previously, the triple complement regulators-expressing pigs have been produced by CD46 gene and CD59 gene construct microinjections to a single CD55-tg pig. Unfortunately, the transgenes are not ubiquitously or abundantly expressed [79]. Demerit of the study was also in that transgenic integration at different genomic locus is not efficient due to segregation-decreased multi-tg offspring proportion. Previously, CD46 itself reduces the activation and cascade levels of complement coagulation and fibrinolysis [80] but is not enough. Therefore, additional transgenic expressions in donor pigs are desired to protect the donor organs from dysregulated coagulation events and thrombosis-caused microangiopathy involved in activation of ECs.

For such available transgene candidates, thrombomodulin protein known as a human anticoagulant protein, CD39, and the endothelium protein C receptor (EPCR) are reported [81–83]. Further, to suppress the next step of adaptive responses of immunity, immunosuppression is also required. Because *GGTA1* KO disruption and CD46 tg upregulation suppress the in vitro immune responses of human T cells to xenoantigenic cells of pigs [84, 85], recently developed CTLA4-Ig or LEA29Y60 as regulators of T cells can be considered with additional transgenes. Those developed pig donors reflect an advanced progression for pig-to-human xenotransplantation at the clinical levels, as genetic modification of livestock is now reached matured. The continued improvements are progressed in genomic sequencing, nuclear transferring and synthetic endonuclease technologies [86, 87] toward xenotransplantation.

On the other hand, recently, genetic modified animals have been developed by targeted insertion of specific transgenes into pig genomic locus of GGTA1 gene using FokI-dCas9 [88]. This is based on the fact that the initially developed wild-type Cas9 nuclease has potentially a disadvantage in introduction of directional off-targeting mutation [89], consequently producing an unidentified mutation. Furthermore, CRISPR has a substantial demerit that it cannot be used to insert protection transgenes to detrimentally harmful pig genes. Then, to improve such CRISPR-Cas9 gene editing quality, recently, the improved Cas9 has been made with high Cas9 fidelity [90]. The developed FokI-dCas9 contains a nuclease-defective Cas9, which the non-specific FokI endonuclease subunit is fused. FokI-dCas9 utilizes two RNA guides to cleave 20-nucleotide site targets that are spaced, showing excellently 140-fold higher specificity than the original Cas9 wild type [9]. A similarly developed fdCas9 as a chimeric nuclease exhibits on-target effect, almost corresponding to Cas9 wild-type specificity, without wild-type Cas9 off-target site activity [91]. Although the FokI-dCas9 technology has been applied to create various KO mice at that time [92], its KO utilization is now being waited for pigs. However, further, the CRISPR gene editing is not yet used to create pig knock-in, which is different from KO pigs, for pig-to-human xenotransplantation [93, 94].

Recently, the gene-modified pigs have been created by so-called second-generation version of FokI-dCas9 nuclease, which shows extremely low rates of off-target mutations compared the original Cas9 nuclease [60]. In the pig GGTA1 gene-targeting, FokI-dCas9 is used for a 7.1 kb-transgene integration via two RNA guides to the GGTA1 gene exon 9 region in fibroblasts of pig fetal. Thus, the modification-produced pig fetal fibroblasts lack the α1,3Gal-xenoantigenic epitopes, instead, expressed an CD2-specific mAbs. The pig fibroblast cells, utilized for nuclear transfer of somatic cells, produce healthy knock-in piglets that α1,3Gal antigens are deficient but secrete serum CD2-specific Abs. The high-fidelity knock-in transgene system to the genomes of pigs can be utilized in pig-to-human xenotransplantation. To explain the observed gaps in efficacy, the gene editing protocol of a FokI-dCas9 system was developed to efficiently increase the successful knock-in rate of a target transgene of the CD2-specific monoclonal antibody named diliximab, as an immunomodulatory molecule, into the pig GGTA1 gene locus. Secreted immune-modulating molecule by engineered pig cells can protect the pig cells from "local" grafts-reactive immune responses. For the rationale of the targeted insertion of the anti-CD2 mAb gene, a CD2-specific monoclonal antibody, termed diliximab, recognizes human, New World primate and Old World primate T cells [95]. Similarly, this anti-CD2 mAb gene-expressed adenovirus-transduced pig islets protect themselves from human T-cell infiltration in a humanized model of mice [96].

In addition, two other examples of knocked-in anti-CD2 and hTBM genes are also successfully expressed. For example, the CD2-specific mAb gene knock-in cells were produced by SCNT and resulting in generation of two healthy male pigs. Currently, the 7.1 kb CD2-specific mAb transgene has been regarded as the largest sized transgene case for pig CRISPR-based gene knock-in. The α1,3Gal-T KO pigs express CD2-specific monoclonal antibody in the pig sera, probably because the mAbs specifically bind to CD2 protein expressed in primate and human [95, 96] and they don't deplete T cells of pigs. Xenograft-specified expression of diliximab is anticipated to the prolonged survivals of pig xenografts by local depletion of T-cell infiltration of recipient hosts such as human and primate. Therefore, the GGTA1-based knock-in strategy can be further applied to incorporate a multiple cloning site to potentiate easy transgene exchange and interchange for gene expression-regulating factors and coding genes.

References

1. Groenen MA, Archibald AL, Uenishi H, Tuggle CK, Takeuchi Y, Rothschild MF, et al. Analyses of pig genomes provide insight into porcine demography and evolution. Nature. 2012;491(7424):393–8.
2. Prather RS. Pig genomics for biomedicine. Nat Biotechnol. 2013;31(2):122–4.
3. Hammer RE, Pursel VG, Rexroad CE Jr, Wall RJ, Bolt DJ, Ebert KM, et al. Production of transgenic rabbits, sheep and pigs by microinjection. Nature. 1985;315(6021):680–3.
4. Dai Y, Vaught TD, Boone J, Chen SH, Phelps CJ, Ball S, et al. Targeted disruption of the alpha1,3-galactosyltransferase gene in cloned pigs. Nat Biotechnol. 2002;20(3):251–5.
5. Lai L, Kolber-Simonds D, Park KW, Cheong HT, Greenstein JL, Im GS, et al. Production of alpha-1,3-galactosyltransferase knockout pigs by nuclear transfer cloning. Science. 2002;295 (5557):1089–92.
6. Rogers CS, Stoltz DA, Meyerholz DK, Ostedgaard LS, Rokhlina T, Taft PJ, et al. Disruption of the CFTR gene produces a model of cystic fibrosis in newborn pigs. Science. 2008;321(5897): 1837–41.

7. Ryu J. Lee K.CRISPR/Cas9-mediated gene targeting during embryogenesis in swine. Methods Mol Biol. 2017;1605:231–44.
8. Watanabe M, Nagashima H. Genome editing of pig. Methods Mol Biol. 2017;1630:121–39.
9. Cowan PJ. The use of CRISPR/Cas associated technologies for cell transplant applications. Curr Opin Organ Transplant. 2016;21:461–6.
10. Hai T, Teng F, Guo R, et al. One-step generation of knockout pigs by zygote injection of CRISPR/Cas system. Cell Res. 2014;24:372–5.
11. Higginbotham L, Mathews D, Breeden CA, et al. Pretransplant antibody screening and anti-CD154 costimulation blockade promote long-term xenograft survival in a pig-to-primate kidney transplant model. Xenotransplantation. 2015;22:221–30.
12. Iwase H, Hara H, Ezzelarab M, et al. Immunological and physiological observations in baboons with life-supporting genetically engineered pig kidney grafts. Xenotransplantation. 2017;24: e12293. https://doi.org/10.1111/xen.12293. Epub2017 Mar 17
13. Hering BJ, Wijkstrom M, Graham ML, et al. Prolonged diabetes reversal after intraportal xenotransplantation of wild-type porcine islets in immunosuppressed nonhuman primates. Nat Med. 2006;12:301–3.
14. Cardona K, Korbutt GS, Milas Z, et al. Long-term survival of neonatal porcine islets in nonhuman primates by targeting costimulation pathways. Nat Med. 2006;12:304–6.
15. van der Windt DJ, Bottino R, Casu A, et al. Long-term controlled normoglycemia in diabetic nonhuman primates after transplantation with hCD46 transgenic porcine islets. Am J Transplant. 2009;9:2716–26.
16. Bottino R, Wijkstrom M, van der Windt DJ, et al. Pig-to-monkey islet xenotransplantation using multitransgenic pigs. Am J Transplant. 2014;14:2275–87.
17. Shin JS, Kim JM, Kim JS, et al. Long-term control of diabetes in immunosuppressed nonhuman primates (NHP) by the transplantation of adult porcine islets. Am J Transplant. 2015;15:2837–50.
18. Mohiuddin MM, Singh AK, Corcoran PC, et al. Chimeric 2C10R4 anti-CD40 antibody therapy is critical for long-term survival of GTKO.hCD46.hTBM pigto-primate cardiac xenograft. Nat Commun. 2016;7:11138.
19. Jinek M, Chylinski K, Fonfara I, Hauer M, Doudna JA, Charpentier E. A programmable dual-RNA-guided DNA endonuclease in adaptive bacterial immunity. Science. 2012;337:816–21.
20. Wiedenheft B, Sternberg SH, Doudna JA. RNA-guided genetic silencing systems in bacteria and archaea. Nature. 2012;482:331–8.
21. Wei C, Liu J, Yu Z, Zhang B, Gao G, Jiao R. TALEN or Cas9-rapid, efficient and specific choices for genome modifications. J Genet Genomics. 2013;40:281–9.
22. Bassett AR, Liu JL. CRISPR/Cas9 and genome editing in Drosophila. J Genet Genomics. 2014;41:7–19.
23. Shen B, Zhang J, Wu H, Wang J, Ma K, Li Z, Zhang X, Zhang P, Huang X. Generation of gene-modified mice via Cas9/RNA-mediated gene targeting. Cell Res. 2013;23:720–3.
24. Tan W, Carlson DF, Lancto CA, Garbe JR, Webster DA, Hackett PB, Fahrenkrug SC. Efficient nonmeiotic allele introgression in livestock using custom endonucleases. Proc Natl Acad Sci U S A. 2013;110:16526–31.
25. Wang H, Yang H, Shivalila CS, Dawlaty MM, Cheng AW, Zhang F, Jaenisch R. One-step generation of mice carrying mutations in multiple genes by CRISPR/Cas-mediated genome engineering. Cell. 2013;153:910–8.
26. Hai T, Teng F, Guo R, Li W, Zhou Q. One-step generation of knockout pigs by zygote injection of CRISPR/Cas system. Cell Res. 2014;24:372–5.
27. Ma Y, Zhang X, Shen B, Lu Y, Chen W, Ma J, Bai L, Huang X, Zhang L. Generating rats with conditional alleles using CRISPR/Cas9. Cell Res. 2014;24:122–5.
28. Niu Y, Shen B, Cui Y, Chen Y, Wang J, Wang L, Kang Y, Zhao X, Si W, Li W. Generation of gene-modified cynomolgus monkey via Cas9/RNA-mediated gene targeting in one-cell embryos. Cell. 2014;156:836–43.

29. Chen F, Wang Y, Yuan Y, Zhang W, Ren Z, Jin Y, Liu X, Xiong Q, Chen Q, Zhang M, Li X, Zhao L, Li Z, Wu Z, Zhang Y, Hu F, Huang J, Li R, Dai Y. Generation of B cell-deficient pigs by highly efficient CRISPR/Cas9-mediated gene targeting. J Genet Genomics. 2015;42(8): 437–44.
30. Schnieke AE, et al. Human factor IX transgenic sheep produced by transfer of nuclei from transfected fetal fibroblasts. Science. 1997;278:2130–3.
31. McCreath KJ, et al. Production of gene-targeted sheep by nuclear transfer from cultured somatic cells. Nature. 2000;405:1066–9.
32. Li P, Estrada JL, Burlak C, et al. Biallelic knockout of the alpha-1,3 galactosyltransferase gene in porcine liver-derived cells using zinc finger nucleases. J Surg Res. 2013;181:e39–45.
33. Lutz AJ, Li P, Estrada JL, et al. Double knockout pigs deficient in N-glycolylneuraminic acid and galactose alpha-1,3-galactose reduce the humoral barrier to xenotransplantation. Xenotransplantation. 2013;20:27–35.
34. Carlson DF, Tan W, Lillico SG, et al. Efficient TALEN-mediated gene knockout in livestock. Proc Natl Acad Sci U S A. 2012;109:17382–7.
35. Paris LL, Estrada JL, Li P, et al. Reduced human platelet uptake by pig livers deficient in the asialoglycoprotein receptor 1 protein. Xenotransplantation. 2015;22:203–10.
36. Xin J, Yang H, Fan N, et al. Highly efficient generation of GGTA1 biallelic knockout inbred mini-pigs with TALENs. PLoS One. 2013;8:e84250.
37. Fischer K. Efficient production of multi-modified pigs for xenotransplantation by 'combineering', gene stacking and gene editing. Sci Rep. 2016;6:29081. https://doi.org/10.1038/srep29081.
38. Wells KD, Prather RS. Genome-editing technologies to improve research, reproduction, and production in pigs. Mol Reprod Dev. 2017 Sep;84(9):1012–7. https://doi.org/10.1002/mrd.22812.
39. Whyte JJ, Zhao J, Wells KD, Samuel MS, Whitworth KM, Walters EM, et al. Gene targeting with zinc finger nucleases to produce cloned eGFP knockout pigs. Mol Reprod Dev. 2011;78(1):2.
40. Carlson DF, Tan W, Lillico SG, Stverakova D, Proudfoot C, Christian M, et al. Efficient TALEN-mediated gene knockout in livestock. Proc Natl Acad Sci U S A. 2012;109(43): 17382–7.
41. Lillico SG, Proudfoot C, Carlson DF, Stverakova D, Neil C, Blain C, et al. Live pigs produced from genome edited zygotes. Sci Rep. 2013;3:2847.
42. Hai T, Teng F, Guo R, Li W, Zhou Q. One-step generation of knockout pigs by zygote injection of CRISPR/Cas system. Cell Res. 2014;24(3):372–5.
43. Whitworth KM, Lee K, Benne JA, Beaton BP, Spate LD, Murphy SL, et al. Use of the CRISPR/Cas9 system to produce genetically engineered pigs from in vitro-derived oocytes and embryos. Biol Reprod. 2014;91(3):78.
44. Reyes LM, Estrada JL, Wang ZY, et al. Creating class I MHC null pigs using guide RNA and the Cas9 endonuclease. J Immunol. 2014;193:5751–7.
45. Martens GR, Reyes LM, Butler JR, et al. Humoral reactivity of renal transplant-waitlisted patients to cells from GGTA1/CMAH/B4GalNT2, and SLA class I knockout pigs. Transplantation. 2017;101:e86–92.
46. Petersen B, Frenzel A, Lucas-Hahn A, et al. Efficient production of biallelic GGTA1 knockout pigs by cytoplasmic microinjection of CRISPR/Cas9 into zygotes. Xenotransplantation. 2016;23:338–46.
47. Estrada JL, Martens G, Li P, et al. Evaluation of human and nonhuman primate antibody binding to pig cells lacking GTA1/CMAH/beta4GalNT2 genes. Xenotransplantation. 2015;22: 194–202.
48. Li P, Estrada JL, Burlak C, et al. Efficient generation of genetically distinct pigs in a single pregnancy using multiplexed single-guide RNA and carbohydrate selection. Xenotransplantation. 2015;22:20–31.

49. Reyes LM, Estrada JL, Wang ZY, et al. Creating class I MHC-null pigs using guide RNA and the Cas9 endonuclease. J Immunol. 1950;2014(193):5751–7.
50. Butler JR, Martens GR, Estrada JL, et al. Silencing porcine genes significantly reduces human-antipig cytotoxicity profiles: an alternative to direct complement regulation. Transgenic Res. 2016;25(5):751–9.
51. Butler JR, Skill NJ, Priestman DL, et al. Silencing the porcine iGb3s gene does not affect Galalpha3Gal levels or measures of anticipated pig-to-human and pig-to-primate acute rejection. Xenotransplantation. 2016;23:106–16.
52. Yeom HJ, Koo OJ, Yang J, et al. Generation and characterization of human heme oxygenase-1 transgenic pigs. PLoS One. 2012;7:e46646.
53. Klymiuk N, van Buerck L, Bahr A, et al. Xenografted islet cell clusters from INSLEA29Y transgenic pigs rescue diabetes and prevent immune rejection in humanized mice. Diabetes. 2012;61:1527–32.
54. Wheeler DG, Joseph ME, Mahamud SD, et al. Transgenic swine: expression of human CD39 protects against myocardial injury. J Mol Cell Cardiol. 2012;52:958–61.
55. Hara H, Witt W, Crossley T, et al. Human dominant-negative class II transactivator transgenic pigs – effect on the human antipig T-cell immune response and immune status. Immunology. 2013;140:39–46.
56. Budorf L, Rybak E, Zhang T, et al. Human EPCR expression in GalTKO. hCD46 lungs extends survival time and lowers PVR in a xenogenic lung perfusion model. J Heart Lung Transplant. 2013;32(4 Supplement):S137.
57. Sato M, Miyoshi K, Nagao Y, et al. The combinational use of CRISPR/Cas9-based gene editing and targeted toxin technology enables efficient biallelic knockout of the alpha-1,3-galactosyltransferase gene in porcine embryonic fibroblasts. Xenotransplantation. 2014;21: 291–300.
58. Wolf E, Wunsch A, Bahr A, et al. Genetic modification of source pigs-1 section 5-2 genetically engineered pigs as donors of pancreatic islets and as large animal models for testing the functionality of transplanted islets. Xenotransplantation. 2014;21:588–93.
59. Park SJ, Cho B, Koo OJ, et al. Production and characterization of soluble human TNFRI-fc and human HO-1(HMOX1) transgenic pigs by using the F2A peptide. Transgenic Res. 2014;23: 407–19.
60. Li SZ, Qu YC, Liu BQ, et al. Synergistic effects of alpha-1,2-fucosyltransferase, DAF, and CD59 in suppression of xenogenic immunological responses. Xenotransplantation. 2009;16: 27–33.
61. Klymiuk N, Aigner B, Brem G, Wolf E. Genetic modification of pigs as organ donors for xenotransplantation. Mol Reprod Dev. 2010;77:209–21.
62. Lai L, Kolber-Simonds D, Park KW, et al. Production of alpha-1,3-galactosyltransferase knockout pigs by nuclear transfer cloning. Science. 2002;295:1089–92.
63. Tearle RG, Tange MJ, Zannettino ZL, et al. The alpha-1,3-galactosyltransferase knockout mouse. Implications for xenotransplantation. Transplantation. 1996;61:13–9.
64. Butler JR, Santos RMN, Martens GR, et al. Efficient generation of targeted and controlled mutational events in porcine cells using nuclease-directed homologous recombination. J Surg Res. 2017;212:238–45.
65. Gao H, Zhao C, Xiang X, et al. Production of alpha1,3-galactosyltransferase and cytidine monophosphate-N-acetylneuraminic acid hydroxylase gene double-deficient pigs by CRISPR/Cas9 and handmade cloning. J Reprod Dev. 2017;63:17–26.
66. Cowan PJ, Ayares D, Wolf E, Cooper DK. First update of the international xenotransplantation association consensus statement on conditions for undertaking clinical trials of porcine islet products in type 1 diabetes – chapter 2b: genetically modified source pigs. Xenotransplantation. 2016;23:32–7.
67. Mali P, Esvelt KM, Church GM. Cas9 as a versatile tool for engineering biology. Nat Methods. 2013;10:957–63.

68. Hsu PD, et al. DNA targeting specificity of RNA-guided Cas9 nucleases. Nat Biotechnol. 2013;31:827–32.
69. Shen B, et al. Generation of gene-modified mice via Cas9/RNA-mediated gene targeting. Cell Res. 2013;23:720–3.
70. Ma Y, et al. Generating rats with conditional alleles using CRISPR/Cas9. Cell Res. 2014;24: 122–5.
71. Cowan PJ, et al. Knock out of alpha1,3-galactosyltransferase or expression of alpha1,2-fucosyltransferase further protects CD55-and CD59-expressing mouse hearts in an ex vivo model of xenograft rejection. Transplantation. 1998;65:1599–604.
72. McGregor CG, et al. Human CD55 expression blocks hyperacute rejection and restricts complement activation in gal knockout cardiac xenografts. Transplantation. 2012;93:686–92.
73. Lillico SG, Proudfoot C, King TJ, Tan W, Zhang L, Mardjuki R, et al. Mammalian interspecies substitution of immune modulatory alleles by genome editing. Sci Rep. 2016;6:21645.
74. Beaton BP, Wells KD. Compound transgenics: recombinase-mediated gene stacking. In: Pinkert CA, editor. Transgenic animal technology: a laboratory handbook, vol. 21. 3rd ed. Elsevier; 2014. p. 565–79.
75. Phelps CJ, et al. Production of alpha 1,3-galactosyltransferase-deficient pigs. Science. 2003;299:411–4.
76. Wakayama S, et al. Successful serial recloning in the mouse over multiple generations. Cell Stem Cell. 2013;12:293–7.
77. Deng W, et al. Use of the 2A peptide for generation of multi-transgenic pigs through a single round of nuclear transfer. PLoS One. 2011;6:e19986.
78. Cowan PJ, et al. Renal xenografts from triple-transgenic pigs are not hyperacutely rejected but cause coagulopathy in nonimmunosuppressed baboons. Transplantation. 2000;69:2504–15.
79. Lutz AJ, et al. Double knockout pigs deficient in N-glycolylneuraminic acid and galactose α – 1,3-galactose reduce the humoral barrier to xenotransplantation. Xenotransplantation. 2013;20: 27–35.
80. Zhou CY, et al. Transgenic pigs expressing human CD59, in combination with human membrane cofactor protein and human decay-accelerating factor. Xenotransplantation. 2005;12: 142–8.
81. Bongoni AK, et al. Transgenic expression of human CD46 on porcine endothelium: effect on coagulation and fibrinolytic cascades during ex vivo human-to-pig limb xenoperfusions. Transplantation. 2015;99:2061–9.
82. Wuensch A, et al. Expression of human thrombomodulin on the endothelium of pig xenograft donors. Transplantation. 2012;94:787.
83. Taylor FB Jr, Peer GT, Lockhart MS, Ferrell G, Esmon CT. Endothelial cell protein C receptor plays an important role in protein C activation in vivo. Blood. 2001;97:1685–8.
84. Wheeler DG, et al. Transgenic swine: expression of human CD39 protects against myocardial injury. J Mol Cell Cardiol. 2012;52:958–61.
85. Ezzelarab M, et al. Genetically-modified pig mesenchymal stromal cells: xenoantigenicity and effect on human T-cell xenoresponses. Xenotransplantation. 2011;18:183–95.
86. Wilhite T, et al. The effect of gal expression on pig cells on the human T cell xenoresponses. Xenotransplantation. 2012;19:56–63.
87. Petersen B, Niemann H. Molecular scissors and their application in genetically modified farm animals. Transgenic Res. 2015;24:381–96.
88. Groenen MA, et al. Analyses of pig genomes provide insight into porcine demography and evolution. Nature. 2012;491:393–8.
89. Nottle MB, Salvaris EJ, Fisicaro N, McIlfatrick S, Vassiliev I, Hawthorne WJ, O'Connell PJ, Brady JL, Lew AM, Cowan PJ. Targeted insertion of an anti-CD2 monoclonal antibody transgene into the GGTA1 locus in pigs using FokI-dCas9. Sci Rep. 2017;7(1):8383. https://doi.org/10.1038/s41598-017-09030-6.
90. Fu Y, et al. High-frequency off-target mutagenesis induced by CRISPR-Cas nucleases in 95 human cells. Nat Biotechnol. 2013;31:822–6.

91. Guilinger JP, Thompson DB, Liu DR. Fusion of catalytically inactive Cas9 to FokI nuclease improves the specificity of genome modification. Nat Biotechnol. 2014;32:577–82.
92. Aouida M, et al. Efficient fdCas9 synthetic endonuclease with improved specificity for precise genome engineering. PLoS One. 2015;10:e0133373. https://doi.org/10.1371/journal.pone.0133373.
93. Hara S, et al. Generation of mutant mice via the CRISPR/Cas9 system using FokI-dCas9. Sci Rep. 2015;5:11221. https://doi.org/10.1038/srep11221.
94. Lai S, et al. Generation of knock-in pigs carrying Oct4-td tomato reporter through CRISPR/Cas9-mediated genome engineering. PLoS One. 2016;11:e0146562. https://doi.org/10.1371/journal.pone.0146562.
95. Peng J, et al. Production of human albumin in pigs through CRISPR/Cas9-mediated knockin of human cDNA into swine albumin locus in the zygotes. Sci Rep. 2015;5:16705. https://doi.org/10.1038/srep16705.
96. Brady JL, et al. Monoclonal antibodies generated by DNA immunization recognize CD2 from a broad range of primates. Immunol Cell Biol. 2009;87:413–8.

Chapter 20
Solid Xenoorgan Xenotransplantation

20.1 Introduction

Besides the immunological damages, the xenotransplantation also induces several blood and vascular damages through fatal hemorrhagic coagulation, phagocytotic cell death, and activation of platelets as thrombocytopenic consequence. Several biochemical events such as inflammatory response and thrombotic failure occur at xenotransplantation of solid organs. Therefore, to preventively suppress such events, several regulating factors including anti-coagulant factor, anti-inflammatory factor, and cytoprotection factor-expressing pig transgenic strains are strategically considered. From the mice experiments, human EPCR-expressing transgenic mice protect themselves from reperfusion injury of warm renal ischemia. Heart organs of these mice showed the prolonged survival time when the organs were heterotrophic-transplanted in C6-deleted rats and showed less hemorrhage and edema levels [1].

Platelet aggregation develops thrombotic microangiopathy (TM) to xenografted pig organs in primates as recipients. In platelet aggregation event, several biochemical mediators or agonists such as adenosine diphosphate (ADP), collagen, and thrombin are crucial inducing factors in humans stronger than baboons, while ADP and thrombin are more strong inducing factors for aggregation activation of platelets in cynomolgus monkeys than baboons. Among the above agonists, thrombin is the strongest inducing factor of platelets aggregation through all the animal species [2]. Therefore, the formed thrombin activates coagulation, and human TM-expressing α1,3Gal-T KO pig organs can decrease its immune rejection responses. Pig transgenic organ xenotransplantation that expresses complement regulatory factors (CRFs) of human is benefit because the complement cascade is suppressed, showing effective strategy in overcoming HR. Indeed, hearts from human DAF and CD59-expressing transgenic pigs during pig-to-baboon transplantation exhibit decreased levels of harmful vascular injuries and survived for prolonged longer than that of wild-type hearts of pigs [3]. CD59(DAF)-expressing organ hearts of pigs in xenotransplanted baboons with immunosuppressive agents

C.-H. Kim, *Glycoimmunology in Xenotransplantation*,
https://doi.org/10.1007/978-981-99-7691-1_20

such as cyclosporine, leflunomide/mofetil mycophenolate, or methylprednisolone are kept to functions for longer time than wild type, showing the lowered level of HR. However, in the organs, acute vascular rejection (AVR) type still developed with a high accumulation level of IgG type [4].

Interestingly, when baboons and cynomolgus monkeys in the condition of treatment with immunosuppression drugs and human DAF-transgenic kidneys or hearts of pigs were prepared, HR phenotype of the kidney xenografts in cynomolgus monkey has fully been prevented [5]. Heterotopic hearts using the transgenic hCD46-expressing pig-to-baboon xenotransplantation showed no any symptom including cellular infiltration, consumptive coagulopathy, or lymphoproliferative disorders [6]. Although B cells elicit production of non-Galα1,3Gal-specific antibody and DXR [7], cardiac xenografts using heterotopic hCD46/GTKO pigs to baboons significantly prolonged the recipient survival. Peritransplant B-cell depletion also prolonged hCD46/GTKO xenograft survival. Current graft rejections of HR, AXR, and acute CR in the solid genetically modified xenoorgans are partially regulated by immunosuppression, but the xenograft vasculopathy in the CR is still a problem in solid xenoorgan xenotransplantation.

20.2 Islet Xenotransplantation

Recently, IDF, referred from International Diabetes Federation, summarized the total patient number of diabetes mellitus (DM) through the world, with currently over 382 million people. Unfortunately, the DM number is estimated to be increased to near 600 million patients in coming 2035 [8]. Although type I insulin-dependent DM (IDDM) etiology has not well been clarified yet, an autoimmune disease-based B-cell destruction is characteristically depended on HLA alleles. Since the insulin discovery in the 1920s, diabetes is classified to a chronic disease. In 1894, Drs. Williams and Hareshant group reported the first islet transplantation, before Drs. Banting, Best, and Collip group succeeded in 1921 and reported the insulin isolation. In the beginning of the 1900s, Dr. Williams group subcutaneously implanted sheep pancreases into the 15-year-old ketoacidosis male patient tissues without immunosuppression. As expected, the xenograft was rejected. The first islet allotransplantation was succeeded in 1990 with Scharp et al. As the Network of Organ Procurement and Transplantation reports, due to the limited pancreas donors for transplantation, most pancreas donors are not adequate to fulfill the selection criteria [9].

Non-vascularized tissue transplantation does not lead to HR and islet transplantation is not invasive, compared to pancreas transplantation [10]. However, graft survival rates remain low. Despite many positive outcomes and breakthrough in the clinical xenoislet graft transplantation, the islet xenotransplantation is still under study due to different immune and sophisticated system between human and xenoanimals. Thus, as an effort to understand the difference, the rodent model has been subjected to develop the humanized rodent models [11]. The first pig-to-human xenotransplantation case has been reported by pig islet cell grafting to cure type I IDDM. To treat human type 1 IDDM, pig islets have initially been used as the general clinical supplier. Nevertheless partial success in the lasted lives of patients, the transmission of pig microorganisms to the recipient was then concern. Because the number of IDDM patients is increased and alternative therapies such as pluripotent stem cells and cellular reprogramming are just at the beginning stage, pig islet cell xenotransplantation has been accepted for the insulin production as a promising tool [12].

Pig islets provide an alternative replacement of human islets in diabetic patients, because pig islets are similar to the humans in structural and physiological aspects. Furthermore, pig insulin and human insulin are quite similar together, with only a different residue of amino acid sequence [13, 14]. Therefore, pig insulin has long been used for several decades for human diabetic patients [14, 15]. The pig pancreatic intact islets have reverse diabetes capacity upon transplantation to NHPs [16]. Historically, Dr. Williams PW firstly reported xenotransplantation of pancreas was carried out by in 1894 [17]. After 100 years of his action, Groth group firstly carried out clinical xenotransplantation of islet cells of fetal pigs to human, giving survival of grafts to IDDM kidney-transplanted patients [18]. Among the NHPs transplanted with pig islet xenografts, survival period with functional islets in NHPs was more than 6 months when immunosuppression agents were administered [19]. Pig islet cells weakly express the α1,3Gal epitope and are therefore regarded as resistant type of cells to xenoreactive antibodies [20]. When in vitro pig Langerhans islets cells are exposed with human blood, a fast inflammatory reaction occurs with additional coagulation activation, complement system activation, platelets consumption and immune-responsive T-cell infiltration [21]. However, interestingly, IBMIR event, referred to instant blood-mediated inflammatory reaction, is similarly regarded as the phenotypic development of HR, but importantly without antibody deposition in the IBMIR. In the case of pig islets, although the pig islets can survive IBMIR; however, they easily progress to cell-mediated rejection.

As the known immunosuppressive drugs, monoclonal Abs such as IL2R-specific mAbs and CD40L-specific mAbs as well as compounds such as everolimus, fingolimode, leflunomide, and tacrolimus showed the prolonged survival time of islets of wild-type pigs in cynomolgus macaques that were pancreatectomized. The status under temporal depletion of T cells upon thymocyte-specific antibody therapy and the depletion of T cell with combination treatment of alpha-1-antitrypsin of human enable to tolerance induction against islet xenograft [22]. When the encapsulated rat islets with the complement activation inhibitors like soluble form of complement receptor-1 (sCR-1) were grafted to mouse, xenografts showed the prolonged survival for 22 days, which are much longer, compared to the islet controls due to the sCR1 protection efficacy. However, despite the fact that encapsulated pig islets do not directly encounter with the host immune cells, the host macrophage infiltration of grafts was existed and this induced the capsular fibrosis in the encapsulated cavity of the mice [23]. In the human case, mouse macrophages are known to destroy, destruct and collapse the encapsulated human islets in Balb/c mice [24]. As the immunosuppressive strategies, in order to decrease the level of T cells, induced response of immune system is suggested. Thus, the TNF-α-related apoptosis-inducing ligand (TRAIL) of human-expressing animals is genetically engineered in pig DCs [25]. The TRAIL signaling mainly destroys virus-infected cells and tumor cells. Human TRAIL expressed on pig DCs in vitro suppressed the immune responses of primary human T cells [26]. α1,3Gal-T KO pigs exhibited the reduced immunorejection incidence with the improved donor/recipient incompatibility [27, 28]. Produced transgenic pigs improved efficient xenotransplantation levels and encapsulated pig islets also protect the pig islets from the defense immunity of hosts, as both trials are overcoming strategies from the immune rejection of xenografts [29].

The major advantages of pig islets regarding the pig-to-human islet xenotransplantation have well been listed by Dhanasekaran group [29]. They include ethically-usable sources, pathophysiological and structure similarities between pigs and human, non-limited availabilities, breeding and production, rapid growth, low maintenance costs, elective organ availabilities, low zoonotic risks, clean facility, illegal organ trafficking, tolerated immunosuppression, genetic modification and encapsulation [29]. The threat risk of zoonosis is avoided by pathogenic infection-free maintenance and pig breeding. However, integration issue of PERVs that into the pig genome is still a hurdle to overcome and eliminate the virus infection. Despite circumstance, a PERV transmission has been significantly reduced with the improved hypoglycemic and exogenous insulin doses. However, transmission issue of PERVs was not been officially raised in IDDM patients who was subjected to transplantation of pig islets [30]. In addition, other pathogenic viruses such as cytomegalovirus, herpes virus, and lymphotropic herpes virus are also raised to be transmission agents [19]. DPF, referred to designated pathogen-free maintenance of breeding and PERV gene KO pig housing, can reduce the zoonotic infection risks to minimum levels. CRISPR/Cas9 technology created target gene mutation lineages in pigs.

In the intraportal delivery, the forms of transplantation of "naked" islets or microencapsulated islets are also used. Intraportal islet xenotransplantation was in part efficient in preclinical NHP models. Bennet group [31] reported, using adult and fetal pig islets exposed to fresh-prepared sera of humans, that only fetal islets cell cytotoxicity was greatly decreased when anti-α1,3Gal-reactive Abs were used to deplete the sera, but adult islets cell cytotoxicity was not decreased. Therefore, non-α1,3Gal antibodies destroy adult pig islet cells. Recent research trends to avoid rejection are to use the encapsulation and gene editing. Encapsulation of macroencapsulation and microencapsulation in pig islet cells into capsules protects islet cells from host immune attacks [17, 32]. Macrocapsules contain a large islets number [33]. Alginate is used for capsule material for microencapsulation. The genetic-modified and gene-edited islets of pig in capsules promote survival time and functionality of the engrafted islets [34]. Microencapsulated neonatal porcine 5000–20,000 islets using alginic acid intraperitoneally transplanted to 14 diabetic patients in New Zealand and Argentina [35, 36]. This trial was safe in issue of zoonosis. For the issue of inflammatory/immune response, the transplanted porcine islets are still an unsolved problem.

For HR, preformed host antibodies raise the reaction on the pig islets, where the surface proteins are known for α1,3Gal and non-α1,3Gal antigenic proteins [37]. Although the α1,3Gal antigenic xenoepitope is not expressed in organs of apes, old-world monkeys and humans, some bacterial strains, new-world monkeys and NHPs are positive for the α1,3Gal antigenic xenoepitope expression. α1,3Gal antigenic expression of pigs is decreased in pig pancreas of fetal/neonatal births, depending on their growth until adult time, but α1,3Gal antigenic xenoepitope level is downregulated in adults [38, 39]. Humans produce α1,3Gal-specific Abs to 1% of the circulating Abs [38]. Transplanted islets of pigs are attacked by the natural-preformed Abs and rapidly by CDC-based destruction [40], causing islet loss [38, 40]. Specific Abs to other surfaced nonα1,3Gal antigenic epitopes including NeuGc and β1,4 N-acetylgalactosaminyltransferase (B4GALNT2) [40–42], that are associated with CDC-based destruction of xenografted organs [40]. As HR-preventive strategies, the following two methods include: (1) the α1,3Gal and Neu5Gc antigenic epitope genes KO pigs prevent their expression [38] and (2) artificial expression of human CRFs, including human forms of CD46, CD55 and CD59, protects the islet cells [38, 43]. The IBMIR and immunosuppressive agents, which suppress cellular and humoral responses of immune system [44] should be improved. In baboons [45] and monkeys [46], α1,3Gal-T KO with human hCD55 or hCD59 neonatal pig islets were improved for the immune response. Double KO pigs (α1,3Gal-T KO and Neu5Gc) decrease the humoral rejection incidence [38, 47]. α1,3Gal-specific IgG Abs titers are increased in the condition of stopped immunosuppression after islets transplantation of adult pigs [44, 48]. Therefore, α1,3Gal-T KO pigs are effective. In intraportal-infused pig islets experiment, the expressed tissue factors in the islets induce inflammatory responses mediated by blood, contributing to islet loss by complement activation (alternative pathway), coagulation pathways and platelet activation initiated by DCs infiltration [39, 48, 49].

For CR, a $CD4^+$ T-cell-dependent event [50, 51] destructs islet damaged death [17, 52]. Acute CR is occurred early step within 24 h to 20 days by macrophage cells as well as $CD4^+$ T/$CD8^+$ T-cell infiltration. Two different signaling pathways of T-cell receptor and co-stimulatory receptor signaling are directly involved in T-cell activation [50, 53]. T-cell activation needs for TCR double-signaling pathways with additional co-stimulatory molecules [17] and blockades of the known CD870/86-CD28 and/or CD40L (CD154)-CD40, which are known as co-stimulatory molecules present on cell surfaces, will much improve survival level of grafted organs, regardless the presence of immunosuppression [17, 54]. The targeted immunosuppression to molecular blockade helps to preventively protect the xenotransplanted islets from cellular destructions [17]. As a critical event, islet cell revascularization affects to prolong transplanted pig islet survivals. Grafted islets after transplantation acquire nutrients by diffusion, until the host vasculature revascularization within 10–14 days after transplantation [33, 50].

In the pig renal xenotransplantation to human, human patient was xenografted with fetal pig pancreatic islets. The anti-α1,3Gal antibody response was studied in fetal pancreatic islets-grafted recipients in renal transplantation [55, 56]. The recipient sera after xenotransplantation reacted highly to glutaraldehyde-fixed pig endothelial cells. However, adsorption of the sera with Galα1,3Gal antigens reduced the reactivity, and therefore, the initially produced antibodies were suggested to be α1,3Gal antigens-specific [55]. Using pig lymphoblasts-specific ADCC activity [57], it is certain that the patient antibody binding was blocked by a Galα1,3Gal-antigen-based absorption or α-Galactosidase enzyme digestion of the MHC C-I antigens [56]. Hence, the pig MHC reactivity of the patient antibody is not involved to the protein antigens, but caused by glycosylated structures. Generating transgenic pigs designated for long-term survived xenografts will be further used in clinical applications. Pig islet xenotransplantation will give a chance as a translational strategy bridgeable between stem-cell therapy and allotransplantation. Currently, immune tolerance, donor characteristics, and microbial safety are carefully studied for improvement of quality and quantity of pig islets as well as functionality and successful transplantations.

20.3 Kidney Xenotransplantation

The kidney transplant patients are double-increased in number during the last 10 years. Pig kidney is also similarly structured to human kidney, but pig and primate kidneys are physiologically different [58–60]. Humans excrete uric acid, a purine metabolic end-product; however, lower mammal species oxidize uric acid by enzymatic use of uric acid (urate) oxidase [61]. As described above, human species are not capable of oxidization of uric acid, while kidney of pigs is capable of elimination of uric acid, and thus, uric acid-caused disease, termed hyperuricemia, is not problematic in pigs [62]. In the renin–angiotensin–aldosterone system (RAAS), pig-produced renin cannot degrade human angiotensinogen [62]. However, transplanted grafts of pig kidney maintain normal balance of fluids with normal body weights at least for months. Thus, pigs have been suggested to have an alternatively acting mechanism(s) for normal maintenance of healthy fluids balance with lowered activity of renin. Currently, guideline of the USA FDA (Food and Drug Administration) refers to restrict the xenotransplantation application only for "patients with serious or life-threatening diseases for whom adequately safe and effective alternative therapies are not available."

The history of kidney transplantation is well described by Hatzinger group [63] and is ascent to the first dog kidney transplantation by the Austrian E. Ullmann in 1902, in the kidney organ connection to the carotid artery of dogs. The planted organ excreted urine for several days. In 1909, the first human kidney transplantation was tried to monkey and, in 1920, the first human xenotransplantation was completed. In 1939, the first human-to-human transplantation using a deceased human donor has been performed by the Voronoy, giving a survival of several days. In 1953, human kidney transplantation was firstly and successfully performed in Paris by J. Hamburger, Germany. In 1954, a brief guideline for the long-term transplantation of kidney was designed by Joseph Murray and he received the Nobel medicine prize in 1990. Genetically non-related patient kidney transplantation was performed with immunosuppression in 1962 and general kidney transplantation application was performed by R. Nagel and W. Brosig in Germany in 1963. The first xenotransplantation which was clinically applied, in pig kidney-to-human and sheep kidney-to-human patients was failed without any further reports. However, several xenotransplantation to humans was tried with close-related species [64]. Chimpanzee kidneys-to-human patient xenotransplantation was performed by Reemtsma in the 1960s and gave survival for 1 day to 2 months. At that time, six baboon kidney xenotransplantation to patients done by Starzl gave a survival for 19–60 days. From the short survival of kidney xenotransplantation, allotransplantation has been increased.

Xenotransplantation using genetically engineered pig kidneys covers the deceased and living donor shortage of human and renal failure. The above-mentioned gene editing and manipulation are performed for pig xenoantigen gene KO to avoid naturally-preformed Abs and CDC-based destruction or transgenes for protection from the human complement, coagulation, or inflammation. During 1990 and 2015, pig kidney graft survival rate in NHPs was gradually upgraded from 23 days to 10 months, despite the risk of PERVs [65]. Pig xenografted kidney survival showed some progresses in NHPs using the available pigs by genetic modifications and recent development of co-stimulation blockade as immunosuppressive drugs. The modified pig tissues are protected from the immune system of primates and incompatibility between donors and recipients [66, 67]. The CRISPR-Cas9-modified pigs improved major histocompatibility issue. In pig-to-NHP pre-clinical models for kidney xenotransplantation, 7-months survival has been succeeded [68], showing that the recipients had not proteinuria, an indicator of xenotransplantation. Proteinuria is another issue, as it is induced by antibody-mediated inflammation due to rejection or due to incompatibility [69].

For kidney xenotransplantation, the donor pig α1,3Gal-T KO and one or two human CRPs are importantly required with the blockade of basic immune co-stimulatory ligand-receptor. Donor-specific antibodies and human transgene for the anticoagulant, anti-inflammatory EPCR are also important. Genetically-modified donor pigs of α1,3Gal-T KO pigs with human CRPs (CD46, CD59) and human DAF (CD55)-transgenes [70, 75] xenotransplantation to NHPs exhibited lowered antibody-mediated rejections of HR or acute vascular rejection. The longest xenograft survival length obtained in pig to-NHP kidney xenotransplantation was 3 months before 2015 [72]. Using wild-type kidney of pigs, survival time of the xenografted kidney was calculated to be 23 days in 1989, [73, 74]. Since 1990, CD55 kidney xenograft survival was 90 days [75]. In 2005, in the co-transplantation approach of thymus tissues of donor and α1,3Gal-T KO organ kidney of pigs, survival time was calculated to 83 days [76]. Survival of double phenotype of α1,3Gal-T KO/hCRP pig organ kidneys was calculated to 125 days in 2015 [77, 78]. Higginbotham group prolonged the organ kidney survival time over 125 days of rhesus monkeys with α1,3Gal-T KO pig kidneys with transgenic the human CRP CD55 (α1,3Gal-T KO/CD55 pigs) in 2015 [68]. With the α1,3Gal and coagulation resolution improved the TM [79] and coagulopathy [58, 80], which is severe in the kidney [71, 81]. The monkeys with the low anti-pig Abs levels are administered by CD154-specific mAb or belatacept for immunosuppression. The α1,3Gal-TKO/CD46 kidney transplantation of pig in anti-CD154mAb-treated baboons exhibited the coagulopathy [82], whereas the pig kidney transplantation of a α1,3Gal-T KO/CD46/CD55/thrombomodulin (TBM)/endothelial protein C receptor (EPCR)/CD39 organ in a baboon even with high anti-non-α1,3Gal IgM titers allowed a survival of 136 days [83]. An α1,3Gal-T KO/CD46/CD55/TBM/EPCR/CD39 graft transplantation of kidney of pig to baboon was functioned for 136 days [78]. Recently, the human EPCR expression with presence or absence of CD55 in the kidney has been suggested for pig kidney graft survival [66]. In another kidney xenotransplantation into baboons utilizing human CD55/CD59/CD39/FucT (HT) transgenic α1,3Gal-T KO pigs, baboons received immunosuppressive agents and recombinant human C1 inhibitor bortezomib. In the rejected grafts, IgM, C3, and C5b-9 were deposited; however, the C4d was not stained. The results suggest that the classical complement pathway is blocked, while the alternate pathway is activated. Moreover, rejected organs were infiltrated with monocyte/macrophage and lymphocyte [84]. Thus, the major rejection mechanism has been explained by alternative complement pathway, non-α1,3Gal Abs, monocyte activation and PCMV replication as innate mechanisms.

Genetically modified pig with a co-stimulation blockade immunosuppression and anti-inflammatory therapy is suggested to give better outcome of pig kidney xenografts xenotransplantation. Three responses of coagulation, inflammation, and immunity have a strong relationship, suggesting multiple genetic modifications including them protect the xenograft. Such candidates are human A20 and/or HO-1 [85, 86], or pig MHC class II trans-activator, which is dominant-negative [87, 88]. Immunosuppressive agents are also crucial for pig kidney xenotransplantation in NHPs for prolonging kidney xenograft survival. Co-stimulation blockades to suppress the CD40/CD154 or CD28/B7 signaling are also candidates [89]. Co-stimulation blockades of CTLA4-Ig- and CD40-specific monoclonal Abs are also candidates. Anti-CD154 monoclonal antibody prolongs GTKO renal xenograft survival [76]. In dialysis patients, pig xenoantigenic kidney organs have extracorporeally been linked to the circulation system of the dialysis patients. The circulation of human patient by connection of a pig kidney was clinically performed in the UK [90, 91]. The pig-specific Abs of patients are eliminated by immune-adsorbing tool before operation, and then, blood perfusion is carried out. The patient recipient kidney normally excreted urine and plasma creatinine for 4 weeks. Similarly, other two dialysis patients were also subjected to extracorporeal connection to the circulation using the pig kidney organ as the xenografts [92, 93]. However, the cases were dropped simply due to issue of PERV that suggests hot discussion on the transmission concern at that time. Interestingly, a case has been connected to the hDAF transgenic pig kidney what the result was missed.

20.4 Liver Xenotransplantation

Allotransplantation of the liver is also limited by the human donor shortage and functional failure in accident donors. For treating end-stage liver disease, organs supply for transplantation is needed. From inadequate organ supply for patients with liver failure, continuously improved pig-to-NHP LXT survival is regarded as a translational bridge to patient allotransplantation of acute hepatic failure. Liver transplantation has firstly been tried to end-stage hepatic disease by Thomas Earl Starzl, who diseased on March 4, 2017 with his 91 years old, in 1963 [94]. Thomas Earl Starzl MD, PhD was an early pioneer of organ transplantation and also a representative of clinical xenotransplantation of NHP organs (kidneys and livers) [95]. The pig liver utilization for clinical xenotransplantation covers the medical desires by recent progress in genome-editing technologies to generate genetic modified donor pig [96]. Many trials have been attempted during pig-to-baboon xenotransplantation of liver (LXT) and the pig-to-primate LXT was accomplished by Sir Roy Calne in 1968, at the era when usage of immunosuppression drugs was limited to such azathioprine, hydrocortisone, and prednisolone, but his trial gave just short survival time [97]. At that time, the maximum survival was 3.5 days. Since the 1968, many immunosuppression drugs were applied to give prolonged survival.

The donor liver organs of pigs are restricted due to thrombocytopenia issue in pig-to-primate xenotransplantation [98]. Hepatic platelet consumption is recently issued. Sinusoidal ECs and Kupffer cells of pigs eliminate vascularly circulating platelets of human [99], inducing the status of thrombocytopenia as a limiting factor of in the pig-to-primate hepatic xenotransplantation. Pig livers eliminate human platelets by phagocytosis in the absent conditions of tissue injuries caused by immune-mediated responses [100]. Thus, xenogeneic thrombocytopenic elimination in a pig-to-primate xenotransplantation is an actual goal of protection from immune-mediated graft injury [101, 102]. Activated human platelets bind to ECs of pigs and pig livers also remove un-activated platelets [99, 103]. Pig asialoglycoprotein receptor (ASGR) and vWF are directly involved in this interaction [104, 105]. Immunosuppression can prevent thrombocytopenia during pig-to-primate xenotransplantation when pig organs such as hearts and kidneys were grafted [106]. From the substantial phagocytic property of the liver, additional modifications are required to overcome hepatic xenograft-caused thrombocytopenia events.

The pig ASGR1 gene produced ASGR protein clearly eliminates human platelets. $ASGR1^{-/-}$-deficient liver organs remove few platelets level of humans compared to normal WT liver organs [107]. Removing the pig liver ASGR1 greatly decreases the human platelet uptake levels. As silenced GGTA1 and CMAH genes help to avoid early xenogeneic antibody rejection [47], α1,3Gal-T KO/CMAH-KO modifications also decrease the human platelet consumption, which is taken by the pig livers, even in the absent condition of immune-responded injuries. CMAH gene elimination protects the major platelet consumption. At present, although glycan profiles is not defined to hepatic uptake of platelets, the mechanism underlying the relationship between profile and uptake is interesting in explanation how do the carbohydrates on pig organs affect the hepatic recognition and transport. To answer such questions, liver hepatic lectin proteins and cell surfaced glycans are implicated in the hepatic consumption of platelets.

The most crucial demerit in pig-to-primate LXT including human is extremely short survival time that arose from the reasons of lethal blood coagulopathy including thrombocytopenia, bleeding, and TM [108]. These are caused by interspecies incompatibilities. In the orthotopic pig-to-NHP LXT, during 1968–2012, pig-to-NHP LXT survival time was below 9 days only, because of HR and ultimately a thrombocytopenic coagulopathy, TM, and bleeding [109]. They are caused by the interspecies incompatibilities as the lethal barrier like immune rejection. Clinical LXT is an allotransplantation bridge for hepatic patients, if long-term survival is achieved [110]. However, recent LXT of α1,3Gal-T KO and cytomegalovirus-absent pig xenografts into baboons, which are administered with exogenous coagulation factors and co-stimulation blockades, improved survival for a month [111]. The coagulopathy-causing factor was a target and pig liver sinusoidal ECs induce phagocytosis of human and NHP platelets by the ASGR [104]. Inactivation of the pig ASGR1 gene decreased the hepatic platelet consumption.

Livers of ASGR1 KO pigs produced were weak in phagocytosis of human platelets]. The inactivating GGTA1 and CMAH genes consumed xenogeneic platelets of human without immune-responded graft injuries. WT pigs, ASGR1$^{-/-}$ KO pigs, GGTA1$^{-/-}$ KO pigs, and GGTA1$^{-/-}$CMAH$^{-/-}$ double KO pigs showed their efficient hepatic consumption of xenogeneic platelets of human. Human platelet uptake in WT pigs, ASGR1$^{-/-}$ KO pigs, GGTA1$^{-/-}$ KO pigs, and GGTA1$^{-/-}$CMAH$^{-/-}$ double KO pigs were measured by perfusion for the human platelet association with sinusoidal ECs of liver (LSECs). GGTA1$^{-/-}$ KO pigs and CMAH$^{-/-}$ KO pigs LSECs showed decreased the recognition levels of human platelets, in comparison to GGTA1$^{-/-}$ KO pig and WT KO pig LSECs. During continuing perfusion practice, GGTA1$^{-/-}$ CMAH$^{-/-}$ double KO pig livers consume low levels of human platelets compared to GGTA1$^{-/-}$ KO pigs and WT pig liver. GGTA1$^{-/-}$ CMAH$^{-/-}$ double KO pig liver also consume fewer human platelet levels compared to ASGR1$^{-/-}$ KO pig livers. Silenced glycan genes of pigs protect themselves from Abs-caused rejection response in pig-to-human xenotransplantation and decrease the xenogeneic consumption level of human platelets, caused by the pig hepatic function. The above combined gene modifications efficiently reduce the thrombocytopenia events involved in pig liver-to-human xenotransplantation.

The increased Neu5Gc expression in malignant human solid organs induces the tumor-exerted prothrombotic level and Neu5Gc-induced aggregation reflects the supportive tumor metastasis of human platelets. Livers from α1,3Gal-T/CMAH double-KO pigs were also similarly weak in phagocytosis of human platelets [112]. Thus, the pig glycan profile humanization significantly decreases xenogeneic consumption levels of human platelets, as caused by the hepatic livers of pig. Simultaneously silenced GGTA1$^{-/-}$ KO and ASGR1$^{-/-}$ KO livers reduced uptake of human platelets. Combination strategy of glycan and receptor protein KO makes a limitation of the xenogeneic thrombocytopenia involved in pig-to-human transplantation of liver and potentiates clinical approach. To prevent liver xenograft-induced coagulopathy, recombinant human coagulation factors have been treated in baboon recipients. When α1,3Gal-T KO pig donors were used, baboon hosts are treated with immunosuppressive agents such as ATG, CVF, steroids, and tacrolimus and, also with coagulation factor (CF) such as protein C, protein S, CF-II, CF-VII, CF-IX, and CF-X or CF-VIIa of human source [113].

Pig liver may maintain life for a bridge to transplant, using donors genetically modified to prevent thrombocytopenia. To overcome the concern of disordered coagulation following pig-to-baboon LXT experiments, a protocol using exogenous coagulation factor administration with co-stimulation blockade such as belatacept has been suggested in baboon (*Papio hamadryus*) primate [114]. α1,3Gal-T KO pig livers-xenografted baboon recipients displayed the improved thrombocytopenia and TM [113]. Like co-stimulation blockade of belatacept, anti-CD40 mAb (2C10R4) during cardiac xenotransplantation in transgenic α1,3Gal-T KO was effective for prolonged survival from 500 to 945 days [115] without humoral rejection and coagulopathy. In NHP model of hepatic allotransplantation, the similar effect has been observed [116]. Thus, the co-stimulation blockade leads to prolonged xenograft survival in renal, cardiac and liver models, probably due to inhibition of production of non-α1,3Gal baboon anti-pig cytotoxic antibodies.

In pig-to-Tibetan macaque LXT, the inflammatory cytokine profiles with the drastic elevation are evidenced to involve in recruitment of innate immune cells and xenografted organ injuries in recipients of the clinical xenotransplantation [117]. Recently, the 25-day survived success of a pig α1,3Gal-T KO liver xenografted in baboon [118]. However, there are still barriers such as the rapid platelets loss and coagulation disease onset to long-term survival after orthotopic LXT. For example, the total platelet number of hosts falls to 50,000 within 24 h, keeping below <100,000 until xenograft removal [117]. By supplementation of exogenous human coagulation factors, recipient survival was prolonged with constant circulating platelets [119]. Systemic inflammation regulates both allo- and xenotransplantation, because before the graft damage onset, the produced cytokines recruit innate immune cells to induce subsequent early loss of grafts [120]. Xenotransplantation of pig hepatocytes can supply patients with metabolic deficiencies to provide the injured, damaged and failed organ function. When spleens of cynomolgus monkeys were transplanted with 1–2 billion hepatocytes, xenoantigenic hepatocytes were survived in functional status for 80 days more of postinfusion and 253 days more of post-retransplantation with the xenografts' survival [121]. In summary, clinical LXT is the transplantation of NHP livers. Transplantation is not frequent with pig livers, and thus, the therapeutic impact of pig LXT in acute liver failure is not informative from clinical trials. Instead, surgical heterotopic pig LXT is somewhat informative in bridging to overcome the acute liver failure until a suitable liver allograft is supplied [122]. Any hurdles can be overcome by genetic modification technology.

20.5 Heart and Lung Xenotransplantation

To date, the pig-to-NHP xenotransplantation of cardiac organs was frequently tried rather than human cases. The human thrombomodulin (TBM) expression in pigs is importantly preventive for the dysregulated coagulation event. Lungs are low in recovery efficiency and transplantation rate per donor and lung xenotransplantation could be desired for clinical transplantation [123]. Survival of lung xenografts is very limitedly in hours' level, despite "HR" replacement [124]. HR is the initial hurdle to lung xenotransplantation, like other vascularized xenografts. Although α1,3Gal-T KO pig lungs and other pig organs are avoided from HR [125], complement dysfunction caused by non-α1,3Gal Abs induces endothelial injuries, thrombosis coagulation and consequent failures of xenografts [126]. From cross-species molecular incompatibilities, xenogeneic thrombosis is caused [127, 128]. Binding of endothelial TBM of pigs to human thrombin is not effectively capable of protein C activation and leads to endogenous anticoagulant deficiency [129]. Seemingly, human vWF normally binds to human platelet glycoprotein (GP) Ib. However, pig vWF is sufficiently expressed in the pig lung and binds human GPIb to activate platelets and endothelial adhesion [130]. The NeuGc residue is also a second xenoantigen, as in other xenografted organs. Thus, this antigen removement protects from antibody-induced injuries in lung xenotransplantation.

IgM and IgG Abs of human recognize pig antigenic epitopes to injure lung xenografts, as xenoantibody binding activates the classical complement cascade to cause endothelial injuries, recruitment of neutrophils, activation of platelets and microvascular thrombosis event [131, 132]. Activation of complement system and consequent thrombosis events are linked to hyperacute xenograft rejection (HXR). For the pre-clinical animal test, in vivo orthotopic lung xenotransplantation [133] is performed by long-term xenotransplantation and immunosuppression and xenotolerant treatments [130, 133] in a baboon recipient. Mohiuddin group succeeded 1-year survival of heterotopic hearts with α1,3Gal-TKO-human CRP CD46 and TBM of pig donors, by CD40/CD40L co-stimulation blockade with B-cell depletion induction [134]. In an in vivo orthotopic pig-to-NHP xenotransplantation, the lung xenograft survival was ultimately 4.5 days to date [135]. A limitation of this model is NeuGc expression. Another ex vivo xenoperfusion of pig lungs is performed by alternative blood of human donor to study the xenograft injures.

HR of lung xenografts is also mediated by an innate immune response [130, 132]. Human CRPs of hCD46 (MCP) and hCD55 (complement DAF) strongly expressed in the α1,3Gal-T KO system suppress complement activation [124, 130]. Particularly, the α1,3Gal-T KO.hCD46 pig strain decreases activation of complement system, thrombosis, and recruited neutrophils lining in xenografted lung organs [124]. In xenografted organs with α1,3Gal-T KO.hCD46 pig donor, other CRPs transgenic expression seems to be effective. In fact, hCD55 or hCD59 (known as membrane attack complex inhibitor) expression reduces complement activation and thrombosis event with lung function and survival [130]. Different from xenogeneic thromboregulatory dysfunctional failure, pig endothelial CAMs efficiently recognize human leukocytes and consequently potentiate strong attachment. These recognitions are related to leukocyte and endothelial activations, endothelial injuries and inflammatory mediator synthesis. Human CD39 addition to degrade prothrombotic molecule, ADP, to the α1,3Gal-T KO.hCD55.hCD59 basis improves lung function. Likely, human TFPI expression and thrombin inhibitor in pig ECs decrease thrombotic response by pig tissue factors [136]. Multiple human TBM or EPCR transgenic pigs [137] indicate successful function, as transgenic hTBM activates activated protein C synthesis [138]. Expressed hEPCR particularly promise during *ex vivo* xenoperfusion [19]. hTBM and hEPCR co-expression synergistically enhances constitutively activated protein C synthesis and EPCR-induced cytoprotection signaling pathway [130, 138].

Inflammatory mediators represent Abs binding, thrombotic response, innate immune response and tissue injuries in failure of lung xenografts. The lungs are characteristic of high capacity of thromboxane synthesis and sensitive thromboxane effect [139] that occurs to amplify in lung xenotransplantation [140]. In addition, pig lung xenoperfusion generates both human and pig cytokines such as IL-8 in pig as well as IL-6 and IL-8 in human, causing for local and systemic inflammatory responses. In lung xenotransplantation models, organ xenografts are advanced. But innate immune and inflammatory responses-induced acute injuries are restricted in lung xenograft models. Overcoming acute lung xenograft injury is needed in lung xenotransplantation. Combination of α1,3Gal-T KO and human complements such as CD46, CD55 and CD59 or human coagulation regulators of CD39, TBM and EPCR transgenes is effective for improved survival of xenogeneic organs [124, 141]. GalTKO.hCD46 lungs were protected from antibody—and complement-mediated HR of lung [124]. Thrombosis as clotting requires aggregation of platelets by GPIIb/IIIa-fibrinogen binding to cross-link platelets. GPIIb/IIIa activation activates platelets and adhesion downstream from GPIb/VWF. Thus, fibrinogen-binding inhibition attenuates platelet adhesion and activation. Pig lung vasculature largely release vWF. Human platelet GPIb adheres to human vWF in thrombin presence [142], whereas NHP and human GPIb receptors bind pig vWF and consequently platelets adhere to pig endothelium [143]. Blocking this interaction is a strategy to prevent adhesion of primate platelets to pig endothelium. To target the GPIb-vWF interaction has been tried in xenotransplant models. vWF-deficient pigs examined in wild-type pig xenografts showed vWF's importance in xenograft injuries of lung [137], while the bleeding property in phenotype is a significant barrier. A humanized anti-GPIb Fab fragment Abs, named 6B4, inhibit the platelet GPIb-binding site by vWF, as efficiently safety and effective blocker of GPIb-induced activation of platelets in a baboon arterial thrombotic injuries [138].

Recent advance in CRISPR-CAS9 technology allowed multiple functional modification of pigs with the prolonged pig xenograft survival, as recently reported for α1,3GalT KO, hCD46, and hTBM (hCD46/hTBM/α1,3Gal-T KO) heterogeneous pig-to-baboon xenotransplantation of heart organs for more 2 years [139]. The heterotopic heart xenotransplantation showed the different growth curves of heart grafts transplanted even by heterotopic or orthotopic way.

Heterotopic-transplantation of hearts showed the prolonged grafts survival for 146 days. However, heart-orthotopic transplantation has only up to 57 days [124, 140]. α1,3Gal-T KO co-expressed with human complement factors such as CD46, CD55) and human coagulation regulatory factor TBM showed the prolonged xenograft survival in preclinical heart (>2 years) models. However, pig lung xenograft damage is observed within a few minutes to hours upon human blood exposure despite the α1,3Gal-T KO pig modification by regulation of complement and coagulation [139]. Lung Xenotransplantation is believed to be the big issue of solid organ xenotransplantation. However, lung xenografts are still unsuccessful from vascular barrier rejection within hours despite Galα1,3Gal antigen (α1,3Gal-T KO) removal and expression of human membrane cofactor protein (hCD46). In the ex vivo α1,3Gal-TKO.hCD46 pig lung perfusion with human bloods, NK killer cells were rapidly disappeared by pulmonary vascular resistance, reduces histamine elaboration and platelet activation. NK-cell inhibitory ligand HLA-E and β2m of human expressed in pigs inhibit pig α1,3Gal-T KO.hCD46 cell injuries by ADCC and antibody-independent and NK-dependent cytotoxicity with prolonged lung function. Inhibition of the inflammatory responses, complement, coagulation, and innate/adaptive immune responses are prerequisite for better survival of lung xenograft recipients [141].

20.6 Corneal and Tissue Xenotransplantation

Xenotransplantation using pig corneas is another possible alternative replacement of human corneas because corneal blindness is prevalent in elderly peoples [142]. Corneal blindness can be treated by corneal transplantation if biologically regenerative cornea or xenotransplantation is available. Corneal xenotransplantation for human corneas by the pig cornea is also a potential replacement in terms of optical similarity between pigs and humans. As a merit of the corneal *xenotransplantation,* the cornea is not easily vascularized and thus more applicable than other solid xenoorgans *xenotransplantation.* The relative similarity between the pig cornea and human cornea is also high, giving a possibility of saving for humans suffering from corneal blindness by xenocorneal transplantation [143]. Evolutionary adaptation and evolution that privilege the protection of body's important structures, if exemplified, the adrenal cortex, brain, ovaries and testes, from inflammatory responses-injured damages. Simply and conclusively, privilege of immune system is therefore evolution and adaptation. In eye system, the immune-privileged regions or sites include several surrounded tissues such as the anterior chamber, cornea, sub-retinal space and vitreous cavity. Among them, the cornea site is not vascularized without microvascular vessels and it consists of anti-complementary factor molecules [142]. Currently, xenocorneal transplantation is regarded as potential and successful field.

Despite recent development of α1,3Gal-T KO pigs, commercially available xenografts are still long distant from clinical xenotransplantation. Such circumstance invited needs for immune modulatory system toward critical xenotransplantation

[144]. The use of immunosuppressants is likely crucial [145]. In corneal xenotransplantation, such immune modulatory agents are studied. Using immunosuppressive agents, decellularized porcine corneal lamellar grafts can be expected for long-term survival in an anterior lamellar keratoplast [146]. For example, full-thickness pig corneal grafts survived 6 months by an CD154-specific antibody-treated immunosuppression [147]. CD154-specific Abs-induced co-stimulation blockade is known to be effective in NHPs [148, 149]. CD40-specific Abs-induced co-stimulation blockade has now been challenged to pig corneas for a clinical pig-to-NHP corneal xenotransplantation model [145]. CD40/CD154 co-stimulation blockade is beneficial in allotransplantation, xenotransplantation and autoimmune diseases. CD154-specific Abs are also efficiently acted in xenotransplantation trials [150], but CD154-specific Abs are antagonistic to CD40-specific Abs because of the thrombotic complications [148].

A recent case using wild-type Wuzhishan pig corneal xenografts, the non-immunosuppressed rhesus monkey eye received the xenograft by a method of orthotopic penetrating xenotransplantation and shortly rejected it without any HR. However, the rejection response was prolonged to 4 months upon steroid betamethasone conjunctival injection. In addition, eye xenotransplantation of the lamellar corneal xenografts gave the persisted corneal transparency over 3 months without steroid [151]. Corneal xenotransplantation in NHP models gave the good achievement of prolonged survival time (933 days) in pig corneal xenografts with full-thickness xenotransplanted to monkey recipients when the hosts were treated with immunosuppressive CD154-specific Abs. The corneal graft size seems to influence survival of the xenografts, as was confirmed in the case of human corneal grafts in rhesus monkeys [152]. When Wuzhishan pigs corneal graft were xenografted in rhesus monkeys and further, donor BM transplantation was examined to test the chimerism, the survival in the xenografted rhesus monkeys was prolonged longer than the control monkeys without bone marrow transplantation. In addition, immunoglobulin and complement levels as well as inflammatory cell infiltration and eosinophil infiltration were also decreased [153]. The cornea immunogenicity of pigs can also be eliminated by xenoantigenic gene deletion and inactivation [154]. In xenocorneal transplantation, the number of transplanted cells is less than solid organs. In addition, xenocorneal transplantation to human give a longer survival time than others [155].

20.7 Other Tissue/Cellular Xenotransplantations: Pig Red Blood Cells and Replacement of Cruciate Ligament with Pig α1,3Gal Epitope-Lacking Tendon

Pig red blood cells are similar in diameters and counts to those of human red blood cells. However, the average life span of pig red blood cells is short compared to RBCs of human. Pig hemoglobin is also similar in 3D structure by 85% protein sequence homology with the human hemoglobin. Pig RBCs are negative for swine leukocyte antigen (SLA) expression with reduced immune response. Galα1,3Gal epitope removal prevents baboon and human antibody binding to pig RBCs. Wild-type pig RBC treatment with neuraminidase removes the harmful transfusion-mediated hemolysis risk [156]. Negative human or baboon IgM binding is observed on α1,3Gal-T KO pig RBCs while IgG binding is not detected. In vivo α1,3Gal-T KO pig RBCs are rapidly phagocytosed from the primate by a non-α1,3Gal antibodies [157]. GTKO pRBCs show high compatibility compared to human ABO-incompatible RBCs and wild-type pig RBCs. Of course, the compatibility extent of GTKO pRBCs is much lower than ABO-compatible human cases [158].

In the blood transfusion, hCD47-GTKO pigs is thus suggested because the low compatibility with 73% amino acid sequence homology of pig CD47 and human CD47 can lead to xenogeneic cell phagocytosis by macrophages even in antibodies—or complement-caused opsonization absence. SIRPα (signal regulatory protein-α) known as a macrophage receptor interacts with CD47 to prevent autogenic phagocytic process. The human CD47 transgenic pig cells inhibit the SIRPα signaling on macrophages of human [159]. In the stem cell, ADMSC (adipose-derived mesenchymal stem cells) have been emerged as a resource for eye diseases as regenerative medicine. ADMSC of human have been subjected to xenotransplantation to the rat eyes and gave 6-month survival and used to the ocular tissue integration. Blood-retina barrier and spleen analysis suggested the migration issue of stem cells moving to non-targeted tissues [160, 161]. ADMSC of human also cross the blood barrier upon xenotransplantation in case of the rat, requiring is improvement by immune cells.

A patient was also tried to xenograft the α1,3Gal epitope-negative pig patellar tendons to replace anterior cruciate ligaments [162]. The patient increased in anti-non-α1,3Gal antibodies, probably due to elimination of immunogenic pig xenoantigens. The xenografted tendons were stained with glutaraldehyde to secure non-PERV transmission.

References

1. Lee KF, Lu B, Roussel JC, et al. Protective effects of transgenic human endothelial protein C receptor expression in murine models of transplantation. Am J Transplant. 2012;12:2363–72.
2. Iwase H, Ekser B, Zhou H, et al. Platelet aggregation in humans and nonhuman primates: relevance to xenotransplantation. Xenotransplantation. 2012;19(4):233–43.
3. McCurry KR, Kooyman DL, Alvarado CG, et al. Human complement regulatory proteins protect swine-to-primate cardiac xenografts from humoral injury. Nat Med. 1995;1:423–7.
4. Chen RH, Naficy S, Logan JS, et al. Hearts from transgenic pigs constructed with CD59/DAF genomic clones demonstrate improved survival in primates. Xenotransplantation. 1999;6:194–200.
5. Schuurman HJ, Pino-Chavez G, Phillips MJ, et al. Incidence ofhyperacute rejection in pig-to-primate transplantation using organs from hDAF-transgenic donors. Transplantation. 2002;73(7):1146–51.

6. McGregor CG, Davies WR, et al. Cardiac xenotransplantation: recent preclinical progress with 3-month median survival. J Thorac Cardiovasc Surg. 2005;130(3):844–51.
7. Mohiuddin MM, Corcoran PC, Singh AK, et al. B-cell depletion extends the survival of GTKO.hCD46Tg pig heart xenografts in baboons for up to 8 months. Am J Transplant. 2012;12(3):763–71.
8. Guariguata L, Whiting DR, Hambleton I, Beagley J, Linnenkamp U, Shaw JE. Global estimates of diabetes prevalence for 2013 and projections for 2035. Diabetes Res Clin Pract. 2014;103(2):137–49. https://doi.org/10.1016/j.diabres.2013.11.002.
9. Iuamoto LR. Human islet xenotransplantationinrodents: a literature review of experimental model trends. Cinics. 2017;72(4):238–43.
10. Merani S, Shapiro AM. Current status of pancreatic islet transplantation. Clin Sci. 2006;110 (6):611–25. https://doi.org/10.1042/CS20050342.
11. Jacobson S, Heuts F, Juarez J, Hultcrantz M, Korsgren O, Svensson M, et al. Alloreactivity but failure to reject human islet transplants by humanized Balb/c/Rag2gc mice. Scand J Immunol. 2010;71(2):83–90. https://doi.org/10.1111/j.1365-3083.2009.02356.x.
12. Borisov MA, Petrakova OS, Gvazava IG, Kalistratova EN, Vasiliev AV. Stem cells in the treatment of insulin-dependent diabetes mellitus. Acta Nat. 2016;8(3):31–43.
13. Home PD, Alberti KG. The new insulins. Their characteristics and clinical indications. Drugs. 1982;24:401–13.
14. Sonnenberg GE, Berger M. Human insulin: much ado about one amino acid? Diabetologia. 1983;25:457–9.
15. Hu Q, Liu Z, Zhu H. Pig islets for islet xenotransplantation: current status and future perspectives. Chin Med J. 2014;127:370–7.
16. van der Windt DJ, Bottino R, Kumar G, et al. Clinical islet xenotransplantation: how close are we? Diabetes. 2012;61:3046–55.
17. Williams PW. Notes on diabetes treated with grafts of sheeps' pancreas. BMJ. 1894;19:1303–4.
18. Groth C, Korsgren O, Tibell A, et al. Transplantation of porcine fetal pancreas to diabetic patients. Lancet. 1994;344:1402–4.
19. Ellis CE, Korbutt GS. Justifying clinical trials for porcine islet xenotransplantation. Xenotransplantation. 2015;22:336–44.
20. Dor FJ, Cheng J, Alt A, et al. Galalpha1,3Gal expression on porcine pancreatic islets, testis, spleen, and thymus. Xenotransplantation. 2004;11:101–6.
21. Bennet W, Sundberg B, Lundgren T, et al. Damage to porcine islets of Langerhans after exposure to human blood in vitro, or after intraportal transplantation to cynomologus monkeys: protective effects of sCR1 and heparin. Transplantation. 2000;69:711–9.
22. Ashkenazi E, Baranovski BM, Shahaf G, Lewis EC. Pancreatic islet xenograft survival in mice is extended by a combination of alpha-1-antitrypsin and single-dose anti-CD4/CD8 therapy. PLoSOne. 2013;8(5):e63625.
23. Vaithilingam V, Fung C, Ratnapala S, et al. Characterisation of the xenogeneic immune response to microencapsulated fetal pig islet like cell clusters transplanted into immunocompetent C57BL/6 mice. PLoS One. 2013;8(3):e59120.
24. Qi M, Mørch Y, Lacík I, et al. Survival of human islets in microbeads containing high guluronic acid alginate crosslinked with Ca2+ and Ba2+. Xenotransplantation. 2012;19(6): 355–64.
25. Klose R, Kemter E, Bedke T, et al. Expression of biologically active human TRAIL in transgenic pigs. Transplantation. 2005;80(2):222–30.
26. Kemter E, Lieke T, Kessler B, et al. Human TNF-related apoptosisinducing ligand-expressing dendritic cells from transgenic pigs attenuate human xenogeneic T cell responses. Xenotransplantation. 2012;19(1):40–51.
27. Kolber-Simonds D, Lai L, Watt SR, et al. Production of alpha-1,3-galactosyltransferase null pigs by means of nuclear transfer with fibroblasts bearing loss of heterozygosity mutations. Proc Natl Acad Sci U S A. 2004;101:7335–40.

28. Bottino R, Balamurugan AN, Smetanka C, et al. Isolation outcome and functional characteristics of young and adult pig pancreatic islets for transplantation studies. Xenotransplantation. 2007;14:74–82.
29. Dhanasekaran M, George JJ, Loganathan G, Narayanan S, Hughes MG, Williams SK, Balamurugan AN. Pig islet xenotransplantation. Curr Opin Organ Transplant. 2017;22:452–62. https://doi.org/10.1097/MOT.
30. Valdes-Gonzalez R, Dorantes LM, Bracho-Blanchet E, et al. No evidence of porcine endogenous retrovirus in patients with type 1 diabetes after longterm porcine islet xenotransplantation. J Med Virol. 2010;82:331–334.24.
31. Bennet W, Bjorkland A, Sundberg B, et al. A comparison of fetal and adult porcine islets with regard to gal alpha (1,3) gal expression and the role of human immunoglobulins and complement in islet cell cytotoxicity. Transplantation. 2000;69:1711–7.
32. Meyer T, Ho¨cht B, Ulrichs K. Xenogeneic islet transplantation of microencapsulated porcine islets for therapy of type I diabetes: long-term normoglycemia in STZ-diabetic rats without immunosuppression. Pediatr Surg Int. 2008;24:1375–8.
33. Weir GC. Islet encapsulation: advances and obstacles. Diabetologia. 2013;56:1458–61.
34. Cooper DK, Matsumoto S, Abalovich A, et al. Progress in clinical encapsulated islet xenotransplantation. Transplantation. 2016;100:2301–8.
35. Wynyard S, Nathu D, Garkavenko O, Denner J, Elliott R. Microbiological safety of the first clinical pig islet xenotransplantation trial in New Zealand. Xenotransplantation. 2014;21:309–23.
36. Matsumoto S, Abalovich A, Wechsler C, Wynyard S, Elliott RB. Clinical benefit of islet xenotransplantation for the treatment of type 1 diabetes. EBioMedicine. 2016;12:255–62.
37. Ekser B, Ezzelarab M, Hara H, et al. Clinical xenotransplantation: the nextmedical revolution? Lancet. 2012;379:672–83.
38. Denner J. Recent progress in xenotransplantation, with emphasis on virological safety. Ann Transplant. 2016;21:717–27.
39. Ekser B, Ezzelarab M, Hara H, et al. Clinical xenotransplantation: the next medical revolution? Lancet. 2012;379:672–83.
40. Bottino R, Trucco M. Use of genetically-engineered pig donors in islet transplantation. World J Transplant. 2015;5:243.
41. Bouhours D, Pourcel C, Bouhours JE. Simultaneous expression by porcine aorta endothelial cells of glycosphingolipids bearing the major epitope for human xenoreactive antibodies (gal alpha 1-3Gal), blood group H determinant and N-glycolylneuraminic acid. Glycoconj J. 1996;13:947–53.
42. Byrne GW, Du Z, Stalboerger P, et al. Cloning and expression of porcine beta1,4 N-acetylgalactosaminyl transferase encoding a new xenoreactive antigen. Xenotransplantation. 2014;21:543–54.
43. Petersen B, Carnwath JW, Niemann H. The perspectives for porcineto-human xenografts. Comp Immunol Microbiol Infect Dis. 2009;32:91–105.
44. Shin JS, Kim JM, Kim JS, et al. Long-term control of diabetes in immunosuppressed nonhuman primates (NHP) by the transplantation of adult porcine islets. Am J Transplant. 2015;15:2837–50.
45. Hawthorne WJ, Salvaris EJ, Phillips P, et al. Control of IBMIR inneonatal porcine islet xenotransplantation in baboons. Am J Transplant. 2014;14:1300–9.
46. Thompson P, Badell IR, Lowe M, et al. Islet xenotransplantation using gal-deficient neonatal donors improves engraftment and function. Am J Transplant. 2011;11:2593–602.
47. Lutz AJ, Li P, Estrada JL, et al. Double knockout pigs deficient in Nglycolylneuraminic acid and galactose alpha-1,3-galactose reduce the humoral barrier to xenotransplantation. Xenotransplantation. 2013;20:27–35.
48. Park CG, Bottino R, Hawthorne WJ. Current status of islet xenotransplantation. Int J Surg. 2015;23(Pt B):261–6.

49. van der Windt DJ, Marigliano M, He J, et al. Early islet damage after directexposure of pig islets to blood: has humoral immunity been underestimated? Cell Transplant. 2012;21:1791–802.
50. Zhu HT, Wang WL, Yu L, Wang B. Pig-islet xenotransplantation: recent progress and current perspectives. Front Surg. 2014;1:7.
51. Koulmanda M, Laufer TM, Auchincloss H Jr, Smith RN. Prolonged survival of fetal pig islet xenografts in mice lacking the capacity for an indirect response. Xenotransplantation. 2004;11: 525–30.
52. Scalea J, Hanecamp I, Robson SC, Yamada K. T-cell-mediated immunological barriers to xenotransplantation. Xenotransplantation. 2012;19:23–30.
53. Trikudanathan S, Sayegh MH. The evolution of the immunobiology of co-stimulatory pathways: clinical implications. Clin Exp Rheumatol. 2007;25(5 Suppl 46):S12–21.
54. Tian M, Lv Y, Zhai C, et al. Alternative immunomodulatory strategies for xenotransplantation: CD80/CD86-CTLA4 pathway-modified immature dendritic cells promote xenograft survival. PLoS One. 2013;8:e69640.
55. Galili U, Tibell A, Samuelsson B, et al. Increased anti-gal activity in diabetic patients transplanted with fetal porcine islet cell clusters. Transplantation. 1995;59:1549–56.
56. Satake M, Kawagishi N, Rydberg L, et al. Limited specificity of xenoantibodies in diabetic patients transplanted with fetal porcine islet cell clusters. Main antibody reactivity against alpha-linked galactose-containing epitopes. Xenotransplantation. 1994;1:89–101.
57. Satake M, Kumagai-Braesch M, Kawagishi N, et al. Kinetics and character of xenoantibody formation in diabetic patients transplanted with fetal porcine islet cell clusters. Xenotransplantation. 1994;1:24–35.
58. Cowan PJ, Aminian A, Barlow H, Brown AA, Chen CG, Fisicaro N, et al. Renal xenografts from triple-transgenic pigs are not hyperacutely rejected but cause coagulopathy in non-immunosuppressed baboons. Transplantation. 2000;69:2504–15.
59. Soin B, Smith KG, Zaidi A, Cozzi E, Bradley JR, Ostlie DJ, et al. Physiological aspects of pig-to-primate renal xenotransplantation. Kidney Int. 2001;60:1592–7.
60. Ibrahim Z, Busch J, Awwad M, Wagner R, Wells K, Cooper DK. Selected physiologic compatibilities and incompatibilities between human and porcine organ systems. Xenotransplantation. 2006;13:488–99.
61. Oda M, Satta Y, Takenaka O, Takahata N. Loss of urate oxidase activity in hominoids and its evolutionary implications. Mol Biol Evol. 2002;19:640–53. [PubMed]
62. Simmonds HA, Hatfield PJ, Cameron JS, Cadenhead A. Uric acid excretion by the pig kidney. Am J Phys. 1976;230:1654–61.
63. Hatzinger M, Stastny M, Grützmacher P, Sohn M. The history of kidney transplantation. Urologe A. 2016;55(10):1353–9.
64. Cooper DK. A brief history of cross-species organ transplantation. Proc. 2012;25(1):49–57.
65. Wijkstrom M, Iwase H, Paris W, Hara H, Ezzelarab M, Cooper DK. Renal xenotransplantation: experimental progress and clinical prospects. Kidney Int. 2017;91(4):790–6.
66. Iwase H, Hara H, Ezzelarab M, Li T, Zhang Z, Gao B, Liu H, Long C, Wang Y, Cassano A, Klein E, Phelps C, Ayares D, Humar A, Wijkstrom M, Cooper DKC. Immunological and physiological observations in baboons with life-supporting genetically engineered pig kidney grafts. Xenotransplantation. 2017;24(2):e12293. https://doi.org/10.1111/xen.12293.
67. Cooper DK, Satyananda V, Ekser B, et al. Progress in pig-to-nonhuman primate transplantation models (1998-2013): a comprehensive review of the literature. Xenotransplantation. 2014;21:397–419.
68. Higginbotham L, Mathews D, Breeden CA, et al. Pre-transplant antibody screening and anti-CD154 costimulation blockade promote long-term xenograft survival in a pig-to-primate kidney transplant model. Xenotransplantation. 2015;22:221–30.
69. Tasaki M, Wamala I, Tena A, et al. High incidence of xenogenic bone marrow engraftment in pig-to-baboon intra-bone bone marrow transplantation. Am J Transplant. 2015;15:974–83.

70. Menoret S, Plat M, Blancho G, Martinat-Botte F, Bernard P, Karam G, et al. Characterization of human CD55 and CD59 transgenic pigs and kidney xenotransplantation in the pig-to-baboon combination. Transplantation. 2004;77:1468–71.
71. Lin CC, Ezzelarab M, Shapiro R, Ekser B, Long C, Hara H, et al. Recipient tissue factor expression is associated with consumptive coagulopathy in pig-to-primate kidney xenotransplantation. Am J Transplant. 2010;10:1556–68.
72. Baldan N, Rigotti P, Calabrese F, et al. Ureteral stenosis in HDAF pig-to-primate renal xenotransplantation: a phenomenon related to immunological events? Am J Transplant. 2004;4:475–81.
73. Alexandre GPJ, Gianello P, Latinne D, Carlier M, Dewaele A, van Obbergh L. Plasmapheresis and splenectomy in experimental renal xenotransplantation. In: Hardy M, editor. Xenograft, vol. 25. Amsterdam/New York: Excerpta Medica; 1989. p. 259–66.
74. Lambrigts D, Sachs DH, Cooper DK. Discordant organ xenotransplantation in primates: world experience and current status. Transplantation. 1998;66:547–61.
75. Baldan N, Rigotti P, Calabrese F, Cadrobbi R, Dedja A, Iacopetti I, et al. Ureteral stenosis in HDAF pig-to-primate renal xenotransplantation: a phenomenon related to immunological events? Am J Transplant. 2004;4:475–81.
76. Yamada K, Yazawa K, Shimizu A, Iwanaga T, Hisashi Y, Nuhn M, et al. Marked prolongation of porcine renal xenograft survival in baboons through the use of alpha1,3-galactosyltransferase gene-knockout donors and the cotransplantation of vascularized thymic tissue. Nat Med. 2005;11:32–4.
77. Higginbotham L, Mathews D, Breeden CA, Song M, Farris AB 3rd, Larsen CP, et al. Pre-transplant antibody screening and anti-CD154 costimulation blockade promote long-term xenograft survival in a pig-to-primate kidney transplant model. Xenotransplantation. 2015;22:221–30. [PMC free article] [PubMed]
78. Iwase H, Liu H, Wijkstrom M, Zhou H, Singh J, Hara H, et al. Pig kidney graft survival in a baboon for 136 days: longest life-supporting organ graft survival to date. Xenotransplantation. 2015;22:302–9. (In press)
79. Shimizu A, Yamada K, Yamamoto S, Lavelle JM, Barth RN, Robson SC, et al. Thrombotic microangiopathic glomerulopathy in human decay accelerating factor-transgenic swine-to-baboon kidney xenografts. J Am Soc Nephrol. 2005;16:2732–45.
80. Ierino FL, Kozlowski T, Siegel JB, Shimizu A, Colvin RB, Banerjee PT, et al. Disseminated intravascular coagulation in association with the delayed rejection of pig-to-baboon renal xenografts. Transplantation. 1998;66:1439–50.
81. Knosalla C, Yazawa K, Behdad A, Bodyak N, Shang H, Buhler L, et al. Renal and cardiac endothelial heterogeneity impact acute vascular rejection in pig-to-baboon xenotransplantation. Am J Transplant. 2009;9:1006–16.
82. Lin CC, Ezzelarab M, Shapiro R, et al. Recipient tissue factor expression is associated with consumptive coagulopathy in pig-to-primate kidney xenotransplantation. Am J Transplant. 2010;10:1556–68.
83. Iwase H, Liu H, Wijkstrom M, et al. Pig kidney graft survival in a baboon for 136 days: longest life-supporting organ graft survival to date. Xenotransplantation. 2015;22:302–9.
84. Le Bas-Bernardet S, Tillou X, Branchereau J, Dilek N, Poirier N, Châtelais M, Charreau B, Minault D, Hervouet J, Renaudin K, Crossan C, Scobie L, Takeuchi Y, Diswall M, Breimer ME, Klar N, Daha MR, Simioni P, Robson SC, Nottle MB, Salvaris EJ, Cowan PJ, d'Apice AJ, Sachs DH, Yamada K, Lagutina I, Duchi R, Perota A, Lazzari G, Galli C, Cozzi E, Soulillou JP, Vanhove B, Bortezomib BG. C1-inhibitor and plasma exchange do not prolong the survival of multi-transgenic GalT-KO pig kidney xenografts in baboons. Am J Transplant. 2015;15(2):358–70. https://doi.org/10.1111/ajt.12988.
85. Oropeza M, Petersen B, Carnwath JW, Lucas-Hahn A, Lemme E, Hassel P, et al. Transgenic expression of the human A20 gene in cloned pigs provides protection against apoptotic and inflammatory stimuli. Xenotransplantation. 2009;16:522–34.

86. Petersen B, Lucas-Hahn A, Lemme E, Queisser AL, Oropeza M, Herrmann D, et al. Generation and characterization of pigs transgenic for human hemeoxygenase-1 (hHO-1). Xenotransplantation. 2010;17:102–3.
87. Hara H, Witt W, Crossley T, Long C, Isse K, Fan L, et al. Human dominant-negative class II transactivator transgenic pigs – effect on the human anti-pig T-cell immune response and immune status. Immunology. 2013;140:39–46.
88. Iwase H, Satyananda V, Zhou H, Hara H, Bajona P, Wijkstrom M, et al. Initial in vivo experience of pig artery patch transplantation in baboons using mutant MHC (CIITA-DN) pigs. Transpl Immunol. 2015;32:99–108.
89. Miwa Y, Kobayashi T, Nagasaka T, Liu D, Yu M, Yokoyama I, et al. Are N-glycolylneuraminic acid (Hanganutziu-Deicher) antigens important in pig-to-human xenotransplantation? Xenotransplantation. 2004;11:247–53.
90. Cairns TD, Taube DH, Stevens N, et al. Xenografts–future prospects for clinical transplantation. Immunol Lett. 1991;29:167–70.
91. Welsh K, Taube D, Thick M, et al. Human antibodies to pig determinants and their association with hyperacute rejection of xenografts. In: Cooper D, Kemp E, Reemtsma K, White D, editors. Xenotransplantation; the transplantation of organs and tissues between species. Berlin: Springer-Verlag; 1991. p. 501–10.
92. Breimer ME, Bjorck S, Svalander CT, et al. Extracorporeal ("ex vivo") connection of pig kidneys to humans 1. Clinical data and studies of platelet destruction. Xenotransplantation. 1996;3:328–339.92.
93. Rydberg L, Bjorck S, Magnusson S, et al. Extracorporeal ("ex vivo") connection of pig kidneys to humans 2. The anti-pig antibody response. Xenotransplantation. 1996;3:340–53.
94. Starzl TE, Marchioro TL, Vonkaulla KN, Hermann G, Brittain RS, Waddell WR. Homotransplantation of the liver in humans. Surg Gynecol Obstet. 1963;117:659–76.
95. Cooper DKC, Joseph Tector A. Thomas E. Starzl, MD, PhD, 1926-2017. Xenotransplantation. 2017;24(2):e12307. https://doi.org/10.1111/xen.12307.
96. Butler JR, Ladowski JM, Martens GR, Tector M, Tector AJ. Recent advances in genome editing and creation of genetically modified pigs. Int J Surg. 2015;23(Pt B):217–22.
97. Calne RY, White HJ, Herbertson BM, et al. Pig-to-baboon liver xenografts. Lancet. 1968;1: 1176–8.
98. Ekser B, Long C, Echeverri GJ, et al. Impact of thrombocytopenia on survival of baboons with genetically modified pig liver transplants: clinical relevance. Am J Transplant. 2010;10(2): 273–85.
99. Burlak C, Paris LL, Chihara RK, et al. The fate of human platelets perfused through the pig liver: implications for xenotransplantation. Xenotransplantation. 2010;17(5):350–61.
100. Butler JR, Paris LL, Blankenship RL, Sidner RA, Martens GR, Ladowski JM, Li P, Estrada JL, Tector M, Tector AJ. Silencing porcine CMAH and GGTA1 genes significantly reduces xenogeneic consumption of human platelets by porcine livers. Transplantation. 2016;100(3): 571–6.
101. Ezzelarab MB, Ekser B, Azimzadeh A, et al. Systemic inflammation in xenograft recipients precedes activation of coagulation. Xenotransplantation. 2015;22(1):32–47.
102. Buhler L, Yamada K, Kitamura H, et al. Pig kidney transplantation in baboons: anti-gal(alpha) 1-3Gal IgM alone is associated with acute humoral xenograft rejection and disseminated intravascular coagulation. Transplantation. 2001;72(11):1743–52.
103. Iwase H, Ezzelarab MB, Ekser B, Cooper DK. The role of platelets in coagulation dysfunction in xenotransplantation, and therapeutic options. Xenotransplantation. 2014;21(3):201–20.
104. Paris LL, Chihara RK, Reyes LM, et al. ASGR1 expressed by porcine enriched liver sinusoidal endothelial cells mediates human platelet phagocytosis in vitro. Xenotransplantation. 2011;18 (4):245–51.
105. Robson SC, Cooper DK, d'Apice AJ. Disordered regulation of coagulation and platelet activation in xenotransplantation. Xenotransplantation. 2000;7(3):166–76.

106. Schulte Am Esch J 2nd, Robson SC, Knoefel WT, Hosch SB, Rogiers X. O-linked glycosylation and functional incompatibility of porcine von Willebrand factor for human platelet GPIb receptors. Xenotransplantation. 2005;12(1):30–7.
107. Bongoni AK, Kiermeir D, Denoyelle J, et al. Porcine extrahepatic vascular endothelial asialoglycoprotein receptor 1 mediates xenogeneic platelet phagocytosis in vitro and in human-to-pig ex vivo xenoperfusion. Transplantation. 2015;99(4):693–701.
108. Shah JA, Navarro-Alvarez N, DeFazio M, et al. A bridge to somewhere: 25-day survival after pig-to-baboon liver xenotransplantation. Ann Surg. 2016;263:1069–71.
109. Kim K, Schuetz C, Elias N, et al. Up to 9-day survival and control of thrombocytopenia following alpha1,3-galactosyl transferase knockout swine liver xenotransplantation in baboons. Xenotransplantation. 2012;19:256–64.
110. Cooper DK, Wijkstrom M, Hariharan S, et al. Selection of patients for initial clinical trials of solid organ xenotransplantation. Transplantation. 2016;01(7):1551–8. https://doi.org/10.1097/TP.0000000000001582.
111. Patel MS, Louras N, Vagefi PA. Liver xenotransplantation. Curr Opin Organ Transplant. 2017;22:535–40. https://doi.org/10.1097/MOT.0000000000000459.
112. Butler JR, Paris LL, Blankenship RL, et al. Silencing porcine CMAH and GGTA1 genes significantly reduces xenogeneic consumption of human platelets by porcine livers. Transplantation. 2016;100:571–6.
113. Navarro-Alvarez N, Shah JA, Zhu A, et al. The effects of exogenous administration of human coagulation factors following pig-to-baboon liver xenotransplantation. Am J Transplant. 2016;16:1715–25.
114. Shah JA, Patel MS, Elias N, Navarro-Alvarez N, Rosales I, Wilkinson RA, Louras NJ, Hertl M, Fishman JA, Colvin RB, Cosimi AB, Markmann JF, Sachs DH, Vagefi PA. Prolonged survival following pig-to-primate liver xenotransplantation utilizing exogenous coagulation factors and Costimulation blockade. Am J Transplant. 2017;17(8):2178–85.
115. Mohiuddin MM, Singh AK, Corcoran PC, et al. Chimeric 2C10R4 anti-CD40 antibody therapy is critical for long-term survival of GTKO.hCD46.hTBM pig-to-primate cardiac xenograft. Nat Commun. 2016;7:11138.
116. Oura T, Yamashita K, Suzuki T, et al. Long-term hepatic allograft acceptance based on CD40 blockade by ASKP1240 in nonhuman primates. Am J Transplant. 2012;12:1740–54.
117. Zhang Z, Li X, Zhang H, Zhang X, Chen H, Pan D, Ji H, Zhou L, Ling J, Zhou J, Yue S, Wang D, Yang Z, Tao K, Dou K. Cytokine profiles in Tibetan macaques following α-1,3-galactosyltransferase-knockout pig liver xenotransplantation. Xenotransplantation. 2017;24:e12321.
118. Shah JA, Navarro-Alvarez N, DeFazio M, Rosales IA, Elias N, Yeh H, Colvin RB, Cosimi AB, Markmann JF, Hertl M, Sachs DH, Vagefi PA. A bridge to somewhere: 25-day survival after pig-to-baboon liver xenotransplantation. Ann Surg. 2016;263(6):1069–71.
119. Navarro-Alvarez N, Shah JA, Zhu A, Ligocka J, Yeh H, Elias N, Rosales I, Colvin R, Cosimi AB, Markmann JF, Hertl M, Sachs DH, Vagefi PA. The effects of exogenous administration of human coagulation factors following pig-to-baboon liver xenotransplantation. Am J Transplant. 2016;16(6):1715–25.
120. Schneider MK, Seebach JD. Current cellular innate immune hurdles in pig- to- primate xenotransplantation. Curr Opin Organ Transplant. 2008;13:171–7.
121. Nagata H, Nishitai R, Shirota C, et al. Prolonged survival of porcine hepatocytes in cynomolgus monkeys. Gastroenterology. 2007;132(1):321–9.
122. Cooper DK, Dou KF, Tao KS, Yang ZX, Tector AJ, Ekser B. Pig liver xenotransplantation: a review of Progress toward the clinic. Transplantation. 2016;100(10):2039–47.
123. Van Raemdonck D, Neyrinck A, Verleden GM, et al. Lung donor selection and management. Proc Am Thorac Soc. 2009;6:28–38.
124. Burdorf L, Stoddard TA, Zhang T, et al. Expression of human CD46 modulates inflammation associated with GalTKO lung xenograft injury. Am J Transplant. 2014;14:1084–95.
125. Arcasoy SM, Kotloff RM. Lung transplantation. NEJM. 1999;340:1081–91.

126. Chen G, Qian H, Starzl T, et al. Acute rejection is associated with antibodies to non-gal antigens in baboons using gal-knockout pig kidneys. Nat Med. 2005;11:1295–8.
127. Cowan PJ, Roussel JC, D'apice AJ. The vascular and coagulation issues in xenotransplantation. Curr Opin Organ Transplant. 2009;14:161–7.
128. Cowan PJ, D'apice AJ. The coagulation barrier in xenotransplantation: incompatibilities and strategies to overcome them. Curr Opin Organ Transplant. 2008;13:178–83.
129. Roussel J, Moran C, Salvaris E, et al. Pig thrombomodulin binds human thrombin but is a poor cofactor for activation of human protein C and TAFI. Am J Transplant. 2008;8:1101–12.
130. Cooper DK, Ekser B, Burlak C, et al. Clinical lung xenotransplantation–what donor genetic modifications may be necessary? Xenotransplantation. 2012;19:144–58.
131. Pierson RN. Antibody-mediated xenograft injury: mechanisms and protective strategies. Transpl Immunol. 2009;21:65–9.
132. Pierson RN, Kasper-konig W, Tew DN, et al. Hyperacute lung rejection in a pig-to-human transplant model: the role of anti-pig antibody and complement. Transplantation. 1997;63: 594–603.
133. Burdorf L, Azimzadeh AM, Pierson RN. Xenogeneic lung transplantation models. Xenotransplantation. 2012;885:169–89.
134. Mohiuddin M, Singh A, Corcoran P, et al. One year heterotopic cardiac Xenograft survival in a pig to baboon model. Am J Transplant. 2013;14:488–9.
135. Cantu E, Balsara K, Li B, et al. Prolonged function of macrophage, von Willebrand factor-deficient porcine pulmonary Xenografts. Am J Transplant. 2007;7:66–75.
136. Chen D, Riesbeck K, Kemball-Cook G, et al. Inhibition of tissue factor-dependent and-independent coagulation by cell surface expression of novel anticoagulant fusion proteins. Transplantation. 1999;67:467–74.
137. Lau CL, Cantu E 3rd, Gonzalez-stawinski GV, et al. The role of antibodies and von Willebrand factor in discordant pulmonary xenotransplantation. Am J Transplant. 2003;3(9):1065–75.
138. Salvaris E, Fisicaro N, Moran C, Cowan P. In vitro assessment of the compatibility of thrombomodulin and endothelial protein C receptor in the pig-to-human setting. Xenotransplantation. 2013;20:345.
139. Zamora CA, Baron DA, Heffner JE. Thromboxane contributes to pulmonary hypertension in ischemia-reperfusion lung injury. J Appl Physiol. 1993;74:224–9.
140. Azimzadeh A, Zorn G, Blair K, et al. Hyperacute lung rejection in the pig-to-human model. 2. Synergy between soluble and membrane complement inhibition. Xenotransplantation. 2003;10:120–31. [PubMed]
141. Harris DG, Quinn KJ, Dahi S, et al. Lung xenotransplantation: recent progress and current status. Xenotransplantation. 2014;21(6):496–506.
142. Cosemans JM, Schols SE, Stefanini L, et al. Key role of glycoprotein Ib/V/IX and von Willebrand factor in platelet activation-dependent fibrin formation at low shear flow. Blood. 2011;117(2):651–60.
143. Gaca JG, Lesher A, Aksoy O, et al. The role of porcine von Willebrand factor-baboon platelet interactions in disseminated intravascular coagulation associated with pulmonary xenotransplantation. Transplantation. 2002;74:1596–603.
144. Fontayne A, Meiring M, Lamprecht S, et al. The humanized anti-glycoprotein Ib monoclonal antibody h6B4-fab is a potent and safe antithrombotic in a high shear arterial thrombosis model in baboons. Thromb Haemost. 2008;100(4):670–7.
145. Mohiuddin MM, Singh AK, Corcoran PC, Thomas ML 3rd, Clark T, Lewis BG, Hoyt RF, Eckhaus M, Pierson RN 3rd, Belli AJ, Wolf E, Klymiuk N, Phelps C, Reimann KA, Ayares D, Horvath KA. Chimeric 2C10R4 anti-CD40 antibody therapy is critical for long-term survival of GTKO.hCD46.hTBM pig-to-primate cardiac xenograft. Nat Commun. 2016;7:11138.
146. Cai W, Kaiser MS, Dekkers JC. Bayesian analysis of the effect of selection for residual feed intake on growth and feed intake curves in Yorkshire swine. J Anim Sci. 2012;90:127–41.
147. Sachs DH, Leight G, Cone J, Schwarz S, Stuart L, Rosenberg S. Transplantation in miniature swine. I. Fixation of the major histocompatibility complex. Transplantation. 1976;22:559–67.

148. Kubicki N, Laird C, Burdorf L, Pierson RN 3rd, Azimzadeh AM. Current status of pig lung xenotransplantation. Int J Surg. 2015;23(Pt B):247–54.
149. Zeyland J, Lipiński D, Słomski R. The current state of xenotransplantation. J Appl Genet. 2015;56:211–8. https://doi.org/10.1007/s13353-014-0261-6.
150. Hara H, Cooper DK. Xenotransplantation–the future of corneal transplantation? Cornea. 2011;30:371–8.
151. Cooper DK, Bottino R. Recent advances in understanding xenotransplantation: implications for the clinic. Expert Rev Clin Immunol. 2015;11:1379–90.
152. Kim J, Kim DH, Choi HJ, Lee HJ, Kang HJ, Park CG, Hwang ES, Kim MK, Wee WR. Anti-CD40 antibody-mediated costimulation blockade promotes long-term survival of deep-lamellar porcine corneal grafts in non-human primates. Xenotransplantation. 2017;24(3): e12298. https://doi.org/10.1111/xen.12298.
153. Choi HJ, Kim MK, Lee HJ, et al. Efficacy of pig- to- rhesus lamellar corneal xenotransplantation. Invest Ophthalmol Vis Sci. 2011;52:6643–50.
154. Choi HJ, Lee JJ, Kim DH, et al. Blockade of CD40-CD154 costimulatory pathway promotes long-term survival of full-thickness porcine corneal grafts in nonhuman primates: clinically applicable xenocorneal transplantation. Am J Transplant. 2015;15:628–41.
155. Buhler L, Alwayn IP, Appel JZ 3rd, Robson SC, Cooper DK. Anti-CD154 monoclonal antibody and thromboembolism. Transplantation. 2001;71:491.
156. Kawai T, Andrews D, Colvin RB, Sachs DH, Cosimi AB. Thromboembolic complications after treatment with monoclonal antibody against CD40 ligand. Nat Med. 2000;6:114.
157. Kuwaki K, Knosalla C, Dor FJ, et al. Suppression of natural and elicited antibodies in pig-to-baboon heart transplantation using a human anti-human CD154 mAb-based regimen. Am J Transplant. 2004;4:363–72.
158. Pan Z, Sun C, Jie Y, et al. WZS-pig is a potential donor alternative in corneal xenotransplantation. Xenotransplantation. 2007;14:603–11.
159. Kelley CG, Yamaguchi T, Santana E, Kaufman HE. A primate model of human corneal transplantation. Invest Ophthalmol Vis Sci. 1984;25:1061–4.
160. Jie Y, Liu L, Pan Z, Wang L. Survival of pig-to-rhesus corneal xenografts prolonged by prior donor bone marrow transplantation. Mol Med Rep. 2013;7(3):869–74.
161. Lee W, Hara H, Ezzelarab MB, et al. Initial in vitro studies on tissues and cells from GTKO/CD46/Neu GcKO pigs. Xenotransplantation. 2016;23:137–50.
162. Choi HJ, Kim JY, Kim JY, Lee HJ, Wee WR, Kim MK, Hwang ES. Long- term safety from transmission of porcine endogenous retrovirus after pig- to- non- human primate corneal transplantation. Xenotransplantation. 2017;24:e12314.

Chapter 21
Infectious Risk and Protection

21.1 Introduction

Xenotransplantation as an optional strategy can overcome the human transplantable organs and donor shortage and failure of functional organs. The currently outstanding three barriers are existed in clinical xenotransplantation, and they are the immune rejection, the physiological compatibility achievement, and zoonotic microorganism transmission from pigs to human. Among them, recently issued obstacle in xenotransplantation is the problems caused by the inter- and cross-species transmission of pig pathogens, but this is anticipated in improvement as recently CRISPR-Cas9-applied genome-wide inactivation of PERVs has been performed. Xenozoonosis has been recognized as a great interest from the worldwide outbreak of Flu and respiratory syndrome, although exogenous retroviruses are not demonstrated in pigs. Pig-to-human xenotransplantation has been advanced in animals to improve immunological and microbial compatibility. The zoonotic xenotransplantation safety is a risk for infection. Therefore, considering the zoonotic property in xenotransplantation, pig cells and organs should be treated after pig management from conditions of defined pathogen-free (DPF) or specified pathogen free (SPF) state. SPF indicates the completely pathogen-free swine herds from the specified pathogenic bacteria and viruses representatively. SPF piglets are sterilely protected and generated by Caesarean delivery on xenotransplantation request. To examine microbiological safety of pig breeds and to combat unknown risks, many efforts were made. In breeding of pigs, the SPF pig breeding facilities are limited due to short-term pig research history.

Currently, several swine breeds have been established to apply to human islets transplantation, although they are not the intermediate breeds. For example, Auckland island pigs have been approved in pancreatic islet cells of pigs to diabetic patients of humans' clinical transplantation [1]. Likely, the swine breeds named Göttingen minipigs established from crossbreeding between the Vietnamese pot-belly pigs, German Landrace pigs and Minnesota minipigs, are utilized in

C.-H. Kim, *Glycoimmunology in Xenotransplantation*,
https://doi.org/10.1007/978-981-99-7691-1_21

biomedical technology and donor sources of swine pancreatic islet cells to xenotransplantation [2]. In addition, the bred herd, Gottingen mini pig, established via domestic long breeding at Ellegaard is managed as a level of SPF facility (http://minipigs.dk/the-goettingen-minipig/) in Denmark. For clinical aspects in donnish cases, experimental pigs are controlled by genetic characteristics, health status, pathophysiologic properties and are selectively checked twice a year for the required infectious pathogens including 27 bacteria, 3 fungi, 16 viruses and 4 lower eukaryotic parasites (http//www.minipigs.dk/).Minesota mini pig (Sinclair) and Hanford mini pig (Sinclair) are those pig bred to date. So pig zoonotic pathogens are strictly controlled via SPF- or DPF-labeled breeding establishments. At present, results on the results obtained from the consequential zoonoses to apply in pig-to-human xenotransplantation are very limited. Future extensive identification and information of potential zoonotic agents will help to improve the safety acquisition of xenotransplantation. For official leading to xenotransplantation progress, the US FDA and the WHO define human xenotransplantation guidelines for facilities, breed selection, and pathogen analysis of pig animal [1].

The distance in phylogenetic relationship between pigs and humans restricts the infectious prevalence occurred by zoonotic pathogens including bacteria, virus and parasites. During the past decades, any issue on transmission of pig-derived infectious agents to humans has not been reported. Moreover, infectious agent transmission to human cells exposed to cells, tissues organs of live pigs was not reported to date. However, depending on the research data on endogenous viral behavior, several viral agents are suggested to be biohazard in transplantation field. From them, pig endogenous retroviruses (PERVs) are treated as the most interested agent. The PERVs have three distinct subclasses of the PERV-A, PERV-B, and PERV-C, classified by the specific receptor interaction. Although A-PERV and B-PERV types interact with in vitro human cells, but PERV-C type is free of interaction with human cells. However, through genetic recombination between PERV-A and PERV-C, the recombinantly generated PERV virus variants can obtain its infectibility to human cells as host. But the PERV-A/PERV-C fusion type cannot infect humans. Thus, PERV-C-free pig donors with the low PERV-A and PERV-B infectivity are recommended to use for xenotransplantation. For another pig cytomegalovirus (PCMV), PCMV is observed in primate xenotransplantation; however, fortunately the virus does not develop invasive disease in hosts. In the pig lymphotropic viruses (PLHVs), on the other hand, PLHV-1 type only involves in lymphoproliferative syndrome, although the lymphoproliferative syndrome is not documented yet during pig-to-primate xenotransplantation [3].

In the past of xenotransplantation, only several pig viruses have been regarded and suspected to infect human cells, raising the rejection issue of transplants. However, recent progress in next-generation sequencing (NGS) suggested existence of unknown viral sequences in pig genome and its impact in xenotransplantation, leading to new viral discovery and identification of new viruses [4]. Consequently, the new term of "pig virome" has been expressed as the total amount of viruses together with the endogenous retroviruses in and on the pig body [5]. The issue of virus safety has been interested in the recent xenotransplantation from the facts that viral infections are deleted shortly within 24–48 h soon after birth, and thus, viral infection seems to be rapidly post-natal or transplacental transmission. Such examples are the coronaviruses, pig epidemic diarrhea viruses (PEDV) and transmissible gastroenteritis viruses (TGEV) that have been previously reported in Chinese pigs [6]. When the feral pig virome was analyzed, 16 different virus genomes have previously presented [7]. As single-stranded (ss) DNA viruses, anellovirus, circovirus 2, circoviridae, and parvovirus were detected. Torque teno virus was the most frequently observed (73% more). Torque teno virus (or TTV), anelloviridae anellovirus, is a non-enveloped SS DNA virus, widely ranged for its animal hosts. As RNA viruses, kobuvirus, sapelovirus, pestivirus, and orthopneumovirus, were detected [7]. Because domestic pigs are reservoir for zoonotic infection [8, 9], feral pigs are frequently infectious by influenza A virus and its H1N1 subtype of the influenza 2009 virus [8]. H1N1 influenza virus as an influenza A subtype is a communicable viral diseases, causing upper respiratory tract host infections and lower tract at less extent. Notably, human to pig influenza virus transmission is more frequently occurred than pig-to-human zoonosis [10], indicating that pigs-to human xenotransplantation should be take care of monitoring to avoid the influenza viral issue.

21.2 Pig Endogenous Retrovirus (PERV) General Aspects

Currently, in pigs, biotechnological and technical trials can eliminate most dangerous viruses from donor pigs by the following tools including early stepped weaning, drugs-based treatment, anti-infectious immune vaccination, Caesarean delivery, and qualified embryo transfer technology and pig breed selection, except for only PERVs. PERVs are reasonably considered to be the resulting products obtained during retroviral infections in the long past of evolution. Endogenous retroviruses are derived from infection of germ cells with a retrovirus. If oocytes or sperm cells are infected, the proviruses are located in all the developing progeny cells with all the inherited genes. Endogenous retroviruses are well conserved and found in all mammals and humans [11]. Although the human endogenous retroviruses (HERVs) have been known to be mostly active without viral particle production to infect, a lot of retrovirus family is actively infectious. Likely other retrovirus family, PERVs can also cause tumor genesis and immunodeficiency in hosts. Nevertheless, their zoonotic transmission potentials are not known yet because PERV transmission between pigs and humans has not been detected in pig-to-NHP xenotransplantation even in preclinical or initial clinical transplantations [12–14].

Among HERVs, HERV-K type is tested for its characterization. The HERV genetic integration events were started 28 million years ago and thereafter the HERV copy number is consistent conserved in human cells [15]. Although HERVs are inactive in replication, the release of other species' endogenous retroviruses including endogenous γ-retroviruses is observed as active viral particles with inter-species and trans-species infectivities [16]. Developed virus types of retroviruses as the retroviridae family have 2 ssRNA copies in a genome. Upon entry into the hosts, the viral ssRNA genome is reverse transcribed to the corresponded DNA and integrated into the host genomic DNA, calling a provirus form. If retrovirus infection occurs at host germline cells, the proviruses are transmitted to offspring of hosts though Mendelian inheritance law and this is now called an endogenous retrovirus. In the human genomic DNA, retrovirus-like DNA sequence has been known to occupy at least 8% of total DNA sequence. Like this, pig genomes have endogenous retroviruses, as retroviral remnants of their ancestral infections [17]. As a zoonotic retroviral infection in humans, a human familial prostate cancer has been known to bear the retroviral sequences and the retroviral sequences in human is closed to the xenotropic MuLV (murine leukemia virus) sequence. The name of xenotropic MuLV is thus originated [18]. Thus, xenograft endogenous retrovirus can infect the hosts, giving the epidemics.

Pig genome DNA bears beta- and gamma-retroviral sequences, where major difference between the two genera is in the envelope sequence [19]. Among them, only gammaretrovirus genus is infectious [20]. The PERVs are gammaretroviruses, which are closed to feline leukemia virus (FeLV), MuLV, and koala retrovirus (KoRV). Origin of African PERV type is ascended to more than 7.5 million years ago. Another PERV-C type was occurred more than 1.5–3.5 million years ago through homologous recombination events with uncertain DNA homology sequence [21, 22]. The PERV viral copy numbers in each breed, each animal, identical animal-derived different cells and established cells of pigs are not characterized [23]. Limited provirus number and RNA low expression level are known only. In the PERV copy numbers of Iberian pig in Spain were found without any correlated relationship between the heterozygosity and the numbers [24]. The PERV copy numbers were calculated to be 40 PERV copies of each miniature pig, 30 PERV copies of Duroc pigs, 20 PERV copies of Landrace pigs, 15 PERV copies of Korean Jeju pigs and 25 PERV copies of Yorkshire pigs, which were estimated from the pig PBMC DNAs [25]. Different copy numbers of proviruses are found in a single pig and different organs of one pig [26–28], indicating that PERVs are under active replication to form new proviruses. In Banna minipigs, different PERV copy numbers of DNA sequence of the gag gene in different organs have been calculated [27]. *Upon* pig PBMC stimulation with the phytohemagglutinin (PHA) lectin known as a T-cell mitogen elevates expression of viral genes for possible release of viral particles [29, 30]. HERVs in humans are not active, while in pigs, PERVs are active. PERV 7.5 has been known to infect the host pigs 7.5 million year ago, while the HERVs are known to infect the ancestral primates of humans 28 million years ago. In the evolutionary process, primates such as humans developed restriction mechanisms to inactivate the endogenous viruses such as HERVs to protect the host [31].

The PERV's three variants are infectible to human cells as host in vitro. Because three PERV types of A, B and C can be typically integrated into pig genomic DNA, most pigs contain the PERV-A and PERV-B types, and certain pigs, but not all, contain only PERV-C type [32]. PERV-A and PERV-B types are classified to polytropic virus species, infectible for several species cells from humans' cells, while PERV-C as an ecotropic virus type is infectible for only pig-derived cells.

Of interest, PERV-A and PERV-C types can be infectible for immortalized and primary human-derived cells. Ecotropic virus type, PERV-C, is infectible for pig-derived cells only [23, 32, 33]. In vivo PERV replication activity indicates homologous recombination events between the PERV-A as human tropic virus and the PERV-A as human ecotropic virus in pigs, creating the recombinants of PERV-A and PERV-C. However, the recombinants between PERV-A and PERV-C having a high replication capacity cannot integrate into the germ line, although the PERV-A and -C recombinants are observed in the cellular genomic DNAs [34–36]. Because PERV-A and PERV-B types, and also PERV-A/PERV-C fusion types can highly infect in vitro human-derived cells, it is regarded that PERVs are threating risk factor for xenotransplantation [12].

Recombinations between two defective proviruses yield a replication competent virus and the virus can induce tumors. Until now, several types of PERV-A, PERV-B, PERV-NIH and PERV-A/PERV-C recombinant types were known in many cell lines during multiple passages and infection in in vitro human-derived cells [37, 38]. Microorganism stimulation of immune-related cells induces PERV viral expression and homologous recombination events. An important question is whether HERVs homologously recombine with the different viral species, PERVs [39]. But, HERV-homologously recombined DNA sequences were not detected in the different viral species PERV virus of its viral particles of HERVs-expressing human-derived cells [40]. However, in vivo, during the initial pig to NHP xenotransplantations and in the experimental PERV behavior in mice and NHPs, PERV transmission was not detected [41]. Transmission of PERV has not been detected in PERV inoculations into NHP xenotransplantation. Moreover, transmission of PERV has not been reported in the first pig neonatal islets-to-diabetic patients' xenotransplantation in Argentina and New Zealand [13, 14]. Thus, gammaretroviruses are not transmittable to humans. Although PERV from various pig sources has not been transmitted to recipient hosts in vivo [42, 43], in the mice immune-deficiency or NOD (non-obese diabetic)/SCID (severe combined immunodeficiency) cells to pig islet cells xenotransplantation, PERV infection, PERV gene expression, and PERV replication were observed [44, 45]. In the mouse cells-adapted PERV (mPERV)-expressing mouse cell-transplanted mice, the PERV was transmitted to the mice with a high frequency in BALB/c mice, compared to pPERV-generating pig cells [46].

21.3 PERV Transmission

Despite the potential merits, pig-to human xenotransplantation has a concern of the transmission of zoonotic pig microorganisms. Microbiological safety remains a still unresolved problem. Most pig microorganisms develop zoonosis and they are bacteria, fungi, protozoa, and viruses. The zoonosis prevention in xenotransplantation has been recognized as one of the most critical explants—threatening and late-occurring risks for the successful survival [47]. The pig zoonotic microbes and parasites are generally capable of elimination and removal by various efforts such as vaccination and pathogen-free breeding. Although the possible pig fungal and bacterial transmissions can be eliminated using by anti-mycotics and antibiotics, xenotransplantation-based virus transmission may contribute to human recipient diseases, calling “zoonoses” [48]. There are no antivirals and vaccines available for most porcine viruses, although pig viruses can eventually infect humans and cause zoonoses (but the reason is not clear), except for HEV genotype 3 having a zoonotic potential [49, 50]. PCMV is also known as an indirect pathogen, which is not infective to the host cells.

To avoid the infection problems, selection of uninfected pigs, antiviral agents, Cesarean delivery, colostrum deprivation, embryo transfer, antigenic vaccination and weaning, are considered [51]. However, pig-infected PERVs are an exceptional case for the above events to eliminate due to simply their DNA integration into the pig genome DNA. As a counterpart to PERV, HERVs are known to be defective and not competent in replication, although pig cells release PERVs for infectious potentials. All pig genome-integrated PERVs release the infectious viral particles to infect human cells in their specific environmental conditions. Exogenous retroviruses like the HIV infect target cells to integrate the virus genomes, forming a pro-viral DNA form, and the pro-viral DNA again integrates into the target genomic DNA. Endogenous PERV retroviruses are derived by oocytes and sperm cells-infected retrovirus, and consequently all organism cells are integrated with the endogenous retroviruses [52, 53]. The PERV risk degree to infect the human hosts remains unclear yet. However, several experimental evidences are obtained. For example, PERV infection to human cells is clearly observed in the co-cultured human EK-293 cells with the PERV [54]. Transmission of cross-species in pig breeds is another possible concern in pigs-to-SCID mice xenotransplantation [55]. Therefore, they are a threatening risk factor in pigs to human xenotransplantation.

Apart from xenotransplantation, considering the interspecies allotransplantation, the transmission of viruses of human immunodeficiency, rabies, West Nile, human cytomegalo, and human hepatitis viruses as well as other microorganisms of bacterial *Trepanoma pallidum*, fungal *Aspergillus*, and in some case parasites is the issue to the recipients [50, 55–57]. Although the human pathogens in allotransplantation are well characterized for immunodeficiency, rabies and hepatitis viruses, the zoonotic microorganisms in xenotransplantation are partially known. Thus, the zoonotic microorganisms-free pigs are needed to render xenotransplantation, more safely than that of allotransplantation. Several national and regional regulations have been documented for the microbiological test of donors for allotransplantation and xenotransplantation [58, 59]. In allotransplantation, the transmitted microorganisms are adapted to humans; however, pig microorganisms in xenotransplantation are not infectible to human hosts because of the innate/adaptive immune responses of immune system and absence of infection suitable receptors on human cells. Pig hepatitis E virus (HEV) is infectible human cells. Before transplantation, the immediate detection, even in the viral replication condition, is important. In addition, the elimination of specific pig microorganisms is obtained by antiviral drug treatments and vaccination-based prevention. However, it is unfortunate that there are no current antiviral drugs and vaccines to apply for some pathogens. Current prevention of transmission of pig microorganisms includes virus-free animals, anti-viral agents, vaccinations, colostrum deprivations, weaning at early stage, Caesarean delivery, embryo transfer, and target-gene editing. The most promise of elimination strategies is to isolate the cleaned animal breeds without de novo infectious issue. In zoonotic viruses, HEV genotype 3, pig circoviruses (PCV) PERV, PLHV, and PCMV are currently well known. PCMV, PCV, HEV, and PLHV as well as fungi and bacteria can be technologically removed from pig donors; however, PERVs are not eliminated due to integration into the pig genome [60].

The pig xenograft-originated PERV transmission risk to human has been recognized as significant big safety issue. The issue of retroviral transmission from one species to another is important in the PERVs transmission to the human recipients during xenotransplantation. The issue limits the future possibility for clinical xenotransplantation using live pig organs/cells/tissues. The xenograft can be considered to be the delivery mediator of bacterial, fungal, or viral pathogens [61]. Among them, the issue of PERVs is complicated and PERV is carefully treated because they are found as integrated DNA copies in the pig genomic DNA and is capable of infection to human cells, although there is no evidence for pigs to humans or NHPs PERV [48, 60]. In the trans-species viral transmission, various viral subtypes are born, although trans-species transmission of retroviruses is not frequently occurred, but generally seen in HIV-1 that is originated from a trans-species lentivirus transmission from chimpanzees to human hosts. In addition, in HIV-2, it is also observed that is originated from a trans-species SIV (simian immunodeficiency virus) transmission from sooty mangabeys to human hosts [62], as known for the cases of the chimpanzee lentiviral transmission to humans and the sooty mangabey SIV transmission to humans to yield the new HIV-1 and HIV-2 subtypes [63]. These trans-species transmissions are found in the γ-retroviruses such as PERVs as the final product of the mice retroviral transmission to pigs [64], because since similar DNA sequences are detected in mice, and thus, PERVs are considered to derive from mouse endogenous retroviruses. The pig-to-mouse transmission of PERVs is calculated approximately 10 million years ago [65], and the age of PERVs is thought to be 7.6–10^6 years old [66]. However, mechanistic pig-related viral trans-species transmission is still required to be informative.

In issue of the PERV, the downregulation of viral infection to human cells is performed by glycosyltransferases. Because PERV carries some *N*-linked sugar-ligands, controls of glycosylation pathways may regulate PERV infectivity. International Xenotransplantation Association advices the donor pigs and PERV expression to monitor xenograft recipients [58]. In order to exclude the possible PERV risk, donor pig CRISPR/Cas9 technology-based gene editing can be applied to knock out PERV copies and finally generate a PERV-free genome-bearing healthy pigs [67]. Natural anti-α1,3Gal Abs function as the PERV transmission-derived barrier [68–70], and this, reversely, helps to persist PERV life span of α1,3Gal-T KO organ in recipient human circulation system. α1,3Gal-T KO pig-produced PERV will not be subject of anti-α1,3Gal antibody/complement pathway, showing resistance to binding [71, 72]. In xenotransplantation of α1,3Gal-T KO organ into NHPs, PERV RNAs are not observed in recipient PBMC

[73]. However, in the live pig cells clinical trials, transmission of PERV infection was not obtained in non-immunosuppressed [74] and immunosuppressed [74, 75] recipients. The issue of PERV transmission is solved in near future and the α1,3Gal antigens should also be considered during antiviral action. At present, because pig-to-human xenotransplantation requires α1,3Gal antigen elimination, the risk of PERV and any other virus are need to considered together.

21.4 Inactivation of PERVs by Genetic Manipulation

Despite the anti-zoonotic advances, PERV transmission should be secure during xenotransplantation or PERV infection should be prevented. In the case of PERVs, their elimination is not easy due to genomic integration into the pig genomes. For the safety and PERV transmission prevention from pig to human xenotransplantation, PERV-A and PERV-B expression-attenuated pigs are required as donors. In addition, PERV-C negative pigs and PERV-C-free pigs are required, because fusion of PERV-C strain and PERV-A yields the PERV-A and PERV-C fusion type. PERV-A and PERV-C recombination event is easily detectable. The fusion recombinants between PERV-A and PERV-C strains are capable of integration to different organs-derived cells. However, they do not to the animal germ line-derived cells. In screening of such pigs, the multiple available samples of blood, ear biopsies, saliva, or feces are used. When the organ PERV copy numbers are different, the whole animals are examined. For in vitro RNA interference to inactivate the pig genome PERV proviruses [76, 77], in vivo transgenic expression of a PERV-specific small-interfering RNAs (siRNAs) [76] and PERV-specific vaccination [78–80] are also such strategies to combat PERV infection and PERV expression.

In addition, for gene editing technology-engaged elimination of the multiple proviruses, the first attempt was performed by the pol region conserved sequence-specific ZFN [81]. Unfortunately, simply due to the high pro-viral copy number, cytotoxic effects were detected [81]. The other antiretroviral drugs [82–84] or CRISPR/Cas, CRISPR-associated technologies are also used. Genetic modification of pigs through PERV-targeted siRNAs and CRISPR/Cas-used gene editing can be used for the development of effective drugs for PERV. Recent genome editing is rising in the pigs to inactivate all resident PERVs using the CRISPR-Cas9 technology [67]. Among them, due to multiple PERV provirus copies in pig genome, gene editing is considered to be the best way to inactivate PERV. ZFN-based gene editing created in highly expressed ZFN to induce toxicity, cutting the genome DNA at multiple sites [81]. PERVs in pigs have been targeted by using CRISPR/Cas genome editing. Interestingly, the recent CRISPR/Cas9 technology yielded 62 PERV proviruses by inactivation in immortalized PK-15 cells of pigs [67], suggesting that the PERV issue is solved. They did not show any PERVs transmission [85], but with basic question of suitability of PERV-free pig for xenotransplantation [86]. However, the question is arose whether the current CRISPR/Cas9 technology inactivates most PERVs or replicable PERVs in the genome at least, creating usable piglets with good heath [86]. For construction of PERV free pigs, PERV-targeted CRISPR/Cas technology should be used to targeted cells for nuclear transfer. From the limited primary cell division (Hayflick Limit), CRISPR/Cas-inactivated PERV-carrying cells are not easy to clone and select. Because PERV replicated in specific conditions of pig cells, CRISPR/Cas resistance is removed and specific designation of CRISPR/Cas-specific distinct DNA sequences in PERV eliminate such cases [87].

21.5 PERV Life Cycle-Targeting Antiretroviral Agents

As the PERV drug, anti-HIV drugs are used. If inactivation of all proviruses cannot be applicable by genome editing, it is because that strong antiretroviral drugs and PERVs-free pigs are required. Antiretroviral viral reverse transcriptase (RT) and integrase inhibitors suppress in vitro PERV infection capacities, being integrase inhibitors as the biggest potential agents [88]. As anti-PERVs drug, antiretroviral, HIV-1 agent, azidothymidine (AZT) is efficient in inhibition of PERV replication in pig genomes. Inhibitors of virus-derived reverse transcriptase and integrase enzymes are also effectively acting for PERVs. The integrase inhibitors were also effective for PERV. During infection, retroviruses utilize receptor(s) to infect their specific target host cells and transcribe the viral RNA genomes to a single-stranded DNAs by its viral RT, replicate to double-stranded DNA by host-derived DNA polymerases, and finally integrate the cDNA to the host cell genome by its viral integrase. Pro-viral genomic transcription covers a full-length mRNAs that encode for the virus proteins required for virus life cycle, being Gag core proteins, viral enzymes and envelope (Env) proteins.

PERV life cycle is prevented at various stages to target viral proteins responsible for replication [88]. Initially viral entry inhibitors block the vial entry into the host cells. Such entry inhibitors predominantly bind to the host co-receptor proteins and consequently block the viral envelope protein binding. Another type of fusion inhibitors bind to viral structures and block the virus entry. Artificial-designed peptides which recognize the transmembrane envelope protein helical domains to prevent the cellular membrane fusion. Next, the RT inhibitors target the RT enzyme that reverse-transcribe the retrovirus RNA genome to a ssDNAs. There are two RT enzyme inhibitor subclasses: the nucleoside- or nucleotide-based RT inhibitors and the non-nucleoside-based RT inhibitors [89]. Azidothymidine (known as AZT or zidovudine) and stavudine as the thymidine analogs, lamivudine and emtricitabine as the cytidine analogs, didanosine as an adenosine analog, and abacavir as an guanosine analog are the cases. Tenofovir and adefovir recognize the enzyme hydrophobic pocket and shows the allosteric inhibition of viral DNA synthesis. All the mechanism is explained from activity inhibition of RT RNase H and/or inhibition of triphosphate binding capacities. Nevirapin and efavirenz are developed during 1996–1998, etravirin in 2008, and rilpivirin in 2011 [88].

The integrase inhibitors inhibit the integrase activity that mediates the provirus DNA integration event to the chromosomal genome DNA of hosts. Integrase inhibitors known are raltegravir, dolutegravir, and elvitegravir. Protease inhibitors blocks protease required and acted for the late life cycle stage and degrades the precursor form of Gag–polymerase (Pol) polyprotein to each different product subunit, leading to the matures virus particles [90]. The above inhibitors are effective for HIV-1-infection blocking [91] and have been approved by Japan, the USA FDA, Canada agency and the European Medicines Agency (EMEA) and Korea. Due to the virus genomic mutations, a combination of antiretroviral therapy is used for HIV-1 [90–92].

As inhibitors of RT, the first PERV agent is AZT [88]. In 1974, Ostertag group [93] reported AZT to block the Friend murine leukemia virus that is closely related to PERV. In 1987, it has been approved by the USA FDA as HIV infectious blocking agent [94]. AZT selectively inhibits the RT enzymes. Upon administration, host metabolic enzymes catalyze the AZT conversion to its triphosphate derivative as a metabolite and consequently, this specifically blocks RT DNA synthesis activity. The AZT blocks PERV [88]. The PERV-blocking order is AZT, tenofovir, adefovir, and stavudine. Other nucleoside RT inhibitor (NRTI) type AZT, and nucleotide reverse-transcriptase inhibitor (NtRTI) adefovir and tenofovir are also potential inhibitor of PERV activity [88]. HIV-1 and PERV RTs share 22.5% homologies in amino acid sequences and the AZT target domain sequence homology is much higher than other sequences. For inhibitors of PERV integrase and PERV expression, raltegravir and dolutegravir inhibitors of the integrase effectively inhibit PERV replication. The PERV and HIV-1 integrase catalytic domains are well conserved. Rather than AZT, the raltegravir and dolutegravir integrase inhibitors are the most active to inhibit PERV replication. Another PERV replication-blocking strategy in the donor pigs is to use drugs. This strategy can be combined to the PERV-specific siRNA-expressing pig generation, which reduces PERV replication. The currently known 4 HIV-1 gene expression inhibitors suppress PERV replication and expression. For example, K-37 and the EM2487 known as the fluoroquinolone derivatives derived from a *Streptomyces* species selectively inhibited PERV expression [95, 96].

21.6 Pig Cytomegalovirus (PCMV) Transmission

Issues of human organ shortage raised the pig donor-based xenotransplantation strategy, which uses pig-derived cells, tissues, and organs. Xenotransplantation using pig sources is now a replacement tool and an alternative for transplantation [47]. The current clinical xenotransplantations being tried in Argentina and New Zealand seems to be lowly evaluated in biomedical benefits, although pig microorganism transmission was not apparent [13, 14]. The some viruses are prevalent in pig populations. Pigs carry many viruses and other microorganisms as non-harmful phenotype, but sometime the pathogens are zoonotic in the human recipient. The examples are pig cytomegalovirus (PCMV) and HEV. For microbial and parasitic safeties, pig donors should be examined in detailed levels for elimination of zoonotic microbes and parasites. In the pig zoonotic microorganism-threatening xenotransplantation, the transmission of PCMV is another case. PCMV is frequent and latent in adult pigs. The latent viral occurrence raises a difficulty to detect virus from blood or serum. Although the PCMV infection risk of humans is not well reported, the viral propagation is reported in pig transplantation to non-human primates [90, 91]. In preclinical pig kidney xenotransplantation to baboons and cynomolgus monkeys, the PCMV infection causes early failure of transplantation [92]. Due to absence in the obtained evidences for NHP PCMV infection as well as human cell PCMV infection, failures in the organ transplantation seem to be based on viral antigens-induced cytokine production responses [92].

In pig to NHP xenotransplantation, animal transplanted engrafts are infected by PCMV and PCMV is replicated in the recipient [93, 94]. Indeed, in the case of CMV in transplanted grafts of human recipients [95], PCMV is likely to be harmful for human recipients. Eighty-five percent of pigs were found to be PCMV-positive [96]. The PCM DNA has been demonstrated not to present in pig tissues if the relevant pretreatments are done. For example, specialized pathogen-free conditional herd-produced pigs are PCMV DNA-free [47]. In the pig solid organ xenotransplantation to baboons, PCMV DNA was found in only blood of baboons, in which two cases received organs without early PCMV-excluding weaning treatment [13], requiring endless checking the PCMV absence. Although the PCMV-infected nonhuman primate tissue is free of PCMV-derived pathogenicity, PCMV is known to be permissive to human fibroblasts with cytopathic effects, as PCMV proteins were observed in infected human fibroblast cells [14]. Problem is apparently solid in that PCMV transmission to host animal lowers the survival length of the transplanted grafts for 3 times. Such lowered survival is observed in the pig α1,3Gal-T KO kidney xenotransplantation to baboons [97] and cynomolgus monkeys [98]. PCMV infections are symptomless, chronic and non-detectable. PCMV transmission has been known to easily transmit horizontally to hosts, but vertical PCMV transmission is not found [99]. PCMV ubiquitously infects and particularly, causes in pigs a mild disease state at the early younger time, not over the 6 weeks [100]. In contrast, PCMV is latently infected in adult pigs. PCMV transmission and its replication are detected in the recipients in preclinical xenotransplantation [95, 97, 101, 102]. PCMV-infected pig kidney xenotransplantation to NHPs displays the decreased survival level. In pig α1,3Gal-T-KO kidney transplantation to baboons in the absence condition of PCMV, a survival prolonged up to 53 days, but when the animals infected with PCMV gave survival of 14.1 days [90].

Even though PCMV infection is not detected in donor animals, later, the PCMV virus transmission to the heart transplants is still possible and the PCMV virus can actively replicate in the NHP recipients [101]. Therefore, PCMV as a potentially pathogenic zoonose [73, 102, 103] has been regarded. PCMV is a enveloped double-stranded (ds) DNA virus, which has total 128,367 nucleotide base pairs, 73 promoters, 8 putative protein genes, and 79 open reading frames (ORFs), belonging to the herpesviridae family, subfamily beta-herpesviridae, and genus roseolavirus [104]. PCMV function is unclear because until now, one virus species was isolated and identified from pulmonary alveolar macrophage cells and now complete viral genome sequence is elucidated [105]. PCMV strains are similarly closed to human herpes virus (HHV) types including HHV-6A, HHV-6B, and HHV-7, rather than to human CMV (HCMV) and HCMV type of HHV-5. PCMV pig infection is a form of endemic [97, 99, 106]. In newborn and young piglets, PCMV is infectious with rhinitis, while PCMV is latent in adult animals [10]. PCMV infection in human cells is not clearly evidenced yet, although in vitro infection is still controversial, as raised from the results obtained from the PCMV-infected pig cells co-cultivated with two established human cells [106] and primary fibroblast cells [107].

Apart from PCMV, HCMV is still reactivated after allotransplantation, despite HCMV diseases-preventive strategies, requiring immunosuppression in the future pig-to-human xenotransplantation. The reactivated HCMV causes direct damages to human and grafted pig organ. HCMV easily infect pig endothelial cells causing cytopathogenicity [108].

21.7 Hepatitis E Virus (HEV) Transmission

Currently, actual zoonotic activity of many pig-bearing viruses remains still unelucidated, although certain zoonotic virus types of pig-bearing virus are studied. The well-studied zoonotic virus type is the hepatitis E virus (HEV). The HEV transmission to human is very easy and frequent due to of undercooked pork consumption or pig contact. These circumstances can cause HEV-derived diseases and immunosuppressed individual chronic infection as well as liver failure people infection [109]. Pig HEV zoonotic transmission to humans is already evidenced [110]. HEV infection raises chronic infectious diseases in liver and hepatitis in grafts of allotransplant recipients [98, 111]. HEV in pig viruses can be easily transmitted to humans, indicating a threat factor in xenotransplantation [109, 112]. The HEV as representative virus of the HEV elimination-difficult viruses is initially identified from human non-A and non-B hepatitis cases [113]. Non-B hepatitis is also found in pigs and HEV distribution is also broadly present [114]. HEV hepatitis is classified to one of the self-limiting hepatitis diseases in humans and does not induce chronic hepatitis [104, 115]. Nevertheless, HEV pregnancy infection is severely occurred during fulminant hepatitis period, premature delivery and still birth, which shows about 20–25% mortality as relative high mortality rate [50, 104].

Pig herd HEV infection is asymptomatic and prevalent in porcine herds worldwide [2]. HEV species is known as the sole Hepeviridae family and Hepevirus genus. HEV is very small sized having a diameter with 30–35 nm size as ssRNA, positive-sense RNA and non-enveloped form with 7.2 kb in length. Positive-sense viral RNA indicates the viral RNA sequence translation to the viral proteins. The viral RNAs are capped and polyadenylated [2, 116] and the HEV genomic DNAs contain 3 ORFs ranged from ORF-1 to -3, where ORF-2 encodes the capsid proteins. The viral capsid protein size is 71 kDa and glycosylated capsid protein to be 79 kDa [117]. The capsid protein amino acid 394–457 region is the immunologically dominant region to bind Abs [118]. Some new genotypes are described in different species of bats, birds, ferrets, fishes, rabbits and rats [119, 120]. In taxonomical classification, the family Hepeviridae is subdivided into two genus and five species (Piscihepevirus A species and Orthohepevirus A-D species) in humans, pigs, wild boar, and deer. Orthohepevirus A is also name to some animals [119, 120].

HEV is divided to 4 genotypes (gt1-4) in mammals and further classified into 24 subtypes [121]. Differences are observed between gt1 HEV and gt2 HEV on one hand and gt3 and gt4 HEV on the other. Among them, HEV gt1 contains for 5 subtypes and HEV gt2 has 2 subtypes, while HEV gt3 contains 10 subtypes and HEV gt4 has 7 subtypes. HEV gt1 has been found in Asian and other peoples and HEV gt2 is in Mexico and West Africa. HEV gt1 and HEV gt2 belong to human-specific viruses. HEV gt1 and HEV gt2 cause human liver hepatitis. Also, HEV gt1 and HEV gt2 cause pregnant women fatal infections and they are detected in humans. Their transmission takes places by allotransplantations, water, fecal-oral routes, and blood donations [104, 122, 123]. Additionally, HEV chronic infection develops in immunosuppressed patients [124] and consequent HIV infection [125]. Xenotransplantation patients y needs immunosuppressive treatment. HEV infection in pigs is latent [126]. HEV gt3 and gt4 subtypes are relatively low in virulence in humans than virulent in HEV gt1 and HEV gt2 subtypes [50, 123]. HEV gt1 and HEV gt2 subtypes raise light sporadic symptoms of hepatitic disease [104, 127]. In contrast, HEV gt3 and HEV gt4 subtypes are found in humans as well as in animals including pigs. Their subtypes are swine-infectious virus types without causing diseases in pigs, but infecting humans with rare zoonotic diseases. Therefore, it is thought that HEV gt3 and gt4 subtypes belong to pig virus, but may infect humans with mild diseases. Only HEV gt3 and HEV gt4 subtypes are risk factors in xenotransplantation because these subtype gt3 and gt4 transmission is mainly made by pigs with broad distribution in pig herds and human population [128]. HEV subtype gt3 and gt4 RNAs are observed in liver of pigs [129]. HEV gt3 subtype transmission is mediated via transfusion of bloods [130]. Human HEV gt3 infection is observed in Asia, Europe, Australia, Africa and New Zealand, while gt4 human infection is reported mainly in Asian countries [130, 131].

The HEV gt3 and gt4 subtype main carriers are suggested to domestic pigs and to wild boars [132]. The HEV gt3 subtype is frequently found in the European countries and USA, whereas HEV gt4 subtype is seen in Asian countries [122]. In the animal infection of HEV, disease symptom is not observed [110]. Zoonotic human transmission HEV gt3 and HEV gt4 is dependent on consumption of wild boar or deer, wild pork liver, or wild pork meat [133]. HEV gt3 subtype infection including c, e and f is frequent in European peoples [121]. HEV gt3 and gt4 human infection is not symptomatic, but symptomatic in immunocompromised for chronic hepatitis [104, 123, 134]. HEV gt1 and gt2 subtypes are then subdivided to five (a, b, c, d and e) subtypes and two (a and b) subtypes, respectively [104]. HEV gt3 subtype and HEV gt4 subtype are also divided to 10 (a to j) subtypes and 7 (a to g) subtypes [131, 135]. Although HEV gt3 and gt4 human transmission is harmless and attempts are needed to eliminate the virus subtypes from the pig donors, such heterogeneous viral distribution in many HEV subtypes of 10 subtypes of a type to j type in gt3 and 7 subtype a to g in gt4 is hurdle [131, 135]. The development of detection method of pig microorganisms is needed for selection pathogen-free pigs and prevention of pathogen transmission to the hosts. Although multi-transgenic pigs were currently generated to confer the xenotransplants, HEV-eliminated pigs are not made yet. However, non-transgenic pig breeds negative for PERVs and HEV are suitable for encapsulated cell transplantation in pig to non-human primate islets xenotransplantation [136]. Another Ellegaard-generated Gottingen minipig in Denmark is utilized for medical studies [137, 138]. To prevent HEV human transmission, selection of HEV-negative animals and HEV-specific vaccine development are considered [139]. Due to a high prevalence in pigs, HEV monitoring in donor herds is crucial with the removing strategy from the infected animals. Ribavirin agent is reported to eliminate the HEV virus in allotransplants, applicable to xenotransplantation [130].

21.8 Pig Lymphotropic Herpesvirus (PLHV) and Other Herpesviruses

Contrary to PCMV, PLHV is not easily eliminated by the current protocols. The PLHV prevalence and PLHV importance in xenotransplantation remains unknown [138, 140]. It is still not clear whether PLHVs are a life-threatening risk factor in pig-to-human xenotransplantation, although the pathogenic value of PLHV for humans is unclear. The PLHV are classified into the three subclass of PLHV-1, -2, and -3. PLHV-1, -2, and -3 subtypes are commonly found and latent viruses as pig *herpesvirus* pathogens. However, how PLHVs do exactly transmit to pigs are not well known, although PLHV transmission can be occurred by vertical transfer of pre-partum cross-placenta and horizontal transmission of post-partum, although the vertical transfer of pre-partum cross-placenta is relatively rare. PLHV-1 virus DNA is detected in PBMCs and recipient baboon bloods during pig xenotransplantation to baboon of solid organs [141]. Virus productive infections are not observed, and microchimerism is found in all pig cells. PLHV is not found in different herd-originated donor tissues, suggesting that elimination is possible [142]. PLHV can be detected from pig sera using ELISA, although herpesviruses are generally present as latent forms in most animals, and thus, such PLHV viral titers are low [13].

Approximately more than 26–85% of different herd animals are reported to be infected with the PLHV. Early weaning cannot eliminate PHLV. PLHV-1 seems to induce the PTLD (post-transplant lymphoproliferative disease) in mini-pigs during allotransplantation of hematopoietic stem cells. The clinical symptom of pig PTLD resembles human PTLD raised by the Epstein–Barr (EV) virus, which is classified to HHV-4 [143]. Latency seen and specific in herpesviruses is to remain dormant in the host. Because of immune system surveillance over herpesviruses, latent herpesvirus reactivation and transmission occurs via bi-directional mechanism from the recipients to the donors and vice versa. Most mammalians have coevolved with the herpesviridae family by specific bindings of virus to the host cells [144]. For HSV type 1 and type 2, pig cells are not well infected by HSV type 1 and 2 infection due to defected virus entry. Non-heparan sulfate receptor (HVEM) is used as viral receptor for pig cells. Pig cells having the HVEM can be infected with HSV [50]. Different HSV-1 and -2 can cytopathically infect a pig ECs and PK-15 kidney cells, as easily detected.

21.9 Other Exogenous Pathogens

Apart from the prevalent PERV and pig herpesviruses, other pathogens are also highlighted for xenotransplantation [145]. Such pathogens are circovirus, hepatitis virus, parvovirus and rotaviruses present in a pig donors. For parvoviruses, infection with parvovirus contributed to serious diseases in patients [146, 147] and is also observed in allotransplantation. A specific human parvovirus, named parvovirus 4 (PARV-4), was isolated from HIV patient plasma [73], whose showed an acute virus syndrome in 2005 [103, 148]. PARV-4 in human was also found in blood donors and PARV-4 transmission occurs by blood transfusion [149, 150]. The PARV-4 presence in human bloods indicates a viremia condition to potentiate the viral spread to human, although its prevalence is high in NHPs [151]. Such similar parvovirus was also found in pigs [152] and the pig virus induces a transmission by xenotransplantation. As additional problematic virus, pig circovirus (PCV) is also problematic in xenotransplantation. In fact, PCV type 1 and PCV type 2 prevalence is high in healthy pigs during dietary pork meat consumption bu. humans [153]. PCV is classified to the smallest virus, which can replicate by autonomous manner in mammal cells and its distribution is very wide [154]. Higher than 69% of the pork meats have PCV1 or PCV2 virus DNAs. PCV2 transmission is observed between pigs, but the human seroconversion is not detected, indicating non-replicable pig circoviruses in humans. The transmission risk of pig circovirus to human is low. Arenaviruses were identified in allotransplantation in human, causing a grafted patient death [155]. Although the PCV2 causes the severe multi-factorial disease with the related factors, the disease severity is dependent of the immune system and genetic predisposition. In vitro human cell infection with PCV1 showed the prolonged viral existence without any changes, but cytopathogenicity [156]. In addition, the PCV1 infection seems not to not be productive [156, 157]. Details of animal arenaviruses and the zoonosis are discussed [158]. Main carrier of these viruses is the rodent and pig transmission of virus is very low.

As well-characterized zoonotic virus, in paramyxovirus, Nipah virus subtype shows pig to human transmission property [159, 160]. The virus subtype induces severe symptoms of human respiratory tract infection and human encephalitis, being approximately 48% infection death [159, 160]. The Menangle virus subtype is known in an Australia piggery with a dangerous flu-form ill disease [161]. Both subtypes of Nipah and Menangle viruses are fruit bats-originated [162]. The named Rabies as a pig disease is invariably fatal to cause death in most all animal species, which include humans [161, 163]. Additional vesicular stomatitis virus (VSV) as rhabdovirus shows insect infection and other cattle, horse and pig infections, indicating zoonotic property and flu-ill disease of human infection [164]. The Japanese encephalomyelitis virus (JEV) and Eastern equine encephalomyelitis virus (EEEV) as arbovirus infect pigs. The JEV virus amplifying mediums are pigs and mosquitoes are used as delivery vectors between infectious hosts such as cattle, human and horse [165]. They cause stillbirth and abortion in pigs as well as growth retardation in survived pigs [166]. Specific vaccination to EEEV infection can protect pigs [167]. Philippine domestic pigs are known to be carriers of Reston ebolavirus (REBOV) in filoviridae [168], but REBOV does not cause disease in human.

In the pathogenic bacteria in pigs, the multi-drug resistant strain of methicillin-resistant *Staphylococcus aureus* (MRSA) and anaerobic bacterium *Clostridium difficile* are potently concerned to recipients in xenotransplantation [169]. Pig is a MRSA source to cause human MRSA infection from pigs [170]. A MRSA ST398 strain was identified in pig and can cross barriers of specie for human infection [171]. Pig pathogenic *C. difficile* is used for pig skin xenografts [172]. C. difficile identified from the asymptomatic animal feces is utilized for treatment of burns patients. But C. difficile is not identified from the skin. Pig C. difficile genotypes are potentially capable of cross-species infection. Sporogenic *C. difficile* can survive from antibiotic treatment of porcine xenografts and this indicates antibiotics importance of vancomycin or metronidazole control. Thus, before selection of donor pigs, screening with *C. difficile* is needed.

References

1. US Food and Drug Administration. PHS guideline on infectious disease issues in xenotransplantation. Rockville: US Food and Drug Administration; 2006. www.fda.gov/cber/gdlns/xenophs0101.htm. Accessed 5 Feb 2008
2. Morozov VA, Morozov AV, Rotem A, Barkai U, Bornstein S, Denner J. Extended microbiological characterization of Göttingen Minipigs in the context of xenotransplantation: detection and vertical transmission of hepatitis E virus. PLoS One. 2015;10(10):e0139893.
3. Mueller NJ, Sulling K, Gollackner B, Yamamoto S, Knosalla C, Wilkinson RA, Kaur A, Sachs DH, Yamada K, Cooper DK, Patience C, Fishman JA. Reduced efficacy of ganciclovir against porcine and baboon cytomegalovirus in pig-to-baboon xenotransplantation. Am J Transplant. 2003;3(9):1057–64. https://doi.org/10.1034/j.1600-6143.2003.00192.x.
4. Zou S, Caler L, Colombini-Hatch S, Glynn S, Srinivas P. Research on the human virome: where are we and what is next. Microbiome. 2016;4(1):32. https://doi.org/10.1186/s40168-016-0177-y.
5. Denner J. The porcine virome and xenotransplantation. Virol J. 2017;14(1):171. https://doi.org/10.1186/s12985-017-0836-z.
6. Zhang B, Tang C, Yue H, Ren Y, Song Z. Viral metagenomics analysis demonstrates the diversity of viral flora in piglet diarrhoeic faeces in China. J Gen Virol. 2014;95(Pt 7):1603–11.
7. Hause BM, Padmanabhan A, Pedersen K, Gidlewski T. Feral swine virome is dominated by single-stranded DNA viruses and contains a novel Orthopneumovirus which circulates both in feral and domestic swine. J Gen Virol. 2016;97(9):2090–5. https://doi.org/10.1099/jgv.0.000554.

8. Clavijo A, Nikooienejad A, Esfahani MS, Metz RP, Schwartz S, Atashpaz-Gargari E, Deliberto TJ, Lutman MW, Pedersen K, Bazan LR, Koster LG, Jenkins-Moore M, Swenson SL, Zhang M, Beckham T, Johnson CD, Bounpheng M. Identification and analysis of the first 2009 pandemic H1N1 influenza virus from U.S. feral swine. Zoonoses Public Health. 2013;60 (5):327–35. https://doi.org/10.1111/zph.12006.
9. Feng Z, Baroch JA, Long LP, Xu Y, Cunningham FL, Pedersen K, Lutman MW, Schmit BS, Bowman AS, Deliberto TJ, Wan XF, Influenza A. Subtype H3 viruses in feral swine, United States, 2011–2012. Emerg Infect Dis. 2014;20(5):843–6. https://doi.org/10.3201/eid2005.131578.
10. Nelson MI, Vincent AL. Reverse zoonosis of influenza to swine: new perspectives on the human-animal interface. Trends Microbiol. 2015;23(3):142–53. https://doi.org/10.1016/j.tim.2014.12.002.
11. Denner J. Endogenous retroviruses. In: Kurth R, Bannert N, editors. Retroviruses: molecular biologygenomics and pathogenesis. Hethersett: Caister Academic Press; 2010. p. 35–69.
12. Denner J, Tönjes RR. Infection barriers to successful xenotransplantation focusing on porcine endogenous retroviruses. Clin Microbiol Rev. 2012;25(2):318–43. https://doi.org/10.1128/CMR.05011-11.
13. Wynyard S, Nathu D, Garkavenko O, Denner J, Elliott R. Microbiological safety of the first clinical pig islet xenotransplantation trial in New Zealand. Xenotransplantation. 2014;21 (4):309–23. https://doi.org/10.1111/xen.12102.
14. Morozov VA, Wynyard S, Matsumoto S, Abalovich A, Denner J, Elliott R. No PERV transmission during a clinical trial of pig islet cell transplantation. Virus Res. 2017;227:34–40. https://doi.org/10.1016/j.virusres.2016.08.012.
15. Reus K, Mayer J, Sauter M, Zischler H, Müller-Lantzsch N, Meese E. HERV-K(OLD). Ancestor sequences of the human endogenous retrovirus family HERV-K(HML-2). J Virol. 2001;75:8917–26.
16. Denner J. How active are porcine endogenous retroviruses (PERVs). Viruses. 2016;8:215. https://doi.org/10.3390/v8080215.
17. Mattiuzzo G, Scobie L, Takeuchi Y. Strategies to enhance the safety profile of xenotransplantation: minimizing the risk of viral zoonoses. Curr Opin Organ Transplant. 2008;13(2):184–8.
18. Dong B, Kim S, Hong S, et al. An infectious retrovirus susceptible to an IFN antiviral pathway from human prostate tumors. Proc Natl Acad Sci U S A. 2007;104:1655–60.
19. Gemeniano M, Mpanju O, Salomon DR, Eiden MV, Wilson CA. The infectivity and host range of the ecotropic porcine endogenous retrovirus, PERV-C, is modulated by residues in the C-terminal region of its surface envelope protein. Virology. 2006;346(1):108–17. https://doi.org/10.1016/j.virol.2005.10.021.
20. Ericsson T, Oldmixon B, Blomberg J, et al. Identification of novel porcine endogenous betaretrovirus sequences in miniature swine. J Virol. 2001;75:2765–70.
21. Tönjes RR, Niebert M. Relative age of proviral porcine endogenous retrovirus sequences in Sus scrofa based on the molecular clock hypothesis. J Virol. 2003;77:12363–8.
22. Niebert M, Tönjes RR. Evolutionary spread and recombination of porcine endogenous retroviruses in the suiformes. J Virol. 2005;79:649–54.
23. Patience C, Switzer WM, Takeuchi Y, Griffiths DJ, Goward ME, Heneine W, Stoye JP, Weiss RA. Multiple groups of novel retroviral genomes in pigs and related species. J Virol. 2001;75:2771–5.
24. Quereda JJ, Herrero-Medrano JM, Abellaneda JM, García-Nicolás O, Martínez-Alarcón L, Pallarés FJ, Ramírez P, Muñoz A, Ramis G. Porcine endogenous retrovirus copy number in different pig breeds is not related to genetic diversity. Zoonoses Public Health. 2012;9:401–7.
25. Lee D, Lee J, Yoon JK, Kim NY, Kim GW, Park C, Oh YK, Kim YB. Rapid determination of perv copy number from porcine genomic DNA by real-time polymerase chain reaction. Anim Biotechnol. 2011;22:175–80.

26. Yoon JK, Choi J, Lee HJ, Cho Y, Gwon YD, Jang Y, Kim S, Choi H, Lee JH, Kim YB. Distribution of porcine endogenous retrovirus in different organs of the hybrid of a landrace and a Jeju domestic pig in Korea. Transplant Proc. 2015;47:2067–71.
27. Zhang P, Yu P, Wang W, Zhang L, Li S, Bu H. An effective method for the quantitative detection of porcine endogenous retrovirus in pig tissues. In Vitro Cell Dev Biol Anim. 2010;46:408–10.
28. Mazurek U, Kimsa MC, Strzalka-Mrozik B, Kimsa MW, Adamska J, Lipinski D, Zeyland J, Szalata M, Slomski R, Jura J, et al. Quantitative analysis of porcine endogenous retroviruses in different organs of transgenic pigs generated for xenotransplantation. Curr Microbiol. 2013;67:505–14.
29. Dieckhoff B, Kessler B, Jobst D, Kues W, Petersen B, Pfeifer A, Kurth R, Niemann H, Wolf E, Denner J. Distribution and expression of porcine endogenous retroviruses in multi-transgenic pigs generated for xenotransplantation. Xenotransplantation. 2009;16:64–73.
30. Semaan M, Rotem A, Barkai U, Bornstein S, Denner J. Screening pigs for xenotransplantation: prevalence and expression of porcine endogenous retroviruses in Göttingen minipigs. Xenotransplantation. 2013;20:148–56.
31. LaRue RS, Jónsson SR, Silverstein KAT, Lajoie M, Bertrand D, El-Mabrouk N, Hötzel I, Andrésdóttir V, Smith TP, Harris RS. The artiodactyl APOBEC3 innate immune repertoire shows evidence for a multi-functional domain organization that existed in the ancestor of placental mammals. BMC Mol Biol. 2008;9:104.
32. Patience C, Takeuchi Y, Weiss RA. Infection of human cells by an endogenous retrovirus of pigs. Nat Med. 1997;3:282–6.
33. Denner J. Porcine endogenous retrovirus infection of human peripheral blood mononuclear cells. Xenotransplantation. 2015;22:151–2.
34. Denner J. Recombinant porcine endogenous retroviruses (PERV-A/C): a new risk for xenotransplantation? Arch Virol. 2008;153:1421–6.
35. Harrison I, Takeuchi Y, Bartosch B, Stoye JP. Determinants of high titer in recombinant porcine endogenous retroviruses. J Virol. 2004;78:13871–9.
36. Denner J, Specke V, Thiesen U, Karlas A, Kurth R. Genetic alterations of the long terminal repeat of an ecotropic porcine endogenous retrovirus during passage in human cells. Virology. 2003;314:125–33.
37. Scheef G, Fischer N, Krach U, Tonjes RR. The number of a U3 repeat box acting as an enhancer in long terminal repeats of polytropic replication-competent porcine endogenous retroviruses dynamically fluctuates during serial virus passages in human cells. J Virol. 2001;75(15):6933±40. https://doi.org/10.1128/JVI.75.15.6933±6940.2001.
38. Wilson CA, Wong S, Muller J, Davidson CE, Rose TM, Burd P. Type C retrovirus released from porcine primary peripheral blood mononuclear cells infects human cells. J Virol. 1998;72 (4):3082–7.
39. Wildschutte JH, Williams ZH, Montesion M, Subramanian RP, Kidd JM, Coffin JM. Discovery of unfixed endogenous retrovirus insertions in diverse human populations. Proc Natl Acad Sci U S A. 2016;113:E2326–34.
40. Suling K, Quinn G, Wood J, Patience C. Packaging of human endogenous retrovirus sequences is undetectable in porcine endogenous retrovirus particles produced from human cells. Virology. 2003;312:330–6.
41. Popp SK, Mann DA, Milburn PJ, Gibbs AJ, McCullagh PJ, Wilson JD, et al. Transient transmission of porcine endogenous retrovirus to fetal lambs after pig islet tissue xenotransplantation. Immunol Cell Biol. 2007;85(3):238±48.
42. Di Nicuolo G, van de Kerkhove MP, Hoekstra R, Beld MG, Amoroso P, Battisti S, et al. No evidence of in vitro and in vivo porcine endogenous retrovirus infection after plasmapheresis through the AMCbioartificial liver. Xenotransplantation. 2005;12(4):286±92.
43. Hermida-Prieto M, Domenech N, Moscoso I, Diaz T, Ishii J, Salomon DR, et al. Lack of cross-species transmission of porcine endogenous retrovirus (PERV) to transplant recipients and abattoir workers in contact with pigs. Transplantation. 2007;84(4):548±50.

44. van der Laan LJ, Lockey C, Griffeth BC, Frasier FS, Wilson CA, Onions DE, et al. Infection by porcine endogenous retrovirus after islet xenotransplantation in SCID mice. Nature. 2000;407(6800):90±4.
45. Martina Y, Marcucci KT, Cherqui S, Szabo A, Drysdale T, Srinivisan U, et al. Mice transgenic for a human porcine endogenous retrovirus receptor are susceptible to productive viral infection. J Virol. 2006;80(7):3135±46.
46. Kim N, Choi J, Kim S, Gwon Y-D, Cho Y, Yang JM, et al. Transmission of porcine endogenous retrovirus produced from different recipient cells in vivo. PLoS One. 2016;11 (11):e0165156. https://doi.org/10.1371/journal.pone.0165156.
47. Denner J. Recent progress in xenotransplantation, with emphasis on virological safety. Ann Transplant. 2016;21:717–27.
48. Denner J, Mueller NJ. Preventing transfer of infectious agents. Int J Surg. 2015;23(Pt B):306–11. https://doi.org/10.1016/j.ijsu.2015.08.032.
49. Pavio N, Meng XJ, Doceul V. Zoonotic origin of hepatitis E. Curr Opin Virol. 2015;10:34–41. https://doi.org/10.1016/j.coviro.2014.12.006.
50. Denner J. Xenotransplantation and hepatitis E virus. Xenotransplantation. 2015;22(3):167–73. https://doi.org/10.1111/xen.12156.
51. Weiss RA. On the concept and elucidation of endogenous retroviruses. Philos Trans R Soc Lond Ser B Biol Sci. 2013;368:20120494.
52. Dewannieux M, Heidmann T. Endogenous retroviruses: acquisition, amplification and taming of genome invaders. Curr Opin Virol. 2013;3:646–56.
53. Denner J. Endogenous retroviruses. In: Kurth R, Bannert N, editors. Retroviruses: molecular biology, genomics and pathogenesis. Hethersett: Caister Academic Press; 2010. p. 35–69.
54. Yu P, Zhang L, Li SF, Li YP, Cheng JQ, Lu YR, Bu H. Long-term effects on HEK-293 cell line after coculture with porcine endogenous retrovirus. Transplant Proc. 2005;37:496–9.
55. van der Laan LJ, Lockey C, Griffeth BC, et al. Infection by porcine endogenousretrovirus after islet xenotransplantation in SCID mice. Nature. 2000;407:90–4.
56. Vora NM, Basavaraju SV, Feldman KA, Paddock CD, Orciari L, Gitterman S, Griese S, Wallace RM, Said M, Blau DM, Selvaggi G, Velasco-Villa A, Ritter J, Yager P, Kresch A, Niezgoda M, Blanton J, Stosor V, Falta EM, Lyon GM 3rd, Zembower T, Kuzmina N, Rohatgi PK, Recuenco S, Zaki S, Damon I, Franka R, Kuehnert MJ. Transplant-associated rabies virus transmission investigation team. Raccoon rabies virus variant transmission through solid organ transplantation. JAMA. 2013;310(4):398–407.
57. Tandoi F, Romagnoli R, Martini S, Mazza E, Nada E, Cocchis D, Lupo F, Salizzoni M. Outcomes of liver transplantation from hepatitis B core antibody-positive donors in viral cirrhosis patients: the prevailing negative effect of recipient hepatitis C virus infection. Transplant Proc. 2012;44(7):1963–5.
58. Denner J, Tönjes RR, Takeuchi Y, Fishman J, Scobie L. First update of the international xenotransplantation association consensus statement on conditions for undertaking clinical trials of porcine islet products in type 1 diabetes–chapter 5: recipient monitoring and response plan for preventing disease transmission. Xenotransplantation. 2016;23(1):53–9. https://doi.org/10.1111/xen.12227.
59. Cowan PJ, Ayares D, Wolf E, Cooper DK. First update of the international xenotransplantation association consensus statement on conditions for undertaking clinical trials of porcine islet products in type 1 diabetes–chapter 2b: genetically modified source pigs. Xenotransplantation. 2016;23(1):32–7. https://doi.org/10.1111/xen.12224.
60. Enner J, Tönjes RR. Infection barriers to successful xenotransplantation focusing on porcine endogenous retroviruses. Clin Microbiol Rev. 2012;5(2):318–43.
61. Schuurman HJ. Microbiological safety of clinical xenotransplantation products: monitoring strategies and regulatory aspects. A commentary. Xenotransplantation. 2016;23:440–3.
62. Denner J. Transspecies transmission of Gammaretroviruses and the origin of the gibbon ape leukaemia virus (GaLV) and the koala retrovirus (KoRV). Viruses. 2016;8:336.

63. Peeters M, Jung M, Ayouba A. The origin and molecular epidemiology of HIV. Expert Rev Anti-Infect Ther. 2013;11:885–96.
64. Denner J. Transspecies transmissions of retroviruses: new cases. Virology. 2007;369:229–33.
65. Benveniste RE, Todaro GJ. Evolution of type C viral genes preservation of ancestral murine type C viral sequences in pig cellular DNA. Proc Natl Acad Sci U S A. 1975;72:4090–4.
66. Godehardt AW, Rodrigues Costa M, Tönjes RR. Review on porcine endogenous retrovirus detection assays–impact on quality and safety of xenotransplants. Xenotransplantation. 2015 Mar-Apr;22(2):95–101. https://doi.org/10.1111/xen.12154.
67. Yang L, Guell M, Niu D, et al. Genome-wide inactivation of porcine endogenous retroviruses (PERVs). Science. 2015;350:1101–4.
68. Mckane BW, Ramachandran S, Yang J, et al. Xenoreactive anti-Galalpha(1,3)gal antibodies prevent porcine endogenous retrovirus infection of human in vivo. Hum Immunol. 2003;64:708–17.
69. Mckane BW, Ramachandran S, Xu XC, et al. Natural antibodies prevent in vivo transmission of porcine isletderived endogenous retrovirus to human cells. Cell Transplant. 2004;13:137–43.
70. Ramachandran S, Jaramillo A, Xu XC, et al. Human immune responses to porcine endogenous retrovirus-derived peptides presented naturally in the context of porcine and human major histocompatibility complex class I molecules: implications in xenotransplantation of porcine organs. Transplantation. 2004;77:1580–8.
71. Quinn G, Wood JC, Ryan DJ, et al. Porcine endogenous retrovirus transmission characteristics of galactose alpha1-3 galactose-deficient pig cells. J Virol. 2004;78:5805–11.
72. Magre S, Takeuchi Y, Langford G, et al. Reduced sensitivity to human serum inactivation of enveloped viruses produced by pig cells transgenic for human CD55 or deficient for the galactosyl-alpha(1-3) galactosyl epitope. J Virol. 2004;78:5812–9.
73. Issa NC, Wilkinson RA, Griesemer A, et al. Absence of replication of porcine endogenous retrovirus and porcine lymphotropic herpesvirus type 1 with prolonged pig cell microchimerism after pig-to-baboon xenotransplantation. J Virol. 2008;82:12441–8.
74. Leyh RG, Wilhelmi M, Walles T, Kallenbach K, Rebe P, Oberbeck A, Herden T, Haverich A, Mertsching H. Acellularized porcine heart valve scaffolds for heart valve tissue engineering and the risk of cross-species transmission of porcine endogenous retrovirus. J Thorac Cardiovasc Surg. 2003;126(4):1000–4. https://doi.org/10.1016/s0022-5223(03)00353-2.
75. Levy MF, Crippin J, Sutton S, et al. Liver allotransplantation after extracorporeal hepatic support with transgenic (hCD55/hCD59) porcine livers: clinical results and lack of pig-to-human transmission of the porcine endogenous retrovirus. Transplantation. 2000;69:272–80.
76. Semaan M, Kaulitz D, Petersen B, Niemann H, Denner J. Long-term effects of PERV-specific RNA interference in transgenic pigs. Xenotransplantation. 2012;19(2):112–21. https://doi.org/10.1111/j.1399-3089.2012.00683.x.
77. Ramsoondar J, Vaught T, Ball S, Mendicino M, Monahan J, Jobst P, Vance A, Duncan J, Wells K, Ayares D. Production of transgenic pigs that express porcine endogenous retrovirus small interfering RNAs. Xenotransplantation. 2009;16:164–80.
78. Kaulitz D, Fiebig U, Eschricht M, Wurzbacher C, Kurth R, Denner J. Generation of neutralizing antibodies against porcine endogenous retroviruses (PERVs). Virology. 2011;411:78–86.
79. Waechter A, Denner J. Novel neutralising antibodies targeting the N-terminal helical region of the transmembrane envelope protein p15E of the porcine endogenous retrovirus (PERV). Immunol Res. 2014;58:9–19.
80. Denner J, Mihica D, Kaulitz D, Schmidt CM. Increased titers of neutralizing antibodies after immunization with both envelope proteins of the porcine endogenous retroviruses (PERVs). Virol J. 2012;9:260.
81. Semaan M, Ivanusic D, Denner J. Cytotoxic effects during knock out of multiple porcine endogenous retrovirus (PERV) sequences in the pig genome by zinc finger nucleases (ZFN). PLoS One. 2015;10:e0122059.

82. Shi M, Wang X, Okamoto M, Takao S, Baba M. Inhibition of porcine endogenous retrovirus (PERV) replication by HIV-1 gene expression inhibitors. Antivir Res. 2009;83:201–4.
83. Demange A, Yajjou-Hamalian H, Gallay K, Luengo C, Beven V, Leroux A, Confort MP, Al Andary E, Gouet P, Moreau K, et al. Porcine endogenous retrovirus-a/C: biochemical properties of its integrase and susceptibility to raltegravir. J Gen Virol. 2015;96:3124–30.
84. Argaw T, Colon-Moran W, Wilson C. Susceptibility of porcine endogenous retrovirus to antiretroviral inhibitors. Xenotransplantation. 2016;23:151–8.
85. Schuurman HJ. Pig-to-nonhuman primate solid organ xenografting: recent achievements on the road to first-in-man explorations. Xenotransplantation. 2016;23(3):175–8.
86. Denner J. Elimination of porcine endogenous retroviruses from pig cells. Xenotransplantation. 2015;22:411–2.
87. Liang C, Wainberg MA, Das AT, Berkhout B. CRISPR/Cas9: a double-edged sword when used to combat HIV infection. Retrovirology. 2016;13:37.
88. Denner J. Can antiretroviral drugs be used to treat porcine endogenous retrovirus (PERV) infection after xenotransplantation? Viruses. 2017;9:213. https://doi.org/10.3390/v9080213.
89. Sluis-Cremer N, Tachedjian G. Mechanisms of inhibition of HIV replication by nonnucleoside reverse transcriptase inhibitors. Virus Res. 2008;134:147–56.
90. Günthard HF, Saag MS, Benson CA, del Rio C, Eron JJ, Gallant JE, Hoy JF, Mugavero MJ, Sax PE, Thompson MA, et al. Antiretroviral drugs for treatment and prevention of HIV infection in adults: 2016 recommendations of the international antiviral society—USA panel. JAMA. 2016;316:191–210.
91. Kirtane AR, Langer R, Traverso G. Past, present, and future drug delivery systems for antiretrovirals. J Pharm Sci. 2016;105:3471–82. [
92. Lackey P, Mills A, Carpio F, Hsu R, DeJesus E, Pierone G, Henegar C, Fusco J, Fusco G, Wohlfeiler M. Virologic effectiveness of Abacavir/lamivudine with Darunavir/ritonavir versus other protease inhibitors in treatment-experienced HIV-infected patients in clinical practice. Clin Drug Investig. 2017;37:51–60.
93. Boi S, Rosenke K, Hansen E, Hendrick D, Malik F, Evans LH. Endogenous retroviruses mobilized during friend murine leukemia virus infection. Virology. 2016;499:136–43. https://doi.org/10.1016/j.virol.2016.07.009.
94. Brook I. Approval of Zidovudine (AZT) for acquired immunodeficiency syndrome. A challenge to the medical and pharmaceutical communities. JAMA. 1987;258:1517.
95. Wang X, Yamataka K, Okamoto M, Ikeda S, Baba M. Potent and selective inhibition of tat-dependent HIV-1 replication in chronically infected cells by a novel naphthalene derivative JTK-101. Antivir Chem Chemother. 2007;18(4):201–11. https://doi.org/10.1177/095632020701800404.
96. Okamoto M, Okamoto T, Baba M. Inhibition of human immunodeficiency virus type 1 replication by combination of transcription inhibitor K-12 and other antiretroviral agents in acutely and chronically infected cells. Antimicrob Agents Chemother. 1999;43(3):492–7. https://doi.org/10.1128/AAC.43.3.492.
97. Sekijima M, Waki S, Sahara H, et al. Results of life-supporting galactosyltransferase knockout kidneys in cynomolgus monkeys using two different sources of galactosyltransferase knockout swine. Transplantation. 2014;98:419–26.
98. Denner J. Xenotransplantation and porcine cytomegalovirus. Xenotransplantation. 2015;22 (5):329–35.
99. Mueller NJ, Barth RN, Yamamoto S, Kitamura H, Patience C, Yamada K, Cooper DK, Sachs DH, Kaur A, Fishman JA. Activation of cytomegalovirus in pig-to-primate organ xenotransplantation. J Virol. 2002;76:4734–64.
100. Mueller NJ, Livingston C, Knosalla C, Barth RN, Yamamoto S, Gollackner B, Dor FJ, Buhler L, Sachs DH, Yamada K, et al. Activation of porcine cytomegalovirus, but not porcine lymphotropic herpesvirus, in pig-to baboon xenotransplantation. J Infect Dis. 2004;189:1628–33.

101. Ramanan P, Razonable RR. Cytomegalovirus infections in solid organ transplantation: a review. J Infect Chemother. 2013;45(3):260–71.
102. Plotzki E, Keller M, Ivanusic D, Denner J. A new western blot assay for the detection of porcine cytomegalovirus (PCMV). J Immunol Methods. 2016;437:37–42.
103. Garkavenko O, Dieckhoff B, Wynyard S, et al. Absence of transmission of potentially xenotic viruses in a prospective pig to primate islet xenotransplantation study. J Med Virol. 2008;80:2046–52.
104. Whitteker JL, Dudani AK, Tackaberry ES. Human fibroblasts are permissive for porcine cytomegalovirus in vitro. Transplantation. 2008;86:155–62.
105. Yamada K, Tasaki M, Sekijima M, Wilkinson RA, Villani V, Moran SG, Cormac TA, Hanekamp IM, Arn JS, Fishman JA, et al. Porcine cytomegalovirus infection is associated with early rejection of kidney grafts in a pig to baboon xenotransplantation model. Transplantation. 2014;98:411–7.
106. Clark DA, Fryer JFL, Tucker AW, McArdle PD, Hughes AE, Emery VC, Griffiths PD. Porcine cytomegalovirus in pigs being bred for xenograft organs; progress towards control. Xenotransplantation. 2003;10:142–8.
107. Mueller NJ, Denner J. Porcine cytomegalovirus/porcine roseolovirus (PCMV/PRV): a threat for xenotransplantation? Xenotransplantation. 2022;29(5):e12775. https://doi.org/10.1111/xen.12775.
108. Morozov VA, Abicht JM, Reichart B, Mayr T, Guethoff S, Denner J. Active replication of porcine cytomegalovirus (PCMV) following transplantation of a pig heart into a baboon despite undetected virus in the donor pig. Ann Virol Res. 2016;2:1018.
109. Fishman JA, Patience C. Xenotransplantation: infectious risk revisited. Am J Transplant. 2004;4:1383–90.
110. Fishman JA, Scobie L, Takeuchi Y. Xenotransplantation-associated infectious risk: a WHO consultation. Xenotransplantation. 2012;19:72–81.
111. Morozov VA, Heinrichs G, Denner J. Effective detection of porcine cytomegalovirus using non-invasively taken samples from piglets. Viruses. 2017;9:9. https://doi.org/10.3390/v9010009.
112. Gu W, Zeng N, Zhou L, Ge X, Guo X, Yang H. Genomic organization and molecular characterization of porcine cytomegalovirus. Virology. 2014;460–461:165–72.
113. Goltz M, Widen F, Banks M, Belak S, Ehlers B. Characterization of the DNA polymerase loci of porcine cytomegalovirus from diverse geographical origins. Virus Genes. 2000;21:249–55.
114. Liu X, Liao S, Zhu L, Xu Z, Zhou Y. Molecular epidemiology of porcine cytomegalovirus (PCMV) in Sichuan province, China: 2010–2013. PLoS One. 2013;8:e64648.
115. Millard AL, Haberli L, Sinzger C, et al. Efficiency of porcine endothelial cell infection with human cytomegalovirus depends on both virus tropism and endothelial cell vascular origin. Xenotransplantation. 2010;17:274–87.
116. Pischke S, Suneetha PV, Baechlein C, et al. Hepatitis E virus infection as a cause of graft hepatitis in liver transplant recipients. Liver Transpl. 2010;16:74–82.
117. Haagsma EB, Niesters HG, van den Berg AP, et al. Prevalence of hepatitis E virus infection in liver transplant recipients. Liver Transpl. 2009;15:1225–8.
118. Meng XJ. Zoonotic and foodborne transmission of hepatitis E virus. Semin Liver Dis. 2013;33(1):41–9.
119. Balayan MS, Andjaparidze AG, Savinskaya SS, Ketiladze ES, Braginsky DM, Savinov AP, et al. Evidence for a virus in non-a, non-B hepatitis transmitted via the fecal-oral route. Intervirology. 1983;20:23–31.
120. Meng XJ, Purcell RH, Halbur PG, Lehman JR, Webb DM, Tsareva TS, et al. A novel virus in swine is closely related to the human hepatitis E virus. PNAS. 1997;94:9860–5.
121. Kamar N, Dalton HR, Abravanel F, Izopet J. Hepatitis E virus infection. Clin Microbiol Rev. 2014;27:116–38.
122. Scobie L, Dalton HR. Hepatitis E: source and route of infection, clinical manifestations and new developments. J Viral Hepat. 2013;20:1–11.

123. Kumar A, Beniwal M, Kar P, Sharma JB, Murphy MS. Hepatitis E in pregnancy. Int J Gynaecol Obstet. 2004;85:240–4.
124. Meng XJ. Recent advances in hepatitis E virus. J Viral Hepat. 2010;17:153–61.
125. Kabrane-Lazizi Y, Meng XJ, Purcell RH, Emerson SU. Evidence that the genomic RNA of hepatitis E virus is capped. J Virol. 1999;73:8848–50.
126. Guu TSY, Liu Z, Ye Q, Mata DA, Li K, Wang Y, et al. Structure of the hepatitis E virus-like particle suggests mechanisms for virus assembly and receptor binding. PNAS. 2009;106:12992–7.
127. Riddell MA, Li F, Anderson DA. Identification of immunodominant and conformational epitopes in the capsid protein of hepatitis E virus by using monoclonal antibodies. J Virol. 2000;74:8011–7.
128. Johne R, Dremsek P, Reetz J, Heckel G, Hess M, Ulrich RG. Hepeviridae: an expanding family of vertebrate viruses. Infect Genet Evol. 2014;27:212–29.
129. Smith DB, Simmonds P, International Committee on Taxonomy of Viruses Hepeviridae Study Group, Jameel S, Emerson SU, Harrrison TJ, Meng XJ, Van der Poel OH, Purdy MA. Consensus proposals for classification of the family Hepeviridae. J Gen Virol. 2014;95:2223–32.
130. Lu L, Li C, Hagedorn CH. Phylogenetic analysis of global hepatitis E virus sequences: genetic diversity, subtypes and zoonosis. Rev Med Virol. 2006;16:5–36.
131. Meng XJ. Hepatitis E virus: animal reservoirs and zoonotic risk. Vet Microbiol. 2010;140:256–65.
132. Pavio N, Meng X-J, Renou C. Zoonotic hepatitis E: animal reservoirs and emerging risks. Vet Res. 2010;41:46.
133. Ambrosioni J, Mamin A, Hadengue A, Bernimoulin M, Samii K, Landelle C, Negro F, Kaiser L. Long-term hepatitis E viral load kinetics in an immunocompromised patient treated with ribavirin. Clin Microbiol Infect. 2014;20(10):O718–20. https://doi.org/10.1111/1469-0691.12576.
134. Benhamou Y. Hepatitis B in the HIV-coinfected patient. J Acquir Immune Defic Syndr. 2007;45(Suppl 2):S57–65.; discussion S66-7. https://doi.org/10.1097/QAI.0b013e318068d1dd.
135. Casas M, Cortes R, Pina S, Peralta B, Allepuz A, Cortey M, et al. Longitudinal study of hepatitis E virus infection in Spanish farrow-to-finish swine herds. Vet Microbiol. 2011;148:27–34.
136. Purcell RH, Emerson SU. Hepatitis E virus. In: Knipe D, Howley P, Griffin D, Lamb R, Martin M, Roizman, editors. Fields virology. 4th ed. Philadelphia, Lippincott, Williams and Wilkins; 2001. p. 3051–61.
137. Scobie L, Dalton HR. Hepatitis E: source and route of infection, clinical manifestations and new developments. J Virol Hep. 2013;20:1–11.
138. Colson P, Borentai P, Queyriaux B, et al. Pig liver sausage as a source of hepatitis E virus transmission to humans. J Infect Dis. 2010;202:825–34.
139. Mungmunpuntipantip R, Wiwanitkit V. Risk of transfusion-transmitted hepatitis E virus: correspondence. Indian J Hematol Blood Transfus. 2022;38(3):611. https://doi.org/10.1007/s12288-022-01530-4.
140. Di Bartolo I, Ponterio E, Castellini L, Ostanello F, Ruggeri FM. Viral and antibody HEV prevalence in swine at slaughterhouse in Italy. Vet Microbiol. 2011;149:330–8.
141. Abravanel F, Chapuy-Regaud S, Lhomme S, Miedouge M, Peron J-M, Alric L, et al. Performance of anti-HEV assays for diagnosing acute hepatitis E in immunocompromised patients. J Clin Virol. 2013;58:624–8.
142. Lu YH, Qian HZ, Qin X, Jiang QW, Zheng YJ. Subtypes of genotype 3 hepatitis E virus in pigs. Vet J. 2013;197:509–11.
143. Svendsen O. The minipig in toxicology. Exp Toxicol Pathol. 2006;57(5–6):335–9. https://doi.org/10.1016/j.etp.2006.03.003.

144. Simianer H, K€ohn F. Genetic management of the Gottingen Minipig population. J Pharmacol Toxicol Methods. 2010;62:221–6.
145. Kamar N, Rostaing L, Abravanel F, et al. Ribavirin therapy inhibits viral replication on patients with chronic hepatitis E virus infection. Gastroenterology. 2010;139:1612–8.
146. Doucette K, Dor FJ, Wilkinson RA, Martin SI, Huang CA, Cooper DK, Sachs DH, Fishman JA. Gene expression of porcine lymphotrophic herpesvirus-1 in miniature swine with posttransplant lymphoproliferative disorder. Transplantation. 2007;83(1):87–90. https://doi.org/10.1097/01.tp.0000228237.32549.16.
147. Chmielewicz B, Goltz M, Franz T, Bauer C, Brema S, Ellerbrok H, Beckmann S, Rziha HJ, Lahrmann KH, Romero C, Ehlers B. A novel porcine gammaherpesvirus. Virology. 2003;308(2):317–29.
148. Brema S, Lindner I, Goltz M, et al. Development of a recombinant antigen based ELISA for the sero-detection of porcine lymphotropic herpesviruses. Xenotransplantation. 2008;15:357–64.
149. Morozov VA, Plotzki E, Rotem A, Barkai U, Denner J. Extended microbiological characterization of Göttingen minipigs: porcine cytomegalovirus and other viruses. Xenotransplantation. 2016;23(6):490–6.
150. Mueller NJ, Takeuchi Y, Mattiuzzo G, Scobie L. Microbial safety in xenotransplantation. Curr Opin Organ Transplant. 2011;16(2):201–6.
151. Perez A, Li QX, Perez-Romero P, et al. A new class of receptor for herpes simplex virus has heptad repeat motifs that are common to membrane fusion proteins. J Virol. 2005;9:7419–30.
152. Scobie L, Takeuchi Y. Porcine endogenous retrovirus and other viruses in xenotransplantation. Curr Opin Organ Transplant. 2009;14:175–9.
153. Kapoor A, Simmonds P, Slikas E, et al. Human bocaviruses are highly diverse, dispersed, recombination prone, and prevalent in enteric infections. J Infect Dis. 2010;201:1633–43.
154. Mogensen TH, Jensen JM, Hamilton-Dutoit S, et al. Chronic hepatitis caused by persistent parvovirus B19 infection. BMC Infect Dis. 2010;10:246.
155. Jones MS, Kapoor A, Lukashov VV, et al. New DNA viruses identified in patients with acute viral infection syndrome. J Virol. 2005;79:8230–6.
156. Simmonds P, Douglas J, Bestetti G, et al. A third genotype of the human parvovirus PARV4 in sub-Saharan Africa. J Gen Virol. 2008;89:2299–302.
157. Sharp CP, Vermeulen M, Nebie Y, et al. Changing epidemiology of human parvovirus 4 infection in sub-Saharan Africa. Emerg Infect Dis. 2010;16:1605–7.
158. Hourfar MK, Mayr-Wohlfart U, Themann A, et al. Recipients potentially infected with parvovirus B19 by red blood cell products. Transfusion. 2011;51:129–36.
159. Sharp CP, LeBreton M, Kantola K, et al. Widespread infection with homologues of human parvoviruses B19, PARV4, and human bocavirus of chimpanzees and gorillas in the wild. J Virol. 2010;84:10289–96.
160. Cheung AK, Wu G, Wang D, et al. Identification and molecular cloning of a novel porcine parvovirus. Arch Virol. 2010;155:801–6.
161. Li L, Kapoor A, Slikas B, et al. Multiple diverse circoviruses infect farm animals and are commonly found in human and chimpanzee feces. J Virol. 2010;84:1674–82.
162. Denner J, Mankertz A. Porcine circoviruses and xenotransplantation. Viruses. 2017;9:4.
163. Palacios G, Druce J, Du L, et al. A new arenavirus in a cluster of fatal transplant-associated diseases. N Engl J Med. 2008;358:991–8.
164. Hattermann K, Roedner C, Schmitt C, Finsterbusch T, Steinfeldt T, Mankertz A. Infection studies on human cell lines with porcine circovirus type 1 and porcine circovirus type 2. Xenotransplantation. 2004;11(3):284–94. https://doi.org/10.1111/j.1399-3089.2004.00134.x.
165. Dubin G, Toussaint JF, Cassart JP, Howe B, Boyce D, Friedland L, Abu-Elyazeed R, Poncelet S, Han HH, Debrus S. Investigation of a regulatory agency enquiry into potential porcine circovirus type 1 contamination of the human rotavirus vaccine, Rotarix: approach and outcome. Hum Vaccin Immunother. 2013;9(11):2398–408. https://doi.org/10.4161/hv.25973.

166. Charrel RN, de Lamballerie X. Zoonotic aspects of arenavirus infections. Vet Microbiol. 2010;140:213–20.
167. Kirkland PD, Daniels PW, Nor MN, Love RJ, Philbey AW, Ross AD. Menangle and Nipah virus infections of pigs. Vet Clin North Am Food Anim Pract. 2002;18(3):557–71. https://doi.org/10.1016/S0749-0720(02)00038-5.
168. Chua KB. Nipah virus outbreak in Malaysia. J Clin Virol. 2003;26(3):265–75. https://doi.org/10.1016/S1386-6532(02)00268-8.
169. Frölich K, Thiede S, Kozikowski T, Jakob WA. Review of mutual transmission of important infectious diseases between livestock and wildlife in Europe. Ann N Y Acad Sci. 2002;969:4–13. https://doi.org/10.1111/j.1749-6632.2002.tb04343.x.
170. Barr JA, Smith C, Marsh GA, Field H, Wang LF. Evidence of bat origin for Menangle virus, a zoonotic paramyxovirus first isolated from diseased pigs. J Gen Virol. 2012;93(Pt 12):2590–4. https://doi.org/10.1099/vir.0.045385-0.
171. Merriman GM. Rabies in Tennessee swine. J Am Vet Med Assoc. 1966;148(7):809–11.
172. Tesh RB, Peralta PH, Johnson KM. Ecologic studies of vesicular stomatitis virus. I. Prevalence of infection among animals and humans living in an area of endemic VSV activity. Am J Epidemiol. 1969;90(3):255–61. https://doi.org/10.1093/oxfordjournals.aje.a121068.

Chapter 22
Concept of Chimeric Organisms Such as Human/Non-Human Chimera (HNH-Chimera)

As an another direction in the xenotransplantation, chimeric organisms are recently represented in the regenerative medicine using the stem cells (SCs). For definition of chimeric transplantable compatibility, human organs can be generated from pigs usable for xenotransplantation because of the non-new concept of chimeric animals. Since the embryonic chimera from goat and sheep embryos was firstly created in 1984, at present, the ethical issue of the human-derived gamete and animal gamete fertilization is raised with the moral acceptance of human/non-human chimera (HNH-chimera) [1]. In pigs, such chimera is needed to develop animals, which have human organs. In addition, pigs and humans have similar metabolism, physiology, long habitation with human population, and sizes of organs. In addition, the administration of pig cells to humans' xenotransplantation is beneficial for identification of the infectious and immunological risks and barriers [2]. In the chimeric model of rodents, pluripotent stem cell (PSC) injection of rats is made to mice blastocysts as rat-mouse chimeras and the chimera developed into adulthood. Those chimeras can efficiently be constructed by the CRISPR-Cas9 genome editing, as the rodent chimera organs with rat cells can be created [3]. When chimeric competency of human PSCs (hPSCs) was examined in cattle and pigs, the integration has stably been progressed by intermediate hPSCs to the pig embryo progenitors [3]. This inter-species of human–pig blastocyst complementation is in progress.

However, for the controversial prospects, the creation of HNH-chimeras will argue the human dignity status for the morality [4, 5]. Nevertheless, the human–pig chimera creation is not to stop. In France, Germany, Korea, and Japan, construction of human–animal chimeric embryos is forbidden in laws [2]. In 2015, the US NIH (National Institutes of Health) banned the federal research funding of studies on chimeric cells between human stem cells and animal embryos. However, private funding research is regal. In 2016, the NIH started to debate the banned and forbidden chimera research [6]. With the advanced hPSCs and embryo manipulation technology, human–inter-species chimeras showed specific future goal. For example, hSCs integration to pig embryos is implied for human organs-to-animals transplantation in a mode of the human–pig chimera development [6]. Basically,

C.-H. Kim, *Glycoimmunology in Xenotransplantation*,
https://doi.org/10.1007/978-981-99-7691-1_22

the background of chimera concept has been derived from the facts that purely animal model cannot replace the condition and physiology of human, compared to that of a human sample or model. A chimera is defined as the "cellular combinations at the developmental preimplantation blastocyst stage and other entities created by introducing cells at later stages" [7]. Currently, it has been recognized that the chimeric animals have the potential as mouse–rat chimera like animal–animal chimeras are used for human disease animal models [8, 9]. In the translational science, a chimeric organism contains distinct and recognizable cells derived from more than one species, but the chimera is different from a conventional hybrid organism, which is characteristic for the homogenous mixture, and cells in hybrid organism have combined.

References

1. Rashid T, Kobayashi T, Nakauchi H. Revisiting the flight of Icarus: making human organs from PSCs with large animal chimeras. Cell Stem Cell. 2014;15:406–9.
2. Bourret R, Martinez E, Vialla F, Giquel C, Thonnat-Marin A, De Vos J. Human–animal chimeras: ethical issues about farming chimeric animals earing human organs. Stem Cell Res Ther. 2016;7:1–7. https://doi.org/10.1186/s13287-016-0345-9.
3. Wu J, Platero-Luengo A, Sakurai M, et al. Interspecies chimerism with mammalian pluripotent stem cells. Cell. 2017;168:473–86.
4. Karpowicz P, Cohen C, Van der Kooy D. Developing human-nonhuman chimeras in human stem cell research: ethical issues and boundaries. Kennedy Inst Ethics J. 2005;15:107–34.
5. Palacios-González C. Human dignity and the creation of human–nonhuman chimeras. Med Health Care Philos. 2015;18:487–99.
6. Hagan-Brown A, Favaretto M, Borr P. Newspaper coverage of human-pig chimera research: a qualitative study on select media coverage of scientific breakthrough. Xenotransplantation. 2017;24:e12317.
7. Hermeren G. Ethical considerations in chimera research. Development. 2014;142:3–5.
8. Huang Y, Osorno R, Tsakiridis A, Wilson V. In vivo differentiation potential of epiblast stem cells revealed by chimeric embryo formation. Cell Rep. 2012;2:1571–8.
9. Isotani A, Hatayama H, Kaseda K, Ikawa M, Okabe M. Formation of a thymus from rat ES cells in xenogeneic nude mouse↔rat ES chimeras. Genes Cells. 2011;16:397–405.

Chapter 23
Intra-Bone Bone Marrow Transplantation

BM is very specific tissue with an extensive vascular network, and the injected BM cells into the organism easily enter and join the peripheral circulation [1]. Despite successful engraftment of Gal-T$^{+/+}$ mouse BM cells into α1,3Gal-T KO mice recipients [2], α1,3Gal-T KO pig BM cells to NHP xenotransplantation is not yet successful. For example, baboons received with BM exhibit intravenous infusion, and the microchimerism is seen in the BM-received baboons. However, macrochimerism is characteristic of only transient peripheral expression and not in the BM expression. In addition, the responses elicit production of anti-pig ABs [3]. Early loss of grafted BMs in a α1,3Gal-T KO pig-to-NHP xenotransplantation is also observed in the α1,3Gal-T$^{+/+}$ mouse BM to α1,3Gal-T KO mouse transplantation, presumably due to naturally formed non-Gal-specific Abs [4] and CD47 incompatibility-caused phagocytotic property [5, 6]. In mice, donor BM cells with 1–2% occupation are also located at the BM recipients [7], suggesting that the donor BM cell localization at the host BM immunologically is a cause for responses for both engrafted BM and host hematopoiesis reconstitution.

Pig BM α1,3Gal-T KO transplantation to baboons raises loss in macrochimerism within 24 h. For improved engraftment with persisted peripheral chimerism, intrabone BM (IBBM) transplantation designation has been reported. Baboons received α1,3Gal-T KO BM cells in the bilateral tibiae and intravenously (IBBM/BM transplantation). From donor SLA-matched α1,3Gal-T KO kidneys-received recipients in the IBBM and BM transplantation in perioperative period as well as kidney-received recipients after 2 months of IBBM/BM transplantation, peripheral macrochimerism is easily detectable after several days post-IBBM/BM-transplantation and macrochimerism reappeared in animals. Pig CFUs in which pig progenitor cells are engrafted are observed in the recipient host BM area. Anti-pig unresponsiveness is observed. α1,3Gal-T KO/pCMV-kidney survives for more than 60 days, providing for xenogeneic tolerance induction through BM transplantation. The IBBM transplantation increases the pig BM cell engraftment with a higher macrochimerism and higher incidence of hyporesponsively engrafted BM across xenogeneic hurdle in pig-to-baboon xenotransplantation [8]. In addition, IBBM transplantation

C.-H. Kim, *Glycoimmunology in Xenotransplantation*,
https://doi.org/10.1007/978-981-99-7691-1_23

causes chimerism disappearance, baboons kept donor-specifically their hyporesponsiveness and long survival of kidney xenografts when received pCMV negative α1,3Gal-T KO kidneys [8]. Anti-CD154 mAb treatment improved chimerism and engraftment. Increased BM engraftment incidence and sustained survival of pig BM within the host BM were also obtained [8]. Injected pig BM cells to the BM region of recipient host minimize the pig BM cell exposure to macrophage cells and naturally-preformed non-α1,3Gal antigen-specific mAbs. These BM-engrafted microenvironments contribute to donor HSCs survival and mesenchymal SC (MSC) after inoculation into the recipient's BM rather than conventional intravenous BM cells infusion [9].

Because injected BM cells should be retained at the injection sites, a collagen gel has typically been utilized to enhance the cell growth and differentiation, representatively. Cell matrix was used [10, 11]. A part of the intravenous-infused BM cells can prevent naturally formed non-α1,3Gal-specific Abs during IBBM/BM transplantation, allowing the increased donor BM cell survival released from injected bone region. Inflammation also occurs at kidney transplantation as BM transplantation has a negative peripheral chimerism effect. The early graft loss is caused by the pCMV present in the grafts. pCMV present in α1,3Gal-T KO pig kidneys induces the decreased graft survival via early ECs' activation in pig-to-baboon kidney xenotransplantation [12]. Because macrophage cells in the innate immune responses induce xenograft rejection, the GalTKO/hCD47 Tg BM cells or hCD47-transfected BM cells retain valuable potentials. GalTKO/hCD47 Tg pigs can also be used for thymokidney/IBBM transplantation to baboons.

BM transplantation in rodent shows donor-specific tolerance induction [13]. The tolerance induction is also similarly observed in porcine [14], non-human primate [15], and humans [16, 17] as well as concordant rodents [18] and pig-to-NOD/SCID mice [19] xenotransplantation. However, xenoantigenic BM transplantation in pre-clinical pig-to-NHPs is not succeeded [20–23]. When i.v.-infused pig BM cells were ex vivo immunoadsorbed to eliminate naturally preformed anti-α1,3Gal-reactive Abs, transient macrochimerism was pound and the infused pig BM cells are undetected within 24 h [20, 21]. Because naturally preformed anti-α1,3Gal Abs are the major hurdle, α1,3Gal-T KO pigs [24] as BM donors show the limitation in retained peripheral macrochimerism [23]. A few animals show a transient donor-specific hyporesponsiveness during BM transplantation by BM i.v. (IV BM transplantation). I.v.-injected BM cells migrate via the circulatory system, causing for cell loss [16]. In allogeneic models, direct donor BM cell injection to recipient BM spaces (IBBM—transplantation) rapidly reconstitute with a higher survival than the i.v. injection [9]. In pre-clinical pig-to-baboon IBBM transplantation, macrochimerism and BM engraftment across xenogeneic barrier were retained with prolonged peripheral macrochimerism, high BM engraftment at the injection sites, and prolonged α1,3Gal-T KO pig kidney grafts survival [8] of a pig thymic grafts [25].

References

1. Yang Y, Pang D, Hu C, Lv Y, He T, An Y, Tang Z, Deng Z. Nestin positive bone marrow derived cells responded to injury mobilize into peripheral circulation and participate in skin defect healing. PLoS One. 2015;10(12):e0143368. https://doi.org/10.1371/journal.pone.0143368.
2. Ohdan H, Swenson KG, Kitamura H, Yang YG, Sykes M. Tolerization of Gal alpha 1,3Gal-reactive B cells in pre-sensitized alpha 1,3-galactosyltransferase-deficient mice by nonmyeloablative induction of mixed chimerism. Xenotransplantation. 2001;8(4):227–38. https://doi.org/10.1034/j.1399-3089.2001.00006.x.
3. Griesemer A, Liang F, Hirakata A, et al. Occurrence of specific humoral non-responsiveness to swine antigens following administration of GalT-KO bone marrow to baboons. Xenotransplantation. 2010;17:300–12.
4. Galili U. Interacti on of the natural anti-gal antibody with alpha-galactosyl epitopes: a major obstacle for xenotransplantation in humans. Immunol Today. 1993;14:480–2.
5. Wang H, VerHalen J, Madariaga ML, et al. Attenuation of phagocytosis of xenogeneic cells by manipulating CD47. Blood. 2007;109:836–42.
6. Ide K, Wang H, Tahara H, et al. Role for CD47-SIRPalpha signaling in xenograft rejection by macrophages. Proc Natl Acad Sci U S A. 2007;104:5062–6.
7. Cui J, Wahl RL, Shen T, et al. Bone marrow cell trafficking following intravenous administration. Br J Haematol. 1999;107:895–902.
8. Tasaki M, Wamala I, Tena A, Villani V, Sekijima M, Pathiraja V, Wilkinson RA, Pratts S, Cormack T, Clayman E, Arn JS, Shimizu A, Fishman JA, Sachs DH, Yamada K. High incidence of xenogenic bone marrow engraftment in pig-to-baboon intra-bone bone marrow transplantation. Am J Transplant. 2015;15(4):974–83. https://doi.org/10.1111/ajt.13070.
9. Kushida T, Inaba M, Hisha H, et al. Intra-bone marrow injection of allogeneic bone marrow cells: a powerful new strategy for treatment of intractable autoimmune diseases in MRL/lpr mice. Blood. 2001;97:3292–9.
10. Nakano H, Shimada A, Imai K, Takezawa T, Takahashi T, Hashizume K. Bovine trophoblastic cell differentiation on collagen substrata: formation of binucleate cells expressing placental lactogen. Cell Tissue Res. 2002;307(2):225–35. https://doi.org/10.1007/s00441-001-0491-x.
11. Shi M, Adachi Y, Shigematsu A, et al. Intra-bone marrow injection of donor bone marrow cells suspended in collagen gel retains injected cells in bone marrow, resulting in rapid hemopoietic recovery in mice. Stem Cells. 2008;26:2211–6.
12. Sekijima M, Waki S, Sahara H, et al. Results of life-supporting galactosyltransferase knockout kidneys in cynomolgus monkeys using two different sources of galactosyltransferase knockout swine. Transplantation. 2014;98:419–26.
13. Sharabi Y, Sachs DH. Mixed chimerism and permanent specific transplantation tolerance induced by a nonlethal preparative regimen. J Exp Med. 1989;169:493–502.
14. Fuchimoto Y, Huang CA, Yamada K, et al. Mixed chimerism and tolerance without whole body irradiation in a large animal model. J Clin Invest. 2000;105:1779–89.
15. Sommer W, JM O, Pruner KB, Dehnadi A, Ha Huh K, Robinson KA, Hanekamp I, Rosales I, Bean AS, Paster J, Oura T, Neal Smith R, Colvin R, Benichou G, Kawai T, Madsen JC, Allan JS. A Mixed-chimerism protocol utilizing thymoglobulin and Belatacept did not induce lung allograft tolerance, despite previous success in renal allotransplantation. Transplant Direct. 2021;7(6):e705. https://doi.org/10.1097/TXD.0000000000001150.
16. Spitzer TR, Delmonico F, Tolkoff-Rubin N, et al. Combined histocompatibility leukocyte antigen-matched donor bone marrow and renal transplantation for multiple myeloma with end stage renal disease: the induction of allograft tolerance through mixed lymphohematopoietic chimerism. Transplantation. 1999;68:480–4.
17. Kawai T, Cosimi AB, Spitzer TR, et al. HLA-mismatched renal transplantation without maintenance immunosuppression. N Engl J Med. 2008;358:353–61.

18. Sachs DH. Transplantation tolerance through mixed chimerism: from Allo to xeno. Xenotransplantation. 2018;25(3):e12420. https://doi.org/10.1111/xen.12420.
19. Abe M, Qi J, Sykes M, Yang YG. Mixed chimerism induces donor-specific T-cell tolerance across a highly disparate xenogeneic barrier. Blood. 2002;99:3823–9.
20. Sablinski T, Gianello PR, Bailin M, et al. Pig to monkey bone marrow and kidney xenotransplantation. Surgery. 1997;121:381–91.
21. Sablinski T, Emery DW, Monroy R, et al. Long-term discordant xenogeneic (porcine-to-primate) bone marrow engraftment in a monkey treated with porcine-specific growth factors. Transplantation. 1999;67:972–7.
22. Buhler L, Awwad M, Treter S, et al. Pig hematopoietic cell chimerism in baboons conditioned with a non-myeloablative regimen and CD154 blockade. Transplantation. 2002;73:12–22.
23. Tseng YL, Dor FJ, Kuwaki K, et al. Bone marrow transplantation from alpha1,3-galactosyltransferase gene-knockout pigs in baboons. Xenotransplantation. 2004;11:361–70.
24. Lai L, Kolber-Simonds D, Park KW, et al. Production of alpha-1,3-galactosyltransferase knockout pigs by nuclear transfer cloning. Science. 2002;295:1089–92.
25. Yamada K, Yazawa K, Shimizu A, et al. Marked prolongation of porcine renal xenograft survival in baboons through the use of alpha1,3-galactosyltransferase gene-knockout donors and the cotransplantation of vascularized thymic tissue. Nat Med. 2005;11:32–4.

Chapter 24
The Future Prospects of Xenotransplantation

In the past era, gene manipulation techniques were slowly developed and inefficient, making them very difficult to experiment with xenotransplantation. But now it is possible to create genetically modified animals quickly and accurately with CRISPR-Cas9 system. This allows the development of GGTA1 or DKO or TKO to suppress the expression of certain glycan in pigs. Moreover, the expression of hCRPs can be added, thus curbing the HARs and AXHRs that appear by the glycans during the xenotransplantation. This technique enables studies of other barriers while preventing early immune responses as much as possible. They will be SLAs and other immune-related proteins. Suppressing SLAs will significantly reduce the immune response of pigs' APCs. Also, using substances that suppress various proteins involved in the host immunity will enhance the organ viability after implantation. But simply inhibiting multiple molecules at once to transgenic pig organs is a very difficult task. Creation of GTKO pigs and CRISPR-Cas9-based genome-wide inactivation of PERVs allowed the future possibility for xenografts in clinical trials. In addition, immune modulatory drug design will open a new way for better outcome.

Although issues of rejection hurdles and obstacles are overcome and solved, recently developed technological strategy as highly efficient gene-editing and modifying tools has been emerged. The gene-editing CRISPR/Cas9 technology has been improved. The accumulation of molecular characterized evidences on the hurdles and obstacles can justify the success of xenotransplantation. CRISPR/Cas9 generated multiple genetic modifications [1]. For example, the gene-editing tools demonstrated the transgenic utilization for PERV inactivation in the grafts, allowing no more need to be concerned with such hurdles and obstacles. Prolonged and long-term survivals in pig-to-primate xenotransplantation are possible, although we have still powerful hurdle and enemy, more precisely and obviously T-cell-associated adaptive immune responses. In fact, immunosuppression even in allotransplantation is not sufficiently controlled to prolong xenotransplant survivals [2]. Without immune-related molecules, it would be difficult for genetically modified pigs to grow up to be transplanted into humans. Therefore, research should reduce

C.-H. Kim, *Glycoimmunology in Xenotransplantation*,
https://doi.org/10.1007/978-981-99-7691-1_24

expression by uniquely identifying molecules related to a person's immune system while not being fatal to the survival of pigs. Furthermore, for successful xenotransplantation, additional care needs to be provided to those who receive organs. Additional immunosuppressants should be administered for stability, even if non-self-molecules are reduced. And further efforts must be made to overexpression of TBM, EPCR, TFP1, and CD39 to prevent coagulation regulation. Developing these methods will eventually break barriers to all xenotransplantation and successfully transplant them, freeing them from the reality of organ shortages.

References

1. Yang L, Guell M, Niu D, et al. Genome-wide inactivation of porcine endogenous retroviruses (PERVs). Science. 2015;350:1101–4.
2. Hawthorne WJ, Lew AM, Thomas HE. Genetic strategies to bring islet xenotransplantation to the clinic. Curr Opin Organ Transplant. 2016;21:476–83.

Chapter 25
Conclusions

Although many available data describe carbohydrate antigens, it is also true that some discrepancies between each case are seen, indicating current limitation to information on xenotransplantation. The current carbohydrate antigens are α1,3Gal, non-α1,3Gal antigen, NeuGc, Forssman antigen, Galα1-3 Lew^X, α-linked or β-linked GalNAc, β1,3-linked Gal, NeuAc, and SDa-like antigens. However, discrepancies between each antigenic property existed. The most effective immunosuppressive agents are the CD40–CD154 signaling blockers together with the B-cell inhibitors in xenotransplantation models. At present, for better designing of the application in xenogeneic pigs, candidates of genetically modified pig lines, such as SLA proteins and other epitope antigen-deficient pigs, immune-related cell inhibitors, such as macrophage cells, T cells, and NK cells, and also human complement and coagulation-related regulator proteins, in α1,3Gal-T-KO and CMAH-KO lineages, are under investigation.

The development of medical technology has made it possible to overcome serious damage with only donated organs. But fundamentally, there is a lack of organ donation, so we have to think about xenotransplantation. The deceased human donors-caused organ shortage will raise studies on cross-species xenotransplantation using pigs as donors. Moreover, pig breed species are beneficial in the ethical and economic issues. Therefore, pig breeds are the model sources of extensive biomedical and pharmacological researches. Since in 2003, the US FDA outlined a feasible xenotransplantation guideline and prospective reviews, the experimental to clinical approaches are being made using cells and organs xenografts in pig-to-NHS xenotransplantation as models. Hearts, islets, and kidneys are major subjects in NHPs. The potential risks include the infectious microorganism transmission and transfer, as guided by FDA and WHO consensus documents. The two representable hurdles of clinical xenotransplantation would be Galα1,3Galβ1,4GlcNAc-R (α1,3Gal) xenoantigen-displayed immune rejection and PERVs-mediated zoonosis.

C.-H. Kim, *Glycoimmunology in Xenotransplantation*,
https://doi.org/10.1007/978-981-99-7691-1_25

SPRINGER NATURE

GPSR Compliance

The European Union's (EU) General Product Safety Regulation (GPSR) is a set of rules that requires consumer products to be safe and our obligations to ensure this.

If you have any concerns about our products, you can contact us on ProductSafety@springernature.com

In case Publisher is established outside the EU, the EU authorized representative is:

Springer Nature Customer Service Center GmbH
Europaplatz 3
69115 Heidelberg, Germany

The manufacturer's authorised representative in the EU is Springer Nature Customer Service Centre GmbH, Europaplatz 3, 69115 Heidelberg, Germany. If you have any concerns regarding our products, please contact ProductSafety@springernature.com

Printed and bound by CPI Group (UK) Ltd, Croydon, CR0 4YY
01/10/2026
02225899-0012